Mathematical Methods using Python

This advanced undergraduate textbook presents a new approach to teaching mathematical methods for scientists and engineers. It provides a practical, pedagogical introduction to utilizing Python in Mathematical and Computational Methods courses. Both analytical and computational examples are integrated from its start. Each chapter concludes with a set of problems designed to help students hone their skills in mathematical techniques, computer programming, and numerical analysis. The book places less emphasis on mathematical proofs, and more emphasis on how to use computers for both symbolic and numerical calculations. It contains 182 extensively documented coding examples, based on topics that students will encounter in their advanced courses in Mechanics, Electronics, Optics, Electromagnetism, Quantum Mechanics etc.

An introductory chapter gives students a crash course in Python programming and the most often used libraries (SymPy, NumPy, SciPy, Matplotlib). This is followed by chapters dedicated to differentiation, integration, vectors and multiple integration techniques. The next group of chapters covers complex numbers, matrices, vector analysis and vector spaces. Extensive chapters cover ordinary and partial differential equations, followed by chapters on nonlinear systems and on the analysis of experimental data using linear and nonlinear regression techniques, Fourier transforms, binomial and Gaussian distributions. The book is accompanied by a dedicated GitHub website, which contains all codes from the book in the form of ready to run Jupyter notebooks. A detailed solutions manual is also available for instructors using the textbook in their courses.

Key Features:
- A unique teaching approach which merges mathematical methods and the Python programming skills which physicists and engineering students need in their courses.
- Uses examples and models from physical and engineering systems, to motivate the mathematics being taught.
- Students learn to solve scientific problems in three different ways: traditional pen-and-paper methods, using scientific numerical techniques with NumPy and SciPy, and using Symbolic Python (SymPy).

Vasilis Pagonis is Professor of Physics Emeritus at McDaniel College, Maryland, USA. His research area is applications of thermally and optically stimulated luminescence. He taught courses in mathematical physics, classical and quantum mechanics, analog and digital electronics and numerous general science courses. Dr. Pagonis' resume lists more than 200 peer-reviewed publications in international journals. He is currently associate editor of the journal Radiation Measurements. He is co-author with Christopher Kulp of the undergraduate textbook "Classical Mechanics: a computational approach, with examples in Python and Mathematica" (CRC Press, 2020). He has also co-authored four graduate level textbooks in the field of luminescence dosimetry, and most recently published the book "Luminescence Signal analysis using Python" (Springer, 2022).

Christopher Kulp is the John P. Graham Teaching Professor of Physics at Lycoming College. He has been teaching undergraduate physics at all levels for 20 years. Dr. Kulp's research focuses on modelling complex systems, time series analysis, and machine learning. He has published 30 peer-reviewed papers in international journals, many of which include student co-authors. He is also co-author of the undergraduate textbook "Classical Mechanics: a computational approach, with examples in Python and Mathematica" (CRC Press, 2020).

Mathematical Methods using Python

Applications in Physics and Engineering

Vasilis Pagonis and Christopher W. Kulp

CRC Press

Taylor & Francis Group
Boca Raton London New York

CRC Press is an imprint of the
Taylor & Francis Group, an **informa** business

Designed cover image: Vasilis Pagonis

First edition published 2024
by CRC Press
2385 NW Executive Center Drive, Suite 320, Boca Raton FL 33431

and by CRC Press
4 Park Square, Milton Park, Abingdon, Oxon, OX14 4RN

CRC Press is an imprint of Taylor & Francis Group, LLC

Library of Congress Cataloging-in-Publication Data

Names: Pagonis, Vasilis, author. | Kulp, Christopher W., author.
Title: Mathematical physics using Python : applications in physics and
engineering / Vasilis Pagonis and Christopher W. Kulp.
Description: First edition. | Boca Raton : CRC Press, 2024. | Includes
bibliographical references and index.
Identifiers: LCCN 2023052970 | ISBN 9781032278360 (hbk) | ISBN
9781032278384 (pbk) | ISBN 9781003294320 (ebk)
Subjects: LCSH: Mathematical physics--Data processing--Textbooks. | Python
(Computer program language)--Textbooks.
Classification: LCC QC20.7.E4 P34 2024 | DDC
530.15/02855133--dc23/eng/20240209
LC record available at https://lccn.loc.gov/2023052970

ISBN: 978-1-032-27836-0 (hbk)
ISBN: 978-1-032-27838-4 (pbk)
ISBN: 978-1-003-29432-0 (ebk)

DOI: 10.1201/9781003294320

Typeset in Latin Modern font
by KnowledgeWorks Global Ltd.

Publisher's note: This book has been prepared from camera-ready copy provided by the authors.

Dedication

The authors dedicate this book to their students at McDaniel College and Lycoming College.

Contents

Preface

A NEW APPROACH TO TEACHING MATHEMATICAL METHODS FOR SCIENTISTS AND ENGINEERS

In our combined 57 years of experience teaching undergraduate courses, we have found that computation should be front and center in a science education. Programming is a core skill for scientists and engineers, that should be taught alongside advanced mathematical methods. It is in this spirit that we created this textbook.

The use of computers to solve problems has become a fundamental and critical skill in all scientific fields. This book merges instruction in mathematical methods and programming, into a single presentation. In our approach, computer programming and computer algebra systems are treated as simply one more tool for problem solving. Alongside instruction in mathematical methods beyond calculus, we provide instruction in using computers to solve science and engineering problems.

This textbook is intended for students who have had two semesters of calculus and are pursuing degrees in science or engineering, where mathematics is used to solve problems and to model physical systems. We created a book which

1. Provides an overview of the mathematics needed to pursue a degree in science and engineering.

2. Shows students how to use Python to solve mathematical problems.

3. Uses examples and models from physical and engineering systems, to motivate the mathematics being taught.

There are already excellent textbooks being used by instructors to teach mathematical methods for scientists and engineers. For example, the textbook by Boas (Mathematical methods in the physical sciences, Wiley, 2005), is a classic comprehensive mathematical methods textbook, which has been used for many years in universities and colleges. However, there is a lack of undergraduate textbooks in which programming is taught alongside the mathematical methods. We believe that this textbook fills this gap in the literature, by providing a unique teaching approach for both the mathematics and the programming skills which students will need upon graduation.

We chose Python because of its popularity and open source nature. Using Python to perform mathematical calculations is a transferable skill students will find valuable, regardless of what they choose to do with their degree. This book is as much about learning to use Python to solve problems, as it is about learning mathematics. Writing a book intimately tied to a language runs the risk of becoming out of date when the language is updated. However most of the Python commands chosen in this book have been around a long time and are widely used by scientists and engineers. Students who have worked through this book should be comfortable adapting to changes in the Python language as they arise.

TO THE INSTRUCTORS

This textbook can be used for a Mathematical Methods course, or for any course where students need to learn advanced mathematics as well as programming techniques.

It is important for students to see the applications for the mathematics they are learning. In all chapters, we motivate the mathematics with examples from science or engineering.

The Python examples in this book cover topics students will encounter in their advanced courses in Mechanics, Electronics, Optics, Electromagnetism, Quantum Mechanics etc. The end of chapter problems are also chosen with these advanced courses in mind.

Asking students to learn both mathematics and a computer programming language can be a daunting task. Chapter 1 of this textbook gives students a crash course in Python programming and the most often used libraries (SymPy, NumPy, SciPy, Matplotlib). Instructors can use Chapter 1 as an assigned reading, teach the material like any other chapter, or skip it and have students learn the code as they work through the mathematics. This introductory material is followed by chapters dedicated to differentiation, integration, vectors and multiple integration techniques. In the next group of chapters we cover complex numbers, matrices, vector analysis and provide an introduction to vector spaces. Extensive chapters cover ordinary and partial differential equations, followed by a chapter on the analysis of nonlinear systems. Finally, a chapter on the analysis of experimental data introduces students to linear and nonlinear regression techniques, Fourier transforms and Poisson, binomial and Gaussian distributions.

Students and instructors can download the open access codes used in all the chapters of this book from the website *https://github.com/vpagonis/CRCbook* in the form of ready-to-run Jupyter notebooks.

Our goal is for students to learn how to tackle a science problem, by using three complementary and equally important approaches:

- Solve a problem using pen and pencil (the "by hand" approach)

- Use symbolic Python (SymPy) to carry out symbolic analytical calculations (the symbolic approach)

- Use numerical and scientific Python (Numpy and SciPy) for problems requiring a numerical approach (the numerical approach).

This is not a textbook for teaching numerical methods in science, since there are already several very good textbooks covering numerical techniques. Our approach to numerical solutions relies on the numerical integration libraries available in SciPy and NumPy.

TO THE STUDENTS

One of the goals of this book is to teach you how to use computational tools to solve mathematical problems in sciences and engineering. This is an important transferable skill, useful regardless of your career path.

We understand learning both mathematics and programming at the same time may seem daunting. However, you can do it! Search engines and AI are your friend. If you can't remember how to do something in Python, it is okay to search online. For example, a search of "How to solve a differential equation in Python" will result either in the website for a particular Python library (such as SymPy), or to another site which demonstrates the use of the necessary code. Over time, your coding skills will grow and you will find yourself searching for the "simple" things less often.

A word of warning. Do not mindlessly copy and paste code you find online! To learn how to use Python and other languages, it is critical that you write your own codes, and that you understand how the codes are structured and organized. The mathematical methods and the programming skills you will obtain from this book are very valuable skills, which you can carry with you in all of your scientific education! If you simply copy and paste code, you will be cheating yourself out of learning these skills.

ACKNOWLEDGMENTS

We want to thank several people who have helped make this book happen. We thank Rebecca Davies and Danny Kielty at CRC Press and Kumar Shashi at KnowledgeWorks Global, for all of their help in the preparation of this book. While working on this book, Dr Kulp was supported by a Lycoming College Sabbatical Leave and Professional Development Grant.

1 Introduction to Python

In this introductory chapter, we present the basics of using Python, focusing on the most frequently used libraries (SymPy, NumPy, SciPy, Matplotlib). We discuss the various data types, variables and sequences, and give examples of functions, loops and conditional statements. We show how to use functions and methods and how to work with NumPy arrays using indexing and slicing. We introduce the Matplotlib library and how to set up 2D and 3D graphics. Finally, we give examples of symbolic evaluations using SymPy and how to use the lambdify function in Python.

1.1 DATA TYPES AND VARIABLES IN PYTHON

The main numeric types in Python are integers, floating point numbers and complex numbers. A floating point number (or float) is a real number written in decimal form. Python stores floats and integers internally in different ways. Python 3 automatically converts integer to floats as needed. If you are working with an earlier version of Python, you may need to be more careful when working with variables of different types.

In Example 1.1 we assign a value to a variable using the assignment operator $=$, for example a = 0.0003 assigns the value of 0.0003 to the variable a. We can also define variables by using scientific notation to represent floats, for example b = 1.64e-4. Python uses the letter j to represent $\sqrt{-1}$, and the built-in function complex(a,b) creates the complex number a+b*j. We can use the built-in function type() to identify the type of a Python object, and the built-in function print(a) is used to print the value of a variable. Similar to several other computer languages, the line e = 'time' defines a string variable e.

Example 1.1 shows how the function type(a) can be used to find out what class of object the variable a belongs to. In most codes of this book, you will see the line of code

```
print('-'*28,'CODE OUTPUT','-'*29)
```

This line produces a horizontal line labeled CODE OUTPUT and is used to separate the Python code from the output generated by the code.

Example 1.1: Variable types in Python

Assign Python variables to each of the quantities 0.00007, 1.64e-4, 10, 3+4i, time. Use the Python function type(a) to determine what type of variable each of these quantities represents.

Solution:

```
a = 0.00007            #define variables
b = 1.64e-4
c = 10
d = complex(3,4)
e = 'time'

print('-'*28,'CODE OUTPUT','-'*29)

print('\nVariable a = 0.0003 belongs to type:',type(a))
```

DOI: 10.1201/9781003294320-1

1

```
print('Variable b = 1.64e-4 belongs to type:',type(b))
print('Variable c = 10 belongs to type:',type(c),'\n')

print('A complex number created with complex(3,4)=',d)
print('Variable d = complex(3,4) belongs to type:',type(d),'\n')

print("Variable e = 'time' belongs to type:",type(e))

------------------------- CODE OUTPUT -------------------------

Variable a = 0.0003 belongs to type: <class 'float'>
Variable b = 1.64e-4 belongs to type: <class 'float'>
Variable c = 10 belongs to type: <class 'int'>

A complex number created with complex(3,4)= (3+4j)
Variable d = complex(3,4) belongs to type: <class 'complex'>

Variable e = 'time' belongs to type: <class 'str'>
```

Multiple variable assignments can be compressed to one line. For example, the code `a,b,c = 1, 2, 3` is a compressed version of the three individual lines of code `a = 1`, `b = 2` and `c = 3`.

The common arithmetic operators in Python are addition $+$, subtraction $-$, multiplication $*$, division $/$ and exponentiation $**$. In addition, the $\%$ operator is used to find the remainder (or modulo), and the $//$ is used for integer division.

In Python we can use *f-strings* to format the printing of variables. Within the f-string, any variable can be enclosed inside curly brackets; for example, `print(f'{a}')` will print the numerical value of the variable a, and `print(f'{a**2}')` prints the numerical value of a^2. By using the f-string formats `print(f'{a:.3}')`, `print(f'{a:.2e}')`, `print(f'{a:.2f}')` and `print(f'{a:g}')`, we print the variable a using two decimals, using scientific notation, and as a float in generic notation, respectively.

In Example 1.2 we use *f-strings* to format the printing of variables.

Example 1.2; Using f-strings to format Python output

Assign Python variables a, b to the numbers 5.0/3 and 1.645.
(a) Use f-strings to print the variable a using scientific notation with two significant figures.
(b) Print the variable a^2 without any formatting and by using generic notation.
(c) Print the variable b using two decimal points in the output of the code.
(d) What does `round(a,2)` and `round(b,1)` produce in the code?

Solution:

```
a = 5.0/3                #define variables
b = 1.645

print('-'*28,'CODE OUTPUT','-'*29,'\n')
```

```
print(f'Without formatting, the variable a={a}','\n')
print(f'Using scientific notation, the variable a={a:.2e}')

print(f'Without formatting, the square of variable a is {a**2}')
print(f'Using generic notation, the square of variable a is \
{a**2:g}','\n')

print(f'Using two decimals, the variable b=1.645 is      {b:.3}')

print('\nround(a,2) produces ',round(a,2))
print('round(b,1) produces ',round(b,1))

------------------------- CODE OUTPUT ----------------------------

Without formatting, the variable a=1.6666666666666667

Using scientific notation, the variable a=1.67e+00
Without formatting, the square of variable a is 2.777777777777778
Using generic notation, the square of variable a is 2.77778

Using two decimals, the variable b=1.645 is      1.65

round(a,2) produces  1.67
round(b,1) produces  1.6
```

A variable name cannot begin with a number, and there are reserved words in Python which should not be used as variables. Table 1.1 shows a partial list of reserved words in Python.

Table 1.1
Partial list of reserved words in Python.

False	None	def	elif	True
class	for	from	or	if
finally	lambda	nonlocal	else	sum
return	try	and	import	min
is	not	del	break	max
continue	sum	global	list	array

1.2   SEQUENCES IN PYTHON

The main sequence types in Python are *lists*, *tuples* and *range* objects. The main differences between these sequence objects are:

Lists are mutable; i.e. their elements can be modified and are usually homogeneous (i.e. objects of the same type create a list of similar objects)

Range objects are efficient sequences of integers (commonly used in loops). They use a small amount of memory and yield items only when needed

Tuples are immutable; i.e. their elements cannot be modified, and their elements are usually heterogeneous (i.e. objects of different types create a tuple, describing a single structure)

In the next three subsections, we look at examples of these types of sequences.

1.2.1   LISTS

We create a Python list using square brackets, with items separated by commas. Lists can contain data of any type and even other lists. For example

```
a=[2.0,[3,0],5j,[1,1,2],'s',1]
```

defines a list containing floats, integers, complex numbers, other lists and strings.

We can access the elements of a list by their index, and it is important to remember that lists are indexed starting at 0, not at 1. For example, `print(a[0])` prints the first element in list a, and `print(a[2])` prints the third element. We can also use *negative indices* to access elements starting from the end of the list so that `print(a[-1]` prints the last element in the list, `print(a[-2])` prints the second to last element etc.

Since lists are mutable, they can be altered, so we can redefine any of the elements in the list, for example a`[-1]` = -2 sets the last element of the list to -2. We can use *multiple indices* to access several entries in a list of lists; for example, `print(a[1][0])` will print the first element in the element a`[1]`. In the above list a, the element `print(a[1][0])` is 0.

We can define sub-lists called *slices* by using the syntax

```
a[start:end:step]
```

For example, the slice a`[2:5:2]` starts at index 2 and increases by a step of 2 taking every second element in the list, but does *not* include the ending value 5 of the index. We can omit one of the indices or the step parameter in a slice, for example a`[:5]` creates a list of the first five elements of a, and a`[2:]` creates a list of all the elements of a after and including a`[2]`. Likewise, a`[::2]` creates a list which contains every other element in a.

Lists can be *concatenated* using the addition operator +; for example; a`[1]`+a`[3]` creates a new list with the elements from both a`[1]`, a`[3]`.

Example 1.3 presents how to use these general properties of lists and how to access and modify parts of a list.

Example 1.3: Accessing and modifying list elements in Python

Consider the Python list
a = [2,[3,0],5,[1,1,2],'s',[1,4],2,5]
(a) Write a code to add the first and last element in this list.
(b) Modify the third element in the list so that it is equal to -2.
(c) What is the a`[3][2]` element of the list?
(d) Create a new list which starts from the third element in list a, and contains every third element of a.
(e) Create a new list that contains the first five elements of a
(f) What does the code a`[1]`+a`[3]` produce?
(g) What does the code a`[1]`*3 produce?

Solution:

(a) The first and last element in this list are added by `a[0]+a[-1]`

(b)–(c) `a[2]=-2` sets the third element in the list equal to -2. Similarly, `a[3][2]` obtains the third part of the fourth element of the list.

(d)–(e) The line `a[2:len(a):2]` creates a new list which starts from the third element `a[2]` in the list, up to the length of the list which is specified by `len(a)`.

(f)–(g) The code `a[1]+a[3]` produces a new list by concatenating the list $[1, 1, 2]$ to the end of the list $[3, 0]$. Finally, `a[1]*3` produces a new list by repeating the elements of list `a[1]` three times.

```
print('-'*28,'CODE OUTPUT','-'*29,'\n')

a = [2,[3,0],5,[1,1,2],'s',[1,4],2,5] # define list a

print('list a=',a)

print('the sum of the first and last element is: ',a[0]+a[-1])

a[2] = -2
print('\na[2] = -2 modifies list a into \na =',a,'\n')

print('The a[3][2] element is =', a[3][2],'\n')
print('The a[2:len(a):2] sequence is =',a[2:len(a):2])
print('The a[:5] sequence is =',a[:5])

print('\nThe a[1]+a[3] sequence is =',a[1]+a[3])
print('The a[1]*3 sequence is =',a[1]*3)

--------------------------- CODE OUTPUT ----------------------------

list a= [2, [3, 0], 5, [1, 1, 2], 's', [1, 4], 2, 5]
the sum of the first and last element is:  7

a[2] = -2 modifies list a into
a = [2, [3, 0], -2, [1, 1, 2], 's', [1, 4], 2, 5]

The a[3][2] element is = 2

The a[2:len(a):2] sequence is = [-2, 's', 2]
The a[:5] sequence is = [2, [3, 0], -2, [1, 1, 2], 's']

The a[1]+a[3] sequence is = [3, 0, 1, 1, 2]
The a[1]*3 sequence is = [3, 0, 3, 0, 3, 0]
```

Individual elements can be assigned separate variables in a list. For example, a, b, c = [1,2,3] stores the value of 1 in the variable a, the value of 2 in b, and the value of 3 in c. This is sometimes referred to as unpacking a list.

Lists *cannot* be copied like numeric data types by using, for example a statement like b = a. Specifically, a statement like b = a does *not* create a new list b from list a, but simply makes a *reference* to a. By setting a[0] = 1, this automatically sets also the first element of list b to 1. In other words, if we use the statement b = a and then modify any value in list a, the change will also be visible in list b.

This confusion can be avoided by using the b = a.copy() command, which creates two isolated objects whose contents share the same reference. This is referred to as *shallow copying* of an object.

One can also use the b = copy.deepcopy(a) command, which creates two isolated objects whose contents and structure are completely isolated from each other. This is referred to as deep copying of an object. The deep copy is an independent copy of the original object *and all its nested objects*. In other words, if we make changes to any *nested* objects in the original object, we will see no changes to the deep copy.

Example 1.4 shows some of the differences between the shallow and deep copies of a list.

Example 1.4: Shallow and deep copies of objects

What does each of these groups of code produce in the output?

Example A:
```
c = [1,[2,3],4]
d = c
d[1] = 0
print(c,d)
```

Example B:
```
c = [1,[2,3],4]
d = c.copy()
d[1][1] = 0
print(c,d)
```

Example C:
```
c = [1,[2,3],4]
d = c.copy.deepcopy()
d[1][1] = 0
print(c,d)
```

Solution:
The comments in the code describe the result in each of the above codes. Note that this example is the first time we use comments in our code. Any statement following a # is not executed by the Python interpreter. Comments are useful to help a human reader of the code (including yourself) understand what the program is doing. The value of comments cannot be overstated.

```
import copy
print('-'*28,'CODE OUTPUT','-'*29,'\n')


#Example A:
print('Example A: Effect of d = c statement')
c = [1,[2,3],4]
print('old_c = ',c)
d = c
d[1] = 0
print('d[1] = 0')    # modifies both the shallow copy d and the original c
print('new_c = ',c,'  new_d = ',d)

#Example B:
print('\nExample B: Shallow copy using d = c.copy(), modify inner element d[1][1]')
```

```
c = [1,[2,3],4]
print('old_c = ',c)
d = c.copy()
d[1][1] = 0
print('d[1][1] = 0') # modifies the shallow copy d, and the original c
print('new_c = ',c,' new_d = ',d)

#Example C:
print('\nExample C: Deep copy using copy.deepcopy(c), modify inner element d[1][1]')
c = [1,[2,3],4]
print('old_c = ',c)
d = copy.deepcopy(c)
d[1][1] = 0
print('d[1][1] = 0') # modifies the deep copy d, but not the original c
print('new_c = ',c,' new_d = ',d)

-------------------------- CODE OUTPUT --------------------------

Example A: Effect of d = c statement
old_c =  [1, [2, 3], 4]
d[1] = 0
new_c =  [1, 0, 4]   new_d =  [1, 0, 4]

Example B: Shallow copy using d = c.copy(), modify inner element d[1][1]
old_c =  [1, [2, 3], 4]
d[1][1] = 0
new_c =  [1, [2, 0], 4]   new_d =  [1, [2, 0], 4]

Example C: Deep copy using copy.deepcopy(c), modify inner element d[1][1]
old_c =  [1, [2, 3], 4]
d[1][1] = 0
new_c =  [1, [2, 3], 4]   new_d =  [1, [2, 0], 4]
```

1.2.2 RANGE SEQUENCES AND LIST COMPREHENSIONS

The second important type of sequence in Python is a *range* object, created with the built-in function range(a:b:step), where a, b and step are integers. This function creates an object representing the sequence of integers from a to b (excluding b), incremented by the variable step.

An important feature of Python that we use frequently in this book is a list comprehension, with the general syntax

```
[expression for item in sequence]
```

Here sequence is a sequence object (e.g. a range, a list, a tuple etc.), item is a variable name which takes each value in the sequence and expression is a Python expression which is calculated for each value of item. For example

```
[u**2 for u in [1,2,3]]
```

produces a new sequence with the squares of the integer sequence [1,2,3].

As another example, we can use the remainder operator % to create a periodic sequence of (0,1,2,3) and length 12 with the line of code:

```
[x%4 for x in range(0,12)]
```

Example 1.5 presents various properties of the `range()` function. Notice that when we print a range object, this does not display the elements when printed. This is because a range object yields values only when they are needed. However, the function `list()` can convert a range object into a list object whose elements can then be printed. Example 1.5 also demonstrates list comprehensions.

Example 1.5: Using the range() function

(a) What does the line `print(range(1,7,2))` produce in the output of the Python code?

(b) Change the object `range(1,7,2)` into a list by using the Python function `list()`.

(c) Find the maximum element, the sum of the elements and the length of the object `range(1,7,2)`

(d) What will be the result of the code `[u**3 for u in [1,2,3]]`

(e) What will be the result of the code `[x%3 for x in range(0,12)]`

Solution:

The comments in the code describe the result in each case.

```python
print('-'*28,'CODE OUTPUT','-'*29,'\n')

a = range(1,7,2)
print('Define the range sequence a = ',a,'\n')
# print(a) does not print the elements of the range() object

# Use list(a) to convert the range into a list, then print elements
print('list(a) gives: ',list(a),'\n')

print('The length of the range sequence a is = ',len(a))
print('The sum of elements in the range sequence a is = ',sum(a))
print('The maximum of the range sequence a is = ',max(a))

c, d = [3,'s']                    # unpacking a sequence
print("\nUnpack the sequence [3,'s'], to get: c = ",c,'  d = ',d,'\n')

e = [u**3 for u in [1,2,3]]    # a list comprehension
print('The list comprehension e = ',e)

f = [x%3 for x in range(0,12)]  # another list comprehension
print('The list comprehension f = ',f,'\n')

-------------------------- CODE OUTPUT --------------------------

Define the range sequence a =  range(1, 7, 2)

list(a) gives:  [1, 3, 5]

The length of the range sequence a is =  3
The sum of elements in the range sequence a is =  9
```

```
The maximum of the range sequence a is =  5

Unpack the sequence [3,'s'], to get: c =  3   d =  s

The list comprehension e =  [1, 8, 27]
The list comprehension f =  [0, 1, 2, 0, 1, 2, 0, 1, 2, 0, 1, 2]
```

1.2.3 TUPLE SEQUENCES

The third type of Python sequence is a *tuple*. We create a tuple by using *parentheses* instead of square brackets. For example

```
a = (1,3,'s')
```

The major difference between tuples and lists is that tuples are immutable, i.e. their elements cannot be altered. However, the *indexing, slicing* and *concatenating* operations for tuples are essentially the same as for lists. In this book we will be using list and range objects in most examples and will occasionally encounter tuples.

1.2.4 FUNCTIONS ON SEQUENCES

Python has several built-in functions for computing with sequences. For example, `len(a)`, `sum(a)`, `max(a)`, `min(a)`, evaluate the *length, sum, maximum* and *minimum* element in the list. We can also *sort* the list using `sorted(a)`. We also note a major difference between string and list types in Python. Lists are mutable but strings are not, i.e. we can modify the value of an element in a list, but not for a string. However, parts of a string can be extracted using slices; for example, `'hi'[1]` will extract the second letter in the string `'hi'`.

1.3 FUNCTIONS, FOR LOOPS AND CONDITIONAL STATEMENTS

This section is a brief introduction to *functions, for loops* and *conditional statements* in Python. We will demonstrate these over the next four examples. All programming languages have their own way of setting up functions, for loops and conditionals. Python uses indentations and colons for these statements. In the next four examples, pay close attention to how Python uses indentations and colons to delimit blocks of code. There should be no extra white spaces in the beginning of any line, and the line before any indented block must end with a colon character.

We begin with Example 1.6 where we wish to calculate the position of a particle starting at the origin with initial speed v_0, moving under a constant acceleration a at a time t using:

$$y = v_0\, t + \frac{1}{2}a\, t^2 \tag{1.3.1}$$

Because we wish to evaluate (1.3.1) several times in our code, we will define (1.3.1) as a function in Python and we will use a `for` loop for our calculation.

The first three lines of code in Example 1.6 define the function f, by using the `def` keyword. A *function* is generally a named unit of code which can be called from other parts of a program. In Python a function may have one or more variables as arguments, which receive their values from the calling program. In Example 1.6, the function `f` has three arguments vo, a, t, which are listed in parentheses in the line defining `f`. All three arguments must be provided when the function is called.

Inside the indented code of the function definition, the evaluation of (1.3.1) is stored in the variable y. The variable y is a *local variable* used by the function, and its value is not known outside the definition of the function f. By contrast, the variable ypos is a *global variable*, whose numerical value is known to the entire code.

The **return y** statement sends the variable y as the result of the calculation to the calling code. One can specify more than one variable in the return statement by separating them with commas.

Suppose we wanted to calculate the value of the function f for vo = 2, a = 1 and t = 3. This can be done using the syntax f(2,1,3). For example, p = f(2,1,3) would store the value 10.5 in the variable p.

However, Example 1.6 asks us to calculate the particle's position at multiple times. One can construct *for loops* in Python which execute blocks of code repeatedly. In Example 1.6 we create a *for loop* for such a purpose. However, before we begin the *for loop*, the code line ypos = [] initializes an empty list and which will be used to store the position of the particle at various times. Notice that this line of code is unindented because, as a global variable, ypos is defined outside of the function f.

To define a *for loop*, we need two things, a variable for iteration and a range for that iteration. In Example 1.6, the variable of iteration is t which takes on the values in the sequence range(4). The indented block of code following the *for loop* will be repeated several times. In the first iteration of the loop, the variable t has the value of 0, the first element of the sequence range(4). The function f is evaluated using f(1,2,t) which, in this iteration of the loop, is the same as f(1,2,0). The value of f(1,2,0) is then appended to the end of the list ypos by using the ypos.append(f(1,2,t)) command.

In the second iteration of the loop, the variable t has the value of 1 and the process is repeated. Afterwards, the loop has two more iterations, t = 2 and t = 3. The integer 3 is the last element of range(4) and, therefore the loop stops after that iteration.

Example 1.6: A simple function in Python, and a FOR loop

Create a function f(vo,a,t) which evaluates the position y of a particle starting at the origin, moving with initial speed v_0, under a constant acceleration a, and at time t:

$$y = v_0\, t + \frac{1}{2}a\, t^2 \tag{1.3.2}$$

Use a *for loop* so that the function f is called repeatedly to evaluate the position of the particle at times $t = 0, 1, 2$ and 3 seconds.

Solution:
As outlined above, we construct the *for loop* using range(4). The function f is called repeatedly inside the loop, to evaluate the position variable ypos. The value of f is added (appended) to the list ypos by using the ypos.append(f(1,2,t)) command.

```
print('-'*28,'CODE OUTPUT','-'*29,'\n')

def f(vo,a,t):            # define function f
    y = vo*t+a*t**2/2     # functions require indentation
    return y              # return the variable y to the calling code

ypos = []                 # create the empty list ypos
for t in range(4):        # for loops require indentation
```

```
    ypos.append(f(1,2,t))      # add the value of f to the list ypos

print('position y(t) = ',ypos)

-------------------------- CODE OUTPUT --------------------------

position y(t) =   [0.0, 2.0, 6.0, 12.0]
```

In some cases the functions we need in our code are simple and they depend only on one variable, so that we can use a simple function structure called a `lambda` function. Example 1.7 defines a `lambda` named function `f`, which evaluates the position `ypos` of the particle based on (1.3.1).

Example 1.7: A lambda function

Repeat the previous example by using a `lambda` function.

Solution:
Note that when using this simpler type of function, the variables v_0 and a are treated as *global* variables with fixed values $v_0 = 1$ and $a = 2$.

```
print('-'*28,'CODE OUTPUT','-'*29,'\n')

vo, a = 1, 2                  # global variables vo, a
f = lambda t: vo*t+a*t**2/2   # define lambda function f(t)

ypos = []                     # empty list ypos

for t in range(4):            # for loops require indentation
    ypos.append(f(t))         # add value of f to the list ypos

print('position y(t) = ',ypos)

-------------------------- CODE OUTPUT --------------------------

position y(t) =   [0.0, 2.0, 6.0, 12.0]
```

In many occasions we need to execute parts of the code only if certain conditions are true. In Python such conditional statements are implemented using the `if`, `elif` and `else` keywords, as in Example 1.8. The conditional statement also uses indentation, just like the functions and `for` loops.

Example 1.8: A conditional loop

Repeat the previous example, however this time use an `if` statement inside the for loop, so that the code stores only values of the position which are smaller than 7.

Solution:
In this example the ypos variable is evaluated and stored only if the conditional statement `f(t)<7` is true, otherwise the code line `ypos.append(f(t))` is ignored.

```
print('-'*28,'CODE OUTPUT','-'*29,'\n')


vo, a = 1, 2               # global variables xo, vo, a
f = lambda t: vo*t+a*t**2/2   # define lambda function f(t)


ypos = []                          # empty list ypos

for t in range(4):         # for loops require indentation
    if f(t)<7:             # if statements require indentation
        ypos.append(f(t))  # if f(t)<7 is true, then add value
                           # of f(t) to the list ypos
                           # if f(t)>=7, then ignore this statement

print('position y(t) = ',ypos)

------------------------- CODE OUTPUT ----------------------------

position y(t) =  [0.0, 2.0, 6.0]
```

Python allows function arguments to have default values. If a function is called without a particular argument, its default value will be taken. By using this feature, the same function can be called with different number of arguments. The arguments *without* default values must appear first in the argument list and they cannot be omitted while invoking the function. Example 1.9 shows how a function `f` can be called using three arguments in the form `f(vo, t, a = 2.0)`, or using two arguments in the form `f(vo, t)`.

Example 1.9: Named variables in functions

In Example 1.6 we used the function `f(vo,a,t)`. Write a code which sets the default value of a to be 2. Using $vo = 1$, calculate f using the default value of a, and also using the value a$= 3$, for the time values of $t = 0, 1, 2$ and 3.

Solution:
Note that for the default parameter value, f is called using only two parameters as `f(vo, t)`, with the third parameter a using the default value of $a = 2$. However, all three parameter values must be specified in the case where the default value of a is no longer used.

```
def f(vo,t,a=2):            # define function f with default a=2
    y = vo*t+a*t**2/2
    return y

ypos, ypos2 = [], []        # empty lists

for t in range(4):          # for loop evaluates function f twice
    ypos2.append(f(1,t))    # call f with 2 arguments and default a=2
    ypos.append(f(1,t,3))   # call f with 3 arguments and a=3

print('-'*28,'CODE OUTPUT','-'*29,'\n')

print('Using three calling parameters, position y(t)=',ypos)
print('Using two calling parameters, position y(t)=  ',ypos2)

-------------------------- CODE OUTPUT --------------------------

Using three calling parameters, position y(t)= [0.0, 2.5, 8.0, 16.5]
Using two calling parameters, position y(t)=   [0.0, 2.0, 6.0, 12.0]
```

1.4 IMPORTING PYTHON LIBRARIES AND PACKAGES

One of the major advantages of Python is the availability of modules, libraries and packages for various scientific applications. In this text, we will use the term *library* as a generic term for all three.

There are several different ways to import Python libraries and the most common methods are explained in Example 1.10, by using the numpy library and the cosine function cos() as an example.

The simplest way to import functions from the NumPy library is using numpy.cos(0.5), where the function is invoked using the form library_name.function_name() .

In the second method, we use an *alias* for the module name, so that we do not have to type repeatedly long module names. In this example import numpy as np, and we call the cosine function in the form np.cos(0.5).

In the third method of importing a function we use from numpy import cos and it is understood that the cos() function in the code will refer to the numpy library.

In the fourth method we use the character * as a wild card for importing all available functions with the code line from numpy import *. In this last method, we do not need to type the name of the alias or library. However, one must be careful when using the * method, since this can cause trouble when more than one modules are imported, which could be using the same name for different functions.

Other common libraries we will work with are SymPy (Symbolic Python), SciPy (Scientific Python) and the graphics library Matplotlib. These libraries are discussed later in this chapter.

Example 1.10: Importing libraries

Write a Python code to demonstrate the four methods described above for importing the numpy package, and then evaluate the cosine function cos(0.5) in each method.

Solution:
The comments in this code explain the differences between the four methods of loading and using the package.

```
print('-'*28,'CODE OUTPUT','-'*29,'\n')

import numpy            # method 1: use name.function() syntax
print('Using NumPy function, result is: ', numpy.cos(0.5))

import numpy as np      # method 2: use alias np, instead of numpy
print('Using np shorthand notation, result is: ', np.cos(0.5))

from numpy import cos  # method 3: import only function cos() from numpy
print('Importing just the NumPy function, result is: ', cos(0.5))

from numpy import *    # method 4: import all necessary from numpy
print('Importing all NumPy functions, result is: ', cos(0.5))

-------------------------- CODE OUTPUT --------------------------

Using NumPy function, result is:  0.8775825618903728
Using np shorthand notation, result is:  0.8775825618903728
Importing just the NumPy function, result is:  0.8775825618903728
Importing all NumPy functions, result is:  0.8775825618903728
```

Libraries are sometimes organized into multiple modules. For example, the module name A.B indicates a *module* B contained within a *library* named A. For example, `numpy.random.normal()` refers to the function `normal()` contained within the module `random` of the library `numpy`. The general format for this type of function is `package.module.function()`.

1.5 THE NUMPY LIBRARY

NumPy is the core Python library for numerical computing. NumPy supports operations on compound data types like arrays and matrices. The first thing to learn is how to create arrays and matrices using the NumPy library. Python lists can be converted into multi-dimensional arrays.

In this book we usually adopt the standard convention and import NumPy using the alias name np, with the code line `import numpy as np`. As mentioned above, one can also import NumPy functions using the syntax `from numpy import *`. If NumPy is the only package being used in the code, then there is no possibility of any function name conflicts.

1.5.1 CREATING NUMPY ARRAYS

The fundamental object provided by the NumPy library is the `ndarray`. We can think of a 1D (1-dimensional) ndarray as a list, a 2D (2-dimensional) ndarray as a matrix, a 3D (3-dimensional) ndarray as a 3-tensor and so on.

Using the function `np.array()` we can create a NumPy array from a Python sequence, such as a list, a tuple or a list of lists. For example

```
a = np.array([1,5,2])
```

creates a 1D NumPy array from a Python list.

There are several other functions that can be used for creating different types of arrays and matrices. Some examples of such array functions are shown in Table 1.2.

Table 1.2

NumPy functions for creating arrays and matrices.

Function	Description
`np.array(a)`	Create N-dimensional NumPy array from sequence *a*
`np.arange(start, stop, step)`	Create an evenly spaced 1D array from *start* to *stop*, excluding the *stop* value.
`np.linspace(start, stop, N)`	Create a 1D array with length equal to N, including the *start* and *stop* value.
`np.zeros(shape)`	Create array of given shape and type, filled with zeros
`np.ones(shape)`	Create an array of given shape and type, filled with ones
`np.random.random(shape)`	Create an array of given shape and type, filled with random float numbers from 0 to 1
`np.reshape(array, newshape)`	Changes the dimensions of a 1D array

Example 1.11 applies some of these array functions to create various types of arrays. In the first two lines of code, notice that when we print a NumPy array it looks a lot like a Python list, except the elements are separated by *spaces*, while in a list the elements are separated by *commas*.

The function `np.arange(start,stop,step)` is very similar to the list function `range(start,stop,step)` that we saw before when discussing lists. In Example 1.11, the code line `np.arange(0,3.5,.5)` creates a 1D NumPy array with values from 0 to 3.5 in steps of .5, *excluding* the end value of 3.5.

The function `np.linspace(start,stop,N)` is slightly different, since we specify the number of points N between the *start* and *end* values, instead of the *step* in the array. In the same example, `np.linspace(0,8,6)` creates an array with exactly 6 elements between 0 and 8, *including* the end value 8.

Similarly `np.zeros(5)` creates a 1D NumPy array of zeros of length 5. Similarly, `np.zeros(2,1)` creates a 2D NumPy array of zeros with 2 rows and 1 column. The code line `np.random.random([1,2])` will generate a matrix with one row and two columns, with random values between 0 and 1.

Example 1.11 demonstrates various types of arrays, and how to use the `np.arange()`, `np.linspace()`, `np.zeros()` and `np.random.random()` functions.

Example 1.11: Creating arrays in Python

Write a simple code to show how to create arrays using the `np.arange()`, `np.linspace()`, `np.zeros()` and `np.random.random()` functions.

Solution:
The comments in the code explain the various commands.

```
import numpy as np
print('-'*28,'CODE OUTPUT','-'*29,'\n')
print('printing the list [1,2,3]:              ',[1,2,3])
print('printing the array np.array([1,2,3]): ',np.array([1,2,3]),'\n')

print('np.arange(5) gives: ',np.arange(5))
print('np.arange(0,3.5,.5) gives: ',np.arange(0,3.2,.5),'\n')
# array from 0 to 3.5 steps 0.5 (excluding 7)

print('np.linspace(0,8,6) gives: ',np.linspace(0,8,6),'\n')
# create array with 6 values between 0 and 8

print('array with zeros: ',np.zeros(5))  # generate array with 5 zeros
print('array with ones: ',np.ones([2,1]),'\n')
# generate two rows and one column, with ones

print('array with random values: ',np.random.random([1,2]))
# generate random numbers in one row and two columns, in interval (0,1)

--------------------------- CODE OUTPUT ---------------------------

printing the list [1,2,3]:              [1, 2, 3]
printing the array np.array([1,2,3]):  [1 2 3]

np.arange(5) gives:  [0 1 2 3 4]
np.arange(0,3.5,.5) gives:  [0.  0.5 1.  1.5 2.  2.5 3. ]

np.linspace(0,8,6) gives:  [0.  1.6 3.2 4.8 6.4 8. ]

array with zeros:  [0. 0. 0. 0. 0.]
array with ones:  [[1.]
 [1.]]

array with random values:  [[0.26177687 0.78577744]]
```

1.5.2 ARRAY FUNCTIONS, ATTRIBUTES AND METHODS

In this section we present a brief overview and examples of working with Python arrays. We will look at how we can access the properties (or *attributes*) of an array and will examine some of the functions that can be used with NumPy arrays.

Table 1.3

Useful commands (*methods*) for extracting the properties or attributes of an array.

Command	Description
A.dtype	prints the data type for NumPy array A
A.ndim	prints the number of dimensions for NumPy array A
A.size	prints the size (total number of elements) for NumPy array A
A.shape	finds the number of rows and columns for NumPy array A

Example 1.12 applies various functions on arrays and also shows how we can extract some of the important properties of arrays. Some examples of useful commands for extracting the properties of arrays are shown in Table 1.3.

Note that the syntax in Table 1.3 is A.dtype, A.size etc. These are examples of using methods objects in Python, as opposed to the functions objects we have looked at so far in this book. Recall also that we use the function type(A) for a list A, but we use the similar command A.dtype for a NumPy array A. This is an example of using a *function* versus using a *method*.

Functions are called by placing argument expressions in parentheses after the function name as in type(a), where a is an object (list, string, float, etc.). The functions that we define in this book will always be called using the function name first, followed in parenthesis by all of the arguments of the function.

Methods are somewhat similar to functions, but they are called using the dot notation, such as A.dtype. Loosely speaking, methods are always attached to a specific object, while functions are isolated. Another example of a function versus a method is the implementation of the dot and cross products of two vectors A and B in SymPy and NumPy. In SymPy the dot product is implemented as a method A.dot(B), while in NumPy the dot product is represented as a function dot(A,B). We will later see several more different types of functions and methods within the NumPy, SymPy and SciPy packages used in this book.

Example 1.12 shows how we can use some of these functions with arrays. All entries in a NumPy array are of the *same* data type. We will mostly work with numeric arrays which contain integers, floats, complex numbers or booleans. We will also mostly be working with the default integer type numpy.int64 and the default float type numpy.float64. We can access the datatype of a NumPy array A by its A.dtype attribute. In Example 1.12 we create a 2D NumPy array from a list of lists of integers using the lines of code

```
A = np.array([[1,2,3],[4,5,6]])
A.dtype
```

prints out dtype('int32') i.e the data is of integer type and is represented in the computer's memory by 32-bits. Similarly, A.ndim tells us that A has 2 dimensions, with the first dimension corresponding to the vertical direction counting the *rows*, and the second dimension corresponds to the horizontal direction counting the *columns*.

In Example 1.12, A.shape gives (3, 2), i.e. the result is a tuple (3,2) with 3 rows and 2 columns. Similarly, we create a 1D array using u = np.linspace(0,1,5), and u.dtype prints out dtype('float64') i.e. the data are floating type real numbers represented by 64 bits.

Finally, we can find out the total elements in array A with A.size.

Example 1.12: Attributes of Nympy arrays

Write a simple code to demonstrate how to use `A.dtype`, `A.ndim`, `A.shape` and `A.size` with a NumPy array `A=[[1,2,3],[4,5,6]]`.

***Solution*:**

```
print('-'*28,'CODE OUTPUT','-'*29,'\n')
import numpy as np

A = np.array([[1,2,3],[4,5,6]])
print('A = ',A)
print('Data type of A is: ', A.dtype,'\n')

u = np.linspace(0,1,5)
print('u = ',u)
print('Data type of u is: ', u.dtype,'\n')

A = np.array([[1,2,3],[3,4,5]])
print('A = ',A)
print('dimension of A is:', A.ndim)
print('shape of A is: ', A.shape)
print('size of A is: ', A.size)

--------------------------- CODE OUTPUT ---------------------------

A =  [[1 2 3]
 [4 5 6]]
Data type of A is:  int32

u =  [0.   0.25 0.5  0.75 1.  ]
Data type of u is:  float64

A =  [[1 2 3]
 [3 4 5]]
dimension of A is: 2
shape of A is:  (2, 3)
size of A is:  6
```

In the next section we see how to carry out arithmetic operations with various types of arrays.

1.5.3 ARITHMETIC OPERATIONS WITH NUMPY ARRAYS

Mathematical functions in NumPy are *vectorized*, and a partial list of functions we can use to compute with NumPy arrays is given in Table 1.4.

Vectorized functions operate element-wise on arrays and produce new arrays as output. For example

$$np.sin(2*np.pi*x)$$

computes the values for each elements of the array

```
x = np.arange(0,1.25,0.25)
```

These vectorized functions compute values across arrays very quickly. NumPy also provides mathematical constants such as π (`np.pi`), and e (`np.e`).

We can modify the contents of a NumPy array using *indexing* and *slicing*, just as in the case of lists.

Table 1.4

Partial list of useful array functions in Python.

Array functions in NumPy

np.sum	np.argmax	np.min	np.std
np.max	np.argmin	np.mean	np.prod

Mathematical functions in NumPy

np.sin	np.exp	np.arcsin
np.cos	np.log	np.arccos
np.tan	np.log10	p.arctan

Mathematical constants in NumPy

np.pi	np.e

Arithmetic operations are applied to NumPy arrays element-by-element: these include addition +, subtraction -, multiplication *, division / and exponentiation **. See Example 1.13 for a demonstration of the addition of two NumPy arrays.

Normally we can only add vectors or matrices of the same size. However, NumPy has a set of rules called *broadcasting,* which allows the combination of arrays of different sizes when the process makes sense. For example, we can add a constant to a 1D NumPy array, by using the expression

```
1+np.array([5,6,7])**2
```

In this example of broadcasting, the array `np.array([5,6,7])` is squared, and then we add the array of ones [1. 1. 1.] which has the same shape. Here we are adding the scalar number 1 to a 1D NumPy array of length 3, and the broadcasting rule allows us to use a very simple syntax.

Example 1.13 demonstrates various array operations. For example, to compute the average or mean of the values in an array, we use

```
np.mean([5,6,7])
```

Similarly, to find the index of the maximum element in the array, we use

```
np.argmax([5,6,7])
```

which gives the value of the index 2, i.e. it identifies the third element in the array as the max value.

Array functions apply to multi-dimensional arrays as well, but we have to choose along which axis of the array to apply the function. For example, for the array

```
M = np.array([[2,4,2],[2,1,1],[3,2,0]])),
```

the function

```
np.sum(M,axis=0)
```

will sum the *columns* of M. However, the function

```
np.sum(M,axis=1)
```

will sum the *rows* of M. In a similar manner, a 3D array of size 3 x 3 x 3 can be summed over each of its three axes.

Example 1.13: Arithmetic operations on arrays- Broadcasting

Define two NumPy arrays, a and b from the lists $[1, 2, 3]$ and $[4, 5, 6]$, respectively. In addition, define a 3×3 matrix M of your choosing as a NumPy array.
Write a simple code to demonstrate the following:

1. The use of the functions mean(a), argmax(a) , std(a) for the NumPy array a.

2. The result of a + b and a * b

3. Broadcasting, using $1 + $ a**2

4. The use of the functions np.sum(M,axis=0) , np.sum(M,axis=1) .

Solution:

```
import numpy as np
print('-'*28,'CODE OUTPUT','-'*29,'\n')

a = np.array([1,2,3])
b = np.array([4,5,6])
M = np.array([[2,4,2],[2,1,1],[3,2,0]])

#Part 1
print('mean of elements in array a = ',np.mean(a))
print('std dev of elements in array a = ',np.std(a))
print('max of elements in array a = ',np.max(a))
print('sum of elements in array a = ',np.sum(a),'\n')

#Part 2
print('a + b = ' , a+b)
print('a * b = ', a*b, '\n')

#Part 3
print('1 + a**2 = ', 1 + a**2,'\n')

#Part 4
print('matrix M =',M,'\n')
print('sum of elements of M along axis=0 (rows) = ',np.sum(M,axis=0))
print('sum of elements of M along axis=1 (columns) = ',np.sum(M,axis=1))

-------------------------- CODE OUTPUT --------------------------

mean of elements in array a =  2.0
std dev of elements in array a =  0.816496580927726
```

```
max of elements in array a =   3
sum of elements in array a =   6

a + b =  [5 7 9]
a * b =  [ 4 10 18]

1 + a**2 =  [ 2  5 10]

matrix M = [[2 4 2]
 [2 1 1]
 [3 2 0]]

sum of elements of M along axis=0 (rows) =   [7 7 3]
sum of elements of M along axis=1 (columns) =   [8 4 5]
```

1.5.4 INDEXING AND SLICING OF NUMPY ARRAYS

NumPy arrays can be indexed, sliced and copied, just like Python Lists.

Example 1.14 shows how to use slicing and indexing in multidimensional NumPy array. For example, the fourth element in array v is accessed with v[3] and we can use two indexes v[1,2] to find the element at row index 1 and column index 2. Similarly v[-1,-1] accesses the last element. The syntax v[2,:] addresses the *row* with index 2, and v[:,3] addresses the *column* with index 3.

We can also build bigger arrays by *stacking* smaller arrays along different dimensions, using the horizontal and vertical stacking functions np.hstack and np.hstack.

Example 1.14: Indexing and slicing of NymPy arrays

(a) Define a 3×3 matrix v of your choosing and demonstrate the use of indexing with v[3], v[1,2] and v[-1,-1].
(b) What elements of the matrix correspond to v[2,:] and v[:,3] ?
(c) Define two 1×3 matrices, and demonstrate the use of the horizontal and vertical stacking functions np.hstack() and np.hstack().

Solution:
In the Python code, the pprint package is imported and used to format the code output for the printed matrix.

```
import numpy as np
print('-'*28,'CODE OUTPUT','-'*29,'\n')

import pprint
pp = pprint.PrettyPrinter(width=41, compact=True)

# define 3x3 matrix using NumPy array
v = np.array([[1,2,3],[4,5,6],[7,8,9]])

print('array v = ')
pp.pprint(v)
```

```
print('The third element v[2] = ',v[2],'\n')

print('The element at row index 1 and column index 2=',v[1][2])
print('The element v[1,0] = ',v[1,0])
print('The last element v[-1,-1] = ',v[-1,-1],'\n')

print('The column with index 0 = ',v[:,0])
print('The row with index 2 = ',v[2,:],'\n')

x = np.array([1,1,1])
y = np.array([2,2,2])
print('array x = ',x)
print('array y = ',y)
print('vertical stack of x,y,x = ',np.vstack((x,y,x)))
print('horizontal stack of x,y,x = ',np.hstack((x,y,x)))


-------------------------- CODE OUTPUT ----------------------------

array v =
array([[1, 2, 3],
       [4, 5, 6],
       [7, 8, 9]])
The third element v[2] =  [7 8 9]

The element at row index 1 and column index 2= 6
The element v[1,0] =  4
The last element v[-1,-1] =  9

The column with index 0 =  [1 4 7]
The row with index 2 =  [7 8 9]

array x =  [1 1 1]
array y =  [2 2 2]
vertical stack of x,y,x =  [[1 1 1]
 [2 2 2]
 [1 1 1]]
horizontal stack of x,y,x =  [1 1 1 2 2 2 1 1 1]
```

1.6 THE MATPLOTLIB MODULE

Matplotlib is a Python library that can generate many different types of plots and graphical representations of data, using just a few lines of code. The user has full control of line styles, font properties, axes properties, etc. Often data points for the plotting functions are supplied as Python lists or as NumPy arrays.

In the next two subsections we will see how to create two and three-dimensional plots using Matplotlib. We will also briefly discuss two different methods of using Matplotlib, namely object oriented programming versus procedural programming.

1.6.1 2D PLOTS USING MATPLOTLIB

The general procedure to create a 2D plot is to first import the Python library *Matplotlib* and its submodule *pyplot*. Typically we will import the `matplotlib.pyplot` submodule using the short-hand alias `plt`. Next we create two sequences of x-values and y-values, and then use the general plot command

```
plt.plot(x,y,[fmt],**kwargs)
```

Here `[fmt]` is an optional string format and `**kwargs` are optional keyword arguments specifying the properties of the plot. To improve the appearance of the plot, we use `pyplot()` functions to add a figure title, legend, grid lines, etc. Finally, the line `plt.show()` displays the figure.

Let's begin with a basic example, solving a simple physics problem that you are already familiar with. A physicist throws a ball straight up with an initial speed $v_0 = 5$ m/s. We want to plot the position $y(t)$ as a function of time, and also find numerically the maximum height reached by the ball. You recall from Introductory Physics that the kinematic equations for this problem are:

$$y = v_0\, t - \frac{1}{2} g\, t^2 \tag{1.6.1}$$

$$v(t) = v_0 - g\, t \tag{1.6.2}$$

where $g = 9.8$ m/s^2 is the acceleration of gravity. Example 1.15 shows how we would solve this problem using the numerical capabilities of Python.

Example 1.15: Example of plot and numerical evaluation

Write a code to evaluate and plot the kinematic equation (1.6.1), from $t = 0$ to $t = 1$ s with $vo = 5$ m/s.

Solution:
The first line in the code `import numpy as np` causes Python to import the NumPy library, with the alias `np`. We import the Python library *Matplotlib* and its submodule *pyplot*, in order to plot the position $y(t)$. The code imports the `matplotlib.pyplot` submodule using the short-hand alias `plt`. After the libraries are imported, we define the initial condition variable vo and the acceleration of gravity g, in a single line vo, g = 5, 9.8.
We need to tell Python for which values of time t we will be computing the position $y(t)$. In this case, we use `t=np.linspace(0,1,100)` to define the array with the values of times t, which will be used to compute $y(t)$. We next use a *list comprehension* in the form

```
y=[vo*u-g*u**2/2 for u in t]
```

to obtain the array y containing the values of $y(t)$.
The pyplot commands `plot`, `title`, `ylabel`, `xlabel` are used to improve the appearance of the plot in Figure 1.1, and finally `plt.show()` prints the plot. The last 4 lines in the code show how to evaluate numerically the maximum height $ymax$ and the corresponding time $tmax$, by applying the functions `max()` and `argmax()` respectively.

```
print('-'*28,'CODE OUTPUT','-'*29,'\n')
import numpy as np
import matplotlib.pyplot as plt
vo, g = 5, 9.8                          # define values of vo, g
```

```
t = np.linspace(0,1.03,100)      # create sequence of times
y = [vo*u-g*u**2/2 for u in t] # create sequence of positions, then plot y(t)

plt.plot(t,y)

plt.title('Ball thrown straight up')  # add title to plot
plt.ylabel('Vertical distance y(t)')  # add labels for x and y axes
plt.xlabel('Time t [s]')

ymax = max(y)                    # find position ymax
tmax = t[np.argmax(y)]           # find time tmax

print('Max height reached = ',f'{ymax:.4f}', ' m')
print('Time to reach ymax = ',f'{tmax:.4f}',' s')

plt.show()

------------------------- CODE OUTPUT -------------------------

Max height reached =   1.2755  m
Time to reach ymax =   0.5098  s
```

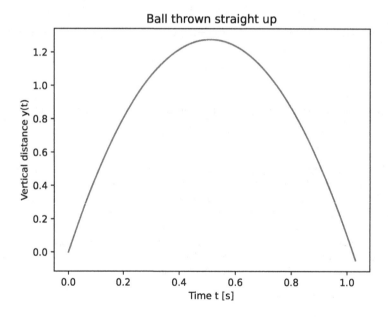

Figure 1.1 Plot of the kinematic equation (1.6.1), from $t = 0$ to $t = 1$ s with $vo = 5$ m/s.

We can improve further the plots by specifying various options inside the `plt.plot()` function. For example, we specify the *color* e.g. by name as in `color='red'`, or by a RGB tuple such as (1,0,1). We can also specify the type of line to be used with `linestyle='dashed'`, the type of marker for data points with `marker='o'`, the linewidth with `linewidth=2` etc.

It is very convenient to use combined format strings as in the command `plt.plot(x,y,'rd--')` where the string `'rd--'` signifies a red line ('r'), a diamond marker ('d') and a dashed line (--).

Examples of *colors* are b blue, g green, r red, c cyan, m magenta, y yellow, k black, w white. Examples of character *Markers* are. point, o circle, v triangle down, ^ triangle up, s square, p pentagon, * star, + plus, x x, D diamond. Finally, examples of *linestyles* are - solid line style, -- dashed line, -. dash-dot line, : dotted line.

The graphical commands in this example are an example of procedural programming. Next, we will use the Matplotlib module to produce 3D plots using object oriented programming.

1.6.2 3D PLOTS USING MATPLOTLIB

Example 1.16 demonstrates how to plot a surface $z = f(x, y)$ in 3D.

Example 1.16: Graphics: Plotting a surface $z = f(x, y)$

Write a Python code to plot the surface function $z = f(x, y) = x+y+3$ in 3D, when $-3 \leq x \leq 3$ and $-3 \leq y \leq 3$.

Solution:
The 3D plot in the example is created using the `plot_surface()` function within the `matplotlib` library. The lines `fig = plt.figure()` and and `fig.add_subplot(projection ='3d')` create the 3D plot, and the wireframe style is used for the plot.
The NumPy command `meshgrid(x, x)` creates double arrays X,Y from the single NumPy array x. It is necessary to created double arrays which are required as inputs for the function `plot_surface(X,Y,Z)`.
The plotting function `plot_wireframe(X,Y,Z)` plots the 3D surface of the plane $z = f(x, y) = 3 + x + y$ with the result shown in Figure 1.2.

```python
import numpy as np
import matplotlib.pyplot as plt


# surface plot for z=3+x+y
x = np.arange(-3, 3, 0.6)       # grid of points on x-axis
X, Y = np.meshgrid(x, x)        # grid of points on xy-plane
Z= 3+X+Y                        # values of z for points on the xy-grid

# plot 3D surface using object oriented programming commands
# define the objects fig and axes for 3D plotting
fig = plt.figure()
axes = fig.add_subplot(projection ='3d')

# plot 3D surface using wireframe style
axes.plot_wireframe(X, Y, Z, color='skyblue')

axes.set_xlabel('X')            # labels for x,y,z axes
axes.set_ylabel('Y')
axes.set_zlabel('Z')
```

```
axes.text(-2.6,2,6,'z=f(x,y)=3+x+y')   # add text to the plot

plt.show()
```

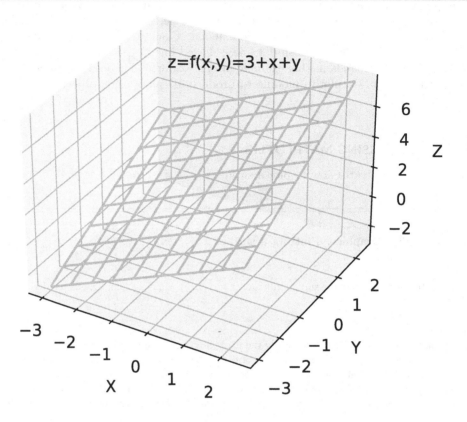

Figure 1.2   Plot of a 3D surface $z = f(x, y)$ from Example 1.16, using MatPlotLib.

1.7 SYMBOLIC COMPUTATION WITH SYMPY

Symbolic Computation involves using a computer to help find closed-form solutions to differential equations, integrals, eigenvalues, and many more types of symbolic evaluations. For example, we can enter the equation or integral we want solved, and the computer program returns a closed form solution, if it exists.

Python cannot perform symbolic manipulations by itself, so we need to import the SymPy library in order to expand Python's capabilities. This is sometimes done in the first line of the code with **import sympy as sym**, which causes Python to import the SymPy library with the alias **sym**. Most often, we will import from SymPy only the functions we need.

As an example of a symbolic type of computation, we will use SymPy to solve again *symbolically* the simple physics problem of throwing a ball straight up with an initial speed $v_0 = 5$ m/s. We want to evaluate symbolically the maximum height y_{max} reached by the ball (neglecting air resistance), and how long it takes to reach this maximum height.

As mentioned previously in this chapter, the kinematic equations for this problem are:

$$y = v_0\, t - \frac{1}{2} g\, t^2 \qquad (1.7.1)$$

$$v(t) = v_0 - g\, t \qquad (1.7.2)$$

By setting the speed $v(t) = v_0 - g\, t = 0$ at the maximum height reached by the ball, we find the corresponding time $t = v_0/g$. Substituting this value of t into (1.7.1) we find the maximum height:

$$y_{\text{max}} = v_0 \left(\frac{v_0}{g}\right) - \frac{1}{2} g \left(\frac{v_0}{g}\right)^2 = \frac{v_0^2}{2g} \qquad (1.7.3)$$

Example 1.17 shows how we would solve this problem using the symbolic capabilities of Python.

Example 1.17: Example of symbolic evaluation

Use SymPy to solve again the problem of throwing a ball straight up with an initial speed $v_0 = 5$ m/s. Evaluate symbolically the maximum height y_{max} reached by the ball (neglecting air resistance), and how long it takes to reach this maximum height.

Solution:
The following Python code evaluates the time t_{max} and the maximum height y_{max}, starting from (1.7.1) and (1.7.2). In this example, the SymPy library includes the function solve and the method .subs which can help us solve algebraic equations.
Notice that we tell Python that the variables vo, t, g need to be defined, using the symbols command, and that we also include the option real=True to signify that they are to be treated as real variables.
The code line solve(vo-g*t,t) solves symbolically the equation $v_0 - g\, t = 0$ for the variable t, while the line y.subs(t,tmax) substitutes symbolically the value of t with tmax.

```
from sympy import  symbols, solve
print('-'*28,'CODE OUTPUT','-'*29,'\n')

vo, t,  g = symbols('vo, t,  g ',real=True)   # define variables

# solve equation for tmax symbolically using solve() function
tmax = solve(vo-g*t,t)[0]
print('Time to reach max height=',tmax)

y = vo*t-g*t**2/2
# substitute tmax in y(t) to find ymax, using method .subs
print('Max height reached = ',y.subs(t,tmax))

---------------------------- CODE OUTPUT ----------------------------

Time to reach max height= vo/g
Max height reached =  vo**2/(2*g)
```

In the next section we provide an example which shows how to use the `lambdify` function in Python.

1.8   THE LAMBDIFY() FUNCTION IN PYTHON

Example 1.18 shows how one can combine the results from NumPy and SymPy libraries, by using the `lambdify()` function. The `lambdify()` function can calculate numerical values from SymPy expressions.

In Example 1.18, we are asked to evaluate a symbolic derivative and then plot its result. We use SymPy to evaluate the symbolic derivative of a function. However, `lambdify` must then be used, so that we can numerically evaluate the result before we create the plot.

Example 1.18: Example of using the lambdify function

Use SymPy to evaluate the symbolic derivative of the function $f = a\sin(bt)$ with respect to time t, and plot the function $f(t)$ and its derivative in the interval $t = 0$ to $t = 6$. Use the numerical values $a = 1$ and $b = 2$.

Solution:
The Python code imports the `lambdify` function with the code line

```
from sympy.utilities.lambdify import lambdify
```

Notice that we tell Python that the variables a, b, t need to be defined using the `symbols` command, and that we also include the option `real=True` to signify that they are to be treated as real variables. The `diff(f,t)` command is used to evaluate the symbolic derivative of the function $f = a\sin(bt)$ and the result is stored in the variable `deriv`.

Next, `f.subs(a:1,b:2)` substitutes the numerical values $a = 1$ and $b = 2$ into the symbolic derivative, and code line

```
y = lambdify(t,f.subs(a:1,b:2))
```

creates a function $y = f(t)$ which can be plotted using Matplotlib.
`plt.subplot(1,2,1)` creates the first of two subplots, arranged in 1 row and 2 columns, and the last index 1 refers to the first subplot. Similarly, `plt.subplot(1,2,2)` creates the second of the two subplots in Figure 1.3.

```python
import matplotlib.pyplot as plt
import numpy as np
from sympy import symbols, sin, diff
from sympy.utilities.lambdify import lambdify

print('-'*28,'CODE OUTPUT','-'*29,'\n')

a, b, t = symbols('a,b,t',real=True) # define symbols

f = a*sin(b*t)
deriv = diff( f,t)                    # evaluate symbolic derivarive of f
print('Symbolic Derivative v=dy/dt: ',deriv)

tims = np.linspace(0,6,50)            # sequence of times tims

# substitute a=1 and b=2 in f, using the .subs method
y = lambdify(t,f.subs({a:1,b:2}))

# plot y(t) and v(t)
plt.subplot(1,2,1)    # first subplot, positions y(t)
```

```
plt.plot(tims,y(tims))
plt.xlabel('Time [s]')
plt.ylabel('y(t)')

u = deriv.subs({a:1,b:2})      # substitute a=1 and b=2 in deriv
v = lambdify(t,u)              # lambdify creates derivative function v(t)

plt.subplot(1,2,2)            # second subplot, speed v(t)
plt.plot(tims,v(tims))
plt.xlabel('Time [s]')
plt.ylabel('Speed v(t)=dy/dt')
plt.tight_layout()            # create the 2 subplots with thight layout
plt.show()

--------------------------- CODE OUTPUT ----------------------------

Symbolic Derivative v=dy/dt:    a*b*cos(b*t)
```

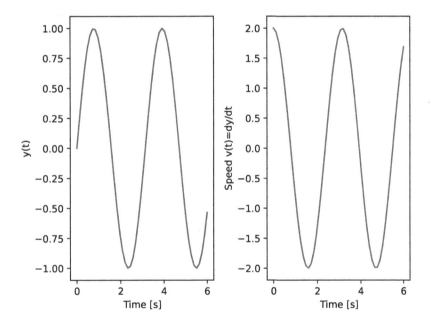

Figure 1.3 Graphical output of Example 1.18, showing subplots of a function $f = a\sin(b\,t)$ and its derivative.

1.9   END OF CHAPTER PROBLEMS

1. **A uniform distribution of random numbers** – Create 100 random numbers x from a uniform distribution using the **random()** function in Numpy. Plot these 100 random numbers on the xy-plane, and also in the form of a histogram. Find the average and the standard deviation of these 100 numbers.

2. **A Gaussian distribution of random numbers** – Create 100 random numbers from a Gaussian distribution using the `random.normal()` function in Numpy, and plot them as a histogram. Find their average and standard deviation.

3. **Example of using functions in Python** – Define the function $f(x, m, k) = e^{-mx} \cos(2\pi k x)$ and compute $f(x)$ for three different combinations of values of m and k.
   (a) Plot the 3 functions on the same plot, using different colors and line types for each curve.
   (b) Annotate the graph with the values of m, k for each plot.

4. **Using a grid of subplots** – Plot three functions from the previous problem with different values of k, m . Create a grid of three subplots. Label the axes and provide a title for each graph, and choose different colors and line types for each curve. Save the resulting graph to a .jpg or .png file.

5. **Using lambda functions** – Plot $f(x) = x^2/10$ and $g(x) = \sin x$ using lambda functions, for the range $x = 0$ to $x = 6$. Create a grid of three subplots, one subplot for each function and a third subplot showing both functions. Each function should have its own color and line type.

6. **Conditional statement** – Write a code in which you solve a cubic equation $x^3 + x^2 + 1 = 0$, then the code determines and prints whether the roots are real or complex. Use the `solve()` function in SymPy.

7. **Defining and plotting piecewise functions** – Using the conditional *if* statement, define the piecewise function

$$f(x) = \begin{cases} 1 & x < 5 \\ x^2 & x \geq 5 \end{cases}$$

and evaluate $f(x)$ for $x = 3$ and $x = 7$, and plot $f(x)$ from $x = 0$ to $x = 10$. Use the `piecewise()` function in NumPy to evaluate this piecewise defined function.

8. **The Fibonacci series** – The Fibonacci series of numbers x_n is defined by the recursive sequence $x_n = x_{n-1} + x_{n-2}$. Compute the first 17 values of the Fibonacci sequence. You may want to use `for` loops, and you may also want to save your results into an array using the `append()` function in order to add values to an array.

9. **Solving algebraic equations** – (a) Use symbolic computation to solve each of these equations:

$7x + 5 = 0$
$x^2 - 5x + 2 = 0$
$x^3 + 7x - 5 = 3$

Note that you will get complex roots for some of the solutions.
(b) Use symbolic computation to solve the system of equations:

$2x - 5y = 7$
$x + y = 2$

10. **Polar plots** – Graph the functions $r = |\cos\theta|$ and $r = |\cos\theta\sin\theta|$ using polar plots in Python, for the range of angles $\theta = 0$ to $\theta = 2\pi$.

11. **Parametric plots in 3D** – Write a Python code to plot the parametric line $(x, y, z) = (t, 1 + 2\cos t, -1 + 3\sin t)$ for the range of values $t = 0$ to $t = 9$.

12. **Calculating the powers of a matrix** – Create the 2×2 square matrix $A = \begin{pmatrix} 1 & 2 \\ 1 & 0 \end{pmatrix}$ using a NumPy array, and use a `for` loop to calculate the matrix representing the powers A^2, A^3 and A^4 etc. Two matrices A, B can be multiplied using the matrix multiplication symbol `@`, in the form `A@B`.

13. **Evaluation of product series for $\cos x$** – Define a Python function $f(x, N)$ containing the parameters x and N, and evaluate the following product series for $\cos x$:

$$\cos x = \prod_{k=1}^{N} \left(1 - \frac{4x^2}{\pi^2 (2k-1)^2} \right)$$

Run the code for $N = 10$, 50, 100 terms and for a value of $x = 1$, and compare the value of the above product with the value of $\cos x$ at $x = 1$.

14. **Evaluation of the sum of infinite series** – Use the `Sum()` function and `.doit()` in SymPy to evaluate the infinite sums:

$$\sum_{n=0}^{\infty} \frac{(-1)^n}{(2n+1)!} x^{2n+1} \qquad \sum_{n=0}^{\infty} \frac{1}{(2n)!} x^{2n} \qquad \sum_{n=0}^{\infty} \frac{(-1)^n}{(2n)!} x^{2n} \qquad \sum_{k=0}^{\infty} \frac{x^k}{k!} \qquad \sum_{k=0}^{\infty} \frac{x^k}{k}$$

15. **Evaluation of infinite series for π^2** – Define a Python function $f(x, N)$ containing the parameters x and N, and evaluate the following series: '

$$\frac{\pi^2}{12} = \sum_{n=1}^{N} \frac{(-1)^{n+1}}{n^2}$$

Run the code for $N = 10$, 50, 100 and compare the sum with the value of $\pi^2/12$.

16. **Constructing multidimensional arrays and lists** –
(a) Use a list comprehension method to construct a list with the values of (x, x^2, x^3) from $x = 1$ to 10.
(b) Use the functions `zip()` and the function `list()` in Python to create a list of pairs (y_1, y_2) from two arrays y_1 and y_2.

17. **The Bessel functions** – The Bessel function $J_p(x)$ of order p is defined by the infinite series:

$$J_p(x) = \sum_{n=0}^{\infty} \frac{(-1)^n}{\Gamma(n+1)\,\Gamma(n+1+p)} \left(\frac{x}{2}\right)^{2n+p} \tag{1.9.1}$$

where $p > 0$ and $\Gamma(n)$ is the Gamma function. Write a Python function to evaluate numerically this infinite series using the `summation()` function in SymPy. Evaluate and plot the Bessel functions $J_p(x)$ for $p = 0$, 2 and for the range of values $x = -20$ to $x = +20$.

18. **Plots of the sum of terms in a Fourier series** – Plot the first $N = 5, 10, 20, 30$ terms of the Fourier series for a sawtooth wave:

$$T(x) = \frac{50}{\pi} \sum_{n=0}^{\infty} \frac{(1)^{n-1}}{n} \sin\left(\frac{n\pi}{L} x\right) \tag{1.9.2}$$

which is defined as $f(x) = 5x$ for $0 < x \le 5$, and as $f(x) = -5x$ for $10 \ge x > 5$. The length $L = 5$.

19. **Plots of the sum of terms in a Taylor series** – Plot the first $N = 5, 6, 8, 10$ terms of the Taylor series for $\cos x$ from $x = 0$ to $x = 8$.

$$T(x) = \sum_{n=1}^{\infty} \frac{(-1)^n}{2n+1} x^{2n} \tag{1.9.3}$$

20. **3D plots of random points** – Write a Python code to plot 100 random points (x, y, z) in 3D.

21. **The Pauli matrices** – Look up the 3 Pauli matrices σ_x, σ_y, σ_z, which are used in Quantum Mechanics to described the spin of an electron. Verify the following identities using matrix algebra in Python:

$$\sigma_x^2 = \sigma_y^2 = \sigma_z^2 = -i\sigma_x\sigma_y\sigma_z = \begin{pmatrix} 1 & 0 \\ 0 & 1 \end{pmatrix} = \mathcal{I}$$

$$\sigma_x\sigma_y = i\sigma_z \qquad \sigma_x\sigma_y + \sigma_y\sigma_x = 0 \qquad \sigma_x\sigma_y - \sigma_y\sigma_x = 2i\sigma_z$$

where $\mathcal{I}$ is the identity matrix.

22. **Plotting circles, ellipses, rectangles, arrows** – Look up the `matplotlib.patches` library in Python, and write a code that plots a circle, an ellipse and several arrows inside a rectangle on the xy-plane.

23. **Symbolic matrix multiplication** – The following matrix B represents a counter-clockwise rotation of a point in the xy-plane by the angle ϕ about the z-axis.

$$B = \begin{pmatrix} \cos\phi & -\sin\phi \\ \sin\phi & \cos\phi \end{pmatrix}$$

Rotating a point by an angle ϕ_1 and then by an angle ϕ_2, is carried out by multiplying the corresponding matrices for the two angles. Use SymPy to show that the double rotation is the same as performing a single rotation by the angle $\phi_1 + \phi_2$.

24. **Transpose and inverse of a matrix** – Define a 3×3 matrix M with random integers between 0 and 2 as elements. Write the Python code which finds the transpose of M, and verify that the product of M and its transpose is a symmetric matrix.

25. **Complex numbers in electronics** – Complex numbers are used in the description of AC electrical circuits. Specifically, we use the complex impedance Z to describe the properties of resistors, inductors and capacitors. The impedance of a *resistor* R is a real number given by $Z_R = R$, while an inductors L has a purely imaginary *inductive impedance* $Z_L = i\omega L$, and capacitors C have a purely imaginary *capacitative impedance* $Z_C = -i/(\omega C)$.
    (a) Calculate symbolically the real and imaginary parts of the total impedance $Z = Z_c + Z_R + Z_L$ by using the functions `Z.real` and `Z.imag` commands.
    (b) Obtain the value of Z by using the `.subs` function and the numerical values of $R = 5$ Ohm, $L = 1$ mH, $C = 1$ mF and $\omega = 10$ s^{-1}.

26. **Finding maxima, minima and inflection points** – Given the function

$$f(x) = x^{-6} - x^{-12}$$

with $x > 0$, evaluate its first and second derivatives using the `diff()` command in SymPy. Also evaluate the max/min points and the inflection points (if any) for $f(x)$.

27. **Plotting the Bose Einstein distribution at different temperatures** – The Bose Einstein distribution was introduced for photons by Bose, and generalized to atoms by Einstein. The expected number of particles $n(E)$ in an energy state E and with temperature T (in K) is:

$$n(E, T) = \frac{1}{e^{E/(k_{\mathbf{B}}T)} - 1}$$

where k_B is the Boltzmann constant. Write a Python function that evaluates $n(E, T)$, and plot this function for three different temperatures $T = 2000, 4000, 6000$ K on the same plot.

Use a range of E values from $E = 0$ eV to $E = 30$ eV and label the 3 plots appropriately, and also use a different symbol for each plot.

28. **3D graphics** – Create three-dimensional plots of these objects. In each case label the axes and the plots.
    (a) A 3D line defined by $(x, y, z) = (t, 1 + 2t, 1 + 3\sin t)$
    (b) A cone of radius $R = 1$ and height $H = 1$
    (c) A sphere of radius $R = 1$
    (d) A cylinder of radius $R = 1$ and height $H = 1$
    (e) An ellipsoid with semi-major axes $a = 1$, $b = 0.6$ and $c = 1$

2 Differentiation

Why are derivatives important in science and engineering?

One of the goals of physics, and science in general, is building predictive models. In other words, based on what we know about a given system, can we predict its future state? To build a successful predictive model, one must know how the values which describe the modeled system change with time.

For example, consider Newton's second law, $F = ma$, where a is the acceleration of an object with mass m experiencing a net force F. Acceleration measures the rate at which the velocity of an object changes, i.e. $a = dv/dt$. If we know the value of a, and the object's initial velocity v_0, we can predict the object's velocity at any time t in the future using the constant acceleration equation, $v = v_0 + at$. Put another way, this well-known equation tells us that if we know the rate at which the velocity changes (a) and how long it changes (t), we can predict the object's velocity. Notice the importance of knowing how the object's velocity changes, i.e. the acceleration. Without knowledge of a, we cannot predict v. It should also be mentioned that to accurately predict the object's velocity, we also need v_0. In other words, rates of change alone are not enough to make predictions, we need to also know a past or current state of the system.

Many laws of physics provide us with the means of finding the rate of change of a physical quantity, i.e. its derivative. Hence, laws of physics are often about rates of change, not the value of particular quantities.

In this chapter, we review differentiation of functions of single and multiple variables, and show how to use Python to evaluate derivatives and total differentials, how to compute Taylor and Maclaurin series, and to find extrema of functions. A section is devoted to power series approximations of functions, and we give examples of how to compute Taylor and Maclaurin series of functions. We also discuss numerical calculations of derivatives from data.

2.1 DERIVATIVES OF SINGLE-VARIABLE FUNCTIONS

We begin with Newton's basic definition of the derivative. Consider a single-variable function $f(x)$ where x is the independent variable. The derivative of $f(x)$ at the point $x = a$ can be found using

$$\left.\frac{df}{dx}\right|_{x=a} = \lim_{\Delta x \to 0} \frac{f(a + \Delta x) - f(a)}{\Delta x}. \tag{2.1.1}$$

Ignoring the limit in (2.1.1), we see that the right-hand side is of the form of a rise over a run:

$$m = \frac{\Delta y}{\Delta x} = \frac{f(a + \Delta x) - f(a)}{\Delta x} \tag{2.1.2}$$

where Δx is the independent variable. In other words, Newton's definition of a derivative provides the slope of a line. We can interpret the slope m as the average rate of change of f between points $x = a$ and $x = a + \Delta x$. Figure 2.1 illustrates how to interpret the limit $\Delta x \to 0$ in (2.1.1) as a graphical representation of the derivative of f with respect to x evaluated at $x = a$.

The slopes of the solid black lines eventually converge to the line tangent to $f(x)$, at the point $x = a$ (the gray line in Figure 2.1). Hence, in the limit $\Delta x \to 0$, (2.1.1) gives the slope of the tangent line. As can be seen in Figure 2.1, the tangent line can be used to make reasonable predictions about the value of $f(x)$ near $x = a$.

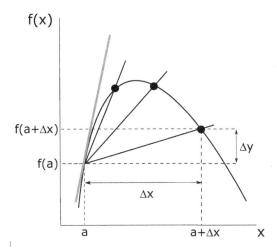

Figure 2.1 Geometrical representation of the derivative of function $f(x)$ at $x = a$. At the limit of $\Delta x \to 0$, the slopes of solid black lines converge to the derivative $f'(a)$, represented by the slope of the tangent gray line.

In the rest of this chapter, we will assume that you are familiar with basic derivatives such as the derivative of x^n, $\cos x$, $\sin x$, e^x, and so on. We will focus on how we can perform more complex calculations by hand, as well symbolically and numerically using Python.

2.1.1 RULES FOR DIFFERENTIATION

Newton's definition of the derivative is an important formula to know. It forms the basis for defining the derivatives of many other objects, such as vectors. However, derivatives in practice are most often done using particular rules and applying specific formulas.

Table 2.1 gives some of the most frequently used derivatives and summarizes the basic differentiation rules. In Table 2.1 f, g represent continuous differentiable functions of a single variable x, and a, b represent constants. The primes in this table denote derivatives with respect to x. Proving the rules in Table 2.1 generally involves using only (2.1.1).

In Example 2.1 we show an application from optics, which demonstrates how to evaluate and plot a function and its derivatives in Python. Specifically, we plot the intensity of light from a single slit diffraction experiment, and also evaluate symbolically its first and second derivatives.

Table 2.1

Common derivative formulas

Type of function	Derivative
Constant rule	$\frac{d}{dx}(a) = 0$
Sum rule	$\frac{d}{dx}(a\,f + b\,g) = a\,\frac{df}{dx} + b\,\frac{dg}{dx}$
Ratio rule	$\frac{d}{dx}\left(\frac{f}{g}\right) = \frac{f'\,g - g'\,f}{g^2}$
Product rule	$\frac{d}{dx}(f\,g) = f'\,g + g'\,f$
Chain rule	$\frac{d}{dx}f(u) = \frac{df}{du}\frac{du}{dx}$
Powers	$\frac{d}{dx}(x^n) = n\,x^{n-1}$
	$\frac{d}{dx}(a^x) = a^x \ln(a)$
Exponentials	$\frac{d}{dx}(e^x) = e^x$
Logarithms	$\frac{d}{dx}(\ln x) = 1/x$
Trig functions	$\frac{d}{dx}(\sin x) = \cos x$
	$\frac{d}{dx}(\cos x) = -\sin x$
	$\frac{d}{dx}(\tan x) = \frac{1}{\cos^2 x}$
	$\frac{d}{dx}\left(\sin^{-1} x\right) = \frac{1}{\sqrt{1-x^2}}$
	$\frac{d}{dx}\left(\cos^{-1} x\right) = \frac{-1}{\sqrt{1-x^2}}$
	$\frac{d}{dx}\left(\tan^{-1} x\right) = \frac{1}{1+x^2}$

Example 2.1: Light intensity from a single slit

The intensity of light from a single slit diffraction experiment is given by the expression:

$$I(\phi) = I_0 \frac{\sin^2 \phi}{\phi^2} \tag{2.1.3}$$

where I_0 is the maximum intensity and ϕ is a dimensionless constant which depends on the experimental conditions. Physically ϕ represents a phase angle characterizing the light intensity in the diffraction pattern. Use Python to calculate and plot the three functions $I(\phi)$, $dI/d\phi$ and $d^2I/d\phi$ on the same plot. Use a numerical value of $I_0 = 1$ SI unit, to simplify the calculations.

Solution:
We begin by using the ratio rule to calculate the first and second derivatives of $I(\phi)$:

$$\frac{dI}{d\phi} = \frac{d}{d\phi}\left[\frac{\sin^2 \phi}{\phi^2}\right] = \frac{2\phi^2\,\sin\phi\,\cos\phi - 2\phi\sin^2\phi}{\phi^4} \tag{2.1.4}$$

Similarly:

$$\frac{d^2I}{d\phi^2} = \frac{d}{d\phi}\left[\frac{dI}{d\phi}\right] = \frac{2\cos^2\phi}{\phi^2} - \frac{8\sin\phi\,\cos\phi}{\phi^3} + \frac{6\sin^2\phi}{\phi^4} - \frac{2\sin^2\phi}{\phi^2} \tag{2.1.5}$$

Below is the code for calculating the derivatives and the respective plots. Notice that we calculate the first and second derivatives using the SymPy command functions `diff(f,x)` and `diff(f,x,x)` respectively.

To create the plots in Figure 2.2, we need to use the `lambdify` function. We give Python the range of values for the plot with the command `np.linspace`, and the `np` in front of the command tells Python to use a command from the NumPy library. We need to use SymPy versions of the sine and cosine function, and we import these from SymPy. We use `lambdify()` in order to be able to plot the functions. We map f and its derivatives onto the array `phase`, and to create arrays for plotting.

```python
from sympy import symbols, sin, diff
import numpy as np
import matplotlib.pyplot as plt
from sympy.utilities.lambdify import lambdify
print('-'*28,'CODE OUTPUT','-'*29,'\n')


p = symbols('p',real=True)            # define symbols, p=phase phi


intensity = lambdify(p,sin(p)**2/p**2)              # light intensity I(p)
deriv = lambdify(p,diff(sin(p)**2/p**2,p))          # derivative dI/dp
secondDeriv = lambdify(p,diff(sin(p)**2/p**2,p,p))  # derivative of I(p)


phase = np.linspace(-7,7,100)                        # phase angle p

# plot I(p), dI/dp and second derivative together
plt.plot(phase,intensity(phase),'k-',label='Intensity I')
plt.plot(phase,deriv(phase),'b.',label='Derivative of I')
plt.plot(phase,secondDeriv(phase),'r--',label='2nd Derivative of I')


plt.xlabel(r'$\phi$')
plt.ylabel("I,  I',  I''")
leg = plt.legend()               # place legends of the three plots
leg.get_frame().set_linewidth(0.0) # remove the black box from around the legend

f1 = diff(sin(p)**2/p**2,p)
print('The First derivative dI/dp is:')
print(f1)

f2 = diff(sin(p)**2/p**2,p,p)
print('\nThe Second derivative of I(p) is:')
print(f2)


plt.show()

------------------------- CODE OUTPUT -----------------------------

The First derivative dI/dp is:
2*sin(p)*cos(p)/p**2 - 2*sin(p)**2/p**3

The Second derivative of I(p) is:
2*(-sin(p)**2 + cos(p)**2 - 4*sin(p)*cos(p)/p + 3*sin(p)**2/p**2)/p**2
```

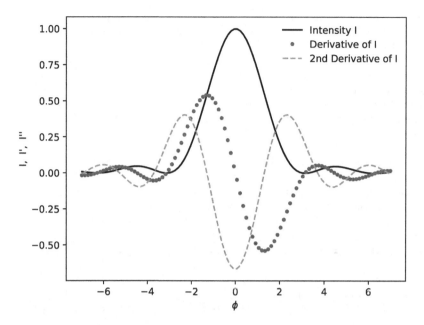

Figure 2.2 The graphical output from Example 2.1, showing plots of the intensity of light and its derivatives, from a single slit diffraction experiment.

Example 2.2 applies the chain rule to evaluate a derivative of a function. We use the `diff` command from SymPy to evaluate the result of the chain rule.

Example 2.2: The chain rule

The electric field E describing a light wave from a single slit diffraction experiment can be written in terms of the diffraction angle θ as:

$$E(\theta) = E_0 \frac{\sin (b \sin \theta)}{(b \sin \theta)} \qquad (2.1.6)$$

where E_0 is the maximum electric field and b is a dimensionless constant which depends on the experimental conditions. Calculate the derivative $dE/d\theta$ by hand using the chain rule, and verify the result by using SymPy. Use a numerical value of $E_0 = 1$ SI unit, to simplify the calculations.

Solution:
We begin by defining $u = b \sin \theta$, such that the function I becomes

$$I(u) = E_0 \frac{\sin u}{u} \qquad (2.1.7)$$

Using the chain rule:

$$\frac{dE}{d\theta} = \frac{dE}{du} \frac{du}{d\theta} \qquad (2.1.8)$$

Using the ratio rule:

$$\frac{dE}{du} = \frac{d}{du}\left[E_0 \frac{\sin u}{u}\right] = E_0\left(\frac{u\cos u - \sin u}{u^2}\right) \tag{2.1.9}$$

$$\frac{du}{d\theta} = \frac{d\,(b\,\sin\theta)}{d\theta} = b\,\cos\theta \tag{2.1.10}$$

Substituting (2.1.9) and (2.1.10) into (2.1.8):

$$\frac{dE}{d\theta} = E_0\left[\frac{(b\,\sin\theta)\cos(b\,\sin\theta) - \sin(b\,\sin\theta)}{(b\,\sin\theta)^2}\right]b\,\cos\theta \tag{2.1.11}$$

The SymPy code for the same calculation is done below. Notice the variables (x, b) need to be defined using the symbols command. The command diff is used to calculate the derivative.

```
from sympy import  symbols, diff, sin

x, b = symbols('x, b',real=True)
print('-'*28,'CODE OUTPUT','-'*29,'\n')


s = diff(sin(b*sin(x))/(b*sin(x)),x)


print('Derivative of E-field  is = ')
print(s)

---------------------------- CODE OUTPUT ----------------------------

Derivative of E-field  is =
cos(x)*cos(b*sin(x))/sin(x) - sin(b*sin(x))*cos(x)/(b*sin(x)**2)
```

2.2   DIFFERENTIATING ANALYTICAL FUNCTIONS IN PYTHON

In this section we present specific applications of Python differentiation methods from various areas of physics. We start with an example from Classical Mechanics.

In Example 2.3 we use SymPy to evaluate analytically the speed and acceleration of a falling object near the surface of the Earth, in the presence of air resistance which depends linearly on the speed of the object.

Example 2.3: Falling object with air resistance

Consider the motion of a falling object with mass m near the surface of the Earth when the air resistance varies linearly with the speed, $F_d = -bv$. By integrating Newton's law $F = m\,a$ we obtain the following expression for the position $y(t)$ of the object as a function of the elapsed time t:

$$y(t) = y_0 + \frac{mg}{b}t + \left(\frac{m^2 g}{b^2} - \frac{mv_0}{b}\right)\left(e^{-\frac{bt}{m}} - 1\right) \tag{2.2.1}$$

where y_0 is the initial position, v_0 is the initial speed and g is the acceleration of gravity.
(a) Find the analytical expressions for the velocity $v(t)$ and the acceleration $a(t)$, both by hand and using SymPy.
(b) Use reasonable values of the parameters y_0, v_0, b, m to plot $x(t)$, $v(t)$ and $a(t)$, and discuss the physical meaning of the graphs.

Solution:

In this example the speed $v(t)$ is found by differentiating with respect to time:

$$v(t) = \frac{dx}{dt} = \frac{mg}{b} + \left(\frac{m^2 g}{b^2} - \frac{m v_0}{b}\right) e^{-\frac{bt}{m}} \left(\frac{-b}{m}\right) \tag{2.2.2}$$

$$v(t) = \frac{mg}{b} + \left(v_0 - \frac{mg}{b}\right) e^{-\frac{bt}{m}} \tag{2.2.3}$$

and the acceleration $a(t)$ is found by differentiating $v(t)$ with respect to time:

$$a(t) = \frac{dv}{dt} = \left(-\frac{b\, v_0}{m} + g\right) e^{-\frac{bt}{m}} \tag{2.2.4}$$

We can also obtain these results by using the `diff` command in SymPy. Once more, the variables t, y_0, m, v_0, b, g need to be defined using the `symbols` command, and we used the option `real=True`. The analytical result from Python agrees with the results from the differentiation by hand.

In order to plot the results, we define NumPy arrays `tims`, `yt`, `vt` containing the values of t, y, v in the time interval from $t = 0$ to $t = 7$ s. The numerical values used in this example are $b = 1$ Ns/m, $g = 9.8$ m/s^2, $y_0 = v_0 = 0$, $m = 1$kg.

Figure 2.3 shows that the speed of the object in the plot increases, until at large times $t > 6$ s it reaches the terminal velocity $v_t = mg/b = 9.8$ m/s. Similarly, the position of the object increases linearly with time at large times $t > 6$ s, since it has reached terminal velocity.

```
from sympy import symbols, exp, diff, simplify
import numpy as np
import matplotlib.pyplot as plt
print('-'*28,'CODE OUTPUT','-'*29,'\n')

t, y0, m, v0, b, g = symbols('t,y0,m,v0,b,g',real=True)   # symbols

v = diff(y0+m*g*t/b+(m**2*g/b**2-m*v0/b)*(exp(-b*t/m)-1),t) # speed v(t)
print('v(t)=',simplify(v),'\n')

a = diff(v,t)                # acceleration a(t)
print('a(t)=',simplify(a))

b, m, g, v0, y0 = 1.0, 1, 9.8, 0, 0 # numerical values for the plots

tims = np.linspace(0,7,50)  # times t to be evaluated

yt = y0+m*g*tims/b+(m**2*g/b**2-m*v0/b)*(np.exp(-b*tims/m)-1) #array y(t)
vt = (b*v0 + g*m*np.exp(b*tims/m) - g*m)*np.exp(-b*tims/m)/b  #array v(t)

plt.subplot(1,2,1)
plt.plot(tims, yt, 'b.')
plt.xlabel('Time t [s]')
plt.ylabel('y(t) [m]')

plt.subplot(1,2,2)
plt.plot(tims, vt, 'k')
plt.xlabel('Time t [s]')
plt.ylabel('v(t) [m/s]')
plt.tight_layout()
plt.show()
```

```
------------------------ CODE OUTPUT ------------------------

v(t)= (b*v0 + g*m*exp(b*t/m) - g*m)*exp(-b*t/m)/b

a(t)= (-b*v0 + g*m)*exp(-b*t/m)/m
```

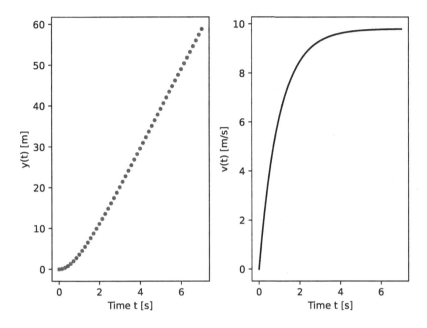

Figure 2.3 Graphical output from Example 2.3, for the speed and acceleration of a falling object near the surface of the Earth.

Example 2.4 is a second example from Classical Mechanics, in which we find the equilibrium points in a 1D potential $V(x)$. Recall from your Introductory Physics course that physical systems can be in three types of equilibrium: stable, unstable, and neutral. A *stable equilibrium* corresponds to values for which the potential energy $V(x)$ is a *local minimum*. An *unstable equilibrium* corresponds to values for which the potential energy $V(x)$ is a *local maximum*. A final type of equilibrium is called *neutral equilibrium*, for which the potential energy $V(x)$ is a *constant*, and hence its derivative is zero.

Example 2.4: The Lennard-Jones potential

The Lennard-Jones potential function, also referred to as the 6–12 potential, is a simple yet realistic model of intermolecular interactions between two particles a distance x apart:

$$V(x) = \left(\frac{\sigma}{x}\right)^{12} - \left(\frac{\sigma}{x}\right)^{6}$$

where all physical quantities σ, x, and V are in appropriate SI units. Determine the type of equilibrium possible in this potential, and plot the potential and the force $F = -dV/dx$, for $\sigma = 1$.

Solution:
With the given value of σ the potential $V(x) = x^{-12} - x^{-6}$ and we find the equilibrium point by setting the derivative of $V(x)$ equal to zero and solving for x:

$$\frac{dV(x)}{dx} = 12x^{-13} - 6x^{-7} = 0$$

$$x = 2^{1/6} = 1.122$$

We can easily verify that this is a stable equilibrium point by substituting $x = 1.122$ in the second derivative d^2V/dx^2, which gives a value of $d^2V/dx^2 > 0$ indicating a local minimum in $V(x)$.

The Python code evaluates the position of stable equilibrium using the `fsolve()` function from the `scipy.optimize` library in the line:

`fsolve(force,1)`. The parameter 1 here provides an initial estimate of the position of stable equilibrium to the code.

The graphical output in Figure 2.4 shows subplots of the potential $V(x)$ and of the force $F = -dV/dx$.

```
from sympy import symbols, diff
import numpy as np
import matplotlib.pyplot as plt
from scipy.optimize import fsolve
print('-'*28,'CODE OUTPUT','-'*29,'\n')


x = symbols('x',real=True)      # define symbols
F = -diff((((x**-12)-(x**-6)),x)  # evaluate F=-dV/dx symbolically
print('Force=',F)


force = lambda x: -12/x**13 + 6/x**7  # create function F(x)
print("Force=0 at x=",np.round(fsolve(force,1)[0],3)," m")


x = np.linspace(1,2,50)         # evaluate positions x

# plot V(x) and F(x)using subplots
plt.subplot(1,2,1)
plt.plot(x, (x**-12)-(x**-6), 'k')
plt.xlabel('Distance x ')
plt.ylabel('Potential V(x) ')


plt.subplot(1,2,2)
plt.plot(x,-12/x**13 + 6/x**7, 'b--')
plt.xlabel('Distance x ')
plt.ylabel('Force F(x)  ')
plt.tight_layout()
plt.show()

------------------------- CODE OUTPUT ----------------------------

Force= -6/x**7 + 12/x**13
Force=0 at x= 1.122  m
```

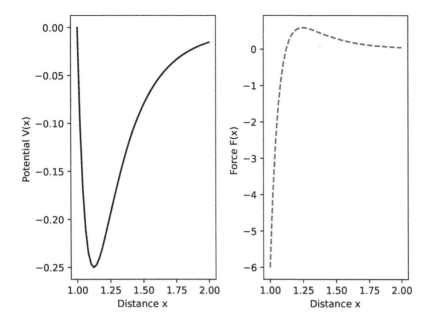

Figure 2.4 Graphical output from Example 2.4, finding the equilibrium points in a 1D potential $V(x)$.

The Maxwell-Boltzmann distribution (MB) is used in Statistical Mechanics to describe the distribution $f(v)$ of the speeds v of atoms or molecules in an ideal gas:

$$f(v) = \sqrt{\frac{2}{\pi} \left(\frac{m}{kT}\right)^3} \, v^2 \exp\left(\frac{-mv^2}{2kT}\right) \tag{2.2.5}$$

where T is the temperature of the gas in K, k is the Boltzmann constant, and m is the mass of the atoms or molecules. Example 2.5 shows how to evaluate the most probable speed of the particles in the MB distribution, and also plots the probability distribution $f(v)$ of an ideal Helium gas.

Example 2.5: Most probable speed in the Maxwell-Boltzmann distribution

(a) Show by hand and by using symbolic Python that the most probable speed for this distribution is

$$v_p = \sqrt{\frac{2kT}{m}} \tag{2.2.6}$$

(b) For Helium at room temperature, plot $f(v)$ and its derivative df/dv.

Solution:

We calculate the most probable speed v_p by evaluating the derivative df/dv using the product and chain rules, and then setting it equal to zero at $v = v_p$:

$$\frac{df}{dv} = \sqrt{\frac{2}{\pi} \left(\frac{m}{kT}\right)^3} \left[2v \exp\left(\frac{-mv^2}{2kT}\right) + v^2 \exp\left(\frac{-mv^2}{2kT}\right) \frac{d}{dv}\left(\frac{-mv^2}{2kT}\right)\right] \qquad (2.2.7)$$

$$\frac{df}{dv} = \sqrt{\frac{2}{\pi} \left(\frac{m}{kT}\right)^3} \exp\left(\frac{-mv^2}{2kT}\right) \left[2v - \frac{mv^3}{kT}\right] \qquad (2.2.8)$$

The derivative $df/dv = 0$ when $v = v_p$. Hence, the term inside the square brackets gives

$$2v_p + \left(\frac{-mv_p^3}{kT}\right) = 0$$

and

$$v_p = \sqrt{\frac{2k\,T}{m}}$$

Using the `solve()` command in SymPy, we obtain the same analytical result $v_p = \sqrt{2kT/m}$, and a numerical value of $v_p = 1116.405$ m/s. We also obtain the same result for the numerical value of v_p using the `fsolve()` function from the `scipy.optimize` library, as in the previous example.

The graphical output in Figure 2.5 shows subplots of the distribution $f(v)$ and its derivative df/dv.

```
from sympy import symbols, pi, sqrt, exp, diff, Eq, solve
import matplotlib.pyplot as plt
import numpy as np
from scipy.optimize import fsolve
print('-'*28,'CODE OUTPUT','-'*29,'\n')


v, m, k, T = symbols('v,m,k,T',positive=True)  # define symbols


f1 = sqrt(2/pi*((m/(k*T))**3))*v**2*exp(-m*v**2/(2*k*T))


deriv = diff(f1,v)
eq1 = Eq(deriv,0)
vp = solve(eq1,v)[0]  # solve symbolically df/dv=0
print("Symbolic vp from SymPy = ",vp)


# Plot MB distribution of speeds for Helium
m = 6.6464731e-27  # atomic mass Helium in kg
T = 300            # room temperature in K
k = 1.380649e-23   # Boltzmann constant in J/K
print("Numerical vp from SymPy = ",round(np.sqrt(2*k*T/m),3)," m/s")


v1 = np.linspace(1,3000,100)  # define values of v for x-axis


a = np.sqrt(2/np.pi*((m/(k*T))**3))
b = m/(2*k*T)
f = lambda v: a*v**2*np.exp(-b*v**2)   # define function for f(v)


# define function for derivative df/dv
der = lambda v: np.sqrt(2)*m**(3/2)*v*(2*T*k - m*v**2)*\
```

```
          np.exp(-m*v**2/(2*T*k))/(np.sqrt(np.pi)*T**(5/2)*k**(5/2))

plt.subplot(1,2,1)
plt.plot(v1,f(v1))
plt.title('(a)')
plt.xlabel('Speed [m/s]')
plt.ylabel('Distribution of speeds f(v)')

plt.subplot(1,2,2)
plt.plot(v1,der(v1))
plt.title('(b)')
plt.xlabel('Speed [m/s]')
plt.ylabel('Derivative df/dv')
plt.plot(v1,[0]*len(v1))

vp=np.round(fsolve(der,1000)[0],3)
print("Numerical vp using fsolve command:",vp," m/s")
plt.tight_layout()
plt.show()

------------------------- CODE OUTPUT -----------------------------

Symbolic vp from SymPy =  sqrt(2)*sqrt(T)*sqrt(k)/sqrt(m)
Numerical vp from SymPy =  1116.405  m/s
Numerical vp using fsolve command: 1116.405  m/s
```

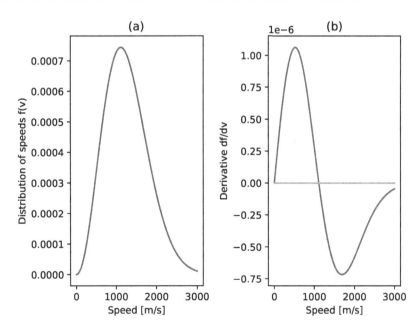

Figure 2.5 Graphical output from Example 2.5 for (a) The probability distribution $f(v)$ of speeds in an ideal Helium gas and (b) The derivative df/dv.

Example 2.6 is an application of symbolic differentiation using Python in Quantum Mechanics. Here we use again the `diff` command in SymPy to evaluate symbolically the derivatives of the wave function $\psi(x)$ describing a simple harmonic oscillator (SHO). We use these derivatives to verify that $\psi(x)$ satisfies the 1D Schrodinger equation (SE).

Example 2.6: The 1D Schrodinger equation

The wave function of a SHO in Quantum Mechanics of mass m and frequency ω is:

$$\psi(x) = A \, \exp\left(-\frac{m\,\omega}{2\hbar}\,x^2\right) \tag{2.2.9}$$

where A is the normalization constant of the wave function and $\hbar$ is the reduced Planck constant. The position x can vary anywhere along the x-axis.
Show that this function is a solution of the one-dimensional Schrödinger equation:

$$-\frac{\hbar^2}{2m}\,\frac{d^2\psi}{dx^2} + \frac{1}{2}m\,\omega^2\,x^2\,\psi = E\,\psi \tag{2.2.10}$$

where the energy $E = \hbar\,\omega/2$.

Solution:
We evaluate the first and second derivatives of the wave function using the chain and product rules:

$$\frac{d\psi}{dx} = A \, \exp\left(-\frac{m\,\omega}{2\hbar}\,x^2\right)\left(-\frac{m\,\omega}{\hbar}\,x\right) \tag{2.2.11}$$

$$\frac{d^2\psi}{dx^2} = A \, \exp\left(-\frac{m\,\omega}{2\hbar}\,x^2\right)\left(-\frac{m\,\omega}{\hbar}\,x\right)^2 + A \, \exp\left(-\frac{m\,\omega}{2\hbar}\,x^2\right)\left(-\frac{m\,\omega}{\hbar}\right) \tag{2.2.12}$$

$$\frac{d^2\psi}{dx^2} = A \, \exp\left(-\frac{m\,\omega}{2\hbar}\,x^2\right)\left[\frac{m^2\,\omega^2}{\hbar^2}\,x^2 - \frac{m\,\omega}{\hbar}\right] \tag{2.2.13}$$

Substituting (2.2.13) into the SE (2.2.10):

$$-\frac{\hbar^2}{2m}\,A \, \exp\left(-\frac{m\,\omega}{2\hbar}\,x^2\right)\left[\frac{m^2\,\omega^2}{\hbar^2}\,x^2 - \frac{m\,\omega}{\hbar}\right] + \frac{1}{2}m\,\omega^2\,x^2\,A \, \exp\left(-\frac{m\,\omega}{2\hbar}\,x^2\right) = E\,\psi \tag{2.2.14}$$

Canceling terms from both sides, we obtain the energy $E = \hbar\omega/2$. The corresponding Python code for evaluating the left-hand side (LHS) and right-hand side (RHS) of the Schrödinger equation is straightforward.

```
from sympy import symbols, exp, diff, simplify
print('-'*28,'CODE OUTPUT','-'*29,'\n')


x, m, omeg, hbar, A = symbols('x,m,omeg,hbar,A',positive=True)   # symbols


psi = A*exp(-m*omeg*x**2/(2*hbar))   # wave function


lhs = -(hbar**2)/(2*m)*diff(psi,x,x)+m*omeg**2*x**2*psi/2
print('LHS of Scrodinger Equation = ',simplify(lhs))


rhs = hbar*omeg*psi/2
print('RHS of Scrodinger Equation = ',simplify(rhs))

-------------------------- CODE OUTPUT --------------------------


LHS of Scrodinger Equation =  A*hbar*omeg*exp(-m*omeg*x**2/(2*hbar))/2
RHS of Scrodinger Equation =  A*hbar*omeg*exp(-m*omeg*x**2/(2*hbar))/2
```

2.3 A DETAILED EXAMPLE: DERIVATION OF WIEN'S DISPLACEMENT LAW

In this section we present a comprehensive example of using Python in Optics, specifically in the analysis of black-body radiation.

The study of black-body radiation was key in the development of quantum mechanics in the early 20th Century. In the simplest case, a black-body is an object that absorbs all of the radiation that falls upon it and reflects none. A common example of a black-body is a hole in the wall of a hollow container. A small fraction of the light entering the hole will be re-emitted. If the black-body is in thermal equilibrium, it will also be an excellent emitter of radiation. Another example of objects that approximates a black-body are stars.

The classical theory of electromagnetism struggled to explain the peak-shaped dependence of the intensity (power per unit area) of the emitted radiation as a function of the wavelength λ of the emitted light. The wavelength of maximum intensity λ_{max} depends on the temperature of the black-body, and as the black-body's temperature T increases, the maximum of the intensity shifts to lower wavelengths. This is the well-known Wien's displacement law:

$$\lambda_{\mathrm{max}} T = 2.898 \times 10^{-3} \mathrm{m} \cdot \mathrm{K} \qquad (2.3.1)$$

In 1900 Max Planck derived the equation for the black-body curve known as Planck's law:

$$I(\lambda) = \frac{2\pi c^2 h}{\lambda^5} \left(\frac{1}{e^{hc/\lambda kT} - 1} \right) \qquad (2.3.2)$$

where T is the black-body's temperature, $c = 2.998 \times 10^8$ m/s is the speed of light, $k = 1.381 \times 10^{-23}$ J/K is the Boltzmann constant and $h = 6.626 \times 10^{-34}$ J·s is Planck's constant.

In Example 2.7 we derive Wien's displacement law (2.3.1) from Planck's radiation law (2.3.2). Here we use again the `diff` command in SymPy to evaluate symbolically the derivative $dI/d\lambda$ of the intensity, and then set it equal to zero in order to find the λ_{max}.

Example 2.7: Planck's law of black-body radiation

(a) Using the SI numerical values of the constants h, c, and k, plot Planck's radiation law (2.3.2), $I(\lambda)$ as a function of the wavelength λ for a back-body with temperature $T = 3000$ K.

(b) Use SymPy to evaluate the derivative $dI/d\lambda$, and derive Wien's displacement law (2.3.1). It is possible to simplify the algebra significantly by introducing a dimensionless variable $x = hc/(kT\lambda)$.

Solution:

(a) By introducing a dimensionless variable $x = hc/(kT\lambda)$, Planck's law becomes:

$$I(x) = A\, x^5 \left(\frac{1}{e^x - 1} \right) \qquad (2.3.3)$$

where A is a constant. The Python code below plots $I(\lambda)$ and $I(x)$ for a black-body with a temperature $T = 3000$ K. The plot appears in Figure 2.6.

(b) Using the product and chain rules:

$$\frac{dI}{dx} = \frac{d}{dx}\left[A\,x^5\left(\frac{1}{e^x-1}\right)\right] = 5A\,x^4\left(\frac{1}{e^x-1}\right) - A\,x^5\frac{e^x}{(e^x-1)^2} = 0 \qquad (2.3.4)$$

$$A\,x^4\left[5\,(e^x-1) - x\,e^x\right] = 0 \qquad (2.3.5)$$

$$x\,e^x - 5\,(e^x-1) = 0 \qquad (2.3.6)$$

Solving $(2.3.6)$ for x is challenging, because it is a transcendental equation which needs to be solved *numerically* using Python.

In the Python code, we begin by using SymPy to differentiate $(2.3.3)$. We can use SymPy to find $(2.3.6)$ by isolating the numerator of `dfdv`. The `[0]` after the `fraction` command extracts the numerator after the derivative has been simplified. Note that without the `simplify` command, a numerator will not be provided by `fraction`, because $(2.3.4)$ is not expressed as a single fraction. The variable `numerator` stores the next result in the code (which we have set equal to zero).

To solve $(2.3.5)$ numerically, we need an initial guess. The plot of $I(x)$ provides a reasonable starting guess near $x = 5$. We can get a better approximation using the SymPy command `fsolve`, which uses as its third argument, the initial guess $x = 6$.

```python
from sympy import symbols, exp, diff, fraction, simplify
import numpy as np
import matplotlib.pyplot as plt
from scipy.optimize import fsolve
print('-'*28,'CODE OUTPUT','-'*29,'\n')


x = symbols('x',real=True)                # define symbols


fv = (x**5)/(exp(x)-1)                     # light intensity I(x)
dfdv = -diff(fv,x)                         # evaluate dI/dx
print('Derivative df/dv=',dfdv,'\n')


numerator = fraction(simplify(dfdv))[0]    # find numerator only
print('Numerator of derivative df/dv=',numerator,'\n')


num = lambda x: x**4*(x*np.exp(x) - 5*np.exp(x) + 5)


# set function df/dv=0 and find x of maximum intenisty
print("df/dv=0 at x=",np.round(fsolve(num,6)[0],4))


h = 6.626e-34    # Planck constant in J s
k = 1.380649e-23 # Boltzmann constant in J/K
c = 2.998e8      # speed of light m/s
T = 3000         # Temperature in K


# define I(wavelength)
f = lambda x: (2*h*c**2.0/x**5.0)*1/(np.exp(h*c/(x*k*T))-1)


x1 = np.linspace(3e-8,3e-6,100)  # wavelength values for plot


plt.subplot(1,2,1)
plt.plot(x1,f(x1),'r+',label='T=3000 K')
plt.xlabel('Wavelength $\lambda$ [m]')
plt.ylabel(r'Intensity I(${\lambda}$)')
plt.title('(a)')
leg = plt.legend()
```

```
leg.get_frame().set_linewidth(0.0)

f = lambda x: (x**5)/(np.exp(x)-1)  # define I(x)
x = np.linspace(1,15,50)            # x values for plot

plt.subplot(1,2,2)
plt.plot(x, f(x), 'k')
plt.xlabel('Parameter x')
plt.ylabel('Distribution f(x)')
plt.title('(b)')
plt.tight_layout()
plt.show()

------------------------- CODE OUTPUT -------------------------

Derivative df/dv= x**5*exp(x)/(exp(x) - 1)**2 - 5*x**4/(exp(x) - 1)

Numerator of derivative df/dv= x**4*(x*exp(x) - 5*exp(x) + 5)

df/dv=0 at x= 4.9651
```

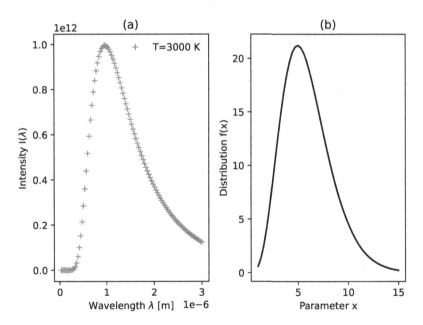

Figure 2.6 Graphical output from Example 2.7: (a) The black-body intensity spectrum $I(\lambda)$. (b) The distribution $I(x)$.

The following short code shows an alternative method of numerically solving (2.3.5), by using the SymPy command `nsolve`.

```
from sympy import nsolve, symbols, exp
x = symbols('x')                        # define symbol x

root = nsolve(5*(exp(x)-1)-x*exp(x),x,5)  # equation to solve
```

```
# another method for solving the transcedental equation

print('-'*28,'CODE OUTPUT','-'*29,'\n')
print('Solution of transcedental equation:   x = ',root)

------------------------ CODE OUTPUT ---------------------------

Solution of transcedental equation:   x =   4.96511423174428
```

Hence, we obtain the condition for maximum intensity:

$$x = \frac{h\,c}{k\,T\,\lambda_{\max}} = 4.9651 \qquad (2.3.7)$$

and after rearranging and inserting the SI values for h, c, and k, we obtain:

$$\lambda_{\max} T = \frac{h\,c}{4.9651\,k} = \frac{\left(6.626 \times 10^{-34}\right)\left(2.998 \times 10^{8}\right)}{4.9651 \times 1.3806 \times 10^{-23}} = 2.898 \times 10^{-3}\,\text{m} \cdot \text{K} \qquad (2.3.8)$$

which represents Wien's law (2.3.1).

2.4   DERIVATIVES OF MULTIVARIABLE FUNCTIONS

In this section, we will address the question of differentiation of multivariable functions. The method is called *partial differentiation* and is of central importance in fields such as thermodynamics and electromagnetism.

There are many quantities in science which depend on more than one variable. For example, consider the pressure of an ideal gas

$$P = \frac{nRT}{V} \qquad (2.4.1)$$

where P, V and T are the pressure, volume, and temperature of the gas, n is the number of moles, and $R = 8.31$ J $K^{-1}\text{mol}^{-1}$ is the ideal gas constant. We can ask the question, how does the pressure of the ideal gas P change with volume? This question is clearly about derivatives, but now we have two variables to account for, T and V. Recall that in thermodynamics, we are often interested in understanding how one quantity changes while others remain constant. For example, we can change our question to, How does the pressure of the gas change with volume when the temperature is held fixed? Such a question can be answered using partial differentiation.

2.4.1   INTRODUCTION TO PARTIAL DIFFERENTIATION

Consider a multivariable function $f(x, y, z)$. For example, $f(x, y, z)$ could describe the temperature of the air in a room at a point with coordinates (x, y, z). For simplicity, let's assume the air has reached a thermal equilibrium such that its temperature does not depend on time. Suppose we want to find the rate of change of f along the x-axis. We could imagine ourselves as walking along the room's x-axis taking temperature measurements. During the walk, our y and z coordinates are held constant and we are finding how $f(x, y, z)$ changes with x. We use the notation $\partial f / \partial x$ which is often read as the partial derivative of f with respect to x, with all other variables held fixed.

In analogy to the derivative of a function of a single variable $f(x)$, we can define the partial derivative $\partial f / \partial x$ (holding the values of y and z fixed) as a limit:

$$\frac{\partial f}{\partial x}(x, y, z) = \lim_{\Delta x \to 0} \frac{f(x + \Delta x, y, z) - f(x, y, z)}{\Delta x} \qquad (2.4.2)$$

Notice that (2.4.2) is the same as Newton's definition of the derivative (2.1.1). Hence, all of the methods of calculating total derivatives can be used to calculate partial derivatives. We can create similar definitions as (2.4.2) for $\partial f/\partial y$ and $\partial f/\partial z$.

As an example, let us consider a function $T(x, y, z) = cx^2y + bz$, where c and b are constants. Then

$$\frac{\partial T}{\partial x} = 2c\,x\,y \qquad (2.4.3)$$

Note that the derivative with respect to x is done just as in the single-variable case (i.e. we use the power rule) treating y and z as constants.

In addition, we can examine second derivatives and higher

$$\frac{\partial}{\partial x}\frac{\partial T}{\partial x} = \frac{\partial^2 T}{\partial x^2}, \qquad \frac{\partial}{\partial y}\frac{\partial T}{\partial x} = \frac{\partial^2 T}{\partial y \partial x}, \qquad \frac{\partial}{\partial z}\frac{\partial^2 T}{\partial x \partial y} = \frac{\partial^3 T}{\partial z \partial x \partial y}, \qquad \text{and so on.}$$

For most common functions in physics, the order in which the derivatives are taken is irrelevant. In other words,

$$\left(\frac{\partial}{\partial x}\right)\left(\frac{\partial T}{\partial y}\right) = \left(\frac{\partial}{\partial y}\right)\left(\frac{\partial T}{\partial x}\right) = \frac{\partial^2 T}{\partial y \partial x} = \frac{\partial^2 T}{\partial x \partial y} \qquad (2.4.4)$$

Example 2.8: Calculating partial derivatives

The temperature of a large two-dimensional plate is modeled using the following equation

$$T(x, y) = c\sin(xy) \qquad (2.4.5)$$

where c is a constant with the appropriate SI units, and the origin is located at the center of the plate. Calculate the partial derivatives:

$$\frac{\partial^2 T}{\partial x^2} \qquad \frac{\partial^2 T}{\partial x \partial y} \quad \text{and} \quad \frac{\partial^3 T}{\partial^2 x \partial y} \qquad (2.4.6)$$

Solution:
Evaluating the partial derivatives:

$$\frac{\partial^2 T}{\partial x^2} = -cy^2\sin(xy)$$

$$\frac{\partial^2 T}{\partial x \partial y} = \frac{\partial}{\partial x}\left[cx\cos(xy)\right] = c\left[\cos(xy) - xy\sin(xy)\right]$$

$$\frac{\partial^3 T}{\partial^2 x \partial y} = c\frac{\partial}{\partial y}\left[-y^2\sin(xy)\right] = -cy\left[2\sin(xy) + xy\cos(xy)\right]$$

The code for computing partial derivatives in SymPy is below. The `diff` command is used just as before.

```
from sympy import symbols, sin, diff
x, y, c = symbols('x,y, c')


T = c*sin(x*y)


print('-'*28,'CODE OUTPUT','-'*29,'\n')
```

```
print('Second partial with respect to x = ', diff(T,x,2))
print('Second partial with respect to x,y = ',diff(T,x,y))
print('3rd partial with respect to x,x,y = ',diff(T,x,x,y))



----------------------------- CODE OUTPUT -----------------------------


Second partial with respect to x =  -c*y**2*sin(x*y)
Second partial with respect to x,y =  c*(-x*y*sin(x*y) + cos(x*y))
3rd partial with respect to x,x,y =  -c*y*(x*y*cos(x*y) + 2*sin(x*y))
```

In some textbooks, a subscript is used to denote which variables are held fixed. Consider the example of the ideal gas law which we used to start this subsection. The partial derivative of P with respect to V, with T held fixed, can be written as

$$\left(\frac{\partial P}{\partial V}\right)_T \tag{2.4.7}$$

The subscript can be helpful when we want to explicitly state which variable is being held constant, such as when describing a thermodynamic process. In this textbook, we will generally not include the subscript, unless it is important to a particular example.

2.4.2 TOTAL DIFFERENTIALS

Consider a single-variable function $f(x)$. How does f change when we change its independent variable x? Figure 2.7 contains a graph of a function $f(x)$ and a line tangent at the point $x = x_0$. We can think of the tangent line as making a linear prediction of the value of the function $f(x)$, as we change its independent variable x by an amount Δx.

We *define* the differential dx of the independent variable x as:

$$dx = \Delta x \tag{2.4.8}$$

Notice that in Figure 2.7 the lengths df and Δf are not the same. If we change the independent variable x by an amount $dx = \Delta x$, there is a difference between the actual

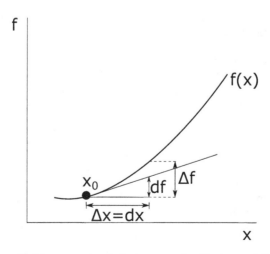

Figure 2.7 The function $f(x)$ is approximated at $x = x_0$ by the tangent line. Note that $dx = \Delta x$ by definition, however the two lengths along the y-axis are not equal ($df \neq \Delta f$).

change of the function's value Δf and the value of the function df predicted by the tangent line.

Since the derivative df/dx represents the slope of the line tangent to f at x_0, we can write for the lengths df and dx in Figure 2.7:

$$df = \frac{df}{dx}\,dx \qquad (2.4.9)$$

Equation (2.4.9) says that the amount df by which the function changes, is equal to the product of the change in its independent variable (dx) and the rate at which f changes (df/dx). This is similar to the $x = v\,t$ formula you learned in basic physics (distance equals speed $v = dx/dt$ multiplied by the time t).

We call the quantity df the *differential of the function f*.

Next we examine functions of two variables $f(x, y)$. The function $z = f(x, y)$ represents a surface in three dimensions, and the partial derivatives $\partial f/\partial x$ and $\partial f/\partial y$ are the slopes of two lines tangent to this surface, just like df/dx represented the slope of the tangent line in Figure 2.7. These two lines, tangent to the surface $z = f(x, y)$, form a plane which is also tangent to the surface. We can then consider small displacements $dx = \Delta x$ and $dy = \Delta y$ along the x and y axes respectively, and we can define the total differential df of the function f as

$$df = \frac{\partial f}{\partial x}dx + \frac{\partial f}{\partial y}dy \qquad (2.4.10)$$

Finally, we can generalize the concept of a total differential to n variables,

$$df = \frac{\partial f}{\partial x_1}dx_1 + \frac{\partial f}{\partial x_2}dx_2 + \cdots \frac{\partial f}{\partial x_n}dx_n \qquad (2.4.11)$$

A specific application common in physics is when the function f depends both on spatial coordinates and time, then

$$df = \frac{\partial f}{\partial x}dx + \frac{\partial f}{\partial y}dy + \frac{\partial f}{\partial z}dz + \frac{\partial f}{\partial t}dt \qquad (2.4.12)$$

Example 2.9: The entropy of a monatomic ideal gas

The entropy S of a monatomic ideal gas is given by the Sackur-Tetrode equation

$$S = Nk\left\{\ln\left[\frac{V}{N}\left(\frac{4\pi mU}{3Nh^2}\right)^{3/2}\right] + \frac{5}{2}\right\} \qquad (2.4.13)$$

where N is the number of atoms, k is Boltzmann's constant, V is the volume of the gas, U is the internal energy of the gas, and h is Planck's constant. Compute dS, assuming that the number of particles is fixed $(dN = 0)$.

Solution:
Notice that the entropy depends on three variables $S = S(V, U, N)$. Hence, we can write

$$dS = \frac{\partial S}{\partial V}dV + \frac{\partial S}{\partial U}dU + \frac{\partial S}{\partial N}dN \qquad (2.4.14)$$

However since the number of particles N is fixed, $dN = 0$, we can ignore the last term. For simplicity, we will find it easier to rewrite (2.4.13) as:

$$S = Nk \ln V + \frac{3}{2} Nk \ln U + f(N) \tag{2.4.15}$$

where $f(N)$ is a constant function of N which will disappear in the total differential. Next we compute the partial derivatives

$$\frac{\partial S}{\partial V} = \frac{Nk}{V} \quad \text{and} \quad \frac{\partial S}{\partial U} = \frac{3Nk}{2U} \tag{2.4.16}$$

and therefore:

$$dS = \frac{Nk}{V} dV + \frac{3Nk}{2U} dU \tag{2.4.17}$$

```
from sympy import symbols, diff, pi, log
print('-'*28,'CODE OUTPUT','-'*29,'\n')


N, k, V, m, U, h, dV, dU= symbols('N, k, V, m, U, h, dV, dU',\
real=True)
#  define symbols

S = N*k* (log((V/N)* ((4*pi*m*U/(3*N*h**2))**(3/2))) +5/2)

dS = diff(S,V)*dV +diff(S,U)*dU
print('dS=', dS)

-------------------------- CODE OUTPUT ----------------------------

dS= 1.0*N*dV*k/V + 1.5*N*dU*k/U
```

2.4.3   TOTAL DERIVATIVE OF A FUNCTION - THE CHAIN RULE REVISITED

Recall that the chain rule provides a method of differentiating a function of a function. We can extend that concept to multivariable functions using the concept of a *total differential* df from the previous section. Physicists sometimes think of this as dividing (2.4.11) by the quantity dt. For example, the total derivative of $f(x_1, x_2, \ldots, x_n)$ with respect to time can be found as:

$$\frac{df}{dt} = \frac{\partial f}{\partial x_1} \frac{dx_1}{dt} + \frac{\partial f}{\partial x_2} \frac{dx_2}{dt} + \cdots \frac{\partial f}{\partial x_n} \frac{dx_n}{dt} \tag{2.4.18}$$

In the case where the function depends explicitly both on spatial coordinates and time,

$$\frac{df}{dt} = \frac{\partial f}{\partial x} \frac{dx}{dt} + \frac{\partial f}{\partial y} \frac{dy}{dt} + \frac{\partial f}{\partial z} \frac{dz}{dt} + \frac{\partial f}{\partial t} \tag{2.4.19}$$

Notice that in general the total derivative df/dt will contain the term $\partial f/\partial t$, unless the function f does not depend explicitly on time.

Example 2.10: Temperature changes in a rectangular plate

The temperature of a rectangular plate located on the xy-plane is described by $T = a\,x^2\,t\,\sin(y)$, where a is constant, and t is the time variable. Find dT/dt along the path $x = ct^2$, $y = be^{2t}$, where a and b are constants.

Solution:
We begin by applying (2.4.18):

$$\frac{dT}{dt} = \frac{\partial T}{\partial x}\frac{dx}{dt} + \frac{\partial T}{\partial y}\frac{dy}{dt} + \frac{\partial T}{\partial t} \tag{2.4.20}$$

Next, we compute all derivatives,

$$\frac{\partial T}{\partial x} = 2axt\sin(y) \quad \frac{\partial T}{\partial y} = ax^2 t\cos(y) \quad \frac{\partial T}{\partial t} = ax^2\sin(y)$$

$$\frac{dx}{dt} = 2ct \qquad\qquad \frac{dy}{dt} = 2be^{2t} \tag{2.4.21}$$

and inserting into (2.4.20)

$$\frac{dT}{dt} = 5ac^2 t^4\sin\left(be^{2t}\right) + 2abc^2 t^5 e^{2t}\cos\left(be^{2t}\right) \tag{2.4.22}$$

Note that we leave the final result as a function of t, the variable used for differentiation. The code for solving the problem in SymPy is below. Note that unlike doing the problem by hand, we substitute the functions for x and y into T before taking the derivative, by using the substitution method .subs in SymPy. We could have done this when solving the problem by hand, but using the method above can sometimes save some algebra. When a computer is doing the algebra for us, we no longer have to worry about that!

```
from sympy import symbols, diff, sin, exp
print('-'*28,'CODE OUTPUT','-'*29,'\n')


x, y, a, b, c, t = symbols('x,y,a,b,c,t')


T = a*x**2*t*sin(y)
x_fcn = c*t**2
y_fcn = b*exp(2*t)

print('The total derivative dT/dt is:')
diff(T.subs([(x,x_fcn),(y,y_fcn)]),t)

--------------------------- CODE OUTPUT ---------------------------

The total derivative dT/dt is:
2*a*b*c**2*t**5*exp(2*t)*cos(b*exp(2*t)) + 5*a*c**2*t**4*sin(b*exp(2*t))
```

Suppose we had not specified $x(t)$ and $y(t)$. SymPy can evaluate a total derivative using the Derivative command, along with the .doit method. In this code we define $x(t)$ as a general function using Function('x')(t). Notice that the final answer in this code now contains dx/dt and dy/dt symbolically, as Derivative(x(t),t) and Derivative(y(t),t) respectively.
We also demonstrate a way to fix the width of a code output, using the textwrap library. Note that we first need to store the output parameter deriv as a string, using the str(deriv) function.

```
from sympy import Function, symbols, Derivative, sin
import textwrap

a, t = symbols('a, t')
x = Function('x')(t)
y = Function('y')(t)

T = a*x**2*t*sin(y)
deriv = str(Derivative(T,t).doit())

print('-'*28,'CODE OUTPUT','-'*29,'\n')
print('The total derivative dT/dt is: ')
print(textwrap.fill(deriv,80))

------------------------- CODE OUTPUT ----------------------------

The total derivative dT/dt is:
a*t*x(t)**2*cos(y(t))*Derivative(y(t), t) +
2*a*t*x(t)*sin(y(t))*Derivative(x(t), t) + a*x(t)**2*sin(y(t))
```

2.4.4   MAXIMUM AND MINIMUM PROBLEMS

In the calculus of single variables, the maximum or minimum of a function can be found by setting $df/dx = 0$, and the concavity of f determines if the extrema is a local maximum or minimum. In multivariable calculus, we have a similar procedure. A point is an extremum of $f(x, y)$ if

$$df = \frac{\partial f}{\partial x}dx + \frac{\partial f}{\partial y}dy = 0 \qquad (2.4.23)$$

This implies that at the extremum we have two conditions $\partial f/\partial x = 0$ and $\partial f/\partial y = 0$. It is often difficult to know if the extremum is a maximum or a minimum. Sometimes, one can tell from the geometry of the problem. A method of identifying the nature of the extremum exists, but it is complicated and we rarely need it.

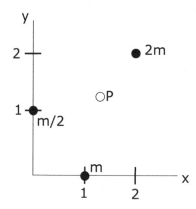

Figure 2.8   Mass configuration for Example 2.11.

Example 2.11 shows how to evaluate the location of the point P which minimizes the moment of inertia of a system of three masses shown in Figure 2.8, using SymPy.

Example 2.11: Moment of inertia

Consider the configuration of three masses $(m, m/2, 2m)$ shown in Figure 2.8. Find the location of the point P which minimizes the moment of inertia of the system, with respect to a rotational axis perpendicular to the xy-plane and passing through point P.

Solution:
Recall that the moment of inertia for a system of N particles is found using

$$I = \sum_{i=1}^{N} m_i r_i^2 \tag{2.4.24}$$

where m_i is the mass of each particle and r_i is the distance between m_i and the axis of rotation going through the point P. Let (x, y) represent the unknown coordinates of point P. For the mass m with coordinates $(x_1, y_1) = (1, 0)$ we write:

$$r_1^2 = (x - x_1)^2 + (y - y_1)^2 = (x - 1)^2 + y^2$$

By repeating this process for the system of three masses $(m, m/2, 2m)$, (2.4.24) becomes

$$I = m \left[(x - 1)^2 + y^2 \right] + \frac{m}{2} \left[x^2 + (y - 1)^2 \right] + 2m \left[(x - 2)^2 + (y - 2)^2 \right] \tag{2.4.25}$$

Next, we compute $\partial I / \partial x$ and $\partial I / \partial y$ and set both equal to zero:

$$\frac{\partial I}{\partial x} = 2m(x - 1) + 4m(x - 2) + mx = 0$$

$$\frac{\partial I}{\partial y} = 2my + m(1 - y) + 4m(y - 2) = 0 \tag{2.4.26}$$

We then solve for x and y. We will let SymPy handle the algebra for us, and we solve system of equations $\partial I / \partial x = 0$ and $\partial I / \partial y = 0 =$ using the `solve()` command in SymPy. Note that the expressions in `solve` must be equal to zero.

```
from sympy import symbols, diff, solve
print('-'*28,'CODE OUTPUT','-'*29,'\n')


x, y, m = symbols('x,y,m')  # define symbols


# I=sum of three terms M*r**2, one term for each mass
I = m*((x-1)**2+y**2)+(m/2)*(x**2-(y-1)**2)+2*m*((x-2)**2+(y-2)**2)

diff_I_x = diff(I,x)      # find partial dI/dx
diff_I_y = diff(I,y)      # find partial dI/dy


# solve system of equations dI/dx=0, dI/dy=0 usin solve()
print('The coordinates of location P are:')
solve([diff_I_x, diff_I_y])

------------------------- CODE OUTPUT ---------------------------

The coordinates of location P are:
[{x: 10/7, y: 7/5}]
```

2.5   POWER SERIES APPROXIMATIONS OF FUNCTIONS

Besides identifying rates of change and local extrema, differentiation provides a powerful tool for approximating single-variable functions (near a point $x = a$) as a power series. A power series is a series of the form:

$$\sum_{n=0}^{\infty} c_n \left(x - a\right)^n \tag{2.5.1}$$

where a and the coefficients c_n are constants.

Let us begin by assuming that a function $f(x)$ can be represented by a series.

$$f\left(x\right) = c_0 + c_1 \left(x - a\right) + c_2 \left(x - a\right)^2 + c_3 \left(x - a\right)^3 + \cdots \tag{2.5.2}$$

We can compute the coefficients c_n by differentiating both sides of (2.5.2) multiple times, and evaluating the derivative at $x = a$. Note that in what follows, the notation $f^{(n)}\left(a\right)$ represented the nth derivative of f with respect to x, evaluated at $x = a$.

$$f\left(a\right) = c_0 \tag{2.5.3}$$

$$f^{(1)}\left(a\right) = c_1 \tag{2.5.4}$$

$$f^{(2)}\left(a\right) = 2!\, c_2 \tag{2.5.5}$$

$$f^{(3)}\left(a\right) = 3!\, c_3 \tag{2.5.6}$$

and we can continue this process to find:

$$c_n = \frac{f^{(n)}\left(a\right)}{n!} \tag{2.5.7}$$

The result is the Taylor series expansion of the function $f\left(x\right)$ about $x = a$:

$$f\left(x\right) = \sum_{n=0}^{\infty} \frac{f^{(n)}\left(a\right)}{n!} \left(x - a\right)^n \tag{2.5.8}$$

A Taylor series about the origin ($a = 0$) is called a Maclaurin series

$$f\left(x\right) = \sum_{n=0}^{\infty} \frac{f^{(n)}\left(0\right)}{n!} x^n \tag{2.5.9}$$

Taylor and Maclaurin series are very useful in physics. They provide an nth-degree polynomial approximation of a function near the point $x = a$ (or near $x = 0$, in the case of Maclaurin series), when the series is truncated at the nth term.

Table 2.2 shows several examples of power series which are often used for approximating analytical functions.

Table 2.2

List of power series used for approximating commonly used analytical functions.

Function	Description	Power series
e^x	exponential	$e^x = \sum\limits_{n=0}^{\infty} \frac{x^n}{n!} = 1 + x + \frac{x^2}{2} + \frac{x^3}{6} + \frac{x^4}{24} + \cdots$
$\cos x$	cosine	$\cos x = \sum\limits_{n=0}^{\infty} \frac{(-1)^n x^{2n}}{(2n)!} = 1 - \frac{x^2}{2!} + \frac{x^4}{4!} - \cdots$
$\sin x$	sine	$\sin x = \sum\limits_{n=0}^{\infty} \frac{(-1)^n x^{2n+1}}{(2n+1)!} = x - \frac{x^3}{3!} + \frac{x^5}{5!} - \cdots$
$\tan x$	tangent	$\tan x = x + \frac{x^3}{3} + \frac{2x^5}{15} +$
$\arcsin x$	arcsinx	$\arcsin x = \sum\limits_{n=0}^{\infty} \frac{(2n)!}{4^n (n!)^2 (2n+1)} x^{2n+1} = x + \frac{x^3}{6} + \frac{3x^5}{40} + \cdots$
$(1+x)^a$	binomial series	$(1+x)^\alpha = \sum\limits_{n=0}^{\infty} \binom{\alpha}{n} x^n = \sum\limits_{n=0}^{\infty} \frac{\alpha(\alpha-1)\cdots(\alpha-n+1)}{n!} x^n$
$\sinh x$	hyperbolic sine	$\sinh x = \sum\limits_{n=0}^{\infty} \frac{x^{2n+1}}{(2n+1)!} = x + \frac{x^3}{3!} + \frac{x^5}{5!} - \cdots$
$\cosh x$	hyperbolic cosine	$\cosh x = \sum\limits_{n=0}^{\infty} \frac{x^{2n}}{(2n)!} = 1 + \frac{x^2}{2!} + \frac{x^4}{4!} - \cdots$
$\ln(1-x)$	logarithm	$\ln(1-x) = -\sum\limits_{n=1}^{\infty} \frac{x^n}{n} = -x - \frac{x^2}{2} - \frac{x^3}{3} - \cdots$
$\ln(1+x)$	logarithm	$\ln(1+x) = \sum\limits_{n=1}^{\infty} (-1)^{n+1} \frac{x^n}{n} = x - \frac{x^2}{2} + \frac{x^3}{3} - \cdots$

Often in physics, processes are described by complicated equations. and the complexity of the equation can interfere with interpreting the physics. A polynomial approximation can often be more easily understood. If a further calculation such as an integration needs to be done, it may be much easier to perform this operation on the polynomial approximation of $f(x)$, than on the function f itself. It is difficult to understate the utility of Taylor series approximations.

You may be wondering if (2.5.2) always holds. It does not. Some functions such as $1/x$ and $\ln x$ are infinite at the origin, and therefore a Maclaurin series for them would not exist. However, for most practical applications in physics, (2.5.2) does indeed hold true. In this book, we focus on the applications of mathematics that scientists routinely encounter. We recommend that the reader consult an introductory calculus book for more information about the cases where Taylor series provide spurious results.

Example 2.12 shows an example of evaluating the Maclaurin series for a logarithmic function.

Example 2.12: Calculating a Maclaurin series

Calculate the first five terms of the Maclaurin series for the following function, by hand and using SymPy:

$$f(x) = \ln\left(\frac{1}{1+x}\right) \tag{2.5.10}$$

Solution:
We begin by calculating the derivatives of $f(x)$

$$f^{(1)}(x) = -(1+x)^{-1}$$
$$f^{(2)}(x) = (1+x)^{-2}$$
$$f^{(3)}(x) = -2(1+x)^{-3}$$
$$f^{(4)}(x) = 6(1+x)^{-4}$$

We find that, $f(0) = 0$, $f^{(1)}(0) = -1$, $f^{(2)}(0) = 1$, $f^{(3)}(0) = -2$, and $f^{(4)}(0) = 6$. Therefore,

$$f(x) = -x + \frac{1}{2!}x^2 - \frac{2}{3!}x^3 + \frac{6}{4!}x^4 + \cdots$$
$$= -x + \frac{1}{2}x^2 - \frac{1}{3}x^3 + \frac{1}{4}x^4 + \cdots$$

We can use `series(f, x, x0, n)` in SymPy to calculate the series expansion around $x_0 = 0$ and keep the first $n = 5$ terms of the expansion. Note that O(x**5) in the output indicates the presence of higher order terms in the series.

```
from sympy import log, series, Symbol

x = Symbol('x')
maclaurin = series( log( 1/(1+x) ), x, x0 = 0, n = 5)

print('-'*28,'CODE OUTPUT','-'*29,'\n')
print('The Maclaurin series for f is: ',maclaurin)

------------------------- CODE OUTPUT ----------------------------

The Maclaurin series for f is:  -x + x**2/2 - x**3/3 + x**4/4 + O(x**5)
```

We now consider an application of series expansion in Classical Mechanics, specifically the motion of a body falling under the influence of air resistance.

Example 2.13: The displacement of a falling body experiencing air resistance

Consider a particle of mass m starting at rest, falling in the Earth's gravitational field. The particle experiences a drag force with a magnitude $f = -bv$ which opposes the particle's velocity.

As we saw earlier in this chapter, using Newton's second law, $F_{\text{Net}} = ma$ and the conditions $y_0 = v_0 = 0$ at time $t = 0$, the particle's position y as a function of time is:

$$y(t) = \frac{mg}{b}t - \frac{m^2 g}{b^2}\left(1 - e^{-bt/m}\right) \tag{2.5.11}$$

where we defined the positive y-direction to be pointing down, toward the Earth's center. Find the series expansion of the particle's displacement as a function of time, and show that air resistance can be ignored early in the particle's motion.

Solution:
We expand the exponential function $e^{-bt/m}$ using a power series:

$$e^x = \sum_{n=0}^{\infty} \frac{x^n}{n!} = 1 + x + \frac{x^2}{2} + \frac{x^3}{6} + \frac{x^4}{24} + \cdots$$

with $x = bt/m$ to obtain:

$$y(t) = \frac{mg}{b}t - \frac{m^2 g}{b^2}\left(1 - e^{-bt/m}\right) = \frac{mg}{b}t - \frac{m^2 g}{b^2}\left(\frac{bt}{m} + \frac{b^2 t^2}{2m^2} - \frac{b^3 t^3}{6m^3}\right) \tag{2.5.12}$$

$$y(t) = \frac{gt^2}{2} - \frac{b\,g\,t^3}{6m} \tag{2.5.13}$$

The leading order term $gt^2/2$ is identical to the displacement predicted by free fall kinematics in introductory physics. Notice that it does not include the drag force coefficient b. The next term in the series, $(bg/6m)t^3$, contains the drag coefficient, but it is small compared to the first term when t is small. Hence, the effect of the drag force is negligible early in the particle's motion.
It is important to remember the argument in the paragraph above, because it is common in physics. Small positive numbers, i.e. numbers much less than one, get smaller as they are exponentiated. In our example $gt^2/2 >> (bg/6m)t^3$, and we can ignore terms of t^3 and higher. In addition, we can look at plots of the original function $y(t)$ and its power series as shown in Figure 2.9. We see that as we add more terms, the power series provides a better approximation for $y(t)$ for longer periods of time.
We can also obtain this result using SymPy to find $y(t)$ near $t = 0$.

```
from sympy import exp, symbols, series

m, g, b, t = symbols('m, g, b, t')
y = m*g/b * t - g*m**2/b**2 *(1 - exp(-b*t/m))

maclaurin = series(y, t, x0 = 0, n = 4)

print('-'*28,'CODE OUTPUT','-'*29,'\n')
print('The Maclaurin series for x(t) is: ',maclaurin)

---------------------------- CODE OUTPUT ----------------------------

The Maclaurin series for x(t) is:  g*t**2/2 - b*g*t**3/(6*m) + O(t**4)
```

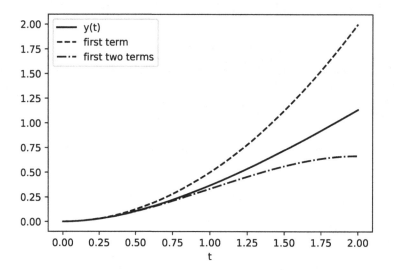

Figure 2.9 A graph of the position $y(t)$ from (2.5.11) and its power series approximations.

It is common in physics to calculate a power series for a function of two variables. Suppose we want to approximate the function $f(x, y)$ about the point $(x, y) = (a, b)$. Then the multivariable power series of f can be found using:

$$f(x, y) = f(a, b) + \frac{\partial f}{\partial x}(a, b) \ (x - a) + \frac{\partial f}{\partial y}(a, b) \ (y - a) +$$

$$+ \frac{1}{2!} \left[\frac{\partial^2 f}{\partial x^2}(a, b) \ (x - a)^2 + 2 \frac{\partial^2 f}{\partial x \partial y}(a, b) \ (x - a) \ (y - b) + \frac{\partial^2 f}{\partial y^2}(a, b) \ (y - b)^2 \right] + \dots \quad (2.5.14)$$

where $\frac{\partial f}{\partial x}(a, b)$ and $\frac{\partial^2 f}{\partial y^2}(a, b)$ indicate the values of the derivatives $\partial f / \partial x$ and $\partial^2 f / \partial y^2$ at the point $(x, y) = (a, b)$.

Example 2.14: The temperature of a metal disk

Consider a large metal circular disk centered at the origin. Suppose the temperature of the plate is described by the function:

$$T(x, y) = a \, e^{-\sigma(x^2 + y^2)} \quad (2.5.15)$$

where a and σ are constants. Find the Maclaurin series of $T(x, y)$.

Solution:
We use (2.5.14) with $a = b = 0$:

$$T(x, y) = T(0, 0) + \frac{\partial T}{\partial x} x + \frac{\partial T}{\partial y} y + \frac{1}{2} \left(\frac{\partial^2 T}{\partial x^2} x^2 + 2 \frac{\partial^2 T}{\partial x \partial y} xy + \frac{\partial^2 T}{\partial y^2} y^2 \right) + \dots \quad (2.5.16)$$

where each derivative is evaluated at the point $(0,0)$. The derivatives are:

$$\frac{\partial T}{\partial x} = -2a\sigma x e^{-\sigma(x^2+y^2)} \qquad \frac{\partial^2 T}{\partial x^2} = -2a\sigma e^{-\sigma(x^2+y^2)} + 4a\sigma^2 x^2 e^{-\sigma(x^2+y^2)}$$

$$\frac{\partial T}{\partial y} = -2a\sigma y e^{-\sigma(x^2+y^2)} \qquad \frac{\partial^2 T}{\partial y^2} = -2a\sigma e^{-\sigma(x^2+y^2)} + 4a\sigma^2 y^2 e^{-\sigma(x^2+y^2)} \qquad (2.5.17)$$

$$\frac{\partial^2 T}{\partial x \partial y} = 4a\sigma^2 xy e^{-\sigma(x^2+y^2)}$$

Most of the derivatives at the point $(0,0)$ are equal to zero, and $T(0,0) = a$. The Maclaurin series is therefore,

$$T(x,y) = a - a\sigma x^2 - a\sigma y^2 + \cdots \qquad (2.5.18)$$

The code below verifies the partial derivatives calculated above.

```
from sympy import symbols, series, exp, diff

a, sigma, x, y = symbols('a,sigma,x,y')

T = a*exp(-sigma*(x**2+y**2))

print('-'*28,'CODE OUTPUT','-'*29,'\n')
print('Partial of T with respect to x = \n', diff(T,x))
print('\nPartial of T with respect to y = \n', diff(T,y))
print('\nSecond Partial of T with respect to x = \n', diff(T,x,x))
print('\nSecond Partial of T with respect to y = \n', diff(T,y,y))
print('\nSecond Partial of T with respect to x and y = \n', diff(T,x,y))


---------------------------- CODE OUTPUT ----------------------------

Partial of T with respect to x =
 -2*a*sigma*x*exp(-sigma*(x**2 + y**2))

Partial of T with respect to y =
 -2*a*sigma*y*exp(-sigma*(x**2 + y**2))

Second Partial of T with respect to x =
 2*a*sigma*(2*sigma*x**2 - 1)*exp(-sigma*(x**2 + y**2))

Second Partial of T with respect to y =
 2*a*sigma*(2*sigma*y**2 - 1)*exp(-sigma*(x**2 + y**2))

Second Partial of T with respect to x and y =
 4*a*sigma**2*x*y*exp(-sigma*(x**2 + y**2))
```

2.6 NUMERICAL EVALUATION OF DERIVATIVES

We can use the Taylor series to derive a numerical method of calculating derivatives. Suppose we want to calculate the Taylor series of a function $f(x)$ about the point x. Then the value of f at a point $x + h$ near x is

$$f(x+h) = f(x) + f'(x)h + \frac{1}{2}f''(x)h^2 + \cdots \qquad (2.6.1)$$

We can solve for the first derivative $f'(x)$:

$$f'(x) = \frac{f(x+h) - f(x)}{h} - h\frac{f''(x)}{2} - \cdots \qquad (2.6.2)$$

If h is small, then we can drop the higher terms and obtain the forward *finite-difference* *formula*

$$f'(x) = \frac{f(x+h) - f(x)}{h} \qquad (2.6.3)$$

which is exact if f is linear. The term $hf''(x)/2$ in (2.6.2) is the leading order term of the *truncation error*. Notice that (2.6.3) is just like (2.1.1) but without the limit. Further note that one can derive a backward finite-difference and a centered-finite difference formula. See Problems 25 and 26, respectively.

Although we will not derive it here, a second derivative can be found using the forward finite-difference method:

$$f''(x) = \frac{f(x+2h) - 2f(x+h) + f(x)}{h^2} \qquad (2.6.4)$$

For more information on second (and higher) derivatives, we recommend consulting a book on numerical methods. See the Further Reading section at the end of this text for a list of such texts.

It is often useful to differentiate data you measure in a laboratory. We can use (2.6.3) to calculate the derivative of experimental data. For example, it is common in the physics lab to use a motion sensor to measure the position of a cart as a function of time. From this information, how do you obtain the car's velocity? We know that the car's instantaneous velocity is the derivative of the car's position. We will need to apply (2.6.3) to experimental data.

In the case of numerical data, we are no longer working with a function of a continuous variable. Instead, we have values $\{x_0, x_1, \ldots, x_n\}$ measured at discrete times $\{t_0, t_1, \ldots, t_n\}$, where the value x_i is measured at the time t_i. However we can use (2.6.3) to calculate the derivative of the data, if the step size $h = t_{i+1} - t_i$ is the same for all values of $i = 0, 1, \ldots, n$. The derivative, using the forward finite-difference formula is

$$x'(t_i) = \frac{x_{i+1} - x_i}{h} \qquad (2.6.5)$$

Let us consider an experiment where a cart is placed along a linear track as shown in Figure 2.10. Suppose that the motion sensor in Figure 2.10 measures the position x of the cart every $\Delta t = 0.1$ second. The cart starts at rest at the origin. The goal of the experiment is to find the velocity and the acceleration of the car as a function of time.

To find the velocity as a function of time, we need to calculate the derivative of the numerical data using (2.6.3):

$$v(t) = \frac{x(t + \Delta t) - x(t)}{\Delta t} \qquad (2.6.6)$$

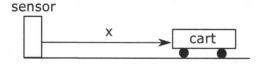

Figure 2.10 The schematic for an experiment which measures the displacement of a cart as a function of time.

where Δt is the time between measurements, similar to h in (2.6.5). Finally, we cannot rewrite $x(t)$ because we do not have continuous functions, but a sampling of the cart's position. We write x_i as the value of the position at time $i\Delta t$ where $i = 0, \ldots 6$. Therefore,

$$v_i = \frac{x_{i+1} - x_i}{\Delta t} \tag{2.6.7}$$

Notice that we need two data points to calculate each value of the velocity in this example. We have 7 data points, so we will only be able to compute 6 values of the velocity. Likewise, we can get the cart's acceleration using (2.6.4)

$$a_i = \frac{x_{i+2} - 2x_{i+1} + x_i}{(\Delta t)^2} \tag{2.6.8}$$

or

$$a_i = \frac{v_{i+1} - v_i}{\Delta t} \tag{2.6.9}$$

The cart's velocity is found in Example 2.15.

Example 2.15: Calculating derivatives from data

Calculate the velocity of the cart whose position was recorded in Table 2.3.

Solution:
The following Python code computes the velocity using (2.6.7) and (2.6.8). We use the variable x for the position, v for velocity.
Notice that the cart's velocity in Figure 2.11 is increasing with time. Further notice that when calculating v, we needed to stop at the second to last element of x. Because the velocity data were shorter than the position data, we could not use the complete time array when creating the velocity plot.
    We use a list comprehension to evaluate the velocity, using (2.6.7).

```
import numpy as np
import matplotlib.pyplot as plt

t = [0.0, 0.1, 0.2, 0.3, 0.4, 0.5, 0.6]  # time data t
x = [0.00119, 0.00088, 0.0331, 0.0770, 0.159, 0.232, 0.381]
# position data x

delta_t = 0.1    # time interval

# use a list comprehension to find v
v = [(x[i+1] - x[i])/delta_t for i in range(0,len(x)-1)]
```

Table 2.3

The position x of the cart from Figure 2.10 as a function of time t.

time [s]	0.0	0.1	0.2	0.3	0.4	0.5	0.6
position x [m]	0.00119	0.00088	0.0331	0.0770	0.159	0.232	0.381

```
plt.subplot(1,2,1)
plt.plot(t,x, 'k.')
plt.xlabel('Time t [s]')
plt.ylabel('Position x [m]')
plt.title('(a)')

plt.subplot(1,2,2)
plt.plot(t[:-1],v, 'k+')
plt.xlabel('Time t [s]')
plt.ylabel('Velocity v [m/s]')
plt.title('(b)')
plt.tight_layout()
plt.show()
```

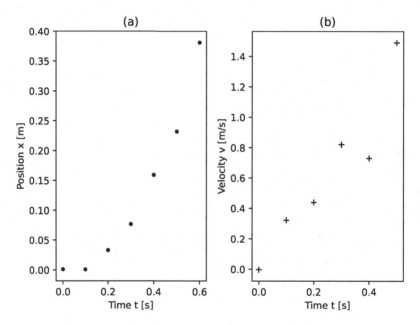

Figure 2.11 The graphical output in Example 2.15 for (a) the position $x(t)$ and (b) the velocity $v(t)$ of the cart.

2.7 END OF CHAPTER PROBLEMS

1. **The kinematics of toy cars and using derivatives to predict future values** – The position of a toy cart can be described by the formula $x(t) = \frac{1}{2}at^2$, where $a = 2.0 \text{ m/s}^2$ is the car's acceleration and t is time measured in seconds.

    a. Plot $x(t)$ and the line tangent to $x(t)$ at $t = 1.5$ seconds.

    b. How well does the line tangent to $x(t)$ at $t = 1.5$ seconds predict the car's location during the next two seconds? Compute and plot the difference between $x(t)$ and the position as predicted by the line tangent to $x(t)$ at $t = 1.5$, as a function of time. Why does the difference increase with time?

2. **Equilibria of the Two-Body Problem** – In classical mechanics, you will study the problem of two bodies interacting via a central force. A common example of this problem is two bodies interacting via the force of gravity. Consider the Earth orbiting the Sun. The effective potential for this system is

$$V(r) = \frac{\ell^2}{2m_E r^2} - \frac{Gm_E m_S}{r} \tag{2.7.1}$$

where G is the universal gravitational constant, r is the distance of the Earth from the Sun, ℓ is the orbital angular momentum of the Earth, and m_E and m_S are the mass of the Earth and Sun, respectively. Using an average orbital angular momentum of $\ell = 2.7 \times 10^{40}$ kgm^2/s, compute the value of r which minimizes $V(x)$. How does that value compare to Earth's mean orbital radius?

3. **Practice with partial derivatives** – Given $f = ax^2 e^{-\lambda(x^2+y^2)}$, where a and λ are constants and $x = x(t)$ and $y = y(t)$. Both by hand and using Python, compute

   a. $\left(\frac{\partial f}{\partial x}\right)_y$

   b. $\left(\frac{\partial f}{\partial y}\right)_x$

   c. $\frac{\partial f}{\partial t}$

   d. $\frac{df}{dt}$

4. **Partial differentiation and the chain rule** – If $z = x \sin y$ and $x = te^{-s}$ and $y = \cos s$. Find $\partial z/\partial s$ and $\partial z/\partial t$. Use Python to help with any algebra. *Hint: Calculate dz in terms of the necessary variables.*

5. **Cartesian coordinates as a function of polar coordinates** – The formulas connecting Cartesian coordinates to polar coordinates are, $x = r\cos\phi$, $y = r\sin\phi$. Calculate $(\partial y/\partial\phi)_x$ and $(\partial y/\partial\phi)_r$.

6. **Chain rule and coupled equations** – If $xs^3 + yt = x^2 t$ and $xs + yt^2 = 1$, calculate $\partial x/\partial s$, $\partial x/\partial t$, $\partial y/\partial s$, and $\partial y/\partial t$ at $(x, y, s, t) = (1, -1, 0, 2)$. Use Python to help with the algebra. It will help to insert the values after you compute the differentials.

7. **Partial derivatives and coupled equations** – Given $x^2 s + yt^2 = 2$ and $x + y = st$, find $(\partial x/\partial s)_t$ and $(\partial x/\partial s)_y$.

8. **Center of mass** – Consider three particles m_1, m_2, m_3, located at (x_1, y_1), (x_2, y_2), and (x_3, y_3), respectively. Find the point P about which their total moment of inertia is minimized. Show that the result is the system's center of mass.

9. **Constrained particle motion** – A particle is constrained to move along the plane, $x + 3y - 4z = 3$. Find the shortest distance between the particle and the origin. Use Python to help with any partial derivatives.

10. **The dimensions of an aquarium** – A glass aquarium with rectangular sides, bottom, and no top is to hold a certain volume of water. Find its dimensions so it will use the least amount of glass.

Maclaurin Series – Find the Maclaurin series for the following functions. Do these problems by hand, using the well-known series for $\sin x$, $\cos x$, $\tan x$, $\exp x$, $\ln x$ and $(1 + x)^n$, and verify the results using Python.

11. $\sin\left(x^2\right)$

12. $x\ln\left(x^2 + 1\right)$

13. $\left(\frac{x+1}{x-1}\right)$

14. $\tan^2 x$

15. $\int_0^x e^{-2t^2}\,dt$

16. $\cosh\left(x\right)$

Taylor Series – Find the Taylor series for each function about the point listed. Do these problems by hand and verify the results using Python.

17. $\sin\left(x\right),\quad x = \pi/2$

18. $1/x,\quad x = 1$

19. $\sqrt{x},\quad x = 16$

20. $\ln\left(e^x - 1\right),\quad x = 1$

21. $\sec\left(x\right),\quad x = \pi/4$

22. $\frac{x}{1-x},\quad x = 2$

23. **Maclaurin series approximation of electric field due to a charged disk** – The electric potential of a charged disk with surface charge density σ and radius R is

$$V_P = \frac{\sigma}{2\,\epsilon_0}\left(\sqrt{z^2 + R^2} - z\right)$$

where z is the distance above the disk along the its central axis and ϵ_0 is the permittivity of free space (a constant). Compute the first three terms of the Maclaurin series for (a) the potential V_p and (b) the corresponding electric field $E = -dV_p/dz$, and interpret the results.

24. **Falling object experiencing quadratic air resistance** – Consider a particle of mass m falling in the Earth's gravitational field and experiencing a drag force with a magnitude $f = cv^2$. Newton's second law gives the net force as $F_{\text{Net}} = mg - cv^2$ if the positive y-direction is defined to be toward the Earth's surface. By integrating Newton's second law twice, we can find the particle's displacement as a function of time

$$y(t) = \frac{m}{c}\ln\left(\cosh\left(\frac{cv_t}{m}t\right)\right)$$

where $v_t = \sqrt{mg/c}$ is the terminal velocity of the mass. Find and plot several terms of the Maclaurin series for $y(t)$ and interpret the result.

25. **Backward Finite-Difference and numerical evaluation of derivatives** –

    a. Derive the backwards finite-difference formula

$$f'(x) = \frac{f(x) - f(x - h)}{h} \qquad (2.7.2)$$

    where h is the step size.

    b. The position of a toy car was measured from $t = 0$ to $t = 1$ s in intervals of 0.1 s, and the results were as shown in Table 2.4. Using the backward finite-difference numerical method, calculate and plot the velocity and acceleration of the car as a function of time.

Table 2.4

The position x of the cart at the time instants t for Problem 25.

t [s]	0	0.1	0.2	0.3	0.4	0.5	0.6	0.7	0.8	0.9	1.0
x [m]	0	0.0008	0.0043	0.0088	0.0161	0.0245	0.0370	0.048	0.061	0.8	0.11

26. **Centered Finite-Difference and numerical evaluation of derivatives** –

    a. Derive the centered finite-difference formula

$$f'(x) = \frac{f(x + h) - f(x - h)}{2h} \qquad (2.7.3)$$

    where h is the step size.

    b. The position of a toy car was measured and the results are shown in Table 2.5. Using the centered finite-difference numerical method, calculate and plot the velocity and acceleration of the car as a function of time.

Table 2.5

The position of the cart for Problem 26.

t [s]	0	0.1	0.2	0.3	0.4	0.5	0.6	0.7	0.8	0.9
x [m]	4.000	3.970	3.940	3.912	3.886	3.862	3.841	3.824	3.811	3.802

27. **Partial derivatives in 3D in Cartesian coordinates** – Show using SymPy and by hand that

$$V(x, y, z) = \sin(3x)\sin(4y)e^{-5z} \qquad (2.7.4)$$

is a solution of the following equation, which is known as the Laplace equation in Cartesian coordinates in three dimensions:

$$\frac{\partial^2 V}{\partial x^2} + \frac{\partial^2 V}{\partial y^2} + \frac{\partial^2 V}{\partial z^2} = 0 \qquad (2.7.5)$$

3 Integration

In this chapter we summarize the concepts of indefinite and definite integrals of a single variable function $f(x)$, and the rules for calculating integrals. We review and give examples of the all-important integration by parts, and provide an overview of symbolic and numerical integration methods in Python, using the SymPy, NumPy and SciPy libraries. We discuss both symbolic and numerical integration methods, and provide examples from undergraduate courses in Mechanics, Electromagnetism and Quantum Mechanics. We demonstrate the evaluation of improper integrals and integrals of special functions, and discuss Simpson's rule and the trapezoidal integration method.

3.1 INTEGRALS

Given a function f of a real variable x and an interval $[a, b]$ along the x-axis, the definite integral

$$\int_a^b f(x)\, dx \tag{3.1.1}$$

is defined informally to be the net (or signed) area of the region in the xy-plane bounded by the graph of $f(x)$, the x-axis, and the vertical lines $x = a$ and $x = b$. This is shown schematically in Figure 3.1a. Integrals of the type shown in (3.1.1) are termed *definite integrals*. The term net area here means that an area above the x-axis is considered positive, and area underneath the x-axis is negative.

The $\int$ sign represents integration, a and b are the *lower limit* and *upper limit of integration* which define the *domain of integration* [a,b], and $f(x)$ is the *integrand* i.e. the function to be integrated. The variable of integration dx has different interpretations. For example, it can be seen as strictly a notation indicating that x is a dummy *variable of integration*, or as an independent mathematical quantity called a *differential*, as discussed in Chapter 2.

A more formal definition of the definite integral is as the limit of an infinite sum, which is obtained by dividing the interval [a,b] into N sub-intervals $[x_{i-1}, x_i]$ and evaluating the function $f(x)$ at an arbitrary point x_i^* located in this sub-interval :

$$\int_a^b f(x)\, dx = \lim_{N \to \infty} \sum_{i=1}^{N} f(x_i^*)\, (x_i - x_{i-1}) \tag{3.1.2}$$

As the number of sub-intervals $N \to \infty$, the width of the sub-interval $\Delta x = x_i - x_{i-1}$ will become negligibly small i.e. $\Delta x \to 0$, and the infinite sum will converge to the value of the definite integral. The summation on the right-hand side of (3.1.2) is known as a *Riemann sum*. The process is shown schematically in Figure 3.1b.

DOI: 10.1201/9781003294320-3

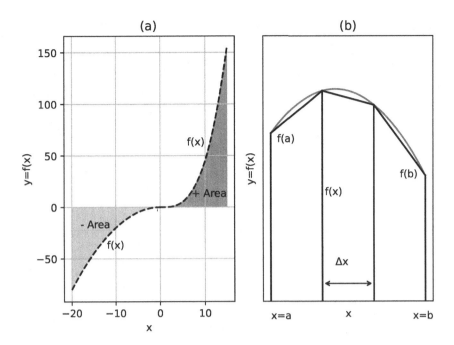

Figure 3.1 (a) A definite integral of a function can be represented as the *net or signed area* of the region bounded by its graph. The part of the area above the x-axis is considered positive, and the part below it is considered negative. (b) The integral can also be defined mathematically as the limit of the infinite sum of the areas of rectangles with base Δx and height $f(x)$, as the width $\Delta x \to 0$.

More generally, the *indefinite integral* $\int f(x)\,dx$ of a function $f(x)$ is defined as a new function $F(x)$ called the *antiderivative* of $f(x)$, such that the derivative dF/dx is equal to the given function $f(x)$. Symbolically we write:

$$F(x) = \int f(x)\,dx \tag{3.1.3}$$

$$f(x) = \frac{d}{dx}\left[\int f(x)\,dx\right] \tag{3.1.4}$$

3.2   REVIEW OF ELEMENTARY INTEGRALS

Table 3.1 gives some of the most frequently used definite integrals and summarizes the basic integration rules. In this table f, g represent any functions of x, and a, b represent constants.

Table 3.1

List of several commonly used integrals and integration rules.

Type of function	Integral
Constant rule	$\int (af)\, dx = a \int f\, dx$
Sum rule	$\int (a\, f + b\, g)\, dx = a\, \int f\, dx + b \int g\, dx$
Integration by parts	$\int f\, g'\, dx = f\, g - \int f'\, g\, dx$
Powers	$\int x^a\, dx = \frac{x^{a+1}}{a+1}$
Rational Functions	$\int (ax + b)^n\, dx = \frac{(ax+b)^{n+1}}{a\,(n+1)}$
	$\int \frac{c}{ax+b}\, dx = \frac{c}{a} \ln |ax + b|$
Exponentials and logarithms	$\int e^x\, dx = e^x$
	$\int \frac{1}{x}\, dx = \ln |x|$
	$\int \ln x\, dx = x \ln x - x$
	$\int a^x\, dx = \frac{a^x}{\ln a}$
Trig functions	$\int \sin x\, dx = -\cos x$
	$\int \cos x\, dx = \sin x$
	$\int \tan x\, dx = -\ln |\cos x|$
	$\int \cot x\, dx = \ln |\sin x|$
	$\int \sec^2 x\, dx = \tan x$
	$\int \csc^2 x\, dx = -\cot x$
	$\int \sin^2 x\, dx = \frac{1}{2}\left(x - \frac{\sin 2x}{2}\right) = \frac{1}{2}(x - \sin x \cos x)$
	$\int \cos^2 x\, dx = \frac{1}{2}\left(x + \frac{\sin 2x}{2}\right) = \frac{1}{2}(x + \sin x \cos x)$
	$\int \sin^n x\, dx = -\frac{\sin^{n-1} x \cos x}{n} + \frac{n-1}{n} \int \sin^{n-2} x\, dx$
	$\int \cos^n x\, dx = \frac{\cos^{n-1} x \sin x}{n} + \frac{n-1}{n} \int \cos^{n-2} x\, dx$
	$\int x \cos(n\, x)\, dx = \frac{x \sin(n\, x)}{n} + \frac{\cos(n\, x)}{n^2}$
	$\int x \sin(n\, x)\, dx = -\frac{x \cos(n\, x)}{n} + \frac{\sin(n\, x)}{n^2}$
Error function integral	$\int_{-\infty}^{\infty} e^{-x^2 t}\, dx = \frac{\sqrt{\pi}}{\sqrt{t}}$

3.3 OVERVIEW OF INTEGRATION METHODS IN PYTHON

This section provides a brief overview of different symbolic and numerical integration methods available in Python.

In general, there are two broad types of integration functions available in Python. The first type of Python function is used to integrate analytical functions $f(x)$, and several examples are given in this chapter from various areas of science. This type of integration is carried out using two different methods:

- *Method #1:* Using the `sympy.integrate` library, we carry out symbolic integration of both proper and improper integrals. In addition, by using the `scipy.special` library we will learn how to evaluate integrals of special functions frequently encountered in physics (Error function, Bessel function, Hermite polynomials etc.).

- *Method #2:* Using the `scipy.integrate` library we can carry out numeric integration of functions from other libraries, or functions defined by the user (see Table 3.2), or samples of a function which are often stored as NumPy arrays (see Table 3.3).

Table 3.2

Integration functions available in the `scipy.integrate` library for given functions.

Python function	Description
quad	General purpose integration
dblquad	General purpose double integration
tplquad	General purpose triple integration
fixed_quad	Integrate func(x) using Gaussian quadrature of order n
quadrature	Integrate with given tolerance using Gaussian quadrature
romberg	Integrate func using Romberg integration

Of the methods in Table 3.2, we will use the `quad` command in this chapter to evaluate 1D integrals, while the functions `dblquad` and `tplquad` will be used in a later chapter to evaluate two-dimensional and three-dimensional integrals.

Table 3.3

Integration functions available in the `scipy.integrate` library for sampled functions.

Python function	Description
trapezoid	Use the trapezoidal rule to compute an integral
cumulative_trapezoid	Use the trapezoidal rule to cumulatively compute an integral
simpson	Use Simpson's rule to compute an integral from samples of the function
romb	Use Romberg Integration to compute integral from $(2^k + 1)$ evenly-spaced samples

For a more complete listing of integration routines in SciPy, and details on how each algorithm works, see the online manual and tutorial for `scipy.integrate`.

3.4 INTEGRATION BY PARTS

Integration by parts is a rule that transforms the integral of products of functions into simpler integrals.

Consider an integral of the product of a function $f(x)$ and the derivative of another function $g(x)$:

$$\int f(x)g'(x)dx \tag{3.4.1}$$

where $g'(x) = dg/dx$. We can integrate (3.4.1) if we use the product rule of differentiation:

$$\int (fg)' \, dx = \int f'g dx + \int fg' dx \tag{3.4.2}$$

This leads to the following two *integration by parts rules* for indefinite and definite integrals:

$$\int f(x)g'(x) \, dx = f(x)g(x) - \int f'(x)g(x) \, dx \tag{3.4.3}$$

$$\int_a^b f(x)g'(x) \, dx = f(x)g(x) \Big|_a^b - \int_a^b f'(x)g(x) \, dx \tag{3.4.4}$$

where

$$f(x)g(x) \Big|_a^b = f(b)g(b) - f(a)g(a) \tag{3.4.5}$$

In your calculus course, you probably performed the variable substitutions, $u = f(x)$, and $v = g(x)$ with the differentials $du = f'(x)dx$ and $dv = g'(x)dx$. Then integration by parts is written in the easy to remember formula

$$\int u \, dv = uv - \int v \, du \tag{3.4.6}$$

Example 3.1 applies integration by parts to evaluate an indefinite integral which appears often in physics. We also use the `sympy.integrate` command to evaluate the indefinite integral.

Example 3.1: An example of integration by parts

Calculate these integrals using integration by parts

$$\text{(a)} \quad \int x \cos(n \, x) \, dx \qquad \text{(b)} \quad \int x \sin(n \, x) \, dx$$

Solution:
(a) We substitute $u = x$ then $du = dx$, and also let $dv = \cos(n \, x) \, dx$, then

$$v = \int \cos(n \, x) \, dx = \frac{\sin (n \, x)}{n}$$

By applying the formula for integration by parts $\int u \, dv = uv - \int v \, du$:

$$\int x \cos(n \, x) \, dx = \frac{x \sin(n \, x)}{n} - \int \frac{\sin (n \, x)}{n} \, dx$$

$$\int x \cos(n \, x) \, dx = \frac{x \sin(n \, x)}{n} + \frac{\cos(n \, x)}{n^2} + C$$

(b) In a similar manner we obtain:

$$\int x \sin(n\,x)\,dx = -\frac{x\cos(n\,x)}{n} + \frac{\sin(n\,x)}{n^2} + C \tag{3.4.7}$$

where C is an arbitrary constant of integration. The following Python code to evaluate the definite integral gives the same result. After defining x as a symbol, we use the symbols command to tell Python that n is not zero by including the option positive=True. We also use the symbolic form of cos() for the cosine function, instead of the numerical form of numpy.cos().

```
from sympy import symbols, integrate, cos, sin
x = symbols('x')
n = symbols('n', positive=True)  # define symbols

print('-'*28,'CODE OUTPUT','-'*29,'\n')
print("The indefinite cosine integral = ",integrate(x*cos(n*x),x))
print("The indefinite sine integral = ",integrate(x*sin(n*x),x))

-------------------------- CODE OUTPUT ----------------------------

The indefinite cosine integral =  x*sin(n*x)/n + cos(n*x)/n**2
The indefinite sine integral =  -x*cos(n*x)/n + sin(n*x)/n**2
```

In science and engineering we often need to evaluate integrals of the type:

$$\int x^3 \sin(x)\,dx \quad \text{and} \quad \int x^2 e^x\,dx$$

These types of integrals can be evaluated by using repeated integration by parts. Each application of the integration parts rule lowers the power of x by one.

3.5   PARAMETRIC INTEGRATION FOR DEFINITE INTEGRALS

In the previous section we mentioned we can evaluate integrals of the type $\int x^3 \sin(x)\,dx$ by using repeated integration by parts. Another very useful technique for evaluating *definite* integrals of this type is by using parametric integration.

If $f(x,t)$ is a function of two parameters x and t, the parametric integration rule states that:

$$\frac{d}{dt}\left(\int_a^b f(x,t)dx\right) = \int_a^b \frac{\partial}{\partial t}f(x,t)dx \tag{3.5.1}$$

This rule also holds for non-finite bounds a, b.

Example 3.2 uses the sympy.integrate command to symbolically evaluate definite integrals, by using the parametric integration method.

Example 3.2: An example of the parametric integration method

Evaluate symbolically the following definite integrals

$$\text{(a)} \quad \int_0^\infty x e^{-x\,t}\,dx \qquad \text{(b)} \quad \int_0^\infty x^2 e^{-x\,t}\,dx$$

using the parametric integration method, with the assumption $t > 0$.

Solution:
(a) We start by applying rule (3.5.1) using $f = e^{-x\,t}$:

$$\frac{d}{dt}\int_0^\infty e^{-tx}\,dx = \int_0^\infty \frac{\partial}{\partial t}\left(e^{-tx}\right)\,dx = -\int_0^\infty x\,e^{-tx}\,dx \tag{3.5.2}$$

The left-hand side of (3.5.2) can be evaluated easily:

$$\frac{d}{dt}\int_0^\infty e^{-tx}\,dx = \frac{d}{dt}\left(\frac{e^{-tx}}{-t}\bigg|_{x=0}^{x=\infty}\right) = \frac{d}{dt}\left[0 - \left(\frac{e^{-0}}{-t}\right)\right] = \frac{d}{dt}\left(\frac{1}{t}\right) = -\frac{1}{t^2} \tag{3.5.3}$$

Substituting (3.5.3) in (3.5.2):

$$\int_0^\infty x\,e^{-tx}\,dx = \frac{1}{t^2} \tag{3.5.4}$$

(b)

$$\frac{d}{dt}\int_0^\infty x\,e^{-tx}\,dx = \int_0^\infty \frac{\partial}{\partial t}\left(x\,e^{-tx}\right)\,dx = -\int_0^\infty x^2\,e^{-tx}\,dx \tag{3.5.5}$$

and using part (a):

$$\frac{d}{dt}\left(\frac{1}{t^2}\right) = -\int_0^\infty x^2\,e^{-tx}\,dx = \frac{-2}{t^3} \tag{3.5.6}$$

$$\int_0^\infty x^2\,e^{-tx}\,dx = \frac{2}{t^3}$$

The Python code below evaluates these definite integrals. We need to specify again that the variables x, t are real and that t is a positive. We use the symbolic form of the exponential function, sympy.exp(). The symbol for infinity in SymPy is oo.

```
from sympy import symbols, integrate, exp, oo
x, t = symbols('x,t', real=True,positive=True)   # define symbols

int1 = integrate(x*exp(-x*t),(x,0,oo))
int2 = integrate(x**2.0*exp(-x*t),(x,0,oo))

print('-'*28,'CODE OUTPUT','-'*29,'\n')
print("The definite integral of x*exp(-x*t) =      ",int1,'\n')
print("The definite integral of x**2.0*exp(-x*t) =",int2)


------------------------- CODE OUTPUT ----------------------------

The definite integral of x*exp(-x*t) =       t**(-2)


The definite integral of x**2.0*exp(-x*t) = 2.0/t**3.0
```

Another common example of using the parametric integration method is the evaluation of the definite Gaussian Integrals of the type:

$$\int_{-\infty}^{\infty} x^{2n} e^{-x^2 t}\, dx \qquad (3.5.7)$$

These are evaluated by starting with the integral

$$\int_{-\infty}^{\infty} e^{-x^2 t} dx = \frac{\sqrt{\pi}}{\sqrt{t}} \qquad (3.5.8)$$

and by taking the derivative with respect to t of both sides. Each time the parametric integration rule is applied, an extra factor of x appears inside the Gaussian integrals (see Problem 6).

3.6 INTEGRATING ANALYTICAL FUNCTIONS IN PYTHON

In this section we present specific applications of Python integration methods from various areas of physics. We start with an example from Classical Mechanics.

In Example 3.3 we use the `integrate` command from SymPy to evaluate an indefinite integral (or antiderivative) of a function. The same command can also be used to evaluate the definite integral of the function in any interval $[a, b]$.

Example 3.3: Integration of Newton's law

The force acting on a mass m is decreasing with time according to $F(t) = F_0/(t^2 + 1)$ where t is the elapsed time. Use Newton's law $F = ma = m\, dv/dt$ to find the velocity $v(t)$ as a function of time.

Solution:
In this example the time dependent force $F = F(t)$ and Newton's law $F = ma$ becomes:

$$F(t) = \frac{F_0}{t^2 + 1} = m\frac{dv}{dt}$$

$$dv = \frac{F_0}{m\,(t^2 + 1)}dt$$

Integrating both sides:

$$v(t) = \frac{1}{m}\int F(t)\,dt = \frac{1}{m}\int \frac{F_0}{t^2 + 1}\,dt = \frac{F_0}{m}\tan^{-1} t + C \qquad (3.6.1)$$

where C is a constant.

We can also obtain these results by using the `integrate` command from SymPy. Notice the variables $t, F0, m$ need to be defined using the `symbols` command. The option `real=True` tells Python that the variables in this example are real numbers. The result from Python agrees with the result from the integration by hand.

```
from sympy import symbols, integrate
t, F0, m = symbols('t, F0, m',real=True)     # define symbols

print('-'*28,'CODE OUTPUT','-'*29,'\n')
int1=integrate(F0/m*(1/(t**2+1)),t)
print('The indefinite integral of F0/m*(1/(t**2+1))=',int1)
```

```
------------------------- CODE OUTPUT ----------------------------

The indefinite integral of F0/m*(1/(t**2+1))= F0*atan(t)/m
```

Example 3.4 is an application of *numerical* integration using Python, in Quantum Mechanics. Here we use again symbolic integration with the `integrate` command. We also show how to numerically evaluate definite integrals using SymPy's `integrate.quad` command.

Example 3.4: Expectation value of the position in Quantum Mechanics

In Quantum Mechanics we often need to evaluate the expectation value $< x >$ of the position of a particle when we know the wave function $\psi(x)$ which describes the particle. The expectation value $< x >$ for a particle moving along the x-axis can be evaluated from:

$$< x >= \int_{-\infty}^{\infty} x\, \psi^2(x) dx \tag{3.6.2}$$

Consider the following wave function for a particle

$$\psi(x) = \begin{cases} x & 0 \le x \le 1 \\ 2 - x & 1 \le x \le 2 \\ 0 & \text{otherwise} \end{cases} \tag{3.6.3}$$

Find the expectation value $< x >$ for the position.

Solution:
We break up the integral into two parts corresponding to the two given ranges along the x axis:

$$< x >= \int_{0}^{2} x\, \psi^2(x) dx = \int_{0}^{1} x\, x^2 dx + \int_{1}^{2} x\, (2-x)^2\, dx$$

$$< x >= \frac{x^4}{4}\Big|_{0}^{1} + \int_{1}^{2} x\, \left(4 + x^2 - 4x\right) dx = \frac{1}{4} + \left(2x^2 + \frac{x^4}{4} - 4\frac{x^3}{3}\right)\Big|_{1}^{2} = \frac{2}{3}$$

The Python code evaluates the two definite integrals, using two different methods, symbolically using SymPy's `integrate` function, and numerically using SciPy's `integrate.quad` function. The line `f=lambda x:  x**3` defines a simple function $f = x^3$ by using the `lambda` method. Furthermore, we included the `print` command to show how formatting with 4 decimal places can be done in Python.

Note that the result of the `integrate.quad` function has two parts, the numerical value of the integral and the corresponding error in the evaluation of the integral. In order to print only the value of the integral we use `int1[0]+int2[0]`.

Further note that in this example we use the `integrate()` functions from *both* SymPy and SciPy. We avoid any conflicts by importing the integrate function in SymPy as `sym.integrate()`, and the SciPy function as `integrate.quad`.

```
import sympy as sym
# symbolic integration with sympy.integrate
from scipy import integrate      # numerical integration with quad()
```

```
x = sym.symbols('x')                   # define symbols
int0 = sym.integrate(x**3,(x,0,1))+sym.integrate(x*((2-x)**2),(x,1,2))
print('-'*28,'CODE OUTPUT','-'*29,'\n')
print("The Integral using SymPy is: ",int0)  # integral using sympy

f = lambda x: x**3                     # function for interval 0<x<1
int1 = integrate.quad(f,0,1)           # first integral
print("The first Integral and its error: ",int1)    # integral and error

g = lambda x: x*((2-x)**2)             # function for interval 1<x<2
int2 = integrate.quad(g,1,2)           # second integral and error
print("The second Integral and its error: ",int2,'\n')  # integral and error

print("The sum of the  Integrals: ", f'{(int1[0]+int2[0]):.4f}')  # sum of integrals

------------------------- CODE OUTPUT --------------------------

The Integral using SymPy is:  2/3
The first Integral and its error:  (0.25, 2.7755575615628914e-15)
The second Integral and its error:  (0.41666666666666674, 4.625929269271486e-15)

The sum of the  Integrals:  0.6667
```

A common application of integration techniques is the evaluation of the center of mass of objects. For a thin rod of constant cross-sectional area, we define the linear mass density $\lambda = dm/dx$ where dm is the mass of a small element with length dx. The SI units of the linear mass density λ are kg/m. In Example 3.5 we find the center of mass of a rod with length L and a variable linear mass density proportional to the distance from the end of the rod, i.e. $\lambda(x) = 2x$.

Example 3.5: Center of mass of a rod

Consider a thin rod of length L and mass M, with a non-uniform linear density $\lambda = 2x$ (in SI units), where x is the distance from the end of the rod. Compute the location X_{cm} of the center of mass for $L = 1$ m, by using the definition of the center of mass:

$$X_{cm} = \frac{1}{M} \int x\,dm = \frac{1}{M} \int_{x=0}^{L} x\,\lambda dx = \frac{\int_{x=0}^{L} x\,\lambda dx}{\int_{x=0}^{L} dm} \tag{3.6.4}$$

Solution:
From the definition $\lambda = dm/dx$ we can replace $dm = \lambda dx = 2x dx$ in (3.6.4) and evaluate the two integrals:

$$M = \int dm = \int_{x=0}^{L} \lambda dx = \int_{x=0}^{L} 2x\,dx = x^2 \Big|_0^L = L^2$$

$$X_{cm} = \frac{1}{M} \int_{x=0}^{L} x\,\lambda dx = \frac{1}{L^2} \int_{x=0}^{L} x\,2x\,dx = \frac{1}{L^2} \frac{2x^3}{3} \Big|_0^L = \frac{2}{3}L$$

For $L = 1$ m we obtain $X_{cm} = 2/3$ m.

We also perform the integration using the numerical integration routine `integrate.quad`, which is imported from SciPy, and we also verify the symbolic result $X_{cm} = 2/3$ by using the `integrate` function imported from SymPy. Notice that in this example we import SymPy using the alias sym, so that there is no confusion between the SymPy and NumPy `integrate` commands.

```python
print('-'*28,'CODE OUTPUT','-'*29,'\n')

import sympy as sym
from scipy import integrate
x, L = sym.symbols('x,L')

int1 = integrate.quad(lambda x: 2*x**2, 0,1) # 1st integral numerically
int2 = integrate.quad(lambda x: 2*x, 0,1)    # second integral

# X_cm is the ratio of the two integrals
print("Numerically evaluated X_cm: ", f'{(int1[0]/int2[0]):.4f}')

# integrate also symbolically in SymPy using integrate()
symb=sym.integrate(2*x**2,(x,0,L))/sym.integrate(2*x,(x,0,L))
print("\nSymbolically evaluated X_cm: ",symb)

-------------------------- CODE OUTPUT ----------------------------

Numerically evaluated X_cm:  0.6667

Symbolically evaluated X_cm:  2*L/3
```

3.7 FOURIER SERIES

The expansion of a periodic function into a Fourier series is a very important technique which allows us to write a periodic function as the linear combination of sine and cosine terms. Let us consider a function $f(t)$ of the real variable t, which is periodic with a period τ, i.e. $f(t)=f(t+\tau)$. Fourier's theorem states that this periodic function can be written as the sum of sine and cosine terms in the form of a *Fourier series*:

$$f(t) = \frac{a_0}{2} + \sum_{n=1}^{N} a_n \cos(n\omega t) + b_n \sin(n\omega t) \qquad (3.7.1)$$

where the lowest angular frequency is $\omega = 2\pi/\tau$, and the higher frequencies are integer multiples of the fundamental angular frequency ω. The coefficients a_n and b_n represent the amplitudes of the various waves and are known as *Fourier coefficients*. The Fourier coefficients a_n and b_n are computed by using the following equations which are known as *Fourier's trick* to many generations of physicists:

$$a_n = \frac{2}{\tau} \int_{-\tau/2}^{\tau/2} f(t) \cos(n\omega t) dt \qquad (3.7.2)$$

$$b_n = \frac{2}{\tau} \int_{-\tau/2}^{\tau/2} f(t) \sin(n\omega t) dt \qquad (3.7.3)$$

The Fourier coefficients a_n and b_n are computed by using the following general integral relationships:

$$\int_{-\tau/2}^{\tau/2} \cos(n\omega t) \cos(m\omega t) \, dt = \begin{cases} 0 & m \neq n \\ \tau/2 & m = n \neq 0 \end{cases}$$

$$\int_{-\tau/2}^{\tau/2} \sin(n\omega t) \sin(m\omega t) \, dt = \begin{cases} 0 & m \neq n \\ \tau/2 & m = n \neq 0 \end{cases} \qquad (3.7.4)$$

$$\int_{-\tau/2}^{\tau/2} \cos(n\omega t) \sin(m\omega t) \, dt = 0 \qquad \text{for all integers } n \text{ and } m$$

The calculation of the Fourier coefficients involves multiplying both sides of (3.7.1) with either $\cos(2\pi mt/\tau)$ or $\sin(2\pi mt/\tau)$, and integrating over the whole period.

We will use Fourier series extensively in Chapter 11, when we discuss the solution of partial differential equations (PDE's).

In Example 3.6 we demonstrate how to evaluate a_n and b_n using integration by parts.

Example 3.6: Evaluation of Fourier coefficients for a periodic function

Find the Fourier series for the following periodic saw tooth function:

$$f(x) = x \quad \text{for } -\pi \leq x \leq \pi$$

$$f(x + 2\pi k) = f(x) \quad \text{for } x < -\pi \text{ and } x > \pi$$

where k is an integer.

Solution:
The period of the function is $\tau = 2\pi$ and therefore we can compute the Fourier coefficients, with $\omega = 2\pi/\tau = 1$. First we find the coefficients a_n :

$$a_n = \frac{1}{\pi} \int_{-\pi}^{\pi} x \cos(\tfrac{2\pi nx}{2\pi}) \, dx = \frac{1}{\pi} \int_{-\pi}^{\pi} x \cos(nx) \, dx$$

We now notice that the integrand $f(x) = x \cos(nx)$ is an odd function of x, since $f(-x) = -f(x)$ for all values of x, and therefore its integral must be zero, i.e. $a_n = 0$ for any integer value of n. Next, we compute the b_n coefficients:

$$b_n = \frac{1}{\pi} \int_{-\pi}^{\pi} x \sin(\tfrac{2\pi nx}{2\pi}) \, dx = \frac{1}{\pi} \int_{-\pi}^{\pi} x \sin(nx) \, dx$$

and using (3.4.7) from Example 3.1:

$$b_n = \frac{1}{\pi} \left[-\frac{x \cos(n\,x)}{n} \bigg|_{-\pi}^{\pi} + \frac{\sin(n\,x)}{n^2} \bigg|_{-\pi}^{\pi} \right] = \frac{2}{n} \cos(n\pi) = -\frac{2\,(-1)^n}{n}$$

where we used $\sin(n\pi) = 0$ and $\cos(n\pi) = (-1)^n$ for all integer values of n.
The resulting Fourier series is therefore:

$$f(x) = 2 \sum_{n=1}^{\infty} \frac{(-1)^{n+1}}{n} \sin(nx) = 2\sin(x) - \sin(2x) + \frac{2}{3}\sin(3x) - \dots$$

In the code evaluating the definite integrals for a_n and b_n we specify that n is an integer, by using the option `integer = True`. Once more, we import cos and sin for the cosine and sine functions in SymPy. The plot of the individual Fourier components and their sum, is left as an exercise in Problem 12.

```
from sympy import symbols, integrate, sin, cos, pi
x = symbols('x')          # define symbols
n = symbols('n', integer = True,positive=True)

print('-'*28,'CODE OUTPUT','-'*29,'\n')

print("Fourier coefficients an=",integrate(x*cos(n*x),(x,-pi,pi))/pi)

print("\nFourier coefficients bn=",integrate(x*sin(n*x),(x,-pi,pi))/pi)

--------------------------- CODE OUTPUT ---------------------------

Fourier coefficients an= 0

Fourier coefficients bn= -2*(-1)**n/n
```

3.8   IMPROPER INTEGRALS AND INTEGRALS OF SPECIAL FUNCTIONS

In this section we look at improper integrals and integrals involving special functions. These types of integrals occur frequently in science, and they can be evaluated either symbolically or numerically using Python.

In SymPy the symbolic infinity symbol is implemented as `sympy.oo`. We start with an example from Quantum Mechanics, on the wave function of simple harmonic oscillator of mass m and frequency ω. Example 3.7 is an example of how the error function is used in Quantum Mechanics, and the mathematical description of the quantum mechanical simple harmonic oscillator (SHO).

Example 3.7: Using the error function in Quantum Mechanics

The wave function of a SHO in Quantum Mechanics of mass m and frequency ω is:

$$\psi(x) = A \, \exp\left(-\frac{m\omega}{2\hbar} \, x^2\right) \tag{3.8.1}$$

where A is the normalization constant of the wave function and $\hbar$ is the reduced Planck constant. The position x can vary anywhere along the x-axis.

(a) Calculate the constant A so that the wave function is normalized, i.e. by requiring that the integral

$$\int_{x=-\infty}^{\infty} \psi^2(x)dx = 1 \tag{3.8.2}$$

(b) Find the expectation value $< x >$ of the position of this particle along the x-axis:

$$< x >= \int_{-\infty}^{\infty} x \, \psi^2(x)dx$$

(c) A nitrogen molecule N_2 provides a simple example of molecular vibration. Look up the vibrational frequency ω and the mass m, and plot the wave function for the nitrogen molecule.

Solution:
(a) We normalize the wave function:

$$\int_{-\infty}^{\infty} \psi^2(x)dx = \int_{-\infty}^{\infty} A^2 \exp\left(-\frac{m\omega}{\hbar} x^2\right) dx = 1 \qquad (3.8.3)$$

In general, the definite integral

$$\text{erf}(x) = \int_0^{\infty} e^{-x^2} dx \qquad (3.8.4)$$

has no explicit analytical formula and is known as the error function $\text{erf}(x)$. The function $\exp\left(-x^2\right)$ is called a Gaussian function centered at $x = 0$. Using a table of integrals or checking the internet, you can find the general formula for Gaussian functions

$$\int_{-\infty}^{\infty} e^{-a\,x^2} dx = \sqrt{\pi/a}$$

so (3.8.3)with $a = m\omega/(\pi\hbar)$ yields:

$$A^2 \sqrt{\frac{\pi\hbar}{m\omega}} = 1 \Longrightarrow A = \left(\frac{m\omega}{\pi\hbar}\right)^{1/4}$$

and the complete normalized wave function is:

$$\psi(x) = \left(\frac{m\omega}{\pi\hbar}\right)^{1/4} \exp\left(-\frac{m\omega}{2\hbar} x^2\right)$$

(b) The expectation value $< x >$ is found from:

$$< x >= \int_{-\infty}^{\infty} x\,\psi^2(x)dx = \int_{-\infty}^{\infty} x \left(\frac{m\omega}{\pi\hbar}\right)^{1/2} \exp\left(-\frac{m\omega}{\hbar} x^2\right) dx$$

Notice the integrand $f(x) = x \exp\left(-\frac{m\omega}{\hbar} x^2\right)$ is an odd function, i.e. $f(-x) = -f(x)$, therefore the integral from $-\infty$ to $+\infty$ must be zero: $< x >= 0$.
In the code we obtain the same results using `sympy.integrate` to carry out the integrations from $-\infty$ to $+\infty$. We also use SymPy's `solve` command to solve the equation $A^2 = 1$ for the value of A. When using `solve`, the right-hand side of the equation must be zero.

```
from sympy import symbols, integrate, exp, oo, solve, sqrt

x, m, omega, hbar, A = symbols('x, m, omega, hbar, A',positive=True)

# Evaluate total probability Ptotal and set =1
lhs = integrate((A*exp(-m*omega/(2*hbar)*x**2))**2,(x,-oo,oo))

# symbolically solve equation Ptotal=1
A1 = solve(lhs-1,A)

print('-'*28,'CODE OUTPUT','-'*29,'\n')
print("Normalization constant A = ",A1[0])

# find <x>
expectx=integrate(x*(A1[0]*exp(-m*omega/(2*hbar)*x**2))**2,\
(x,-oo,oo))
print('Expectation value of x: ',expectx)
```

```
-------------------------- CODE OUTPUT --------------------------
Normalization constant A =  m**(1/4)*omega**(1/4)/(pi**(1/4)*hbar**(1/4))
Expectation value of x:  0
```

(c) The molecular vibration frequency of the nitrogen molecule is $\omega = 5.16 \times 10^{14}$ s^{-1} and the reduced mass $m = 1.163 \times 10^{-26}$ kg.

Figure 3.2 is a plot of the wave function $\psi(x)$ for the nitrogen molecule in this example.

```python
import numpy as np
import matplotlib.pyplot as plt


m = 1.163e-26     # effective vibrational mass of nitrogen molecule in kg
omega = 5.16e14   # molecular vibration frequency in s**-1
hbar = 1.0545e-34 # hbar in J.s

#normalization constant A
A = m**(1/4)*omega**(1/4)/(np.pi**(1/4)*hbar**(1/4))

# define qavefunction f
f = lambda x: A*np.exp(-m*omega/(2*hbar)*x**2)


x1=np.linspace(-.3e-10,.3e-10,100)

# plot wave function
plt.plot(x1,f(x1))
plt.title('Wavefunction of quantum oscillator')
plt.xlabel('Position x [m]')
plt.ylabel('Wavefunction $\psi$(x)')
plt.tight_layout()
plt.show()
```

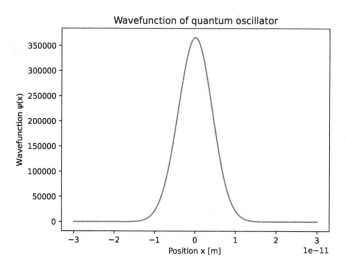

Figure 3.2 The graphical output in Example 3.7, showing the wavefunction of the quantum mechanical simple harmonic oscillator (SHO) for a N_2 molecule.

In Example 3.8 we discuss the Hermite polynomials $H_n(x)$ of degree n, which are used to describe the wavefunction of a SHO in Quantum Mechanics as:

$$\psi(x) = A\, H_n(x) \exp\left(-b\, x^2\right) \tag{3.8.5}$$

where A, b are constants. Specifically, we show how to implement these polynomials in SciPy with `special.hermite(n)`.

Example 3.8: The Hermite polynomials

$H_n(x)$ are the Hermite polynomials, which can be evaluated from the expression:

$$H_n(x) = (-1)^n \, \exp\left(x^2\right) \frac{d^n}{dx^n} \exp\left(-x^2\right) \tag{3.8.6}$$

(a) Use SciPy to show that $H_3(x)$ and $H_4(x)$ obey

$$\int_{-\infty}^{\infty} H_3(x)\, H_4(x) \, \exp\left(-x^2/2\right) dx = 0 \tag{3.8.7}$$

(b) Plot the polynomials $H_3(x)$, $H_4(x)$ on the same graph.

Solution:
We use the function `special.hermite(n,x)` from the library `scipy.special` to define the Hermite polynomials $H_n(x)$ of degree n and argument x. The result is shown in Figure 3.3. Note that in NumPy the infinity symbol is `np.inf`

```python
from scipy import special, integrate
import matplotlib.pyplot as plt
import numpy as np

p4 = special.hermite(4) # Hermite polynomial n=4
p3 = special.hermite(3) # Hermite polynomial n=3

# define integrand function
f = lambda x: special.eval_hermite(3,x)*special.eval_hermite(4,x)*\
np.exp(-x**2)

print('-'*28,'CODE OUTPUT','-'*29,'\n')

# evaluate integral from x=-oo to x=+oo
print('The integral = ',integrate.quad(f,-np.inf,np.inf))

x1 = np.linspace(-2,2, 100)  # define x-values for plot

plt.title('Hermite polynomials for n=3,4')
plt.plot(x1,p3(x1),'b-',label='H$_3$(x)')
plt.plot(x1,p4(x1),'r+',label='H$_4$(x)')
plt.xlabel('x')
plt.plot(x1,[0]*len(x1))
leg = plt.legend()
leg.get_frame().set_linewidth(0.0)
plt.show()
```

```
------------------------- CODE OUTPUT -----------------------------

The integral = (0.0, 0.0)
```

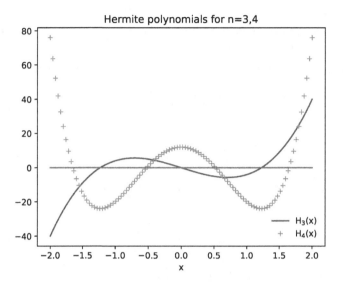

Figure 3.3 The graphical output in Example 3.8, showing plots of Hermite polynomials $H_n(x)$ of degree $n = 3, 4$.

The Maxwell-Boltzmann distribution (MB) is used in Statistical Mechanics to describe the speed v of atoms or molecules in an ideal gas:

$$f(v) = \sqrt{\frac{2}{\pi} \left(\frac{m}{kT}\right)^3} \, v^2 \exp\left(\frac{-mv^2}{2kT}\right) \tag{3.8.8}$$

where T is the temperature of the gas in Kelvin, k is Boltzmann constant and m is the mass of the atoms or molecules. Example 3.9 shows how to evaluate the average speed of the particles in the MB distribution, and also plots the probability distribution $f(v)$ of an ideal nitrogen gas.

Example 3.9: The Maxwell-Boltzmann distribution

(a) Show using symbolic Python that the mean speed or mathematical average of the speed for this distribution is

$$\langle v \rangle = \int_0^\infty v \, f(v) \, dv = \frac{2}{\sqrt{\pi}} \sqrt{\frac{2kT}{m}} \tag{3.8.9}$$

(b) Use reasonable values of the constants m and T, and plot $f(v)$ for an ideal nitrogen gas.

Solution:

The average speed is evaluated in the code symbolically using the `integrate` and `simplify` functions in SymPy.

The plot of the probability distribution $f(v)$ for the speeds is shown in Figure 3.4. Note that its is important to use SI units for all physical quantities in the equations, before attempting to plot the distribution $f(v)$.

```
from sympy import symbols, sqrt, pi, exp, integrate, oo, simplify
import matplotlib.pyplot as plt
import numpy as np

v, m, k, T = symbols('v,m,k,T',positive=True)  # define symbols

# find average speed
f1 = sqrt(2/pi*((m/(k*T))**3))*v**3*exp(-m*v**2/(2*k*T))
int1 = integrate(f1,(v,0,oo))

print('-'*28,'CODE OUTPUT','-'*29,'\n')
print('average speed v: ',simplify(int1))

# Plot MB distribution of speeds for nitrogen
m = 6.6464731e-27   # atomic mass nitrogen in kg
T = 300             # room temeprature in K
k = 1.380649e-23    # Boltzmann constant in J/K

v1 = np.linspace(1,4000,100)         # speed values for plot

a = np.sqrt(2/np.pi*((m/(k*T))**3))
b = m/(2*k*T)

f = lambda v: a*v**2*np.exp(-b*v**2)  # define function for MB

plt.plot(v1,f(v1))
plt.title('Maxwell-Boltzmann distribution of speeds')
plt.xlabel('Speed [m/s]')
plt.ylabel('Distribution of speeds f(v)')
plt.show()

-------------------------- CODE OUTPUT ----------------------------

average speed v:  2*sqrt(2)*sqrt(T)*sqrt(k)/(sqrt(pi)*sqrt(m))
```

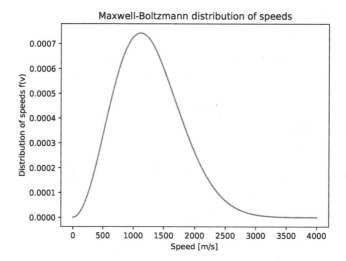

Figure 3.4 The graphical output in Example 3.9, showing a plot of the probability distribution $f(v)$ of speeds for an ideal nitrogen gas.

3.9 INTEGRATING FUNCTIONS DEFINED BY NUMPY ARRAYS

So far in this chapter we have focused on the integration of a known function $f(x)$. In this section we will focus on methods for integrating functions which have been sampled at regular intervals of its independent variable. These functions in Python are often defined by numerical NumPy arrays, whose elements are the value of the function at various evenly spaced points.

This type of numerical integration uses the `scipy.integrate` library and includes the methods in Table 3.3.

The most basic numerical integration algorithm is called the trapezoidal rule. As the name suggests, the trapezoidal rule approximates the area under a curve between $x = x_{i-1}$ and $x = x_i$, by using a single trapezoid as shown in Figure 3.5.

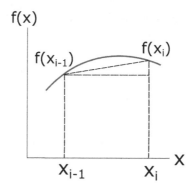

Figure 3.5 The area under the curve $f(x)$ between $x = x_{i-1}$ and $x = x_i$ can be approximated using the trapezoidal rule, which estimates the area using the dashed trapezoid shown in the figure.

The trapezoidal area under the curve between the points $x = x_{i-1}$ and $x = x_i$ is found by taking the sum of the areas of the triangle and rectangle shown in Figure 3.5:

$$\int_{x_{i-1}}^{x_i} f(x)dx \simeq (x_i - x_{i-1})\, f(x_{i-1}) + \frac{1}{2}(x_i - x_{i-1})\,(f(x_i) - f(x_{i-1})) \qquad (3.9.1)$$

$$\int_{x_{i-1}}^{x_i} f(x)dx \simeq \frac{x_i - x_{i-1}}{2}\,[f(x_i) + f(x_{i-1})] \qquad (3.9.2)$$

The trapezoidal rule is the basis of other algorithms, hence it is important to know. In its more general form, one divides the interval $[a, b]$ into $n - 1$ sub-intervals of equal length Δx and the trapezoidal rule is written as:

$$\int_a^b f(x)dx \simeq \sum_{i=1}^n \frac{\Delta x}{2}\,[f(x_i) + f(x_{i-1})] \qquad (3.9.3)$$

As mentioned above, the definite integral

$$\mathrm{erf}(x) = \int_0^\infty e^{-x^2}\,dx \qquad (3.9.4)$$

has no explicit analytical formula and is known as the error function $\mathrm{erf}(x)$, while the function $\exp\left(-x^2\right)$ is called a Gaussian function centered at $x = 0$. In Example 3.10 we evaluate numerically the definite integral of the Gaussian e^{-x^2} between $x = 0$ and $x = 1$ using the trapezoidal rule.

Example 3.10: Error function: applying the trapezoidal rule

(a) Evaluate numerically the definite integral (3.9.4) which defines the error function, between $x = 0$ and $x = 1$ by using the trapezoidal rule.
(b) Compare your answer in (a) using the trapezoidal rule, with the answer obtained using appropriate SciPy functions.

Solution:
We use the general purpose library `integrate.quad` from SciPy to obtain the numerical values. We use `integrate.trapz` to integrate the array `f(x1)`. We use two values of the number of sub-intervals, $N = 10$ and therefore $\Delta x = 0.1$, and $N = 100$ with $\Delta x = 0.01$. As expected, the value obtained with the larger number of intervals $N = 100$ is closer to the result given by `integrate.quad`.

```
from scipy import  integrate
import numpy as np

f = lambda x: np.exp(-x**2/2)    # define function

print('-'*28,'CODE OUTPUT','-'*29,'\n')
print('integrate with quad:  ',integrate.quad(f,0,1)[0],'\n')
x1 = np.linspace(0,1,10)

print('integrate with trapz() and N=10: ',integrate.trapz(f(x1),x1))

x1=np.linspace(0,1,100)
print('integrate with trapz() and N=100:',integrate.trapz(f(x1),x1))
```

```
-------------------------- CODE OUTPUT ----------------------------

integrate with quad:    0.855624391892149

integrate with trapz() and N=10:  0.855000132137889
integrate with trapz() and N=100: 0.8556192348272017
```

3.9.1 SIMPSON'S RULE

Simpson's rule breaks up the range of the integral into more pieces, and also replaces the straight-line top of the trapezoid with a quadratic polynomial. The formula for Simpson's rule is:

$$\int_{x_0}^{x_2} f(x')dx' = \frac{\Delta x}{3} \sum_{i=1}^{N/2} [f(x_{2i-2}) + 4f(x_{2i-1}) + f(x_{2i})] \qquad (3.9.5)$$

where N is the *even* number of sub-intervals, $x_i = a + i\,\Delta x$ and $\Delta x = (b-a)/N$.
In Example 3.11 we evaluate numerically the definite integral of the error function between $x = 0$ and $x = 1$ using Simpson's rule implemented in `integrate.simps`.

Example 3.11: Error function: applying Simpson's rule

Evaluate numerically the definite integral of the error function between $x = -1$ and $x = 1$ by using Simpson's rule.

Solution:
We use `simps` to integrate the array `f(x1)`, using two values of the number of sub-intervals, $N = 10$ and therefore $\Delta x = 0.1$, and $N = 100$ with $\Delta x = 0.01$. As expected, the value obtained with the larger number of intervals $N = 100$ is closer to the answer given by `integrate.quad`.

```
from scipy import  integrate
import matplotlib.pyplot as plt
import numpy as np

f = lambda x: np.exp(-x**2/2)  # define function

print('-'*28,'CODE OUTPUT','-'*29,'\n')
print('integrate with quad:              ',integrate.quad(f,-1,1)[0],'\n')
x1 = np.linspace(-1,1,10)

print('integrate with simps() and N=10: ',integrate.simps(f(x1),x1))
x1 = np.linspace(-1,1,100)
print('integrate with simps() and N=100:',integrate.simps(f(x1),x1))

------------------------- CODE OUTPUT ---------------------------

integrate with quad:              1.7112487837842973

integrate with simps() and N=10:  1.7111544649714445
integrate with simps() and N=100: 1.7112487775824874
```

3.10   END OF CHAPTER PROBLEMS

1. **The normal distribution** – In probability theory, the *normal distribution* or *Gaussian distribution* is a continuous probability distribution that has a bell-shaped probability density function. It is also known as the Gaussian function or informally as the bell curve

$$f = \frac{1}{\sigma\sqrt{2\pi}}e^{-\frac{1}{2}\left(\frac{x}{\sigma}\right)^2} \qquad (3.10.1)$$

The parameter σ is known as the *standard deviation*, and σ^2 is called the *variance*. A normal distribution is often used as a first approximation to describe real-valued random variables that cluster around a single mean value.

   a. Plot several Gaussian functions on the same graph, for values of $\sigma = 1, 2, 3$.

   b. Show that the integral underneath the Gaussian function is always equal to 1, independently of the values of σ.

   c. Show that the integral underneath the Gaussian function between $x = -\sigma$ and $x = +\sigma$ is always equal to 0.67, independently of the values of σ. What is the physical meaning of this result?

   d. Find the integral underneath the Gaussian function between $x = -2\sigma$ and $x = +2\sigma$, and between $x = -3\sigma$ and $x = +3\sigma$. What is the physical meaning of these results?

2. **General form of Gaussian** – A more general form of the Gaussian distribution is

$$f = \frac{1}{\sigma\sqrt{2\pi}}e^{-\frac{1}{2}\left(\frac{x-\mu}{\sigma}\right)^2} \qquad (3.10.2)$$

   a. Plot several Gaussian functions of this type on the same graph, for values of $\mu = 1, 2, 3$ and for the same value of $\sigma = 1$. What is the physical meaning of the extra parameter μ?

   b. Find symbolically the total area under the distribution, and the average position $< x >$.

3. **Newton's law for time dependent force F(t)** – Consider again Example 3.3 of this chapter, in which the force acting on a mass m varied with time according to $F(t) = F_0/(t^2 + 1)$ where t is the elapsed time.

   a. Find the position $x(t)$ with the initial conditions $x(0) = x_0 = 0$ and $v(0) = v_0 = 0$.

   b. Use reasonable values of the parameters F_0, m to plot $x(t)$ and $v(t)$, and discuss the physical meaning of the graphs.

   c. Find symbolically the limits $\lim_{t\to\infty} x(t)$ and $\lim_{t\to\infty} v(t)$, and discuss the physical meaning of these results.

4. **Linear drag force in the presence of gravity** – Consider the case of a falling object near the surface of the Earth. Suppose in this case, the air resistance is linear with the speed $f = -b\,v$. Then the net force on the falling body depends on the velocity v according to

$$F(v) = -bv + mg \qquad (3.10.3)$$

where we have defined $+y$ to be vertically downward so that the body's motion is in the positive direction.

    a. Find the displacement $y(t)$ and the velocity $v(t)$, with the initial conditions $v(0) = v_0$ and $y(0) = y_0$.

    b. Use reasonable values of the parameters F_0, m, b to plot $x(t)$ and $v(t)$, and discuss the physical meaning of the graphs.

    c. Find symbolically the limits $\lim\limits_{t \to \infty} x(t)$ and $\lim\limits_{t \to \infty} v(t)$, and discuss the physical meaning of these results.

5. **Simple harmonic motion** – A mass m is under the influence of a position dependent force, $F = F(x) = -kx$, which is proportional to the distance x from an equilibrium located at $x = 0$. The mathematical form of F is, of course, the well-known Hooke's law for springs.

    a. Starting from the Work-Kinetic energy theorem,

$$\Delta K = \frac{1}{2}mv^2 - \frac{1}{2}mv_0^2 = W = \int_0^x F(x)\, dx$$

    where W is the work done by the force and ΔK is the change in kinetic energy, find $x(t)$ and $v(t)$ with the initial conditions $v(0) = 0$ and $x(0) = x_0 \neq 0$ at $t = 0$ using Symbolic Python.

    b. Plot $x(t)$ and $v(t)$ on the same graph, and verify that they satisfy the initial conditions.

6. **Exponential and Gaussian Integrals** –

    a. Starting from the following integral:

$$\int_0^\infty x^2\, e^{-t\,x}\, dx = \frac{2}{t^3} \tag{3.10.4}$$

evaluate the integral

$$\int_0^\infty x^3 e^{-t\,x}\, dx$$

using the parametric integration method.

    b. Show that the integral

$$\int_{-\infty}^\infty e^{-x^2 t} dx = \frac{\sqrt{\pi}}{\sqrt{t}} \tag{3.10.5}$$

    c. Starting with the integral

$$\int_{-\infty}^\infty e^{-x^2 t} dx = \frac{\sqrt{\pi}}{\sqrt{t}} \tag{3.10.6}$$

show that:

$$\int_{-\infty}^\infty x^2 e^{-x^2 t} = \frac{\sqrt{\pi}}{2} t^{-\frac{3}{2}} \tag{3.10.7}$$

7. **The Maxwell-Boltzmann distribution** – Consider again the Maxwell-Boltzmann distribution

$$f(v) = \sqrt{\frac{2}{\pi} \left(\frac{m}{kT}\right)^3} v^2 \exp\left(\frac{-mv^2}{2kT}\right)$$

a. Show by using parametric integration that the mean speed or mathematical average of the speed for this distribution is

$$\langle v \rangle = \int_0^\infty v\, f(v)\, dv = \frac{2}{\sqrt{\pi}} \sqrt{\frac{2\,k\,T}{m}} \tag{3.10.8}$$

b. Evaluate the standard deviation σ of the Maxwell-Boltzmann distribution, which Maxwell evaluated in 1859:

$$\sigma = \sqrt{\overline{v^2} - (\bar{v})^2} = \sqrt{\left(3 - \frac{8}{\pi}\right)\frac{k\,T}{m}} \tag{3.10.9}$$

8. **Planck's law** – In Chapter 2 we saw that Planck's law for a black-body states that the power emitted per unit solid angle and per unit of area, is given by:

$$u_\lambda(\lambda, T) = \frac{2\,h\,c^2}{\lambda^5} \frac{1}{e^{h\,c/(\lambda\,k\,T)} - 1} \tag{3.10.10}$$

The following equivalent expression for Planck's law can be obtained by transforming from a wavelength variable λ to the frequency variable ν:

$$B_\nu(\nu, T) = \frac{8\pi\,\nu^2}{c^3} \frac{h\,\nu}{e^{h\,\nu/(k\,T)} - 1} \tag{3.10.11}$$

where $B_\nu(T)$ has units of power per unit area, per solid angle and per frequency, T is the temperature of the black-body at thermal equilibrium, h is the Planck constant, c is the speed of light in a vacuum, k is the Boltzmann constant, and ν is the frequency of the electromagnetic radiation.

a. Plot $u_\lambda(\lambda, T)$ as a function of the wavelength λ for a black-body with temperatures $T = 3000$, 4000, 5000 K.

b. Plot $B_\nu(\nu, T)$ as a function of the frequency ν for a black-body with temperature $T = 3000$, 4000, 5000 K.

9. **The Electric field of a finite line of charge** – A uniformly charged rod of length L carries a total charge Q, as shown in Figure 3.6. A small length element dx of the rod is located at a distance x from the origin creates an electric potential dV at a point P located at a distance b from the origin, given by:

$$dV = \frac{dq}{4\pi\,\epsilon_0} \frac{1}{r} = \frac{dq}{4\pi\,\epsilon_0} \frac{1}{(x^2 + b^2)^{1/2}}$$

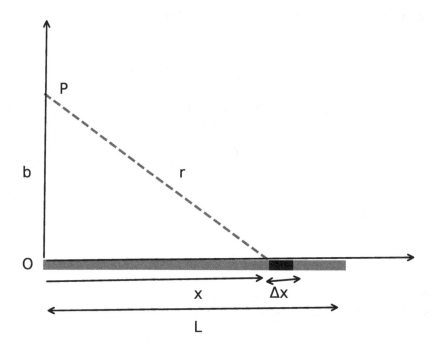

Figure 3.6 A uniformly charged rod of length L creates an electric field E at a point P located on the y-axis, at a distance b from the origin.

     a. Find the electric potential V_P at point P, with respect to the origin, by hand and also using SymPy.

     b. Plot the potential V_P as a function of the distance b, using reasonable estimates for the physical quantities involved. Does the shape of the plot make sense?

10. **The Electric field of a uniformly charged disk** – A uniformly charged disk of negligible thickness carries a total charge Q and has a radius R. The electric field at a point P located at a distance x perpendicularly from the center of the disk is given by:

$$E(x) = \frac{\sigma}{2\,\epsilon_0} \left\{ 1 - \frac{x}{(x^2 + R^2)^{1/2}} \right\}$$

where σ is the surface charge density of the ring.

     a. The corresponding potential function $V(x)$ is found by integrating the electric field:

$$V = -\int_x^\infty E(x)\, dx$$

Calculate $V(x)$ and verify the result using Python.

     b. Find the limits of $V(x)$ and $E(x)$ when very far above the surface of the disk, and also the limits on the surface of the disk.

11. **The Legendre polynomial** – The Legendre polynomials $P_l(x)$ play an important role in Electricity and Magnetism, where they are used to describe the potential from

a continuous charge distribution. They are defined in the interval $x = -1$ to $x = +1$, and can be evaluated by using the `special.eval_legendre` command imported from SciPy.

a. Write a Python code to evaluate the Legendre polynomials $P_l(x)$ for two integer values of $l = 3$ and $l = 4$, in the interval $x = -1$ to $x = +1$. These Legendre polynomials are:

$$P_3(x) = \frac{1}{2}\left(5x^3 - 3x\right) \tag{3.10.12}$$

$$P_4(x) = \frac{1}{8}\left(35x^4 - 30x^2 + 3\right) \tag{3.10.13}$$

b. Show that $P_3(x)$ and $P_4(x)$ are orthogonal, i.e:

$$\int_{-1}^{1} P_3(x)P_4(x)\,dx = 0$$

12. **Fourier series** – Consider the periodic square wave function:

$$f(x) = 1 \quad \text{for} \quad 0 < x < 2$$

$$f(x + 2\pi k) = -1 \quad \text{for} \quad 2 < x < 4$$

where k is an integer.

a. Use the symbolic capabilities of Python, to find analytical expressions for the Fourier coefficients for the periodic square wave function.

b. Plot the sum of the first 10 terms in the Fourier series and discuss how the plot compares with the periodic square wave.

13. **Relativity** – The relativistic linear momentum p of a mass m moving with speed v along the x-axis is given by:

$$p = m\,\gamma\,v = m\frac{1}{\sqrt{1 - \frac{v^2}{c^2}}}v \tag{3.10.14}$$

The mass m moves from $x = x_1$ to $x = x_2$ and is accelerated from rest to some speed v. Show by hand, and by using SymPy, that the work done along this motion is

$$W = \int_{x_1}^{x_2} F\,dx = \int_{x_1}^{x_2} \frac{dp}{dt}\,dx = \frac{m\,c^2}{\sqrt{1 - \frac{v^2}{c^2}}} - m\,c^2 \tag{3.10.15}$$

where $F = dp/dt$ is the force acting on the mass m during the motion.

14. **Integrating Newton's law** – Find the velocity $v(t)$ and position $x(t)$ of a particle starting at position x_0 with speed v_0, experiencing the following forces (assume all constants are positive):

a. $F = a + b\,t$

b. $F = A\,\cos(\omega\,t)$

15. **Speed dependent forces** – An object experiences a force, $F(v) = -cv^2$. If the particle starts at the origin with an initial velocity, $v(0) = v_0$, find the particle's velocity and position as a function of time. Use SymPy to verify your results.

16. **A ball is thrown upwards** – A ball is thrown upwards with an initial velocity of 10 m/s. If the ball has a mass of 0.10 kg and experiences a drag force $F = -bv$ that scales linearly with the ball's velocity such that the magnitude of $b = 0.2$ N s/m.

    a. What is the maximum height reached by the ball?

    b. Compare your answer to the one you would get in the case of no air resistance.

17. **Integrate Newton's law when the force F=F(x)** – Find the velocity $v(x)$ and position $x(t)$ of a particle starting at rest at the origin experiencing a force $F = a + bx$, where a, b are positive constants.

    Use the following equations which are derived from conservation of energy:

$$v(x) = \frac{2}{m} \sqrt{\int F(x)dx}$$

$$t = \int \frac{dx}{v(x)} = \frac{m}{2} \int \frac{dx}{\sqrt{\int F(x)dx}}$$

18. **The Gamma function** – The Gamma function $\Gamma(x)$ is an important mathematical function which appears in many branches of physics. It is defined as:

$$\Gamma(x) = \int_0^\infty t^{x-1} e^{-t} dt \qquad (3.10.16)$$

    Using parametric differentiation, we can show the following simple recursion relation:

$$x\,\Gamma(x) = \Gamma(x+1) \qquad (3.10.17)$$

    This relationship tells us that if we know the value of the function $\Gamma(x)$ 'in the interval between $x = 1$ and $x = 2$, we can evaluate $\Gamma(x)$ anywhere along the x-axis.

    a. Evaluate $\Gamma(1/2)$ and $\Gamma(3/2)$ using the built-in function $\Gamma(x)$ in SymPy.

    b. Verify (3.10.17) for $x = 1/2$ using SymPy.

    c. Evaluate $\Gamma(1/2)$ by numerically integrating (3.10.16) and compare with (a).

    d. Plot $\Gamma(x)$ between $x = 1$ and $x = 2$

19. **The error function erf(x)** – The error function $\text{erf}(x)$ is defined as the integral:

$$\text{erf}(x) = \frac{2}{\sqrt{\pi}} \int_0^x e^{-t^2} dt$$

    a. Plot $\text{erf}(x)$ along the x-axis using Python.

    b. Evaluate $\text{erf}(3)$.

20. **Trapezoidal integration** – Use trapezoidal integration to evaluate $\text{erf}(2)$, and compare with the exact numerical value.

21. **Area and circumference of a circle** –

   a. Evaluate by hand and symbolically the area $A = \int y\,dx$ and the perimeter $P = \int \sqrt{(dx)^2 + (dy)^2}$ of a circle with radius R and center at the origin. Do this problem by hand and by using SymPy.

   b. Plot a circle with radius R and center at the origin, using the `matplotlib.patches` library in Python.

22. **Area and circumference of an ellipse** –

   a. Evaluate by hand and symbolically the area $A = \int y\,dx$ and the perimeter $P = \int \sqrt{(dx)^2 + (dy)^2}$ of an ellipse with semi-major axes a, b and its center at the origin. Do this problem by hand and by using SymPy.

   b. Plot the ellipse using the `matplotlib.patches` library in Python.

23. **The cycloid** – The cycloid is a curve defined by the parametric equations: $x = a\,\theta - a\,\sin\theta$ and $y = a - a\,\cos\theta$.

   a. Plot the cycloid between $\theta = 0$ and $\theta = 2\pi$.

   b. Evaluate the area $A = \int y\,dx$ between $\theta = 0$ and $\theta = 2\pi$.

   c. Find also the arc length of the cycloid between $\theta = 0$ and $\theta = 2\pi$. Do this problem by hand and by using Python.

24. **Expectation values of simple harmonic oscillator in Quantum Mechanics** – In Quantum Mechanics we often need to evaluate the expectation value $<p>$ of the momentum of a mass m, when we know the wavefunction $\psi(x)$ which describes the particle. The expectation value $<p>$ for a particle moving along the x-axis can be evaluated from:

$$<p> = \int_{-\infty}^{\infty} \psi^* \frac{d\psi}{dx} dx \qquad (3.10.18)$$

The wavefunction for a particle in the ground state of a simple harmonic oscillator of mass m and frequency ω is:

$$\psi(x) = A \exp\left(-\frac{m\omega}{2\hbar} x^2\right) \qquad (3.10.19)$$

where A is the normalization constant of the wavefunction and $\hbar$ is the reduced Planck constant. The position x can vary anywhere along the x-axis.

   a. Find the normalization constant a such that the total probability of finding the particle along the x-axis is 100%, i.e.:

$$\int_{-\infty}^{\infty} \psi^* \psi\, dx = 1 \qquad (3.10.20)$$

   b. Find the expectation value $<p>$ of the momentum.

   c. Find the expectation value $<V(x)>$ of the potential energy $V(x) = m\,\omega^2\,x^2/2$:

$$<V(x)> = \int_{-\infty}^{\infty} \psi^* V(x)\,\psi\,dx \qquad (3.10.21)$$

d. Find the expectation value of the kinetic energy $< K >$ of the simple harmonic oscillator:

$$< K >= -\frac{\hbar^2}{2m} \int_{-\infty}^{\infty} \psi^* \frac{d^2\psi}{dx^2} \, dx \tag{3.10.22}$$

and discuss the physical meaning of the results.

25. **Expectation values of particle in an infinite square well in Quantum Mechanics** – The wavefunction for a particle of mass m in an infinite square well of width L:

$$\psi(x) = a \, \sin\left(\frac{n\,\pi}{L}x\right) \tag{3.10.23}$$

where A is the normalization constant of the wavefunction and $n = 1, 2, 3...$ The wavefunction is zero outside the square well, i.e. $\psi(x) = 0$ for $x < 0$ and $x > L$.

a. Find the normalization constant a such that the total probability of finding the particle along the x-axis is 100%, i.e.:

$$\int_{-\infty}^{\infty} \psi^* \, \psi \, dx = 1 \tag{3.10.24}$$

b. Find the expectation value $< p >$ of the momentum.

$$< p >= \int_{-\infty}^{\infty} \psi^* \frac{d\psi}{dx} dx \tag{3.10.25}$$

c. Find the expectation value of the kinetic energy $< K >$:

$$< K >= -\frac{\hbar^2}{2m} \int_{-\infty}^{\infty} \psi^* \frac{d^2\psi}{dx^2} \, dx \tag{3.10.26}$$

and discuss the physical meaning of the results.

4 Vectors

Vectors are important mathematical objects in physics. In this chapter, we will begin with a review of the basic concept of a vector and vector addition. We will then discuss the derivatives of time-dependent vectors and follow with a review of the concepts of the dot product and cross product operations on vectors that are very useful in physics. After establishing the basic rules for Cartesian vectors, we will present non-Cartesian coordinate systems and their vectors. We will then discuss the derivatives of vectors in non-Cartesian systems and will return to Cartesian coordinates to discuss how to develop parametric equations for lines and planes.

4.1 VECTOR BASICS

Vectors are used to describe physical quantities that cannot be fully expressed by a single number (i.e. scalar). Instead, vectors are described using a *magnitude* and a *direction*. In this section we review the basic properties of vectors, and learn how to draw and how to add vectors graphically with Python in 3D.

Let us begin by considering the unit vectors, $\hat{\mathbf{i}}$, $\hat{\mathbf{j}}$, and $\hat{\mathbf{k}}$, which have a length of one unit and point along the x, y, or z axes, respectively. For now, we will focus on Cartesian coordinates, later we will discuss non-Cartesian coordinates.

Consider the simple vector $\mathbf{A} = \hat{\mathbf{i}}$ which has a length of one unit and lies along the x-axis. If we multiply $\mathbf{A}$ by the scalar 2, then we get a new vector $2\mathbf{A}$ which has twice the length but the direction has not changed. If we multiplied $\hat{\mathbf{i}}$ by the scalar -2 we could write $-2\hat{\mathbf{i}} = -\left(2\hat{\mathbf{i}}\right)$. As before, the new vector has a magnitude of 2 (twice its original magnitude). However the minus sign tells us that the new vector now points to the left. Multiplying a vector by a negative scalar changes both the length and direction of a vector. The new vector points in the direction opposite of the original.

Let us now generalize to two dimensions. First, we will need to add two vectors $A_x\hat{\mathbf{i}}$ and $A_y\hat{\mathbf{j}}$ using the head-to-tail method where we place the tail of the vector $A_y\hat{\mathbf{j}}$ at the head of the vector $A_x\hat{\mathbf{i}}$, as you learned in introductory physics. The head to tail method is illustrated in Figure 4.1, where the resulting vector $\mathbf{A}$ is created.

$$\mathbf{A} = A_x\hat{\mathbf{i}} + A_y\hat{\mathbf{j}} \tag{4.1.1}$$

which can also be represented as (A_x, A_y). We denote A_x as the x-component of $\mathbf{A}$, and A_y as the y-component of $\mathbf{A}$ (similarly for a z-component if $\mathbf{A}$ had a term $A_z\hat{\mathbf{k}}$). Notice the importance of using unit vectors. For example, A_x is the length of $\mathbf{A}$ along the x-axis.

From Figure 4.1, we can see that the length or magnitude of $\mathbf{A}$ can be found using the Pythagorean theorem. It is typical to denote the magnitude of $\mathbf{A}$ using either $|\mathbf{A}|$ or A. We will use both notations interchangeably.

In three dimensions,

$$A = \sqrt{A_x^2 + A_y^2 + A_z^2} \tag{4.1.2}$$

In the case of Figure 4.1, $A_z = 0$.

It is often useful to know the direction of a vector, especially when working in two dimensions. From Figure 4.1, we see that

$$\theta = \tan^{-1}\frac{A_y}{A_x} \tag{4.1.3}$$

DOI: 10.1201/9781003294320-4

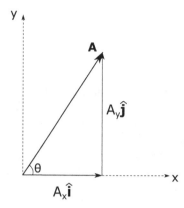

Figure 4.1 The vector $\mathbf{A}$ is created by adding the vectors $A_x\hat{\mathbf{i}}$ and $A_y\hat{\mathbf{j}}$ using the head to tail method.

Returning to the concept of multiplication by a scalar, suppose we multiply $\mathbf{A}$ by a scalar $c > 0$:

$$cA = c\left(A_x\hat{\mathbf{i}} + A_y\hat{\mathbf{j}}\right) = (cA_x)\hat{\mathbf{i}} + (cA_y)\hat{\mathbf{j}} \tag{4.1.4}$$

Because we multiply each component by the same factor c, we do not change the overall direction of $\mathbf{A}$. Hence, multiplication by a scalar does not change the direction of the vector. While the direction of $\mathbf{A}$ is not changed, its magnitude is.

$$|c\mathbf{A}| = \sqrt{(cA_x)^2 + (cA_y)^2} = c\sqrt{A_x^2 + A_y^2} = cA \tag{4.1.5}$$

Hence, the magnitude of $\mathbf{A}$ is multiplied by the scalar c.

If we multiply $\mathbf{A}$ by a negative constant, we create a new vector $\mathbf{B}$

$$\mathbf{B} = -c\mathbf{A} = (-cA_x)\hat{\mathbf{i}} + (-cA_y)\hat{\mathbf{j}} \tag{4.1.6}$$

Each component now contains a negative sign. In other words, each component now points in the opposite direction of its corresponding component in $\mathbf{A}$. Hence, multiplying $\mathbf{A}$ by a negative scalar produces a vector that points in the opposite of $\mathbf{A}$. However, if we calculate $|\mathbf{B}|$, we would get cA. An important example of this is the vector $\mathbf{B} = -\mathbf{A}$, which has the same magnitude as $\mathbf{A}$ but points in the opposite direction.

Let us now return to the concept of vector addition. Consider two vectors $\mathbf{A}$ and $\mathbf{B}$ which, when added, produce a new vector $\mathbf{C} = \mathbf{A} + \mathbf{B}$. As mentioned previously, the head-to-tail method of adding $\mathbf{A}$ and $\mathbf{B}$ is to place the tail of $\mathbf{B}$ at the head of $\mathbf{A}$. The resulting vector $\mathbf{C}$ has its tail at the origin and its head at the head of $\mathbf{B}$. This method is most useful when creating diagrams, but is generally not useful for a quantitative analysis. The head to tail-to-tail method of vector addition is illustrated in Figure 4.2.

The analytical method of vector addition is useful for solving problems involving vector addition. In this case, we add terms of similar components and treat unit vectors like algebraic symbols.

$$\mathbf{A} = A_x\hat{\mathbf{i}} + A_y\hat{\mathbf{j}} \tag{4.1.7}$$

$$\mathbf{B} = B_x\hat{\mathbf{i}} + B_y\hat{\mathbf{j}} \tag{4.1.8}$$

$$\mathbf{C} = (A_x + B_x)\hat{\mathbf{i}} + (A_y + B_y)\hat{\mathbf{j}} \tag{4.1.9}$$

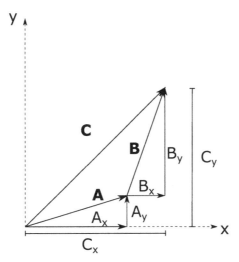

Figure 4.2 The addition of the vectors **A** and **B** resulting in the vector **C**.

Notice that the x-component of **C** is $C_x = A_x + B_x$ and the y-component is $C_y = A_y + B_y$. Figure 4.2 shows that the analytic method of adding vectors is equivalent to the head-to-tail method.

Example 4.1 demonstrates how to plot the two vectors $\mathbf{A} = (Ax, Ay, Az)$ and $\mathbf{B} = (Bx, By, Bz)$ in 3D and the corresponding sum $\mathbf{A} + \mathbf{B}$.

Example 4.1: Plotting two vectors and their sum in 3D

Write a Python code to plot the two vectors $\mathbf{A} = (1, 2, 2)$ and $\mathbf{B} = (1, 1, 1)$ in three dimensions, and their vector sum $\mathbf{A} + \mathbf{B}$.

Solution:

The 3D plot in the example is created within the Matplotlib library. The lines `fig = plt.figure()` and `fig.add_subplot(projection ='3d')` create the 3D plot.
The function `quiver(0,0,0,Ax,Ay,Az)` plots the 3D arrow for vector **A** in Figure 4.3 starting at the origin, while `quiver(Ax,Ay,Az,Bx,By,Bz)` plots the vector in the head-to-tail method of adding the vectors $\mathbf{A} + \mathbf{B}$.

```python
import matplotlib.pyplot as plt

fig = plt.figure()
ax = fig.add_subplot(projection='3d')     # create 3D figure

Ax, Ay, Az = 1, 2, 2                       # components of A vector
Bx, By, Bz = 1, 1, 1                       # components of B vector

# plot arrows for A, B, A+B and head-to-tail arrow from A to B
ax.quiver(0,0,0,Ax,Ay,Az,color='blue',length=1,arrow_length_ratio=.1)
ax.quiver(0,0,0,Bx,By,Bz,color='red',arrow_length_ratio=.1)

ax.quiver(0,0,0,Ax+Bx,Ay+By,Az+Bz,linestyle='--',arrow_length_ratio=.1)
```

```
ax.quiver(Ax,Ay,Az,Bx,By,Bz,color='black',arrow_length_ratio=.2)

ax.set_xlim(0,2.3)
ax.set_ylim(0,3.1);
ax.set_zlim(0,3.1);
ax.set_xlabel('X')
ax.set_ylabel('Y')
ax.set_zlabel('Z')

ax.text(.9,2,2.1,'A')
ax.text(1.1,.3,1.,'B')
ax.text(.9,2,1.,'A+B')

plt.show()
```

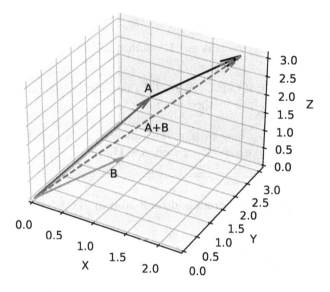

Figure 4.3 The graphical output from Example 4.1, showing the two vectors $\mathbf{A} = (Ax, Ay, Az)$ and $\mathbf{B} = (Bx, By, Bz)$ in 3D, and their vector sum $\mathbf{A} + \mathbf{B}$.

Vector subtraction is similar to vector addition. For example, the vector $\mathbf{A} - \mathbf{B}$ is the same as $\mathbf{A} + (-\mathbf{B})$. Analytically, this is simply subtracting the components of the two vectors. In the head to tail method, at the head of $\mathbf{A}$ we would place a vector which the same length at $\mathbf{B}$, but pointing in the opposite direction. Example 4.2 shows how to find the distance between two position vectors $\mathbf{r}_1$ and $\mathbf{r}_2$ in 3D space, by using subtraction of vectors.

Example 4.2: Displacements as vectors

Two points A, B on the xy-plane correspond to the vectors $\mathbf{r}_1 = 2\hat{\mathbf{i}} + 3\hat{\mathbf{j}}$ and $\mathbf{r}_2 = 4\hat{\mathbf{i}} + 2\hat{\mathbf{j}}$, respectively. What is the magnitude and direction of the vector connecting point A to point B?

Solution:

We begin with a diagram of each vector as shown in Figure 4.4, where each tick on the axes represents one unit of length. We want to find the displacement vector $\Delta\mathbf{r}$.

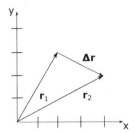

Figure 4.4   The vectors $\mathbf{r}_1$, $\mathbf{r}_2$, and the difference $\Delta\mathbf{r} = \mathbf{r_2} - \mathbf{r_1}$ for Example 4.2.

Notice from the head-to-tail method
$$\mathbf{r}_1 + \Delta\mathbf{r} = \mathbf{r}_2$$

Hence,

$$\Delta\mathbf{r} = \mathbf{r}_2 - \mathbf{r}_1$$
$$= 4\hat{\mathbf{i}} + 2\hat{\mathbf{j}} - \left(2\hat{\mathbf{i}} + 3\hat{\mathbf{j}}\right)$$
$$= (4 - 2)\,\hat{\mathbf{i}} + (2 - 3)\,\hat{\mathbf{j}} = 2\hat{\mathbf{i}} - \hat{\mathbf{j}}$$

The analytical result matches Figure 4.4, as we see that the vector $\Delta\mathbf{r}$ points to the right (positive x-direction) and down (negative y-direction). We can also calculate the magnitude and direction of $\Delta\mathbf{r}$ using (4.1.2) and (4.1.3), respectively.

$$\Delta r = \sqrt{2^2 + (-1)^2} = \sqrt{5}$$
$$\theta = \tan^{-1}\left(\frac{-1}{2}\right) = -26°$$

where θ is the angle $\Delta\mathbf{r}$ makes with the positive x-axis.

Let us examine how to do this example using Python. When working with vectors in Python, one can use either SymPy or NumPy. Which one should you use? It depends on the goals of your calculations and what is most convenient for the problem you are solving. Since we are interested in numerical solutions in Example 4.2, we will use NumPy.

In NumPy, a vector can be represented by an array where the first element is the vector's x-component, the second element is the vector's y-component (and so on in higher dimensions). Note that we use the `linalg.norm` command from the NumPy library to compute the magnitude of the vector $\Delta\mathbf{r}$. Furthermore, to get the direction of $\Delta\mathbf{r}$, we needed to extract its x and y components, as `delta_r[0]` and `delta_r[1]` respectively.

```
import numpy as np
print('-'*28,'CODE OUTPUT','-'*29,'\n')


r1 = np.array([2,3])   # vector r1
r2 = np.array([4,2])   # vector r2


delta_r = r2 - r1      # vector r2-r1


# find magnitude and direction of r2-r1
```

```
magnitude = np.linalg.norm(delta_r)
direction = np.arctan(delta_r[1]/delta_r[0]) * 180/np.pi

print('Delta r = ', delta_r)
print('The magnitude of Delta r = ', round(magnitude,3))
print('The direction of Delta r = ', round(direction,3), 'degrees')

-------------------------- CODE OUTPUT ----------------------------

Delta r =  [ 2 -1]
The magnitude of Delta r =  2.236
The direction of Delta r =  -26.565 degrees
```

We will demonstrate the use of the SymPy library for finding the magnitude of a vector in Example 4.3.

4.2 SCALAR AND VECTOR FIELDS IN PYTHON

In this section we use the library `sympy.vector` in SymPy to describe the properties of scalar and vector fields. A *vector field* assigns a vector to each point in space. Consider the electric field described by the formula, $\mathbf{E}(x, y, z) = c\,x\,\hat{\mathbf{k}}$ (where c is a constant) which assigns a vector to each point in space based on its x-coordinate. The electric field $\mathbf{E}$ is a vector field. Likewise, a *scalar field* assigns a scalar to each point in space. Consider an object with a non-uniform density $\rho(x, y, z) = axy + bz$ (where a and b are constants). The density function ρ is a scalar field that assigns a density (a scalar) value to each point on the object.

In many of the examples of this chapter we will define a Cartesian coordinate system named R, by using `R = CoordSys3D('R')` from `sympy.vector`. In the coordinate system R, the unit base vectors $\hat{\mathbf{i}}, \hat{\mathbf{j}}, \hat{\mathbf{k}}$ are represented as `R.i`, `R.j`, `R.k`, respectively. In addition to this notation for the unit vectors, we will use also the notation `R.x`, `R.y`, `R.z` for the x, y, z coordinates. Note that there is nothing special about using the variable R in this context, any variable will work.

Using these two notations, the vector $x\,y\hat{\mathbf{i}}+z\hat{\mathbf{j}}+y\,z\hat{\mathbf{k}}$ will be written in the Python codes as `R.x*R.y*R.i + R.z*R.j + R.y*R.z*R.k`. Similarly, a scalar function $\phi(x, y, z) = x\,y + z$ will be written in the Python code as `R.x*R.y + R.z`. The usual SymPy functions can still be applied to these general scalar and vector fields, for example `diff(B, R.x)` means the partial derivative of $\mathbf{B}$ with respect to the coordinate x. Note that in the code it is important to write `diff(B, R.x)` instead of the usual `diff(B, x)`.

Some examples of vector functions and methods are shown in Table 4.1.

Table 4.1
Partial list of vector functions and methods in the package `sympy.vector`.

Function/Method	Description
`R = CoordSys3D('R')`	Define Cartesian coordinate system named R
`R.base_vectors()`	The unit base vectors of the coordinate system R
`v.components`	components of vector **v**
`v.magnitude()`	magnitude of vector **v**
`v.normalize()`	the normalized version of vector **v**
`v.projection(u)`	The vector projection of vector **v** on vector **u**
`v.projection(u, scalar=True)`	The scalar magnitude of the projection of vector **v** on vector **u**
`v.cross(u)`	cross product of vectors **v**, **u**
`v.dot(u).`	dot product of vectors **v**, **u**
`v.to_matrix(C)`	Modify vector **v** into a matrix form C

4.3 VECTOR MULTIPLICATION

There are two methods of multiplying vectors, the *dot product* (also called the *scalar product*) and the *cross product*. In this section, we motivate each method with a physical example. We will then follow with a discussion of the analytical and geometric properties of each operation.

4.3.1 THE DOT PRODUCT

We begin by considering a force **F** acting on a particle along its displacement **r** as shown in Figure 4.5a. The energy transferred to (or from) the particle by the force **F** along the displacement **r** is called the work done by **F**. When **F** and **r** are parallel, we can find the work W using the formula $W = Fr$.

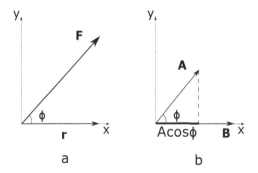

Figure 4.5 (a) A force **F** acting along a displacement **r** and (b) two vectors, **A** and **B**, with an angle ϕ between them.

In the case illustrated in Figure 4.5a, calculating the work done by **F** is not as simple as multiplying F by r because **F** is not parallel to **r**. In other words, the particle isn't being pushed along its path by the full force **F**. Rather, only the component of **F** parallel to **r** is pushing the particle along $\mathbf{r}$. Using trigonometry we see that the component of $\mathbf{F}$ parallel to $\mathbf{r}$ is $F\cos\phi$, where ϕ is the angle between the vectors $\mathbf{F}$ and $\mathbf{r}$. Hence, we compute the work using the formula $W = (F\cos\phi)\,r$, which is more typically written as

$$W = Fr\cos\phi \tag{4.3.1}$$

The important idea behind the development of (4.3.1) is that we needed to multiply the length of one vector $\mathbf{r}$ by the component of a second vector $\mathbf{F}$ parallel to $\mathbf{r}$.

Figure 4.5b shows a similar arrangement for two generic vectors $\mathbf{A}$ and $\mathbf{B}$. In Figure 4.5b we see the component of $\mathbf{A}$ along the direction of $\mathbf{B}$ as a bold line. The length of the bold line is $A\cos\phi$ and is sometimes called the projection of $\mathbf{A}$ on $\mathbf{B}$. The mathematical operation involving the multiplication of the magnitude of $\mathbf{B}$ and the projection of vector $\mathbf{A}$ on $\mathbf{B}$ is called the *dot product* and is found using

$$\mathbf{A} \cdot \mathbf{B} = AB\cos\phi \tag{4.3.2}$$

The dot product produces a scalar. Hence, the dot product is sometimes also called the *scalar product*. We can rewrite (4.3.1) using the dot product to calculate work

$$W = \mathbf{F} \cdot \mathbf{r} \tag{4.3.3}$$

We can use (4.3.2) to calculate the magnitude of a vector. For example, $\mathbf{A} \cdot \mathbf{A} = AA\cos 0 = A^2$. Note that ϕ is the angle between the two vectors. The angle the vector $\mathbf{A}$ makes with itself is 0. Hence, we have

$$A = \sqrt{\mathbf{A} \cdot \mathbf{A}} \tag{4.3.4}$$

We can *normalize* a vector by dividing it by its magnitude.

$$\hat{\mathbf{A}} = \frac{\mathbf{A}}{A} \tag{4.3.5}$$

The result is a vector of unit length that points in the direction of $\mathbf{A}$.

Like the multiplication process in algebra, the dot product is commutative, $\mathbf{A}{\cdot}\mathbf{B} = \mathbf{B}{\cdot}\mathbf{A}$. The product of the magnitude of $\mathbf{B}$ by the projection of $\mathbf{A}$ on $\mathbf{B}$, is the same as the product of the magnitude of $\mathbf{A}$ by the projection of $\mathbf{B}$ on $\mathbf{A}$. Furthermore, the dot product is also distributive

$$\mathbf{A} \cdot (\mathbf{B} + \mathbf{C}) = \mathbf{A} \cdot \mathbf{B} + \mathbf{A} \cdot \mathbf{C} \tag{4.3.6}$$

The distributive property provides a mathematical means of isolating a vector component. For example, if we wanted the x-component of a vector $\mathbf{A}$ we need to calculate $\hat{\mathbf{i}} \cdot \mathbf{A}$

$$\hat{\mathbf{i}} \cdot \mathbf{A} = \hat{\mathbf{i}} \cdot \left(A_x\hat{\mathbf{i}} + A_y\hat{\mathbf{j}} + A_z\hat{\mathbf{k}} \right) \tag{4.3.7}$$

$$= A_x \left(\hat{\mathbf{i}} \cdot \hat{\mathbf{i}} \right) + A_y \left(\hat{\mathbf{i}} \cdot \hat{\mathbf{j}} \right) + A_z \left(\hat{\mathbf{i}} \cdot \hat{\mathbf{k}} \right) \tag{4.3.8}$$

$$= A_x \tag{4.3.9}$$

Using (4.3.4), we find that the term $\hat{\mathbf{i}} \cdot \hat{\mathbf{i}} = 1$. The dot products involving different unit vectors such as $\hat{\mathbf{i}} \cdot \hat{\mathbf{j}}$ and $\hat{\mathbf{i}} \cdot \hat{\mathbf{k}}$ are zero because the unit vectors have an angle of $\pi/2$ radians between them and $\cos(\pi/2) = 0$. In general, the dot product between two perpendicular vectors is zero. If the dot product of two vectors is zero $(\mathbf{A} \cdot \mathbf{B} = 0)$, we say the two vectors $\mathbf{A}$ and $\mathbf{B}$ are *orthogonal*.

The distributive property (4.3.6) also provides a second means of calculating the dot product

$$\mathbf{A} \cdot \mathbf{B} = \left(A_x \hat{\mathbf{i}} + A_y \hat{\mathbf{j}} + A_z \hat{\mathbf{k}} \right) \cdot \left(B_x \hat{\mathbf{i}} + B_y \hat{\mathbf{j}} + B_z \hat{\mathbf{k}} \right) \tag{4.3.10}$$

$$= A_x B_x + A_y B_y + A_z B_z \tag{4.3.11}$$

The step connecting (4.3.10) to (4.3.11) involves distributing the dot product in (4.3.10). This distribution will lead to nine terms. However, only the terms $\hat{\mathbf{i}} \cdot \hat{\mathbf{i}}$, $\hat{\mathbf{j}} \cdot \hat{\mathbf{j}}$ and $\hat{\mathbf{k}} \cdot \hat{\mathbf{k}}$ survive because each of those dot products equals one. Dot products involving different unit vectors such as $\hat{\mathbf{i}} \cdot \hat{\mathbf{j}}$ and $\hat{\mathbf{k}} \cdot \hat{\mathbf{i}}$ are zero as previously discussed.

Using (4.3.2) and (4.3.11), we can compute the angle between two vectors.

$$\cos \phi = \frac{A_x B_x + A_y B_y + A_z B_z}{AB} \tag{4.3.12}$$

In Python, the dot and cross product of two vectors are implemented in both NumPy and SymPy. In Example 4.3 we show how to evaluate in SymPy the dot product of the position and force vectors $\mathbf{F} \cdot \mathbf{r}$, by using the `F.dot(r)` method.

===

Example 4.3: Work done by a force

A force $\mathbf{F} = -3\hat{\mathbf{i}} + 2\hat{\mathbf{j}} - \hat{\mathbf{k}}$ moves a particle along the displacement $\mathbf{r} = \hat{\mathbf{i}} - 3\hat{\mathbf{j}} - 2\hat{\mathbf{k}}$, where all the physical quantities are expressed in SI units. Calculate the work done by the force and find the angle between $\mathbf{F}$ and $\mathbf{r}$.

Solution:

The force is constant, therefore we can calculate the work using (4.3.3)

$$W = \mathbf{F} \cdot \mathbf{r} = F_x r_x + F_y r_y + F_z r_z$$

$$W = (-3)(1) + (2)(-3) + (-1)(-2) = -7$$

We can calculate the angle using (4.3.12). Note that $F = \sqrt{14}$ and $r = \sqrt{14}$.

$$\cos \phi = \frac{F_x r_x + F_y r_y + F_z r_z}{F \, r} = \frac{-7}{14} = -0.5$$

If $\cos \phi = -0.5$, then $\phi = 120°$.

In the Python code, we first import from the SymPy library `sympy.vector`, which is dedicated to the properties of vectors. The command `CoordSys3D` creates a class that represents a coordinate system. In the line R = `CoordSys3D('R')` we create a Cartesian coordinate frame and we name it R.

The Cartesian coordinate frame R has unit vectors which are accessed using the expressions, `R.i`, `R.j`, and `R.k` for $\hat{\mathbf{i}}$, $\hat{\mathbf{j}}$, and $\hat{\mathbf{k}}$, respectively. The magnitude of any vector $\mathbf{F}$ in the SymPy library is evaluated using the method `F.magnitude()`, and the dot product of any two vectors $\mathbf{F} \cdot \mathbf{r}$ is found using the `F.dot(r)` method.

The angle is found by using (4.3.12), by diving the dot product $\mathbf{F} \cdot \mathbf{r}$ by the product $F \, r$, and converting the radians into degrees.

```
from sympy.vector import CoordSys3D
from sympy import acos, pi
print('-'*28,'CODE OUTPUT','-'*29,'\n')


R = CoordSys3D('R')

F = -3 * R.i +  2 * R.j - 1 * R.k
r =  1 * R.i + -3 * R.j - 2 * R.k

F_mag = F.magnitude()
r_mag = r.magnitude()

work = F.dot(r)
cos_angle = work/(F_mag*r_mag)

print('The work = ', work,' Nm (or J)')
print('The cos(phi) = ', cos_angle)
print('phi = ', (acos(cos_angle) * 180/pi).evalf(5),' degrees')

------------------------- CODE OUTPUT ----------------------------

The work =  -7  Nm (or J)
The cos(phi) =  -1/2
phi =  120.00  degrees
```

A more general example of the dot product of two vectors $\mathbf{A} \cdot \mathbf{B}$ in SymPy is shown in Example 4.4. In this case, the components of the two vectors are represented by general variables (Ax, Ay, Az) and (Bx, By, Bz). Here we also show how to verify general identities between vectors, specifically we prove symbolically that $\mathbf{A} \cdot \mathbf{B} = \mathbf{B} \cdot \mathbf{A}$ and that $\mathbf{A} \cdot \mathbf{A} = A^2 = A_x^2 + A_y^2 + A_z^2$.

Example 4.4: The dot product of two vectors in SymPy

Use SymPy to :
(a) Evaluate the general dot product $\mathbf{A} \cdot \mathbf{B}$ of two vectors $\mathbf{A}$ and $\mathbf{B}$, with components (Ax, Ay, Az) and (Bx, By, Bz), respectively.
(b) Show that $\mathbf{A} \cdot \mathbf{B} = \mathbf{B} \cdot \mathbf{A}$ and $\mathbf{A} \cdot \mathbf{A} = A^2 = A_x^2 + A_y^2 + A_z^2$.
(c) Demonstrate that $\hat{\mathbf{i}} \cdot \hat{\mathbf{j}} = 0$ and $\hat{\mathbf{i}} \cdot \hat{\mathbf{i}} = 1$.

Solution:

As in the previous example, we import CoordSys3D, and define the two vectors $\mathbf{A}$ and $\mathbf{B}$ in terms of their components, and in terms of the unit vectors R.i, R.j, and R.k. We use the SymPy dot product method (R.i).dot(R.j) to show that $\hat{\mathbf{i}} \cdot \hat{\mathbf{j}} = 0$ and $\hat{\mathbf{i}} \cdot \hat{\mathbf{i}} = 1$. We verify the identities by using the logical comparison symbol ==.

```
from sympy.vector import CoordSys3D
from sympy import symbols
print('-'*28,'CODE OUTPUT','-'*29,'\n')
```

```
Ax, Ay, Az, Bx, By, Bz = symbols('Ax, Ay, Az, Bx, By, Bz')

R = CoordSys3D('R')

A = Ax * R.i +  Ay * R.j + Az * R.k
B = Bx * R.i +  By * R.j + Bz * R.k

print('The dot product method in SymPy is A.dot(B) = ', A.dot(B))
print('The magnitude of A is A.magnitude() = ', A.magnitude())

print('\nThe identity  A.A = Ax^2+Ay^2+Az^2 is ',\
A.dot(A)==A.magnitude()**2)
print('The identity  A.B = B.A   is ',A.dot(B)==B.dot(A))

print('\nThe dot product of R.i and R.j is ', (R.i).dot(R.j))
print('The dot product of R.i and R.i is ', (R.i).dot(R.i))


-------------------------- CODE OUTPUT ----------------------------

The dot product method in SymPy is A.dot(B) =  Ax*Bx + Ay*By + Az*Bz
The magnitude of A is A.magnitude() =  sqrt(Ax**2 + Ay**2 + Az**2)

The identity  A.A = Ax^2+Ay^2+Az^2 is  True
The identity  A.B = B.A   is    True

The dot product of R.i and R.j is  0
The dot product of R.i and R.i is  1
```

The dot product of two vectors $\mathbf{A} \cdot \mathbf{B}$ is also implemented in NumPy by using the dot(A,B) function, as shown in Example 4.9. In this case, the vectors are represented by NumPy arrays.

Example 4.5: The dot product in NumPy

Use NumPy to :
(a) Evaluate the general dot product $\mathbf{A} \cdot \mathbf{B}$ of two vectors $\mathbf{A}$ and $\mathbf{B}$, with components (Ax, Ay, Az) and (Bx, By, Bz) respectively.
(b) Show that $\mathbf{A} \cdot \mathbf{B} = \mathbf{B} \cdot \mathbf{A}$ and $\mathbf{A} \cdot \mathbf{A} = A^2 = A_x^2 + A_y^2 + A_z^2$.
(c) Demonstrate the distributive property $\mathbf{A} \cdot (\mathbf{B} + \mathbf{C}) = \mathbf{A} \cdot \mathbf{B} + \mathbf{A} \cdot \mathbf{C}$.

Solution:

In the Python code, we first import from the NumPy library the numpy.dot(A,B) function which finds the dot product $\mathbf{A} \cdot \mathbf{B}$, and we also define general arrays $[A_x, A_y, A_z]$ etc.
We prove the distributive property by evaluating the two sides of $\mathbf{A} \cdot (\mathbf{B} + \mathbf{C}) = \mathbf{A} \cdot \mathbf{B} + \mathbf{A} \cdot \mathbf{C}$ and using the .expand() method to expand the parentheses in the results.

```
from numpy import dot, array
from sympy import  expand, symbols
```

```
print('-'*28,'CODE OUTPUT','-'*29,'\n')

# define general symbols for vector components in SymPy
Ax, Ay, Az, Bx, By, Bz, Cx, Cy, Cz = symbols('Ax, Ay, Az, Bx, By, Bz,\
Cx, Cy, Cz')

A = array([Ax,Ay,Az])    # vector A defined as NumPy array
B = array([Bx,By,Bz])
C = array([Cx,Cy,Cz])

print("The dot product A.B in NumPy is\n dot(A,B) = ", dot(A,B))

print('\nThe dot product A.A=',dot(A,A))

print('\nThe identity A.B=B.A is ',dot(A,B)==dot(B,A))

print('\nThe identity A.(B+C)=A.B+A.C is ',\
dot(A,B+C).expand()==dot(A,B)+dot(A,C))


----------------------- CODE OUTPUT ----------------------------

The dot product A.B in NumPy is
 dot(A,B) =  Ax*Bx + Ay*By + Az*Bz

The dot product A.A= Ax**2 + Ay**2 + Az**2

The identity A.B=B.A is  True

The identity A.(B+C)=A.B+A.C is  True
```

It is important to keep in mind that the dot and cross products of vectors are implemented differently in SymPy and NumPy. In SymPy this product is implemented as a *method* `A.dot(B)`, while in NumPy the dot products is represented as a *function* `dot(A,B)`.

4.3.2 THE CROSS PRODUCT

Let us consider the case where a force $\mathbf{F}$ is applied to a door a distance r from its hinge. Figure 4.6a shows a view of the door from above. The force is applied at a location $\mathbf{r}$. The force $\mathbf{F}$ applies a torque $\mathbf{N}$ to the door. Note that $\mathbf{N}$ points out of the page (as represented by the dot enclosed by a circle).

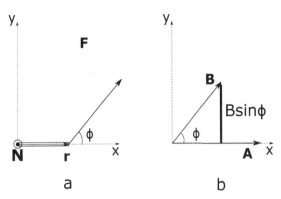

Figure 4.6 (a) A force $\mathbf{F}$ acting at a location $\mathbf{r}$ results in a torque $\mathbf{N}$. (b) Two vectors $\mathbf{A}$ and $\mathbf{B}$ with an angle ϕ between them.

If $\mathbf{F}$ is perpendicular to $\mathbf{r}$, then the torque applied has a magnitude of $N = rF$. However, it is not always the case that the force causing a rotation is perpendicular to the lever arm $\mathbf{r}$. In that case, the component of $\mathbf{F}$ perpendicular to $\mathbf{r}$ is responsible for the door's rotation. We can see in Figure 4.6a that the component of $\mathbf{F}$ perpendicular to $\mathbf{r}$ is $F \sin \phi$. Hence, the torque's magnitude in this situation is

$$N = rF \sin \phi \qquad (4.3.13)$$

Like the case of the dot product, to find the torque we needed to multiply the magnitude of one vector (in this case $\mathbf{r}$) by a component of the second vector. However, when evaluating the torque, we needed the component of the second vector ($\mathbf{F}$) that is perpendicular to the first ($\mathbf{r}$). This type of multiplication is called the *cross product* or *vector product*. For two vectors $\mathbf{A}$ and $\mathbf{B}$ we find the magnitude of the cross product by multiplying the magnitude of $\mathbf{A}$ by the component of $\mathbf{B}$ perpendicular to $\mathbf{A}$

$$|\mathbf{A} \times \mathbf{B}| = AB \sin \phi \qquad (4.3.14)$$

where ϕ is the positive angle ($\leq 180°$) between the vectors $\mathbf{A}$ and $\mathbf{B}$, as shown in Figure 4.6b. Thus, for torque we can rewrite (4.3.13) as

$$N = |\mathbf{r} \times \mathbf{F}| \qquad (4.3.15)$$

Note that if two vectors $\mathbf{A}$ and $\mathbf{B}$ are parallel or anti parallel, then $\phi = 0$ or $\phi = 180°$, and $|\mathbf{A} \times \mathbf{B}| = 0$. This also implies that

$$\mathbf{A} \times \mathbf{A} = 0 \qquad (4.3.16)$$

Unlike the dot product, the cross product produces a vector, hence its name. The vector $\mathbf{A} \times \mathbf{B}$ is perpendicular to the plane defined by $\mathbf{A}$ and $\mathbf{B}$ as illustrated in Figure 4.7a. Figure 4.7a also illustrates that $|\mathbf{A} \times \mathbf{B}|$ is equal to the area of the parallelogram created by $\mathbf{A}$ and $\mathbf{B}$.

Figure 4.7b shows how to find the direction of $\mathbf{A} \times \mathbf{B}$ using the right-hand rule. Using your right hand, point your index finger in the direction of the vector $\mathbf{A}$, your middle finger points in the direction of $\mathbf{B}$, and your thumb points in the direction of $\mathbf{A} \times \mathbf{B}$. A second method for the right-hand rule involves pointing your fingers (of your right hand!) in the direction of $\mathbf{A}$ then curling (or sweeping) your fingers in the direction of $\mathbf{B}$. Your thumb points in the direction of $\mathbf{A} \times \mathbf{B}$. Both methods are equivalent.

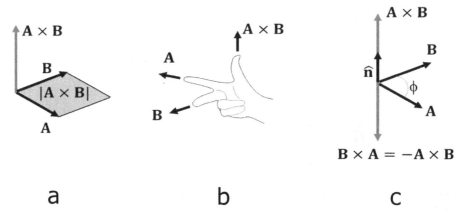

Figure 4.7 The vector nature of the cross product (a), the right-hand rule (b), and the anti-symmetric property of the cross product (c). Image from: Kulp and Pagonis, *Classical Mechanics* (2020), used with permission.

Figure 4.7c illustrates the anti-symmetric nature of the cross product. By using the right-hand rule, you can see that if you point your index finger in the direction of **B** and your middle finger in the direction of **A**, then your thumb would point in the opposite direction as if you began by pointing your index finger in the direction of **A**. Hence

$$\mathbf{B} \times \mathbf{A} = -\mathbf{A} \times \mathbf{B} \tag{4.3.17}$$

The vector $\hat{\mathbf{n}}$ in Figure 4.7c is a unit vector which points perpendicular to the plane made by **A** and **B** and in the same direction as the right-hand rule. The symbol $\hat{\mathbf{n}}$ is often reserved for a unit vector that points outward and perpendicular to a surface. It is commonly used in electromagnetism.

The cross product, like the dot product, obeys the distributive law

$$\mathbf{A} \times (\mathbf{B} + \mathbf{C}) = (\mathbf{A} \times \mathbf{B}) + (\mathbf{A} \times \mathbf{C}) \tag{4.3.18}$$

To calculate the vector $\mathbf{A} \times \mathbf{B}$ we first need to know the cross product relationships between unit vectors. We will use the right-hand rule and (4.3.14) for the Cartesian unit vectors. From (4.3.14), we find

$$\hat{\mathbf{i}} \times \hat{\mathbf{i}} = \hat{\mathbf{j}} \times \hat{\mathbf{j}} = \hat{\mathbf{k}} \times \hat{\mathbf{k}} = 0 \tag{4.3.19}$$

Using the right-hand rule and the fact that each unit vector has a magnitude of one, we have

$$\hat{\mathbf{i}} \times \hat{\mathbf{j}} = \hat{\mathbf{k}} \qquad\qquad \hat{\mathbf{j}} \times \hat{\mathbf{k}} = \hat{\mathbf{i}} \qquad\qquad \hat{\mathbf{k}} \times \hat{\mathbf{i}} = \hat{\mathbf{j}} \tag{4.3.20}$$

The reader should verify the relationships in (4.3.20) using the right-hand rule.

Consider two vectors $\mathbf{A} = A_x\hat{\mathbf{i}} + A_y\hat{\mathbf{j}} + A_z\hat{\mathbf{k}}$ and $\mathbf{B} = B_x\hat{\mathbf{i}} + B_y\hat{\mathbf{j}} + B_z\hat{\mathbf{k}}$. Using (4.3.18), (4.3.19), and (4.3.20) we find

$$\mathbf{A} \times \mathbf{B} = (A_yB_z - A_zB_y)\,\hat{\mathbf{i}} + (A_zB_x - A_xB_z)\,\hat{\mathbf{j}} + (A_xB_y - A_yB_x)\,\hat{\mathbf{k}} \tag{4.3.21}$$

Another method of evaluating the components of the cross product shown in (4.3.21) is by using the determinant:

$$\mathbf{A} \times \mathbf{B} = \begin{vmatrix} \hat{\mathbf{i}} & \hat{\mathbf{j}} & \hat{\mathbf{k}} \\ A_x & A_y & A_z \\ B_x & B_y & B_z \end{vmatrix} \qquad (4.3.22)$$

We will discuss determinants in more detail in a later chapter. For now, we simply state that the expansion of this determinant yields the same expression in (4.3.21).

Example 4.6 shows how to evaluate the torque of a force $\mathbf{N} = \mathbf{r} \times \mathbf{F}$ using Python, as well as its components along a given direction.

Example 4.6: Torque

A force $\mathbf{F} = -3\hat{\mathbf{i}} + 2\hat{\mathbf{j}} - \hat{\mathbf{k}}$ is applied to a particle located at a position $\mathbf{r} = \hat{\mathbf{i}} - 3\hat{\mathbf{j}} - 2\hat{\mathbf{k}}$, where all the physical quantities are measured in SI units.
(a) Calculate the torque applied by the force
(b) Calculate the component of the torque applied by the force along the line in the direction of the vector $\mathbf{B} = -\hat{\mathbf{i}} + \hat{\mathbf{j}} - \hat{\mathbf{k}}$.

Solution:

(a) The force is constant, therefore we can calculate the torque $\mathbf{N}$ using

$$\mathbf{N} = \mathbf{r} \times \mathbf{F}$$
$$= \left(r_y F_z - r_z F_y \right) \hat{\mathbf{i}} + \left(r_z F_x - r_x F_z \right) \hat{\mathbf{j}} + \left(r_x F_y - r_y F_x \right) \hat{\mathbf{k}}$$
$$= [(-3)(-1) - (-2)(2)]\,\hat{\mathbf{i}} + [(-2)(-3) - (1)(-1)]\,\hat{\mathbf{j}} + ((1)(2) - (-3)(-3))\,\hat{\mathbf{k}}$$
$$= 7\hat{\mathbf{i}} + 7\hat{\mathbf{j}} - 7\hat{\mathbf{k}}$$

(b) We know that the component of any vector $\mathbf{A}$ along the direction of vector $\mathbf{B}$ can be found using the dot product $\hat{\mathbf{B}} \cdot \mathbf{A}$, where $\hat{\mathbf{B}}$ is the unit vector along the direction of $\mathbf{B}$. In this case, the unit vector along the direction of the vector $\mathbf{B} = -\hat{\mathbf{i}} + \hat{\mathbf{j}} - \hat{\mathbf{k}}$ is:

$$\hat{\mathbf{B}} = \frac{\mathbf{B}}{B} = \frac{-\hat{\mathbf{i}} + \hat{\mathbf{j}} - \hat{\mathbf{k}}}{|-\hat{\mathbf{i}} + \hat{\mathbf{j}} - \hat{\mathbf{k}}|} = \frac{-\hat{\mathbf{i}} + \hat{\mathbf{j}} - \hat{\mathbf{k}}}{\sqrt{3}}$$

The vector of interest is the torque $\mathbf{N} = \mathbf{r} \times \mathbf{F}$ and since we want to find the component of $\mathbf{N}$ along the direction of $\hat{\mathbf{B}}$, we evaluate the triple scalar product

$$\hat{\mathbf{B}} \cdot \mathbf{N} = \hat{\mathbf{B}} \cdot (\mathbf{r} \times \mathbf{F})$$

Therefore, the component of the torque along $\hat{\mathbf{B}}$ is:

$$\hat{\mathbf{B}} \cdot (\mathbf{r} \times \mathbf{F}) = \left(\frac{-\hat{\mathbf{i}} + \hat{\mathbf{j}} - \hat{\mathbf{k}}}{\sqrt{3}} \right) \cdot \left(7\hat{\mathbf{i}} + 7\hat{\mathbf{j}} - 7\hat{\mathbf{k}} \right) = \frac{7}{\sqrt{3}} \text{ Nm}$$

In the Python code we find the torque using the method `r.cross(F)`. In order to find the unit vector $\hat{\mathbf{B}}$ along the direction of $\mathbf{B}$, we use the `B.normalize()` method in SymPy.
Finally, we combine the `dot()` and `cross()` methods, and use `n.dot(r.cross(F))` to evaluate the triple dot product $\hat{\mathbf{B}} \cdot (\mathbf{r} \times \mathbf{F})$.

We can also obtain the same results for the projection of one vector on another vector, by using the `v.projection(u, scalar=True)` method in SymPy.

```
from sympy.vector import CoordSys3D
print('-'*28,'CODE OUTPUT','-'*29,'\n')


N = CoordSys3D('N')

F = -3 * N.i +  2 * N.j - 1 * N.k   # force vector
r =  1 * N.i + -3 * N.j - 2 * N.k   # position vector

B = -1 * N.i +  1 * N.j - 1 * N.k   # given direction vector B

torque = r.cross(F)
print('The torque = ', torque,' Nm')

B_hat=B.normalize()              # Find unit vector n in direction of B

print('\nThe normal vector is \n B_hat = ', B_hat)

print('\nTorque component along B is:\n', B_hat.dot(r.cross(F)),' N m')
-------------------------- CODE OUTPUT ----------------------------

The torque =  7*N.i + 7*N.j + (-7)*N.k  Nm

The normal vector is
 B_hat =  (-sqrt(3)/3)*N.i + (sqrt(3)/3)*N.j + (-sqrt(3)/3)*N.k

Torque component along B is:
 7*sqrt(3)/3  N m
```

Example 4.7 demonstrates how to plot two vectors and their cross product in 3D.

Example 4.7: Plotting the cross product $\mathbf{A} \times \mathbf{B}$

Write a Python code to plot two vectors $\mathbf{A} = (1, 1, 2)$, $\mathbf{B} = (1, 1, 1)$, and the cross product $\mathbf{A} \times \mathbf{B}$ in three dimensions.

Solution:

The cross product is $\mathbf{A} \times \mathbf{B} = (-1, 1, 0)$. The function `quiver()` is used to plot the vectors and the cross product, with the result shown in Figure 4.8.

```
import matplotlib.pyplot as plt

fig = plt.figure()
ax = fig.add_subplot(projection='3d')   # set up 3D plot

# components of vectors A, B
Ax, Ay, Az = 1, 1, 2
Bx, By, Bz = 1, 1, 1
```

```
# plot 3 vectors A, B, AXB using quiver()
ax.quiver(0,0,0,Ax,Ay,Az,color='blue',length=1,arrow_length_ratio=.1)
ax.quiver(0,0,0,Bx,By,Bz,color='red',arrow_length_ratio=.1)
ax.quiver(0,0,0,Ay*Bz-Az*By,Bx*Az-Bz*Ax,Ax*By-Ay*Bx,color='k',\
arrow_length_ratio=.3)

ax.set_xlim(-1,1.3);
ax.set_ylim(-1,2.1);
ax.set_zlim(-1,2);

ax.set_xlabel('X')
ax.set_ylabel('Y')
ax.set_zlabel('Z')

ax.text(.7,2,1.9,'A')
ax.text(1.1,.3,1.,'B')
ax.text(-.6,-.5,1.,'AXB')

plt.tight_layout()
plt.show()
```

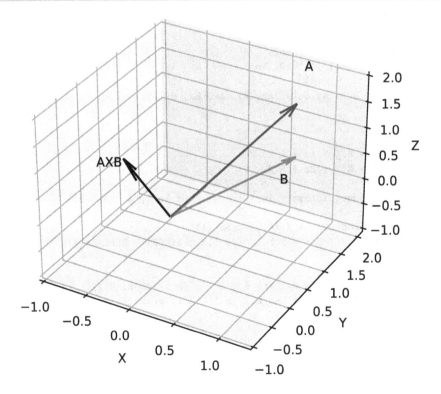

Figure 4.8 The graphical output from Example 4.7, showing the two vectors $\mathbf{A} = (Ax, Ay, Az)$ and $\mathbf{B} = (Bx, By, Bz)$ in 3D, and their cross product $\mathbf{A} \times \mathbf{B}$.

A general example of evaluating the cross product of two vectors $\mathbf{A} \times \mathbf{B}$ in SymPy is shown in Example 4.8. In this case, the components of the two vectors are represented by general variables (Ax, Ay, Az) and (Bx, By, Bz). Here we also show how to verify general identities between vectors, specifically we prove symbolically that $\mathbf{A} \times \mathbf{B} = -\mathbf{B} \times \mathbf{A}$, $\mathbf{A} \times \mathbf{A} = 0$ and that $\hat{\mathbf{i}} \times \hat{\mathbf{j}} = \hat{\mathbf{k}}$ and $\hat{\mathbf{i}} \times \hat{\mathbf{i}} = 0$.

Example 4.8: The cross product of two vectors in SymPy

Use SymPy to:
(a) Evaluate the general cross product $\mathbf{A} \times \mathbf{B}$ of two vectors $\mathbf{A}$ and $\mathbf{B}$, with components (Ax, Ay, Az) and (Bx, By, Bz) respectively.
(b) Show that $\mathbf{A} \times \mathbf{B} = -\mathbf{B} \times \mathbf{A}$ and $\mathbf{A} \times \mathbf{A} = 0$.
(c) Show that $\hat{\mathbf{i}} \times \hat{\mathbf{j}} = \hat{\mathbf{k}}$ and $\hat{\mathbf{i}} \times \hat{\mathbf{i}} = 0$.

Solution:

As in the previous example, we import `CoordSys3D()`, and define the two vectors $\mathbf{A}$ and $\mathbf{B}$ in terms of their components, and in terms of the unit vectors `R.i`, `R.j`, and `R.k`. We use the SymPy cross product method `(R.i).cross(R.j)` to show that $\hat{\mathbf{i}} \times \hat{\mathbf{j}} = 1$ and $\hat{\mathbf{i}} \times \hat{\mathbf{i}} = 0$.

```
from sympy.vector import CoordSys3D
from sympy import acos, pi, symbols
print('-'*28,'CODE OUTPUT','-'*29,'\n')

Ax, Ay, Az, Bx, By, Bz = symbols('Ax, Ay, Az, Bx, By, Bz')

R = CoordSys3D('R')

A = Ax * R.i +  Ay * R.j + Az * R.k
B = Bx * R.i +  By * R.j + Bz * R.k

print('The cross product in SymPy is A.cross(B) is:\n ', A.cross(B))

print('\nThe cross product   A X A=',A.cross(A))
print('The identity   A X B = - B X A   is ',A.cross(B)==-B.cross(A))


print('\nThe cross product in R.i and R.j is ', (R.i).cross(R.j))
print('The cross product in R.i and R.i is ', (R.i).cross(R.i))


---------------------------- CODE OUTPUT ----------------------------

The cross product in SymPy is A.cross(B) is:
 (Ay*Bz - Az*By)*R.i + (-Ax*Bz + Az*Bx)*R.j + (Ax*By - Ay*Bx)*R.k

The cross product   A X A= 0
The identity   A X B = - B X A   is  True

The cross product in R.i and R.j is  R.k
The cross product in R.i and R.i is  0
```

The cross product of two vectors $\mathbf{A} \times \mathbf{B}$ is also implemented in NumPy by using the cross(A,B) function, as shown in Example 4.9. In this case, the vectors are represented by NumPy arrays.

Example 4.9: The cross product in NumPy

Use NumPy to:
(a) Evaluate the general cross product $\mathbf{A} \times \mathbf{B}$ of two vectors $\mathbf{A}$ and $\mathbf{B}$, with components (Ax, Ay, Az) and (Bx, By, Bz) respectively.
(b) Show that $\mathbf{A} \times \mathbf{B} = -\mathbf{B} \times \mathbf{A}$ and $\mathbf{A} \times \mathbf{A} = 0$.
(c) Demonstrate the distributive property $\mathbf{A} \times (\mathbf{B} + \mathbf{C}) = \mathbf{A} \times \mathbf{B} + \mathbf{A} \times \mathbf{C}$.

Solution:

In the Python code, we use the simplify() function to expand the terms in the cross product. Note that the evaluation of $\mathbf{A} \times \mathbf{A}$ in Python produces the zero NumPy array [0,0,0].

```
from numpy import cross, array
from sympy import  symbols, simplify
print('-'*28,'CODE OUTPUT','-'*29,'\n')

Ax, Ay, Az, Bx, By, Bz, Cx, Cy, Cz = symbols('Ax, Ay, Az, Bx, By, Bz,\
Cx, Cy, Cz')

A = array([Ax,Ay,Az])
B = array([Bx,By,Bz])
C = array([Cx,Cy,Cz])

print("The cross product AXB in NumPy is\n cross(A,B) = ", cross(A,B))

print('\nThe cross product AXA =',cross(A,A))

print('\nThe identity  AXB = -BXA  is:',\
simplify(cross(A,B))==simplify(-cross(B,A)))

print('\nThe value of the NumPy array AX(B+C)-AXB-AXC is:\n',\
simplify(cross(A,B+C)-cross(A,B)-cross(A,C)))


------------------------- CODE OUTPUT ----------------------------

The cross product AXB in NumPy is
 cross(A,B) =  [Ay*Bz - Az*By -Ax*Bz + Az*Bx Ax*By - Ay*Bx]

The cross product AXA = [0 0 0]

The identity  AXB = -BXA  is: True

The value of the NumPy array AX(B+C)-AXB-AXC is:
 [0, 0, 0]
```

4.4 TRIPLE PRODUCTS

There are two ways three vectors can be multiplied in physics. These are called *triple products*. There are two triple products we will examine here, the *triple scalar product* and the *triple vector product*.

4.4.1 TRIPLE SCALAR PRODUCT

The triple scalar product multiplies three vectors in such a way that a final result is a scalar. Consider three vectors $\mathbf{A}$, $\mathbf{B}$ and $\mathbf{C}$. Their triple scalar product in Cartesian coordinates is

$$\mathbf{A} \cdot (\mathbf{B} \times \mathbf{C}) = A_x \left(B_y C_z - B_z C_y \right) + A_y \left(B_z C_x - B_x C_z \right) + A_z \left(B_x C_y - B_y C_x \right) \quad (4.4.1)$$

The triple scalar product has a useful geometric interpretation. Consider the parallelopiped created by the vectors $\mathbf{A}$, $\mathbf{B}$ and $\mathbf{C}$ as shown in Figure 4.9. The triple cross product $\mathbf{A} \cdot (\mathbf{B} \times \mathbf{C})$ is equal to the volume of the parallelopiped.

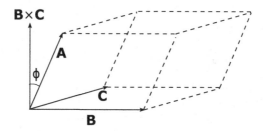

Figure 4.9　A parallelopiped made by the vectors $\mathbf{A}$, $\mathbf{B}$ and $\mathbf{C}$.

To understand the connection between the triple scalar product and the volume, recall that the volume in this case is equal to area times height. The area of the base of the parallelopiped is $|\mathbf{B} \times \mathbf{C}|$. The height of the parallelopiped from Figure 4.9 is $A \cos \phi$. Therefore the volume is

$$|\mathbf{B} \times \mathbf{C}| \, A \cos \phi = \mathbf{A} \cdot (\mathbf{B} \times \mathbf{C}) \quad (4.4.2)$$

Note that to find the volume of the parallelopiped in Figure 4.9, the order in which the sides are multiplied does not matter. Therefore

$$\mathbf{A} \cdot (\mathbf{B} \times \mathbf{C}) = \mathbf{B} \cdot (\mathbf{C} \times \mathbf{A}) = \mathbf{C} \cdot (\mathbf{A} \times \mathbf{B}) \quad (4.4.3)$$

A simple way to remember (4.4.3) is that the alphabetic order (starting at $\mathbf{A}$ and reading to the right) is preserved. In addition, this triple product can be evaluated as a determinant involving the three vectors $\mathbf{A}$, $\mathbf{B}$, $\mathbf{C}$:

$$\mathbf{C} \cdot (\mathbf{A} \times \mathbf{B}) = \begin{vmatrix} A_x & A_y & A_z \\ B_x & B_y & B_z \\ C_x & C_y & C_z \end{vmatrix} \quad (4.4.4)$$

Example 4.10 evaluates the triple scalar products.

Example 4.10: Triple scalar product $\mathbf{A} \cdot (\mathbf{B} \times \mathbf{C})$

Use SymPy to test the identity $\mathbf{A} \cdot (\mathbf{B} \times \mathbf{C}) = \mathbf{B} \cdot (\mathbf{C} \times \mathbf{A})$.

Solution:

We combine the dot and cross product functions in NumPy. Note that we apply the method .expand() and function simplify() to both sides of the identity, in order to simplify the terms in parentheses which are evaluated within the triple scalar product.

```
from numpy import cross, dot, array
from sympy import  symbols, simplify
print('-'*28,'CODE OUTPUT','-'*29,'\n')

Ax, Ay, Az, Bx, By, Bz, Cx, Cy, Cz = symbols('Ax, Ay, Az, Bx, By, Bz,\
Cx, Cy, Cz')

A = array([Ax,Ay,Az])
B = array([Bx,By,Bz])
C = array([Cx,Cy,Cz])

print('\nThe identity A.(B X C) = B.(C X A) is:',\
simplify(dot(A,cross(B,C))).expand()==\
simplify(dot(B,cross(C,A))).expand())


---------------------------- CODE OUTPUT ----------------------------


The identity A.(B X C) = B.(C X A) is: True
```

4.4.2   TRIPLE VECTOR PRODUCT

A second way of multiplying three vectors $\mathbf{A}$, $\mathbf{B}$ and $\mathbf{C}$ is using the *triple vector product*, $\mathbf{A} \times (\mathbf{B} \times \mathbf{C})$. There is a simple expression which can be used to calculate the triple vector product

$$\mathbf{A} \times (\mathbf{B} \times \mathbf{C}) = \mathbf{B}(\mathbf{A} \cdot \mathbf{C}) - \mathbf{C}(\mathbf{A} \cdot \mathbf{B}) \qquad (4.4.5)$$

Equation (4.4.5) is often referred to as the BAC-CAB equation, or the BAC-CAB rule.

Example 4.11: Triple product identity (The BAC-CAB rule)

Use NumPy to demonstrate the triple cross product identity $\mathbf{A} \times (\mathbf{B} \times \mathbf{C}) = \mathbf{B}(\mathbf{A} \cdot \mathbf{C}) - \mathbf{C}(\mathbf{A} \cdot \mathbf{B})$

Solution:

In NumPy we implement the vectors as NumPy arrays and the cross product as the function cross(A,B). The simplify function can be applied directly on the evaluated cross products, to show that the vector $\mathbf{A} \times (\mathbf{B} \times \mathbf{C}) = \mathbf{B}(\mathbf{A} \cdot \mathbf{C}) - \mathbf{C}(\mathbf{A} \cdot \mathbf{B})$ is equal to the zero array [0,0,0] .

```
from numpy import cross, dot, array
from sympy import  symbols, simplify
print('-'*28,'CODE OUTPUT','-'*29,'\n')
```

```
Ax, Ay, Az, Bx, By, Bz, Cx, Cy, Cz = symbols('Ax, Ay, Az, Bx, By, Bz,\
Cx, Cy, Cz')

A = array([Ax,Ay,Az])
B = array([Bx,By,Bz])
C = array([Cx,Cy,Cz])

print('The difference A X (B X C) - (B.(A.C)-C.(A.B)) is: ',\
simplify(cross(A,cross(B,C)) - B*dot(A,C)+C*dot(A,B)))


------------------------ CODE OUTPUT ---------------------------

The difference A X (B X C) - (B.(A.C)-C.(A.B)) is:   [0, 0, 0]
```

Using the asymmetry of the cross product, we find

$$(\mathbf{A} \times \mathbf{B}) \times \mathbf{C} = -\mathbf{C} \times (\mathbf{A} \times \mathbf{B}) = -\mathbf{A}\,(\mathbf{B} \cdot \mathbf{C}) + \mathbf{B}\,(\mathbf{A} \cdot \mathbf{C}) \qquad (4.4.6)$$

which is a different vector from that found in (4.4.5). Therefore, the triple vector product is not associative.

Higher order products can be reduced using (4.4.5). For example

$$\mathbf{A} \times [\mathbf{B} \times (\mathbf{C} \times \mathbf{D})] = \mathbf{B}\,[\mathbf{A} \cdot (\mathbf{C} \times \mathbf{D})] - (\mathbf{A} \cdot \mathbf{B})\,(\mathbf{C} \times \mathbf{D}) \qquad (4.4.7)$$

See Problem 13 for additional higher order products. Example 4.12 shows how to evaluate the triple product in the case of angular momentum vectors.

Example 4.12: Angular momentum

A particle of mass $m = 5$ kg is located at $\mathbf{r} = \hat{\mathbf{i}} - 3\hat{\mathbf{j}} - 2\hat{\mathbf{k}}$ and has an angular velocity of $\omega = \hat{\mathbf{i}} - \hat{\mathbf{j}} + 4\hat{\mathbf{k}}$, where all physical quantities are measured in the SI system. Calculate the particle's angular momentum about the origin.

Solution:

Recall from your introductory physics that the angular momentum $\ell = \mathbf{r} \times \mathbf{p}$, where $\mathbf{p} = m\mathbf{v}$ is the particle's momentum. Furthermore, we can relate the particle's tangential velocity $\mathbf{v}$ to its angular velocity using $\mathbf{v} = \omega \times \mathbf{r}$. Therefore, the angular momentum is an example of the triple cross product

$$\ell = m\,(\mathbf{r} \times \mathbf{v}) = m\,(\mathbf{r} \times (\omega \times \mathbf{r}))$$

We will use the BAC-CAB rule to find $\mathbf{r} \times (\omega \times \mathbf{r})$:

$$\begin{aligned}
\mathbf{r} \times (\omega \times \mathbf{r}) &= \omega\,(\mathbf{r} \cdot \mathbf{r}) - \mathbf{r}\,(\mathbf{r} \cdot \omega) \\
&= \omega\,(1 + 9 + 4) - \mathbf{r}\,(1 + 3 - 8) \\
&= 14\omega + 4\mathbf{r} \\
&= \left(14\hat{\mathbf{i}} - 14\hat{\mathbf{j}} + 56\hat{\mathbf{k}}\right) + \left(4\hat{\mathbf{i}} - 12\hat{\mathbf{j}} - 8\hat{\mathbf{k}}\right) \\
&= \left(18\hat{\mathbf{i}} - 26\hat{\mathbf{j}} + 48\hat{\mathbf{k}}\right)
\end{aligned}$$

The angular momentum is
$$\boldsymbol{\ell} = m\mathbf{r} \times (\boldsymbol{\omega} \times \mathbf{r}) = \left(90\hat{\mathbf{i}} - 130\hat{\mathbf{j}} + 240\hat{\mathbf{k}}\right) \text{ kg m}^2/\text{s}$$

Below, we demonstrate how to solve this example using SymPy.

```
from sympy.vector import CoordSys3D
print('-'*28,'CODE OUTPUT','-'*29,'\n')


N = CoordSys3D('N')  # define coordinate system N

# vectors r and omega, N.i=unit vector i
r     = 1 * N.i - 3 * N.j - 2 * N.k
omega = 1 * N.i - 1 * N.j + 4 * N.k

m = 5

velocity = omega.cross(r)       # cross product omega X r

print('Angular momentum l= ',m*r.cross(velocity) , 'kg m^2/s')

-------------------------- CODE OUTPUT ----------------------------

Angular momentum l=  90*N.i + (-130)*N.j + 240*N.k kg m^2/s
```

4.5 NON-CARTESIAN COORDINATES

Vectors are used to describe quantities such as displacement, velocity, and acceleration in physics. Up to this point, we have examined vectors only in Cartesian coordinates. It is often useful to describe physical quantities using other coordinate systems. For example, if a particle is traveling along a circle of radius r, we can simplify the description of the motion by following its angular displacement $\Delta\theta$ along the circle. In this case, we would only need one variable $\Delta\theta$ to describe the particle's position as opposed to two, x and y. We will need formulas for converting between Cartesian coordinates and non-Cartesian coordinates.

In this section, we will present the three most common non-Cartesian coordinate systems used in physics, plane polar, cylindrical polar, and spherical polar coordinates. These are more commonly referred to as polar, cylindrical, and spherical coordinates, respectively.

4.5.1 POLAR COORDINATES

Polar coordinates are used when dealing with circular motion, or problems with a circular symmetry. We use two variables, ρ which describes the radial distance from the origin and ϕ which describes an angular displacement from the $+x$-axis. The polar coordinates ρ and ϕ are illustrated in Figure 4.10.

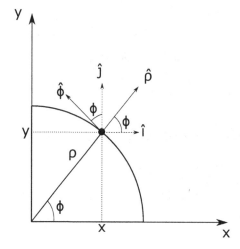

Figure 4.10 Polar coordinates ρ and ϕ and the unit vectors $\hat{\rho}$ and $\hat{\phi}$.

We can use Figure 4.10 to relate the Cartesian coordinates x and y of the black dot to its polar coordinates ρ, ϕ:

$$x = \rho \cos \phi \tag{4.5.1}$$

$$y = \rho \sin \phi \tag{4.5.2}$$

Solving (4.5.1) and (4.5.2) for ρ and ϕ we obtain:

$$\rho^2 = x^2 + y^2 \tag{4.5.3}$$

$$\phi = \tan^{-1}\left(\frac{y}{x}\right) \tag{4.5.4}$$

The unit vectors for polar coordinates are also illustrated in Figure 4.10. Using trigonometry we can find $\hat{\rho}$:

$$\hat{\rho} = |\hat{\rho}| \cos \phi \,\hat{\mathbf{i}} + |\hat{\rho}| \sin \phi \,\hat{\mathbf{j}} = \cos \phi \,\hat{\mathbf{i}} + \sin \phi \,\hat{\mathbf{j}}$$

since $|\hat{\rho}| = \left|\hat{\phi}\right| = 1$. We can repeat a similar trigonometry process for $\hat{\phi}$ and obtain:

$$\hat{\rho} = \cos \phi \,\hat{\mathbf{i}} + \sin \phi \,\hat{\mathbf{j}} \tag{4.5.5}$$

$$\hat{\phi} = -\sin \phi \,\hat{\mathbf{i}} + \cos \phi \,\hat{\mathbf{j}} \tag{4.5.6}$$

Note that unlike the Cartesian unit vectors, the polar unit vectors change orientation based on location. For example, if we place the black dot in Figure (4.10) on the x-axis, then $\hat{\rho} = \hat{\mathbf{i}}$ and $\hat{\phi} = \hat{\mathbf{j}}$. However, if the black dot is placed on the x-axis, then $\hat{\rho} = \hat{\mathbf{j}}$ and $\hat{\phi} = -\hat{\mathbf{i}}$. When taking derivatives of quantities in polar coordinates, we will need to include differentiation of the unit vectors.

By solving (4.5.5) and (4.5.6) for $\hat{\mathbf{i}}$ and $\hat{\mathbf{j}}$ we can obtain the Cartesian unit vectors as a function of the polar unit vectors $\hat{\rho}$ and $\hat{\phi}$:

$$\hat{\mathbf{i}} = \cos \phi \,\hat{\rho} - \sin \phi \,\hat{\phi} \tag{4.5.7}$$

$$\hat{\mathbf{j}} = \sin \phi \,\hat{\rho} + \cos \phi \,\hat{\phi} \tag{4.5.8}$$

Example 4.13 shows how to obtain these equations between the unit vectors $\hat{\mathbf{i}}, \hat{\mathbf{j}}$ and the polar unit vectors $\hat{\rho}, \hat{\phi}$ by using SymPy.

In physics, it is often useful to represent infinitesimal displacements. In two-dimensional Cartesian coordinates, an infinitesimal displacement $d\mathbf{s}$ is calculated using,

$$d\mathbf{s} = dx\,\hat{\mathbf{i}} + dy\,\hat{\mathbf{j}} \tag{4.5.9}$$

We can convert (4.5.9) to polar coordinates by calculating the differentials of (4.5.1) and (4.5.2) and inserting those and (4.5.7) and (4.5.8) into (4.5.9):

$$d\mathbf{s} = d\rho\,\hat{\boldsymbol{\rho}} + \rho\,d\phi\,\hat{\boldsymbol{\phi}} \tag{4.5.10}$$

Note the infinitesimal displacement in polar coordinates is composed of two displacements, the displacement $d\rho$ along the ρ-direction, and $\rho\,d\phi$ along the ϕ-direction, multiplied by the corresponding unit vectors. The resulting magnitude of the arc length is

$$ds^2 = d\rho^2 + \rho^2 d\phi^2 \tag{4.5.11}$$

Example 4.13 shows how to obtain this expression for ds^2 using Python.

Finally, the area element shown in Figure 4.11 is useful for calculations involving surface charge densities and center of mass calculations (to name a few). In Cartesian coordinates, the area element is $dA = dx\,dy$. The area dA is also the product of two infinitesimal displacements, $d\rho$ along the ρ-direction, and $\rho\,d\phi$ along the ϕ-direction. Therefore, the area element dA in polar coordinates is:

$$dA = \rho\,d\rho\,d\phi \tag{4.5.12}$$

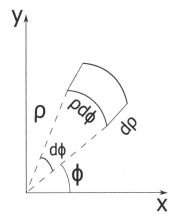

Figure 4.11   The length and area elements for polar coordinates.

Example 4.13 shows how to evaluate the unit vectors $\hat{\mathbf{i}}, \hat{\mathbf{j}}$ in terms of the polar unit vectors $\hat{\boldsymbol{\rho}}, \hat{\boldsymbol{\phi}}$ using SymPy.

Example 4.13: Unit vectors in polar coordinates

Use SymPy to evaluate the unit vectors $\hat{i}, \hat{j}$ in terms of the polar unit vectors $\hat{\rho}, \hat{\phi}$.

Solution:

We can convert Cartesian unit vectors $\hat{i}, \hat{j}$ into polar unit vectors $\hat{\rho}, \hat{\phi}$ using (4.5.7) and (4.5.8). In order to obtain $\hat{i}, \hat{j}$ in terms of the polar unit vectors $\hat{\rho}, \hat{\phi}$ using Python, we use the solve() to solve the system of equations, and store the result in the variable sol. We can print the resulting two equations for $\hat{i}, \hat{j}$ by referring to the two components of the variable sol, using sol[i] and sol[j] respectively. The SymPy result agrees with the analytical equations (4.5.7) and (4.5.8).

```
from sympy import symbols, Eq, diff, sin, cos, simplify, solve

print('-'*28,'CODE OUTPUT','-'*29,'\n')

# define all symbols for real variables
x, y, dx, dy, rho, phi, drho, dphi, i, j, rhat, phihat = \
symbols('x, y, dx, dy, rho, phi, drho, dphi, i, j, rhat,phihat',real = True)

# polar unit vectors as functions of i and j vectors
eq1 = Eq(rhat,cos(phi)*i+sin(phi)*j)
eq2 = Eq(phihat,-sin(phi)*i+cos(phi)*j)

# Use solve to find i,j as functions of the r,phi unit vectors
sol = solve((eq1,eq2),(i,j))

print('i-unit vector = ',simplify(sol[i]))
print('j-unit vector = ',simplify(sol[j]))


-------------------------- CODE OUTPUT ---------------------------

i-unit vector =  -phihat*sin(phi) + rhat*cos(phi)
j-unit vector =  phihat*cos(phi) + rhat*sin(phi)
```

4.5.2 CYLINDRICAL COORDINATES

Problems with cylindrical symmetry appear often in physics. Examples include the electric field surrounding an infinite line of charge and the magnetic field due to a current carrying wire. In both cases, it is easiest to describe these fields using cylindrical coordinates. Cylindrical coordinates are essentially polar coordinates raised a height z above the Cartesian plane.

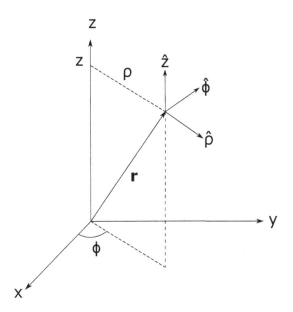

Figure 4.12 Cylindrical coordinates ρ, ϕ, z and their respective unit vectors.

Figure 4.12 illustrates cylindrical coordinates. Suppose the vector $\mathbf{r}$ denotes the location of a particle. Similar to polar coordinates, ρ measures the particle's perpendicular distance from the z-axis and ϕ measures the angle counterclockwise with respect to the positive x-axis. The new coordinate z describes the height of the particle above the xy-plane. Note that z is a Cartesian coordinate. The unit vectors for cylindrical coordinates are the same as polar, but we now include a third unit vector denoted by $\hat{\mathbf{z}}$ which is the same as $\hat{\mathbf{k}}$. However, the symbol $\hat{\mathbf{z}}$ is commonly used to distinguish cylindrical coordinates from Cartesian.

In the previous subsection, we established vector relationships for polar coordinates. Cylindrical coordinates are identical, with the addition of a $z\hat{\mathbf{z}}$ term to describe the third dimension. For example, the relationships between the Cartesian coordinates and cylindrical coordinates are

$$x = \rho\cos\phi \tag{4.5.13}$$

$$y = \rho\sin\phi \tag{4.5.14}$$

$$z = z \tag{4.5.15}$$

and the inverse relationships:

$$\rho^2 = x^2 + y^2 \tag{4.5.16}$$

$$\phi = \tan^{-1}\left(\frac{y}{x}\right) \tag{4.5.17}$$

$$z = z \tag{4.5.18}$$

The relationships between the unit vectors for cylindrical coordinates and Cartesian coordinates are

$$\hat{\boldsymbol{\rho}} = \cos\phi\,\hat{\mathbf{i}} + \sin\phi\,\hat{\mathbf{j}} \tag{4.5.19}$$

$$\hat{\boldsymbol{\phi}} = -\sin\phi\,\hat{\mathbf{i}} + \cos\phi\,\hat{\mathbf{j}} \tag{4.5.20}$$

$$\hat{\mathbf{z}} = \hat{\mathbf{k}} \tag{4.5.21}$$

These unit vectors are shown in Figure 4.12. The inverse relationships are:

$$\hat{\mathbf{i}} = \cos\phi\,\hat{\boldsymbol{\rho}} - \sin\phi\,\hat{\boldsymbol{\phi}} \tag{4.5.22}$$

$$\hat{\mathbf{j}} = \sin\phi\,\hat{\boldsymbol{\rho}} + \cos\phi\,\hat{\boldsymbol{\phi}} \tag{4.5.23}$$

$$\hat{\mathbf{k}} = \hat{\mathbf{z}} \tag{4.5.24}$$

The position vector in cylindrical coordinates is

$$\mathbf{r} = \rho\,\hat{\boldsymbol{\rho}} + z\,\hat{\mathbf{z}} \tag{4.5.25}$$

In three dimensions, the infinitesimal displacement in Cartesian coordinates is

$$d\mathbf{s} = dx\,\hat{\mathbf{i}} + dy\,\hat{\mathbf{j}} + dz\,\hat{\mathbf{k}} \tag{4.5.26}$$

with a resulting arc length of

$$ds^2 = dx^2 + dy^2 + dz^2 \tag{4.5.27}$$

The infinitesimal displacement in cylindrical coordinates can be found using the same method as finding $d\mathbf{s}$ in polar coordinates.

$$d\mathbf{s} = d\rho\,\hat{\boldsymbol{\rho}} + \rho\,d\phi\,\hat{\boldsymbol{\phi}} + dz\,\hat{\mathbf{z}} \tag{4.5.28}$$

The resulting arc length is

$$ds^2 = d\rho^2 + \rho^2 d\phi^2 + dz^2 \tag{4.5.29}$$

In Cartesian coordinates, the volume element is

$$dV = dx\,dy\,dz \tag{4.5.30}$$

which is a product of three infinitesimal displacements. In cylindrical coordinates, we can use (4.5.28) to identify the infinitesimal displacements in each direction. These are $d\rho$, $\rho\,d\phi$, dz and are shown in Figure 4.13, together with the volume elements dV which is:

$$dV = \rho\,d\rho\,d\phi\,dz \tag{4.5.31}$$

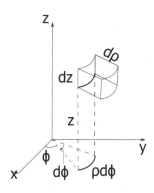

Figure 4.13 The unit lengths $d\rho$, $\rho\,d\phi$, dz in cylindrical coordinates.

Example 4.14 shows how to evaluate the length element ds^2 in polar coordinates.

Example 4.14: Infinitesimal length ds in polar coordinates

Evaluate the length element ds^2 in cylindrical coordinates, both by hand and using SymPy.

Solution:

The differentials of $x = \rho \cos \phi$, $y = \rho \sin \phi$ and $z = z$ are:

$$dx = \frac{\partial x}{\partial \rho}\, d\rho + \frac{\partial x}{\partial \phi}\, d\phi = \cos \phi\, d\rho - \rho \sin \phi\, d\phi \tag{4.5.32}$$

$$dy = \frac{\partial y}{\partial \rho}\, d\rho + \frac{\partial y}{\partial \phi}\, d\phi = \sin \phi\, d\rho + \rho \cos \phi\, d\phi \tag{4.5.33}$$

$$dz = dz \tag{4.5.34}$$

Now that we have the equations for dx, dy and dz, we can carry out the algebra to evaluate the length element ds^2 thus:

$$\begin{aligned} ds^2 = dx^2 + dy^2 + dz^2 &= (\cos \phi\, d\rho - \rho \sin \phi\, d\phi)^2 + (\sin \phi\, d\rho + \rho \cos \phi\, d\phi)^2 + dz^2 \\ &= \cos^2 \phi\ (d\rho)^2 + \rho^2 \sin^2 \phi\ (d\phi)^2 - 2\cos \phi\, d\rho\, \rho \sin \phi\, d\phi + \\ &\quad + \sin^2 \phi\ (d\rho)^2 + \rho^2 \cos^2 \phi\ (d\phi)^2 + 2\cos \phi\, d\rho\, \rho \sin \phi\, d\phi + \\ &\quad + dz^2 \end{aligned}$$

Collecting terms and using $\sin^2 \phi + \cos^2 \phi = 1$ we find:

$$ds^2 = d\rho^2 + \rho^2\, d\phi^2 + dz^2$$

which is the desired result.

The Python code shows how to evaluate the differentials dx, dy, dz. The algebraic evaluation of ds^2 is carried out, before using the `simplify()` function to obtain the final expression for ds^2.

```python
from sympy import symbols, Eq, diff, sin, cos, simplify, solve
import pprint
print('-'*28,'CODE OUTPUT','-'*29,'\n')


x, y, z, dx, dy, dz, rho, phi, drho, dphi = \
symbols('x, y, z, dx, dy, dz, rho, phi, drho, dphi',real=True)
#  define all symbols for real variables

# Equations for (x,y,z) as functions of (r,phi,z)
x = rho*cos(phi)
y = rho*sin(phi)
z = z

# find differentials dx, dy using chain rule
dx = diff(x,rho)*drho+diff(x,phi)*dphi+diff(x,z)*dz
dy = diff(y,rho)*drho+diff(y,phi)*dphi+diff(y,z)*dz
dz = diff(z,rho)*drho+diff(z,phi)*dphi+diff(z,z)*dz

# the length element ds**2 = dx**2+dy**2+dy**2
print('The length element ds**2 in polar coordinates = ')
print(simplify(dx**2+dy**2+dz**2))
```

```
------------------------ CODE OUTPUT ----------------------------

The length element ds**2 in polar coordinates =
dphi**2*rho**2 + drho**2 + dz**2
```

4.5.3 SPHERICAL COORDINATES

Many problems in physics, especially those involving central forces, have spherical symmetry and are most conveniently described using spherical coordinates.

In spherical coordinates, there is one coordinate r which measures the distance from the origin, and two angular coordinates θ and ϕ as illustrated in Figure 4.14. The polar angle θ measures the angle between the vector $\mathbf{r}$ and the positive z-axis. The azimuthal angle ϕ measures the angle between a projection of $\mathbf{r}$ on the xy-plane and the positive x-axis (similar to ϕ in polar coordinates). Note that the radial distance r varies from $r = 0$ to $r = \infty$, the azimuthal angle ϕ varies from $\phi = 0$ to $\phi = 2\pi$ as expected, but the polar angle θ varies from $\theta = 0$ to $\theta = \pi$.

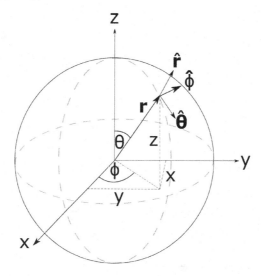

Figure 4.14 Spherical coordinates r, θ, ϕ and their respective unit vectors.

Note that the projection of $\mathbf{r}$ onto the xy-plane has a length $r\sin\theta$. Using the length of this projection, it is possible to find the transformation between spherical and Cartesian coordinates

$$x = r\sin\theta\cos\phi \tag{4.5.35}$$

$$y = r\sin\theta\sin\phi \tag{4.5.36}$$

$$z = r\cos\theta \tag{4.5.37}$$

Solving (4.5.35)–(4.5.37) for r, θ, and ϕ, we find

$$r = \sqrt{x^2 + y^2 + z^2} \tag{4.5.38}$$

$$\theta = \tan^{-1}\left(\sqrt{x^2 + y^2}/z\right) \tag{4.5.39}$$

$$\phi = \tan^{-1}\left(y/x\right) \tag{4.5.40}$$

The unit vectors in spherical coordinates can be expressed in terms of the Cartesian unit vectors. Note that like polar and cylindrical coordinates, the orientation of the unit vectors depends on the direction $\mathbf{r}$.

$$\hat{\mathbf{r}} = \sin\theta\cos\phi\,\hat{\mathbf{i}} + \sin\theta\sin\phi\,\hat{\mathbf{j}} + \cos\theta\,\hat{\mathbf{k}} \tag{4.5.41}$$

$$\hat{\boldsymbol{\theta}} = \cos\theta\cos\phi\,\hat{\mathbf{i}} + \cos\theta\sin\phi\,\hat{\mathbf{j}} - \sin\theta\,\hat{\mathbf{k}} \tag{4.5.42}$$

$$\hat{\boldsymbol{\phi}} = -\sin\phi\,\hat{\mathbf{i}} + \cos\phi\,\hat{\mathbf{j}} \tag{4.5.43}$$

and the inverse relationships:

$$\hat{\mathbf{i}} = \sin\theta\cos\phi\,\hat{\mathbf{r}} + \cos\theta\cos\phi\,\hat{\boldsymbol{\theta}} - \sin\phi\,\hat{\boldsymbol{\phi}} \tag{4.5.44}$$

$$\hat{\mathbf{j}} = \sin\theta\sin\phi\,\hat{\mathbf{r}} + \cos\theta\sin\phi\,\hat{\boldsymbol{\theta}} + \cos\phi\,\hat{\boldsymbol{\phi}} \tag{4.5.45}$$

$$\hat{\mathbf{k}} = \cos\theta\,\hat{\mathbf{r}} - \sin\theta\,\hat{\boldsymbol{\theta}} \tag{4.5.46}$$

Using $\mathbf{r} = x\,\hat{\mathbf{i}} + y\,\hat{\mathbf{j}} + z\,\hat{\mathbf{k}}$ and (4.5.41), one can show that in spherical coordinates:

$$\mathbf{r} = r\,\hat{\mathbf{r}} \tag{4.5.47}$$

Beginning with $d\mathbf{s} = dx\,\hat{\mathbf{i}} + dy\,\hat{\mathbf{j}} + dz\,\hat{\mathbf{k}}$, and using (4.5.44)–(4.5.46), we find the infinitesimal displacement in spherical coordinates

$$d\mathbf{s} = dr\,\hat{\mathbf{r}} + r d\theta\,\hat{\boldsymbol{\theta}} + r\sin\theta\,d\phi\,\hat{\boldsymbol{\phi}} \tag{4.5.48}$$

$$ds^2 = dr^2 + r^2\,d\theta^2 + r^2\,\sin^2\theta\,d\phi^2 \tag{4.5.49}$$

The volume element can be found by multiplying the infinitesimal displacements in each direction:

$$dV = r^2\sin\theta\,dr\,d\theta\,d\phi \tag{4.5.50}$$

For an illustration of differential lengths $r\sin\theta\,d\phi$, dr, $r\,d\theta$ and the volume element dV, see Figure 4.15.

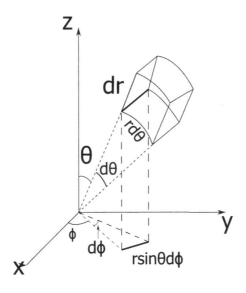

Figure 4.15 The differential lengths $r\sin\theta\,d\phi$, dr, $r\,d\theta$ and the volume element dV in spherical coordinates r, θ, ϕ.

Example 4.15 shows how to evaluate the Cartesian unit vectors $\hat{\mathbf{i}}, \hat{\mathbf{j}}, \hat{\mathbf{k}}$ in terms of the spherical unit vectors $\hat{\mathbf{r}}, \hat{\boldsymbol{\theta}}, \hat{\boldsymbol{\phi}}$, and also how to evaluate the length element ds^2 in spherical coordinates.

Example 4.15: Unit vectors and length element ds in spherical coordinates

Use SymPy to: (a) Evaluate the unit vectors $\hat{\mathbf{i}}, \hat{\mathbf{j}}, \hat{\mathbf{k}}$ in terms of the spherical unit vectors $\hat{\mathbf{r}}, \hat{\boldsymbol{\theta}}, \hat{\boldsymbol{\phi}}$ (b) Evaluate the length element ds^2 in spherical coordinates.

Solution:

(a) We can convert Cartesian unit vectors $\hat{\mathbf{i}}, \hat{\mathbf{j}}, \hat{\mathbf{k}}$ into spherical unit vectors $\hat{\mathbf{r}}, \hat{\boldsymbol{\theta}}, \hat{\boldsymbol{\phi}}$ using (4.5.41)-(4.5.43). As in Example 4.14, we use `solve()` to solve the system of three equations, and store the result in the variable `sol`.
We can print the resulting equations for $\hat{\mathbf{i}}, \hat{\mathbf{j}}, \hat{\mathbf{k}}$ by referring to the components of `sol` using `sol[i]`, `sol[j]` and `sol[k]`, respectively.
(b) The differentials of (4.5.41)–(4.5.43) are:

$$dx = \frac{\partial x}{\partial r}\, dr + \frac{\partial x}{\partial \theta}\, d\theta + \frac{\partial x}{\partial \phi}\, d\phi = \sin\theta\cos\phi\, dr + r\cos\theta\cos\phi\, d\theta - r\cos\theta\sin\phi\, d\phi \qquad (4.5.51)$$

$$dy = \frac{\partial y}{\partial \rho}\, d\rho + \frac{\partial y}{\partial \theta}\, d\theta + \frac{\partial y}{\partial \phi}\, d\phi = \sin\theta\sin\phi\, dr + r\cos\theta\sin\phi\, d\theta - r\cos\theta\cos\phi\, d\phi \qquad (4.5.52)$$

$$dz = \frac{\partial z}{\partial \rho}\, d\rho + \frac{\partial z}{\partial \theta}\, d\theta + \frac{\partial z}{\partial \phi}\, d\phi = \cos\theta\, dr - r\sin\theta\, d\theta \qquad (4.5.53)$$

Once we have the equations for dx, dy and dz, we can carry out the algebra to evaluate the length element $ds^2 = dx^2 + dy^2 + dz^2$. The algebra involved here is rather extensive, so we let Python evaluate the differentials dx, dy, dz and the algebraic evaluation of ds^2. We use the `simplify()` function to obtain the final expression for ds^2.

```
from sympy import symbols, Eq, diff, sin, cos, simplify, solve
import pprint
pp = pprint.PrettyPrinter(width=41, compact=True)
print('-'*28,'CODE OUTPUT','-'*29,'\n')


#  define all symbols
x, y, dx, dy, r, theta, phi, dr, dtheta, dphi, i, j, k, rhat, thetahat,\
phihat = symbols('x, y, dx, dy, r, theta, phi, dr, dtheta, dphi, i, j, k,\
rhat, thetahat, phihat',real=True)


# spherical unit vectors as functions of i, j, k vectors
eq1 = Eq(rhat,sin(theta)*cos(phi)*i+sin(theta)*sin(phi)*j+cos(theta)*k)


eq2 = Eq(thetahat,cos(theta)*cos(phi)*i+cos(theta)*sin(phi)*j-\
sin(theta)*k)


eq3 = Eq(phihat,-sin(phi)*i+cos(phi)*j)


# Use solve to find i,j as functions of the r,phi unit vectors
sol = solve((eq1,eq2,eq3),(i,j,k))


print('i-unit vector = ')
pp.pprint(simplify(sol[i]))
```

```
print('\nj-unit vector = ')
pp.pprint(simplify(sol[j]))
print('\nk-unit vector = ')
pp.pprint(simplify(sol[k]))

# Part b: equations for (x,y,z) as functions of (r,theta,phi)
x = r*sin(theta)*cos(phi)
y = r*sin(theta)*sin(phi)
z = r*cos(theta)

# find differentials dx, dy, dz using chain rule
dx=diff(x,r)*dr+diff(x,phi)*dphi+diff(x,theta)*dtheta
dy=diff(y,r)*dr+diff(y,phi)*dphi+diff(y,theta)*dtheta
dz=diff(z,r)*dr+diff(z,phi)*dphi+diff(z,theta)*dtheta

# The length element ds**2 = dx**2+dy**2+dz**2
print('\nThe length element ds**2 in spherical coordinates = ')
print(simplify(dx**2+dy**2+dz**2))


------------------------- CODE OUTPUT ----------------------------

i-unit vector =
-phihat*sin(phi) + rhat*sin(theta)*cos(phi) + thetahat*cos(phi)*cos(theta)

j-unit vector =
phihat*cos(phi) + rhat*sin(phi)*sin(theta) + thetahat*sin(phi)*cos(theta)

k-unit vector =
rhat*cos(theta) - thetahat*sin(theta)

The length element ds**2 in spherical coordinates =
dphi**2*r**2*sin(theta)**2 + dr**2 + dtheta**2*r**2
```

4.6   DIFFERENTIATION OF VECTORS

In this chapter, we will focus on the derivatives of vectors that depend only on time. In Chapter 8, we will examine vector fields that depend on x, y, and z. Although the following will hold for *any* time dependent vector, we will motivate our discussion with the position of a particle in Cartesian coordinates,

$$\mathbf{r}(t) = x(t)\,\hat{\mathbf{i}} + y(t)\,\hat{\mathbf{j}} + z(t)\,\hat{\mathbf{k}} \tag{4.6.1}$$

We can imagine that (4.6.1) describes the path of a particle in space. For simplicity, we will represent that path on a plane. Figure 4.16 shows the particle's position vector $\mathbf{r}(t)$ in the Cartesian plane ($z = 0$). The path taken by the particle is shown by the dashed line. At a time Δt later, the particle is located at $\mathbf{r}(t + \Delta t)$.

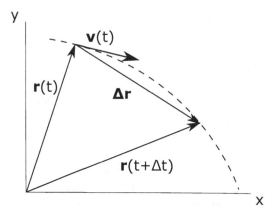

Figure 4.16 The path (dashed line) of a particle in the Cartesian plane.

Using Figure 4.16, we can define the displacement vector $\mathbf{\Delta r}$ using the head-to-tail method of vector addition, $\mathbf{r}(t + \Delta t) = \mathbf{r}(t) + \mathbf{\Delta r}$. Or

$$\mathbf{\Delta r} = \mathbf{r}(t + \Delta t) - \mathbf{r}(t) \tag{4.6.2}$$

The displacement is the distance and direction traveled during the time interval Δt. If we divide the displacement vector by the time interval Δt, we obtain the average velocity,

$$\mathbf{v}_{\text{avg}} = \frac{\mathbf{\Delta r}}{\Delta t} \tag{4.6.3}$$

Notice that the average velocity is the displacement vector multiplied by a scalar $1/\Delta t$. The average velocity gives us a rate at which the particle's position changed over the time interval Δt, and points in the direction of the displacement vector $\mathbf{\Delta r}$. However, the average velocity is just that, an average. It does not tell us the velocity of the particle at any one point in time. For that, we need the instantaneous velocity $\mathbf{v}$. To get $\mathbf{v}$, we take the limit of $\mathbf{v}_{\text{avg}}$ as $\Delta t \to 0$:

$$\mathbf{v}(t) = \lim_{\Delta t \to 0} \frac{\mathbf{\Delta r}}{\Delta t} = \lim_{\Delta t \to 0} \frac{\mathbf{r}(t + \Delta t) - \mathbf{r}(t)}{\Delta t} = \frac{d\mathbf{r}}{dt} \tag{4.6.4}$$

We see that (4.6.4) is Newton's definition of a derivative as presented in Chapter 2. Hence, the instantaneous velocity $\mathbf{v}$ is the derivative of the position vector $\mathbf{r}$ with respect to time. Because the instantaneous velocity is the derivative of the position vector, the instantaneous velocity vector is tangent to the path taken by the particle. In addition, all of the derivative rules that you know also apply to differentiating vectors. For example, the following identities hold:

$$\frac{d}{dt}(\mathbf{A} + \mathbf{B}) = \frac{d\mathbf{A}}{dt} + \frac{d\mathbf{B}}{dt} \tag{4.6.5}$$

$$\frac{d}{dt}(\mathbf{A} \times \mathbf{B}) = \left(\frac{d\mathbf{A}}{dt} \times \mathbf{B}\right) + \left(\mathbf{A} \times \frac{d\mathbf{B}}{dt}\right) \tag{4.6.6}$$

$$\frac{d}{dt}(\mathbf{A} \cdot \mathbf{B}) = \left(\frac{d\mathbf{A}}{dt} \cdot \mathbf{B}\right) + \left(\mathbf{A} \cdot \frac{d\mathbf{B}}{dt}\right) \tag{4.6.7}$$

$$d\mathbf{A}(x, y, z) = \frac{\partial \mathbf{A}}{\partial x}\, dx + \frac{\partial \mathbf{A}}{\partial y}\, dy + \frac{\partial \mathbf{A}}{\partial z}\, dz \tag{4.6.8}$$

For example, let us consider the velocity vector $\mathbf{v}(t)$ as the derivative of the vector (4.6.1):

$$\left. \begin{aligned} \mathbf{v}(t) &= \frac{d\mathbf{r}}{dt} \\ &= \frac{d}{dt}\left(x\hat{\mathbf{i}}\right) + \frac{d}{dt}\left(y\hat{\mathbf{j}}\right) + \frac{d}{dt}\left(z\hat{\mathbf{k}}\right) \\ &= \frac{dx}{dt}\hat{\mathbf{i}} + \frac{dy}{dt}\hat{\mathbf{j}} + \frac{dz}{dt}\hat{\mathbf{k}} \end{aligned} \right\} \tag{4.6.9}$$

Note that the final step in (4.6.9) involves the product rule, however the unit vectors in Cartesian coordinates are constant in both magnitude and direction. That is not true for the unit vectors in other coordinate systems. For example, the unit vectors in polar coordinates always have a length of one, but their direction depends on the particle's location in the Cartesian plane.

Of course, any time-dependent vector can be differentiated similar to what is done in (4.6.9)

$$\frac{d\mathbf{A}}{dt} = \frac{dA_x}{dt}\hat{\mathbf{i}} + \frac{dA_y}{dt}\hat{\mathbf{j}} + \frac{dA_z}{dt}\hat{\mathbf{k}} \tag{4.6.10}$$

Similarly, we evaluate the acceleration $\mathbf{a}(t)$ as the second derivative of the position vector:

$$\mathbf{a}(t) = \frac{d^2\mathbf{r}}{dt^2} = \frac{d^2 x}{dt^2}\hat{\mathbf{i}} + \frac{dy^2}{dt^2}\hat{\mathbf{j}} + \frac{dz^2}{dt^2}\hat{\mathbf{k}} \tag{4.6.11}$$

Example 4.16: Time derivatives of vectors

The position vector of a particle is described by

$$\mathbf{r}(t) = 3c_1\, t^2\hat{\mathbf{i}} - 2c_2\, t^3\hat{\mathbf{j}} + 4c_3 t\,\hat{\mathbf{k}} \tag{4.6.12}$$

where all physical quantities are measured in SI units and the c_i are constants.
(a) Find the instantaneous velocity $\mathbf{v}(t)$ and the instantaneous acceleration $\mathbf{a}(t)$.
(b) Find a unit vector tangent to the curve C defined by the position vector $\mathbf{r}(t)$, at time $t = 0$ s.
(c) Find a unit vector perpendicular to the curve C defined by the position vector $\mathbf{r}(t)$, at time $t = 1$ s.

Solution:

(a) Following (4.6.9), we find the instantaneous velocity by differentiating each vector component individually.

$$\mathbf{v} = \frac{d\mathbf{r}}{dt} = 6c_1 t\,\hat{\mathbf{i}} - 6c_2 t^2\hat{\mathbf{j}} + 4c_3\hat{\mathbf{k}} \tag{4.6.13}$$

Likewise, the instantaneous acceleration is the derivative of the velocity with respect to time, or the second derivative of the position with respect to time:

$$\mathbf{a} = \frac{d^2\mathbf{r}}{dt} = 6c_1\hat{\mathbf{i}} - 12c_2 t\hat{\mathbf{j}} \tag{4.6.14}$$

(b) The velocity vector is always tangent to the curve C, so a unit vector tangent to C is:

$$\hat{\mathbf{n}} = \frac{\mathbf{v}}{|\mathbf{v}|} = \frac{6c_1 t\hat{\mathbf{i}} - 6c_2 t^2\hat{\mathbf{j}} + 4c_3\hat{\mathbf{k}}}{\left|6c_1 t\hat{\mathbf{i}} - 6c_2 t^2\hat{\mathbf{j}} + 4c_3\hat{\mathbf{k}}\right|} = \frac{6c_1 t\hat{\mathbf{i}} - 6c_2 t^2\hat{\mathbf{j}} + 4c_3\hat{\mathbf{k}}}{\sqrt{(6c_1 t)^2 + (-6c_2 t^2)^2 + (4c_3)^2}}$$

At time $t = 0$, we find $\hat{\mathbf{n}} = \hat{\mathbf{k}}$.

(c) There are many vectors perpendicular to the curve C at time $t = 1$ s. All these vectors will also be perpendicular to the tangent velocity vector $\mathbf{v}$. From part (a) we can choose, for example a vector b such that:

$$\mathbf{b} = -4c_3\hat{\mathbf{i}} + 6c_1 t\,\hat{\mathbf{k}}$$

so that the dot product:

$$\mathbf{b} \cdot \mathbf{v} = (-4c_3)(6c_1 t) + (6c_1 t)(4c_3) = 0$$

The corresponding unit vector is

$$\hat{\mathbf{b}} = \frac{\mathbf{b}}{|\mathbf{b}|} = \frac{-4c_3\hat{\mathbf{i}} + 6c_1 t\,\hat{\mathbf{k}}}{\left|-4c_3\hat{\mathbf{i}} + 6c_1 t\,\hat{\mathbf{k}}\right|} = \frac{-4c_3\hat{\mathbf{i}} + 6c_1 t\,\hat{\mathbf{k}}}{\sqrt{(4c_3)^2 + (6c_1 t)^2}}$$

Next we use SymPy to perform the derivatives.

The SymPy command `CoordSys3D()` creates the coordinate system called R. The instantaneous velocity v and instantaneous acceleration a can be found using the `diff()` command as before.

We use the method `v.normalize()` to evaluate the unit vector n̂, and we also use the method `.subs()` to substitute the value of $t = 0$ in the expression for n̂. These are combined in the code line `simplify(v.normalize().subs(t,0))` to evaluate the unit vector at time $t = 0$.

```
from sympy import symbols, diff, simplify
from sympy.vector import CoordSys3D
print('-'*28,'CODE OUTPUT','-'*29,'\n')

R = CoordSys3D('R') #establish a Cartesian coordinate system

c1, c2, c3, t = symbols('c1, c2, c3, t', positive=True)

r = (3*c1*t**2)*R.i - (2*c2*t**3)*R.j + (4*c3*t)*R.k
v = diff(r,t)
a = diff(r,t,t)

print('The instantaneous velocity is: ' + str(v))
print('\nThe instantaneous acceleration is: ' + str(a))

print('\nThe unit vector tangent to the position vetcor at t=0 is:\n',
simplify(v.normalize().subs(t,0)))



------------------------ CODE OUTPUT ----------------------------

The instantaneous velocity is: 6*c1*t*R.i + (-6*c2*t**2)*R.j + 4*c3*R.k

The instantaneous acceleration is: 6*c1*R.i + (-12*c2*t)*R.j

The unit vector tangent to the position vetcor at t=0 is:
 R.k
```

As a final application of derivatives of vectors, we evaluate and interpret the velocity of particle moving on the xy-plane, in polar coordinates. Consider the position vector in polar coordinates, $\mathbf{r} = \rho\hat{\boldsymbol{\rho}}$. Next, we take the derivative of $\mathbf{r}$

$$\mathbf{v} = \frac{d\mathbf{r}}{dt} = \frac{d\rho}{dt}\hat{\boldsymbol{\rho}} + \rho\frac{d\hat{\boldsymbol{\rho}}}{dt} \qquad (4.6.15)$$

Using (4.5.5) we find

$$\frac{d\hat{\boldsymbol{\rho}}}{dt} = -\sin\phi\frac{d\phi}{dt}\hat{\mathbf{i}} + \cos\phi\frac{d\phi}{dt}\hat{\mathbf{j}} = \frac{d\phi}{dt}\left(-\sin\phi\,\hat{\mathbf{i}} + \cos\phi\,\hat{\mathbf{j}}\right) = \frac{d\phi}{dt}\hat{\boldsymbol{\phi}} \qquad (4.6.16)$$

Therefore

$$\mathbf{v} = \frac{d\rho}{dt}\,\hat{\boldsymbol{\rho}} + \rho\,\frac{d\phi}{dt}\,\hat{\boldsymbol{\phi}} \qquad (4.6.17)$$

Notice we could have also found $\mathbf{v}$ by dividing (4.5.10) with the time differential dt

$$\mathbf{v} = \frac{d\mathbf{s}}{dt} = \frac{d\rho}{dt}\,\hat{\boldsymbol{\rho}} + \rho\,\frac{d\phi}{dt}\,\hat{\boldsymbol{\phi}} \qquad (4.6.18)$$

The velocity in polar coordinates has two parts. The first part $(d\rho/dt)\,\hat{\boldsymbol{\rho}}$ corresponds to the velocity of the particle moving radially away from the origin. The second part $\rho\,(d\phi/dt)\,\hat{\boldsymbol{\phi}}$ is the tangential velocity of the particle as it moves around the origin. Recall from introductory physics that the tangential velocity of a particle moving along a circle is the product of the circle's radius and the particle's angular velocity $d\phi/dt$.

The velocity in cylindrical coordinates is

$$\mathbf{v} = \frac{d\mathbf{s}}{dt} = \frac{d\rho}{dt}\,\hat{\boldsymbol{\rho}} + \rho\,\frac{d\phi}{dt}\,\hat{\boldsymbol{\phi}} + \frac{dz}{dt}\hat{\mathbf{z}} \qquad (4.6.19)$$

which is similar to (4.6.18), but includes a component along the z-direction.

To evaluate the particle's velocity in spherical coordinates, we divide (4.5.48) with the time differential dt:

$$\mathbf{v} = \frac{d\mathbf{s}}{dt} = \frac{dr}{dt}\hat{\mathbf{r}} + r\frac{d\theta}{dt}\hat{\boldsymbol{\theta}} + r\sin\theta\frac{d\phi}{dt}\hat{\boldsymbol{\phi}} \qquad (4.6.20)$$

The first term in (4.6.20) is simply the radial component of the particle's velocity, which describes how fast the particle is moving away from the origin. The other two terms in this expression are tangential velocities around circles.

The polar angle θ describes the particle's position along the vertically oriented dashed circle in Figure 4.14. Hence, the second term in (4.6.20) is the particle's tangential velocity $r\,d\theta/dt$ around this circle.

The third term in (4.6.20) describes the particle's tangential velocity along a circle parallel to the xy-plane. If $\theta = \pi/2$, the circle lies in the xy-plane. Otherwise, the plane of the circle lies either above or below the xy-plane. Regardless of the value of θ, the projection of $\mathbf{r}$ onto the plane of the circle has a length of $r\sin\theta$. In other words, the circle has a radius of $r\sin\theta$. Hence, the third term in $r\sin\theta\,(d\phi/dt)$ also has the form of a radius $(r\sin\theta)$ times an angular velocity (in this case $d\phi/dt$).

In this section we discussed the differentiation of vectors. We will postpone the analogous discussion of how to *integrate* vector quantities until the later chapter on Vector Analysis, where we will discuss several different types of vector integration techniques (ordinary vector integrals, line integrals, surface integrals etc.).

4.7   PARAMETRIC EQUATIONS OF LINES AND PLANES

From basic algebra we know the equation for a line is of the form of $y = mx + b$. That is a useful representation of a line when one wants to know the value of the dependent variable y, when the independent variable x is known.

Consider the case where a particle is moving along a line. In that case, we would want to know the particle's location after a time t has passed. Now we need the x and y coordinates as a function of time t, a parameter which gives us the x and y coordinates of the particle while it travels along the line. In this section, we will derive the parametric equation for a line (with parameter t) in two and three dimensions.

In addition, we will derive the equation of a plane. Knowing the equation of a plane can be useful in physics in many applications. For example, the equation of a plane can provide the limits of integration when calculating the center of mass of prisms and parallelograms.

Consider a line in the direction of the vector $\mathbf{v} = v_x\hat{\mathbf{i}} + v_y\hat{\mathbf{j}}$ as shown in Figure 4.17.

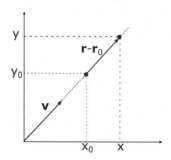

Figure 4.17 A displacement $\mathbf{r} - \mathbf{r}_0$ along a line in the direction of the vector $\mathbf{v}$.

We will derive the equations for a line in two dimensions, for illustrative purposes, but the inclusion of the third dimension is trivial. Suppose a particle on the line starts at the position $\mathbf{r}_0 = x_0\,\hat{\mathbf{i}} + y_0\,\hat{\mathbf{j}}$ and moves along the line to the point $\mathbf{r} = x\,\hat{\mathbf{i}} + y\,\hat{\mathbf{j}}$. The displacement of the particle is then $\mathbf{r} - \mathbf{r}_0$. Because the displacement along the line is described by $\mathbf{v}$, the vector $\mathbf{r} - \mathbf{r}_0$ is parallel to $\mathbf{v}$ and therefore

$$\mathbf{r} - \mathbf{r}_0 = \mathbf{v}\,t \tag{4.7.1}$$

where t is a scalar which serves as a parameter. From (4.7.1) and the definitions of $\mathbf{r}$, $\mathbf{r}_0$, and $\mathbf{v}$ we can solve for x and y as functions of the parameter t, to get the parametric equations for the line.

$$x - x_0 = v_x t$$
$$y - y_0 = v_y t \tag{4.7.2}$$

Solving (4.7.2) for t we obtain

$$\frac{x - x_0}{v_x} = \frac{y - y_0}{v_y} \tag{4.7.3}$$

Equation (4.7.3) can be solved for $y - y_0$ to obtain:

$$y - y_0 = \frac{v_y}{v_x}\left(x - x_0\right) \tag{4.7.4}$$

which is the point-slope formula of a line.

In three dimensions, we would repeat the procedure using the vector $\mathbf{v} = v_x\,\hat{\mathbf{i}} + v_y\,\hat{\mathbf{j}} + v_z\,\hat{\mathbf{k}}$ for the direction of the line, and position vectors $\mathbf{r} = x\,\hat{\mathbf{i}} + y\,\hat{\mathbf{j}} + z\,\hat{\mathbf{k}}$ and $\mathbf{r}_0 = \mathbf{r} = x_0\,\hat{\mathbf{i}} + y_0\,\hat{\mathbf{j}} + z_0\,\hat{\mathbf{k}}$. The resulting equations are

$$\frac{x - x_0}{v_x} = \frac{y - y_0}{v_y} = \frac{z - z_0}{v_z} \tag{4.7.5}$$

$$x - x_0 = v_x\,t$$
$$y - y_0 = v_y\,t$$
$$z - z_0 = v_z\,t \tag{4.7.6}$$

Generalizing the above concepts, any curve C in 3D is represented parametrically by a function $(x(t),\ y(t),\ z(t))$.

Example 4.17 demonstrates how to plot a parametric curve $(x(t),\ y(t),\ z(t))$ in 3D, and how to draw tangent vectors at any point along the curve.

Example 4.17: Plotting a parametric curve (x(t),y(t),z(t)) and tangent vectors

(a) Write a Python code to plot the parametric line $(x, y, z) = (t,\ 1 + 2\cos t,\ -1 + 3\sin t)$.

(b) Find and draw tangent vectors to this line, at the points in space corresponding to the two time instants $t = 2$ s and $t = 5$ s.

Solution:

(a) The 3D plot in the example is created using the `plot3D()` function within the `matplotlib` library. The lines `fig = plt.figure()` and and `fig.add_subplot(projection ='3d')` create the 3D plot.

(b) A vector tangent to the curve is the velocity vector

$$\mathbf{v} = \frac{d\mathbf{r}}{dt} = \frac{d}{dt}(t,\ 1 + 2\cos t,\ -1 + 3\sin t) = (1,\ -2\sin t,\ 3\cos t)$$

Substituting the two time instants $t = 2$ s and $t = 5$ s, we easily find the velocity.

In the Python code we define two functions `pos(t)` and `vel(t)` which calculate the position and velocity arrays `[x,y,z]` and `[vx,vy,vz]`.

The code lines `xo,yo,zo = pos(to)` and `velx, vely, velz = vel(to)` unpack the position and velocity arrays, so that we can plot the arrow. We use the `quiver()` function to plot the corresponding tangent vector **v** at time t=to in Figure 4.18, by starting from the start of the position vector `(xo,yo,zo)`, and ending at the tip of the velocity vector `(velx, vely, velz)`.

```python
import matplotlib.pyplot as plt
import numpy as np

t = np.arange(0, 8.8, 0.1)        # time values from t=0 to t=9 s

def pos(t):
    x = t
    y = 1+2*np.cos(t)             # define x(t), y(t), z(t)
    z = -1+3*np.sin(t)
    return np.array([x,y,z])

def vel(t):
    vx = 1                        # define x(t), y(t), z(t)
    vy = -2*np.sin(t)
    vz = 3*np.cos(t)
    return np.array([vx,vy,vz])

xpos, ypos, zpos = pos(t)         # evaluate the position array

to = 5                            # dot added to graph at to=5 s
xo,yo,zo = pos(to)                # unpack the position array at to=5 s
velx, vely, velz = vel(to)        # unpack the velocity array at to=5 s

# set up 3D plot and plot curve x(t),y(t),z(t)
```

```
fig = plt.figure()
ax = fig.add_subplot(projection='3d')
ax.plot3D(xpos,ypos,zpos)

ax.quiver(xo,yo,zo, velx, vely, velz, color='red',length=1,\
arrow_length_ratio=.2)

ax.scatter(xo,yo,zo)

to = 2                        # dot added to graph at to=2  s
xo,yo,zo = pos(to)            # unpack the position array at t=2 s
velx, vely, velz = vel(to)    # unpack the velocity array at t=2 s
ax.quiver(xo,yo,zo, velx, vely, velz, color='red',length=1,\
arrow_length_ratio=.2)

ax.set_xlabel('X')
ax.set_ylabel('Y')
ax.set_zlabel('Z')

ax.scatter(xo,yo,zo)

ticks=np.arange(-1,4,1)    # define the tick marks on z-axis
ax.set_yticks(ticks)
plt.show()
```

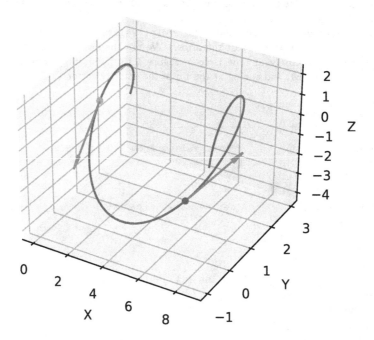

Figure 4.18 The graphical output from Example 4.17, showing the plot of a parametric curve $(x(t),\ y(t),\ z(t))$ in 3D, and the tangent vectors at two points along the curve.

In deriving (4.7.2), we used notation which implies certain physical quantities. From mechanics, we know that the velocity vector gives the instantaneous direction of motion of particle at time t. In this case, the vector $\mathbf{v}$ is behaving as the velocity from mechanics.

However, the derivation of (4.7.2) does not require us to use variables of motion. Any variables sharing a linear relationship would work.

Consider a line pointing in the direction of $\mathbf{v} = a\hat{\mathbf{i}} + b\hat{\mathbf{j}} + c\hat{\mathbf{k}}$. Following the argument above, the equation for a line is then

$$\frac{x - x_0}{a} = \frac{y - y_0}{b} = \frac{z - z_0}{c} \qquad (4.7.7)$$

$$\begin{aligned} x - x_0 &= a\,t \\ y - y_0 &= b\,t \\ z - z_0 &= c\,t \end{aligned} \qquad (4.7.8)$$

Let us return to two dimensions to study a different but related problem. Suppose we want the equation of the line passing through $\mathbf{r}_0$ but perpendicular to the vector $\mathbf{n} = n_x\hat{\mathbf{i}} + n_y\hat{\mathbf{j}}$. How would we find it? Like before, the vector $\mathbf{r} - \mathbf{r}_0$ lies along a line passing through the point $\mathbf{r}_0$. However, we now want $\mathbf{r} - \mathbf{r}_0$ to be perpendicular to $\mathbf{n}$. In other words

$$\mathbf{n} \cdot (\mathbf{r} - \mathbf{r}_0) = 0$$

or

$$n_x\,(x - x_0) + n_y\,(y - y_0) = 0 \qquad (4.7.9)$$

We can rearrange (4.7.9) to obtain the equation of the line perpendicular to $\mathbf{r} - \mathbf{r}_0$

$$y - y_0 = -\frac{n_x}{n_y}\,(x - x_0) \qquad (4.7.10)$$

In two dimensions we are restricted to a single line passing through $\mathbf{r}_0$ which can be perpendicular to $\mathbf{n}$. However, in three dimensions, there are an infinite number of lines which pass through $\mathbf{r}_0$ and are perpendicular to $\mathbf{n} = n_x\hat{\mathbf{i}} + n_y\hat{\mathbf{j}} + n_z\hat{\mathbf{k}}$, forming a plane. Let

$$\mathbf{r} - \mathbf{r}_0 = (x - x_0)\,\hat{\mathbf{i}} + (y - y_0)\,\hat{\mathbf{j}} + (z - z_0)\,\hat{\mathbf{k}} \qquad (4.7.11)$$

Again using $\mathbf{n} \cdot (\mathbf{r} - \mathbf{r}_0) = 0$ we obtain the equation of the plane

$$n_x\,x + n_y\,y + n_z\,z = d \qquad (4.7.12)$$

where $d = n_x\,x_0 + n_y\,y_0 + n_z\,z_0$.

We conclude that the vector (a, b, c) is always perpendicular to the plane $ax + by + cz + d = 0$.

Example 4.18 demonstrates how to plot a plane $a\,x + b\,y + c\,z + d = 0$ and a vector perpendicular to this plane in 3D .

==

Example 4.18: Plotting a plane surface and a vector perpendicular to it

Write a Python code to plot the plane surface function $z = 1 - x - y$ in 3D, when $0 \le x \le 1$ and $0 \le y \le 1$. Plot also a vector perpendicular to this plane.

Solution:

The vector $\mathbf{A} = (a, b, c)$ is perpendicular to the plane $a\,x + b\,y + c\,z + d = 0$. In the case of the plane $x + y + z - 1 = 0$, we have $(a, b, c) = (1, 1, 1)$.

The 3D plot in this example is created using the `plot_surface()` function within the `matplotlib` library. The lines `fig = plt.figure()` and `fig.add_subplot(projection ='3d')` create the 3D plot.

The NumPy command `meshgrid(x, y)` creates double arrays `X,Y` from the single NumPy arrays `x,y`. In this code it is necessary to create double arrays, which are required inputs for the `plot_surface(X,Y,Z)` function.

The plotting function `plot_surface(X,Y,Z)` plots the 3D surface $z = f(x,y) = 1 - x - y$ in Figure 4.19, and `quiver()` is used to plot the arrow.

```python
import matplotlib.pyplot as plt
import numpy as np

fig = plt.figure()
ax = fig.add_subplot(projection='3d')        # setup 3D plot

Ax, Ay, Az = 1, 1, 1               # components of vector A

# set up X, Y, Z for surface plot
# use quiver() to plot the arrow, set limits, add text

# plot the plane z=1-x-y
x = np.arange(0, 1, 0.3)
y = np.arange(0, 1, 0.3)
X, Y = np.meshgrid(x, y)
Z = 1-X-Y
ax.plot_surface(X, Y, Z,alpha=0.4)

# plot arrow for vector A
ax.quiver(0,0,0,Ax,Ay,Az,color='blue',length=1,arrow_length_ratio=.1)

ax.set_xlim(0,1.5);
ax.set_ylim(-1,2.1);
ax.set_zlim(0,1.2);

ax.set_xlabel('X')
ax.set_ylabel('Y')
ax.set_zlabel('Z')

ax.text(1.1,.3,.9,'n=[1,1,1]')   # add text to 3D plot
ax.text(.7,.75,.1,'Plane')
ax.text(.7,.7,-.05,'x+y+z=1')

plt.show()
```

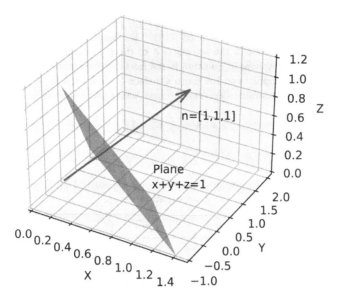

Figure 4.19   The graphical output from Example 4.18, showing a plot of the plane $a\,x + b\,y + c\,z + d = 0$, and of the vector (a, b, c) which is perpendicular to this plane.

In physics, it is common to obtain formulas such as (4.7.1) and (4.7.12). When we do, we can read them to find **v** or **n**, and give a geometric interpretation of our results. An example of this type of procedure is presented in Example 4.19.

Example 4.19: Geodesic on a sphere

A geodesic is the shortest possible path between two points that a particle can take on a curved surface. A common problem in an intermediate mechanics course is to find the geodesic on the surface of a sphere centered at the origin with a radius R. The calculation is done in spherical coordinates and using a branch of mathematics called the calculus of variations, which produces a differential equation whose solution is the formula for the geodesic. The solution to the differential equation (after some manipulation) is

$$R \cos \theta = R\beta \, \cos \alpha \sin \theta \sin \phi - R\beta \, \sin \alpha \sin \theta \cos \phi \qquad (4.7.13)$$

where θ and ϕ are spherical coordinates and β and α are constants. Interpret (4.7.13) and describe the shape of the geodesic.

Solution:

We begin by regrouping (4.7.13)

$$(R \cos \theta) = \beta \, \cos \alpha \, (R \sin \theta \sin \phi) - \beta \, \sin \alpha \, (R \sin \theta \cos \phi) \qquad (4.7.14)$$

Notice that the terms in parentheses are (from left to right) z, y, and x in Cartesian coordinates. Hence, (4.7.13) can be rewritten as

$$z + Bx - Ay = 0 \qquad (4.7.15)$$

where $A = \beta \cos \alpha$ and $B = \beta \sin \alpha$. Therefore, (4.7.13) describes a plane passing through the origin and perpendicular to the vector $\mathbf{n} = B\hat{\mathbf{i}} - A\hat{\mathbf{j}} + \hat{\mathbf{k}}$. Because the particle is constrained to move on the sphere, the geodesic is the intersection of the plane and the sphere. The resulting shape is a circle. When the sphere in question is the Earth, the geodesic is called a great circle. The great circle between two points on the Earth's surface is often used for air and sea navigation. Let us now plot a sample geodesic. Consider the case where $R = A = B = 1$. The sphere can be written as

$$x^2 + y^2 + z^2 = 1 \tag{4.7.16}$$

and the plane as

$$z = y - x \tag{4.7.17}$$

To get the curve of the geodesic, we need to insert (4.7.15) into (4.7.16) to obtain a quadratic equation. The solution of this quadratic is an equation for $y(x)$:

$$y = \frac{x \pm \sqrt{x^2 - 8\left(2x^2 - 1\right)}}{4} \tag{4.7.18}$$

We can then find the path in three dimensional space by choosing an x-value, substituting it into (4.7.18) to get its y-value, and then substituting those values into (4.7.17). We can create the path by repeating the process for multiple x-values as done in the code below. We display only the positive root of (4.7.18) in Figure 4.20. The black curve is the geodesic and it represents the intersection of the plane (4.7.17) with the sphere (4.7.16).

```python
import matplotlib.pyplot as plt
from mpl_toolkits.mplot3d import Axes3D
from matplotlib import cm
import numpy as np

fig=plt.figure()
ax=fig.add_subplot(projection='3d')

#create the 3D plane z=y-x
x = np.arange(0,1,0.3)
y = np.arange(0,1,0.3)
xp, yp = np.meshgrid(x,y)
zp = yp - xp
ax.plot_surface(xp, yp, zp, alpha = 0.5)

# create the sphere u=phi 0-2*pi, v=theta=0-pi
u, v = np.mgrid[0:2*np.pi:20j, 0:np.pi:10j]

xs = np.cos(u)*np.sin(v)
ys = np.sin(u)*np.sin(v)
zs = np.cos(v)
ax.plot_surface(xs, ys, zs, cmap=cm.ocean, alpha = 0.4)

 # create geodesic  using solution of quadratic
xg = np.linspace(0,0.7,100)
yg = (xg + np.sqrt(xg**2 - 8*(2*xg**2-1)))/4
zg = yg - xg
ax.plot3D(xg,yg,zg, 'black', lw = 3)

ticks = np.arange(-1,1.5,0.5)                # graph formatting
ax.set_xticks(ticks)
ax.set_yticks(ticks)
```

```
ax.set_zticks(ticks)
plt.show()
```

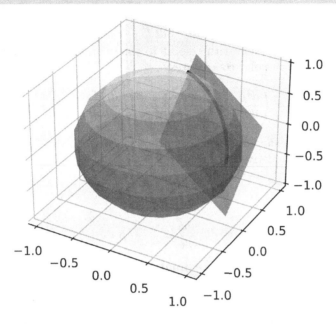

Figure 4.20 The graphical output from Example 4.19, showing the geodesic on a sphere.

4.8 END OF CHAPTER PROBLEMS

1. **Displacement vectors** – Jennifer starts at the origin and walks 2 km in a direction 25° north of east, then turns and walks 1 km 15° west of north. Using a coordinate system which is oriented such that north lies along the positive y-axis and east along the positive x-axis, calculate the magnitude and direction of Jennifer's displacement. Use Python to plot a vector representing each part of the walk.

2. **Derivatives of unit vectors in polar coordinates** – Differentiate the following expressions for the unit vectors $\hat{\rho}$ and $\hat{\phi}$ in polar coordinates with respect to time t,

$$\hat{\rho} = \cos\phi\,\hat{\mathbf{i}} + \sin\phi\,\hat{\mathbf{j}} \tag{4.8.1}$$

$$\hat{\phi} = -\sin\phi\,\hat{\mathbf{i}} + \cos\phi\,\hat{\mathbf{j}} \tag{4.8.2}$$

and show that

$$\frac{d\hat{\rho}}{dt} = \frac{d\phi}{dt}\,\hat{\phi} \tag{4.8.3}$$

$$\frac{d\hat{\phi}}{dt} = -\frac{d\phi}{dt}\,\hat{\rho} \tag{4.8.4}$$

Do this problem by hand and using SymPy.

3. **Acceleration in polar coordinates** – As discussed in this chapter, the velocity in polar coordinates is given by:

$$\mathbf{v} = \frac{d\mathbf{s}}{dt} = \frac{d\rho}{dt}\,\hat{\boldsymbol{\rho}} + \rho\,\frac{d\phi}{dt}\,\hat{\boldsymbol{\phi}} \tag{4.8.5}$$

Differentiate this expression with respect to time t, to prove the following equation for the acceleration in polar coordinates.

$$\mathbf{a} = \left(\frac{d^2\rho}{dt^2} - \rho\left(\frac{d\phi}{dt}\right)^2\right)\hat{\boldsymbol{\rho}} + \left(2\frac{d\rho}{dt}\frac{d\phi}{dt} + \rho\frac{d^2\phi}{dt^2}\right)\hat{\boldsymbol{\phi}} \tag{4.8.6}$$

Do this problem by hand and using SymPy. You may find useful the following relationships which were derived in Problem 2:

$$\frac{d\hat{\boldsymbol{\rho}}}{dt} = \frac{d\phi}{dt}\,\hat{\boldsymbol{\phi}} \tag{4.8.7}$$

$$\frac{d\hat{\boldsymbol{\phi}}}{dt} = -\frac{d\phi}{dt}\,\hat{\boldsymbol{\rho}} \tag{4.8.8}$$

4. **Acceleration in cylindrical coordinates** – Starting with $\mathbf{r} = \rho\,\hat{\boldsymbol{\rho}} + z\,\hat{\mathbf{z}}$ derive the formula for the acceleration in cylindrical coordinates. Use SymPy to help with the calculus and the algebra.

5. **Vector derivatives in polar coordinates** – Consider the vector, $\mathbf{r} = 3ct^2\hat{\boldsymbol{\rho}}$, where c is a constant, $\hat{\boldsymbol{\rho}}$ is the radial unit vector in polar coordinates, and $\mathbf{r}$ is measured in meters and t in seconds. (a) Compute the velocity $\mathbf{v} = d\mathbf{r}/dt$ and the acceleration $\mathbf{a} = d\mathbf{v}/dt$ by hand and using SymPy.

6. **Vector derivatives in spherical coordinates** – Consider the vector, $\mathbf{r} = 3ct^2\hat{\mathbf{r}}$, where c is a constant, $\hat{\mathbf{r}}$ is the radial unit vector in spherical coordinates, and all quantities are in SI units. (a) Compute $\mathbf{v} = d\mathbf{r}/dt$ by hand. (b) Compute $d\mathbf{v}/dt$ using Python. Do the problem by hand and using Python.

7. **Converting between coordinate systems** – A particle is located at $\mathbf{r} = \hat{\mathbf{i}} + 3\hat{\mathbf{j}} + 4\hat{\mathbf{k}}$, find the values of r, θ, ϕ for this vector and the unit vectors $\hat{\mathbf{r}}, \hat{\boldsymbol{\theta}}, \hat{\boldsymbol{\phi}}$ at the position $\mathbf{r}$. In addition, find its components A_r, A_θ and A_ϕ in spherical coordinates, such that $\mathbf{r} = A_r\hat{\mathbf{r}} + A_\theta\hat{\boldsymbol{\theta}} + A_\phi\hat{\boldsymbol{\phi}}$.

8. **Derivatives of unit vectors** – Consider the vector $\mathbf{r} = \mathbf{r}(t)$ which has a length that is always constant (although its direction may vary). Prove that $d\mathbf{r}/dt$ is perpendicular to $\mathbf{r}$ by hand. As a special case, use SymPy to show that the polar unit vector $\hat{\boldsymbol{\rho}}$ and $d\hat{\boldsymbol{\rho}}/dt$ are perpendicular.

9. **Lorentz force** – The force on a particle with charge q and mass m moving with a velocity $\mathbf{v}$ through a magnetic field $\mathbf{B}$ is $\mathbf{F} = q\,(\mathbf{v} \times \mathbf{B})$. The force $\mathbf{F}$ is called the *Lorentz force*. Suppose that the particle moves in the xy-plane through a uniform magnetic field that points in the positive z-direction. Show that the force and the velocity both have constant magnitudes.

10. **Moving charges** – Consider two moving charged particles. They exert forces on each other because each one created a magnetic field through which the other moves. The forces are proportional to $\mathbf{v}_1 \times (\mathbf{v}_2 \times \mathbf{r})$ and $\mathbf{v}_2 \times (\mathbf{v}_1 \times (-\mathbf{r}))$ where $\mathbf{r}$ is the vector joining the two particles. Show that, in general, these forces violate Newton's third law except when $\mathbf{r} \times (\mathbf{v}_2 \times \mathbf{v}_1) = 0$.

11. **Finding the vector tangent to a path** – Find the tangent vector at any point of the curve $(x, y, z) = (t^3, 3t^2 - 1, 2t - 5)$ m where t is time measured in seconds. Plot (a) the path in Python for $t = 0$ to $t = 5$ seconds and the tangent vector at $t = 2$ seconds. (b) The angle between the position and velocity vectors as a function of time.

12. **Torque** – A force $\mathbf{F} = -\hat{\mathbf{i}} + 2\hat{\mathbf{j}} - \hat{\mathbf{k}}$ acts at the point $(2, 1, 3)$. Find the torque due to $\mathbf{F}$ about:

   a. The origin

   b. The x-axis

   c. The line $x/3 = y/2 = -z$

   d. The point $(1, 1, 1)$. Note that in this case $\mathbf{r}$ points from $(1, 1, 1)$ to $(2, 1, 3)$.

13. **Higher order vector products** – Use Python to test the following identities (do not attempt this problem by hand, the algebra becomes very messy):

   a. $\mathbf{A} \times [\mathbf{B} \times (\mathbf{C} \times \mathbf{D})] = \mathbf{B}[\mathbf{A} \cdot (\mathbf{C} \times \mathbf{D})] - (\mathbf{A} \cdot \mathbf{B})(\mathbf{C} \times \mathbf{D})$

   b. $(\mathbf{A} \times \mathbf{B}) \cdot (\mathbf{C} \times \mathbf{D}) = (\mathbf{A} \cdot \mathbf{C})(\mathbf{B} \cdot \mathbf{D}) - (\mathbf{A} \cdot \mathbf{D})(\mathbf{B} \cdot \mathbf{C})$

   c. $\mathbf{A} \times (\mathbf{B} \times \mathbf{C}) + \mathbf{B} \times (\mathbf{C} \times \mathbf{A}) + \mathbf{C} \times (\mathbf{A} \times \mathbf{B}) = 0$.

14. **Derivative of the dot product** – Both by hand and by using Python, prove

$$\frac{d}{dt}(\mathbf{A} \cdot \mathbf{B}) = \left(\frac{d\mathbf{A}}{dt} \cdot \mathbf{B}\right) + \left(\mathbf{A} \cdot \frac{d\mathbf{B}}{dt}\right)$$

   where $\mathbf{A}$ and $\mathbf{B}$ are time-dependent vectors.

15. **Angular velocity** – At a time t, a particle's position is described the vector $\mathbf{r} = 3\hat{\mathbf{i}} + 2\hat{\mathbf{j}} - 5\hat{\mathbf{k}}$ and its velocity by $\mathbf{v} = -\hat{\mathbf{i}} + 3\hat{\mathbf{j}} + 2\hat{\mathbf{k}}$, both in mks units. Using $\mathbf{v} = \boldsymbol{\omega} \times \mathbf{r}$, is it possible to identify the particle's angular velocity $\boldsymbol{\omega}$? Use Python to set up the necessary equations and solve them.

16. **Orthogonal curvilinear coordinate systems** – Cylindrical and spherical coordinates are examples of *orthogonal curvilinear systems*. In such coordinate systems we can obtain expressions for the gradient, divergence and curl by using the following general procedure. Let r be some point in space and q_i represent the coordinate system, we can define three unit vectors $\mathbf{e}_i$ which are orthogonal to each other by evaluating the partial derivatives:

$$\mathbf{e}_i = \frac{\partial \mathbf{r}}{\partial q_i} \Big/ \left|\frac{\partial \mathbf{r}}{\partial q_i}\right| \qquad \mathbf{e}_i \cdot \mathbf{e}_j = 0 \quad (i = 1, 2, 3 \quad i \neq j)$$

   (a) Find expressions for the unit vectors $\mathbf{e}_i = \left\{\hat{\rho}, \hat{\phi}, \hat{\mathbf{k}}\right\}$ in cylindrical coordinates

   (b) Find expressions for the unit vectors $\mathbf{e}_i = \left\{\hat{r}, \hat{\theta}, \hat{\phi}\right\}$ in spherical coordinates.

17. **Particle traveling along a line** – A particle moves along the line

$$\frac{x - 2}{4} = \frac{y + 1}{-3} = \frac{z - 1}{-2}$$

find the velocity of the particle. At what time t_{min} does the particle get the closest to the origin? Plot the path of the particle $\mathbf{r}(t)$. On same graph as the particle's path, plot the vector $\mathbf{r}(t_{min})$. Include dots on the graph to show the origin and the location $\mathbf{r}(t_{min})$.

18. **Distance between two particles** – Particle 1 moves along the line described by $\mathbf{r}_1 = (1,3,2) + (1,1,1)t$. Particle 2 moves along the line $\mathbf{r}_1 = (1 - t^2, 2t, -2 + t^3)$. Plot the path of each particle on the same graph. What is the distance between the two particles as a function of time? At what time are the two particles the closest together (i.e. when is their distance a minimum)? On the same graph as the paths, plot the line that connects the curves at the time of their closest approach.

19. **The angle between two planes** – Find the angle between the two planes $2x - 6y + 3z = 10$ and $-x + y - z = 0$.

20. **A point charge above a plane** – A point charge is located at $(1, 3, 2)$ above the plane $x + y + z = 3$. What is the shortest distance between the point and the plane? Using Python, plot the plane and include a point at the charge's location. Include in the plot, the line alone the shortest distance between the charge and the plane.

21. **Conservation of angular momentum** – The force on a mass m is given by $\mathbf{F} = 3\mathbf{r}$ where $\mathbf{r}$ is the position vector. Show that the angular momentum $\mathbf{L} = \mathbf{r} \times \mathbf{p}$ is conserved, i.e. its time derivative is zero. Here $\mathbf{p}$ is the momentum vector defined by $\mathbf{p} = m\mathbf{v}$ where $\mathbf{v}$ is the velocity vector.

22. **Cross products in cylindrical and spherical coordinates** – Calculate the following cross products between unit vectors. Use Python to help with the algebra.

 a. $\hat{\boldsymbol{\rho}} \times \hat{\boldsymbol{\phi}}$, $\hat{\boldsymbol{\rho}} \times \hat{\mathbf{z}}$, and $\hat{\boldsymbol{\phi}} \times \hat{\mathbf{z}}$ in cylindrical coordinates.

 b. $\hat{\mathbf{r}} \times \hat{\boldsymbol{\phi}}$, $\hat{\mathbf{r}} \times \hat{\boldsymbol{\theta}}$, and $\hat{\boldsymbol{\theta}} \times \hat{\boldsymbol{\phi}}$ in spherical coordinates

5 Multiple Integrals

In this chapter we review integrals which are defined in multiple dimensions, and how to evaluate them by hand and using Python. We present several applications of multiple integrals in various areas of science, and show how the use of polar and spherical coordinates can simplify the evaluation of complex integrals. Specific examples are considered from undergraduate courses in Mechanics (e.g. evaluation of the moment of inertia tensor), in Electromagnetism (e.g. Coulomb forces and potentials for various configurations), and also in gravitational forces and potentials. Multiple integrals are evaluated both symbolically using SymPy, and numerically using SciPy libraries.

5.1 MULTIPLE INTEGRALS

In a previous chapter we discussed the definite integral of a function $f(x)$ of a single real variable x in an interval $[a, b]$ along the x-axis,

$$\int_a^b f(x)\, dx \tag{5.1.1}$$

We saw how the definite integral is defined informally as the net area of the region in the xy-plane bounded by the graph of $f(x)$, the x-axis, and the vertical lines $x = a$ and $x = b$, which define the domain of integration D, such that D is the interval $a \leq x \leq b$.

Mathematically we considered the definite integral as the limit of an infinite sum, which is obtained by dividing the interval [a,b] into N sub-intervals $[x_{i-1},\ x_i]$ and evaluating the function $f(x)$ at an arbitrary point x_i^* located in this sub-interval:

$$\int_a^b f(x)\, dx = \lim_{N \to \infty} \sum_{i=1}^N f(x_i^*)\, (x_i - x_{i-1}) \tag{5.1.2}$$

As the number of sub-intervals $N \to \infty$, the width of the sub-interval $\Delta x = x_i - x_{i-1}$ will become negligibly small i.e. $\Delta x \to 0$, and the infinite sum will converge to the value of the definite integral.

We now extend these concepts to the mathematical and geometrical description of integrals in higher dimensions. The *double integral* is an extension to functions of two variables $f(x, y)$ defined over the domain D, where now D is the area defined by the limits of each variable of integration. Symbolically we write:

$$I = \iint_D f(x, y)\, dx\, dy \tag{5.1.3}$$

By analogy with (5.1.2), we can consider the double integral as the limit of a double infinite sum in the form:

$$\iint_D f(x, y)\, dx\, dy = \lim_{N,\, M \to \infty} \sum_{i=1}^N \sum_{j=1}^M f(x_i^*, y_j^*)\, \Delta x\, \Delta y \tag{5.1.4}$$

where we defined a small rectangular area $\Delta A = \Delta x\, \Delta y$ in the domain of integration D, and (x_i^*, y_j^*) is an arbitrary point located in this sub-interval ΔA. As the number N, M of the intervals $\Delta A = \Delta x\, \Delta y$ goes to infinity, the widths of the sub-interval will become

DOI: 10.1201/9781003294320-5

negligibly small i.e. $\Delta A \to 0$, and the infinite sum will converge to the value of the double definite integral.

The double integral can have a geometrical meaning. Specifically, in analogy to the definite integral $\int_D f(x)\,dx$ representing the *area* of the region between the graph of the positive function $f(x)$ and the x-axis, the double integral $\iint_D f(x,y)\,dx\,dy$ represents the *volume* of the region between the surface defined by the positive function $z = f(x,y)$ and the plane which contains its domain of integration D.

This is shown schematically in Figure 5.1, where we consider a tall box above the small rectangular area $\Delta A = \Delta x\,\Delta y$ located on the domain D on the xy-plane. The box extends to the surface $z = f(x,y)$, and its volume ΔV is given by the product of its height times its base, i.e. $\Delta V = f(x,y)\,\Delta A$. The total volume between the surface $z = f(x,y)$ and the plane containing the domain D is the sum of these tall boxes, and represents the double integral $\iint_D f(x,y)\,dx\,dy$.

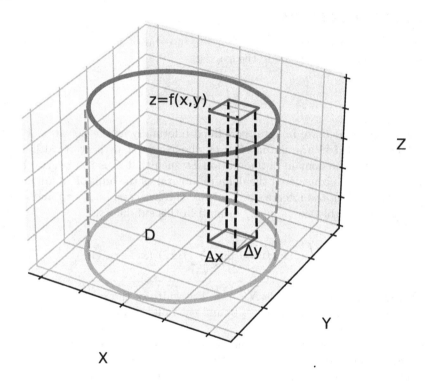

Figure 5.1 A double integral is interpreted geometrically as the volume between the plane containing the domain of integration D and the surface area $z = f(x,y)$. The double integral $\iint f(x,y)\,dx\,dy$ can be calculated as the sum of parallelopipeds of base $\Delta x\,\Delta y$ and height $z = f(x,y)$.

When $f(x,y) = 1$, the double integral gives the *area* of the domain D:

$$\text{Area of } D = \iint_D dx\,dy \tag{5.1.5}$$

Example 5.1 demonstrates how to plot a surface $z = f(x,y)$ in three-dimensions and the corresponding rectangular domain D.

Example 5.1:   Plotting a surface $z = f(x, y)$ **and the rectangular domain** D

Write a Python code to plot the surface function $z = f(x, y) = \sin(x) + \sin(y) + 5$ in three dimensions over the domain D defined by $-3 \leq x \leq 3$ and $-3 \leq y \leq 3$. Plot also the rectangular domain D.

Solution:
The three-dimensional plot in the example is created using the `plot_surface()` function within the `matplotlib` library. The lines `fig = plt.figure()` and `fig.add_subplot(projection ='3d')` create the three-dimensional plot.
The NumPy command `meshgrid(x, y)` creates double arrays X,Y from the single NumPy arrays x and y. It is necessary to create double arrays to be used as inputs for the function `plot_surface(X,Y,Z)`.
The plotting function `plot_surface(X,Y,Z)` plots the three-dimensional surface $z = f(x, y)$, as shown in Figure 5.2. In order to plot the rectangular domain D on the xy-plane in this example, we call `plot_surface(X,Y,Z)` with the Z-variable set to a zero double array.

```python
import numpy as np
import matplotlib.pyplot as plt

# surface plot for z=5+sin(x)+sin(y)
x = np.arange(-3, 3, 0.6)            # x-values
y = np.arange(-3, 3, 0.6)  # y-values

X, Y = np.meshgrid(x,y)              # create grid values (X,Y)
Z = 5+np.sin(X)+np.sin(Y)           # evaluate Z values from equation

fig = plt.figure()
axes = fig.add_subplot(projection ='3d')       # create 3D plot

# plot two 3D surfaces:  using (X,Y,Z) and on xy-plane (Z=0)
axes.plot_surface(X, Y, Z)
axes.plot_surface(X, Y, 0*X)

axes.set_xlabel('X')
axes.set_ylabel('Y')
axes.set_zlabel('Z')

axes.text(-2.6,2,6,'z=f(x,y)')
axes.text(1,-2,0,'D')
plt.show()
```

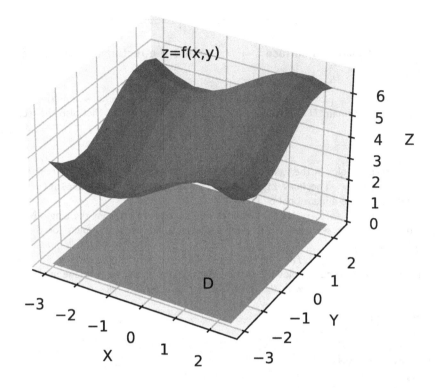

Figure 5.2 The graphical output from Example 5.1, showing the plot of a surface $z = f(x, y)$ in three-dimensions, and the corresponding rectangular domain D.

Multiple integrals have similar properties to integrals of functions of a single variable. For example, when the integrand is a constant function c multiplied by a function, the integral is equal to the product of c and the integral of the function.

$$\iint_D c\, f(x, y)\, dx\, dy = c \iint_D f(x, y)\, dx\, dy \tag{5.1.6}$$

Double integrals also have a linearity property:

$$I = \iint_D [f(x, y) + g(x, y)]\, dx\, dy = \iint_D f(x, y)\, dx\, dy + \iint_D g(x, y)\, dx\, dy \tag{5.1.7}$$

If the domain D can be broken up into two separate regions D_1 and D_2, then we can write:

$$I = \iint_D f(x, y)\, dx\, dy = \iint_{D_1} f(x, y)\, dx\, dy + \iint_{D_2} f(x, y)\, dx\, dy \tag{5.1.8}$$

In the example of Figure 5.1 the domain D is defined in two dimensions, and the double integral of $f(x, y)$ on the domain D was written as:

$$I = \iint_D f(x, y)\, dx\, dy \tag{5.1.9}$$

By generalizing, when the domain D is defined in three-dimensions, the *triple integral* of $f(x, y, z)$ on the domain D is written as:

$$I = \iiint_D f(x, y, z)\, dx\, dy\, dz \tag{5.1.10}$$

When $f(x, y, z) = 1$ this triple integral gives the *volume* of the domain D:

$$\text{Volume of } D = \iiint_D dx\, dy\, dz \tag{5.1.11}$$

More generally, we can define the multiple integral of a function $f(x_1, x_2, \ldots, x_n)$ of n variables over a domain D, as a series of nested (or iterated) integral signs, with the leftmost integral sign computed last.

$$\int \cdots \int_D f(x_1, x_2, \ldots, x_n)\, dx_1 \cdots dx_n \tag{5.1.12}$$

5.2 EVALUATION OF DOUBLE INTEGRALS

In most of cases, evaluating multiple integrals consists of finding a way to reduce them to a series of integrals of one variable. In this section we outline the general method of setting up multiple integrals, and give examples of evaluating double integrals. There are two general cases for multiple integrals common in physics. The first one occurs when the limits of integration are independent of each other. We will refer to this case as having a *rectangular domain*. The second case occurs when one or more of the limits are a function of the others. We will refer to this case as having a *non-rectangular domain*.

5.2.1 THE EVALUATION OF DOUBLE INTEGRALS OVER A RECTANGULAR DOMAIN

In many cases, the value of an integral is independent of the order of integrands. This occurs when the domain of integration D has bounds which are independent of each other. For example, the domain D in Cartesian coordinates defined by the regions $-1 \leq x \leq 1$ and $-2 \leq y \leq 0$ satisfy the condition for being a rectangular domain, because the bounds of each variable are constant. In such cases, we can reverse the order of the variables x, y in a double integral:

$$\iint_D f(x, y)\, dx\, dy = \iint_D f(x, y)\, dy\, dx \tag{5.2.1}$$

Let us now consider the more general case where the domain D is rectangular, defined by the parameter x in the interval $[a, b]$, and the parameter y in $[c, d]$. If the function $f(x, y)$ is continuous on the domain D, then:

$$I = \int_{x=a}^{x=b} \int_{y=c}^{y=d} f(x, y)\, dy\, dx \tag{5.2.2}$$

These nested integrals are also commonly referred to as iterated integrals. We can also change the notation and write:

$$I = \int_{x=a}^{x=b} \left[\int_{y=c}^{y=d} f(x, y)\, dy \right] dx \tag{5.2.3}$$

We can now compute the square bracket by keeping the parameter x constant and integrating with respect to y in the interval $[c, d]$, by following the rules for a integrals of a single variable. This will give a function involving only the parameter x, which can then be integrated in the interval $[a, b]$.

Example 5.2 shows how to set up and evaluate a double integral over a rectangular domain D defined on the xy-plane.

Example 5.2: Evaluating the mass M of an object with non-uniform density

The mass of a two-dimensional object with a non-uniform surface mass density $\rho(x, y)$ defined in a region D of the xy-plane, is given by:

$$M = \int \int_D \rho(x, y) \, dy \, dx \qquad (5.2.4)$$

Evaluate the total mass M of the rectangular region D defined by $2 \leq x \leq 4$ and $3 \leq y \leq 6$, when the density is proportional to the distance y, i.e. $\rho(x, y) = a \, y$, where a is a constant.

Solution:
In this case the domain of integration is rectangular, so we can perform each integration independently of the other. We can attempt the innermost y-integration first, from $y = 3$ to $y = 6$, while keeping x constant. By applying (5.2.3):

$$M = \int \int_D \rho(x, y) \, dy \, dx = a \int \int_D y \, dy \, dx \qquad (5.2.5)$$

$$M = a \int_{x=2}^{x=4} \left[\int_{y=3}^{y=6} y \, dy \right] dx = a \int_{x=2}^{x=4} \left[\frac{y^2}{2} \right]_{y=3}^{y=6} dx$$

$$M = a \int_{x=2}^{x=4} \left[\frac{6^2}{2} - \frac{3^2}{2} \right] dx = a \frac{27}{2} \int_{x=2}^{x=4} dx = a \frac{27}{2} [x]_{x=2}^{x=4} = 27a$$

In the Python code, the `integrate()` command from SymPy is used to evaluate the double integral over the rectangular domain. Note that as expected, the order of integration does not matter, i.e. `integrate(f,(y,3,6),(x,2,4))` produces the same result as `integrate(f,(x,2,4),(y,3,6))`.

```
from sympy import symbols, integrate
print('-'*28,'CODE OUTPUT','-'*29,'\n')
a, x, y = symbols('a, x, y',real=True)    # define symbols


f = a*y
res1 = integrate(f, (y, 3, 6), (x, 2, 4)) # evaluate the integral
print('Double Integral of f(x,y)=y in the intervals [2,4],[3,6] =',res1)

res2 = integrate(f, (x, 2, 4), (y, 3, 6)) # reverse integration order
print('Double Integral in reverse order of integration =',res2)

-------------------------- CODE OUTPUT ---------------------------

Double Integral of f(x,y)=y in the intervals [2,4],[3,6] = 27*a
Double Integral in reverse order of integration = 27*a
```

An interesting special case of double integration in a rectangular domain is when the function $f(x, y)$ is the product of two functions $g(x)$ and $h(y)$, i.e. $f(x, y) = g(x) h(y)$. In this case we can write from (5.2.3):

$$I = \int_{x=a}^{x=b} \left[\int_{y=c}^{y=d} g(x)\, h(y)\, dy \right] dx = \int_{x=a}^{x=b} g(x) \left[\int_{y=c}^{y=d} h(y)\, dy \right] dx \qquad (5.2.6)$$

Since $\int_{y=c}^{y=d} h(y)\, dy$ is constant in x, we can extract it in front of the integral:

$$I = \left(\int_{x=a}^{x=b} g(x)\, dx \right) \left[\int_{y=c}^{y=d} h(y)\, dy \right] \qquad (5.2.7)$$

This equation shows that when $f(x, y) = g(x) h(y)$, we can do each of the two integrals separately, and then multiply them together to evaluate the double integral I.

5.2.2 THE EVALUATION OF DOUBLE INTEGRALS OVER A NON-RECTANGULAR DOMAIN

We can generalize the results of the previous section to double integrals evaluated over a domain D with arbitrary shape. Figure 5.3 shows two general types of shapes for the domain D.

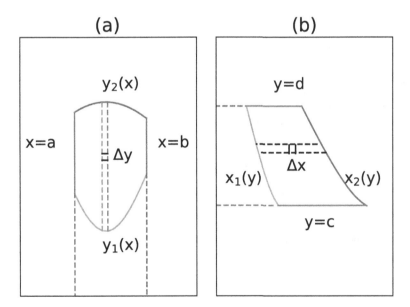

Figure 5.3 Examples of setting up the limits for a double integral. (a) The domain D is defined by the straight lines $x = a$, $x = b$ and the curves $y_1(x)$, $y_2(x)$. (b) D is defined by the straight lines $y = c$, $y = d$ and the curves $x_1(y)$, $x_2(y)$.

An important question when evaluating a double integral $\iint_D f(x,y)\,dx\,dy$ concerns how to choose the order of integration, i.e. is it better to integrate over x first, or over y first? The answer depends on the shape of the domain D. In general we draw small rectangles $dx\,dy$ and we form strips inside the domain D, and then combine the strips to form the shape of the domain D.

For example, for the shapes shown in Figure 5.3a, it is best to choose strips along the y axis, which extend between the curves $y_1(x)$ and $y_2(x)$, and to perform the y-integration first. Symbolically we can write that in these situations:

$$\iint_D f(x,y)\,dx\,dy = \int_a^b dx \int_{y_1(x)}^{y_2(x)} f(x,y)\,dy \qquad (5.2.8)$$

By contrast, in the shapes shown in Figure 5.3b, it is best to choose strips along the x axis, which extend between the curves $x_1(y)$ and $x_2(y)$, and to perform the x-integration first. Symbolically we can write that in these situations:

$$\iint_D f(x,y)\,dx\,dy = \int_c^d dy \int_{x_1(y)}^{x_2(y)} f(x,y)\,dx \qquad (5.2.9)$$

Example 5.3 shows an example of evaluating a double integral over a non-rectangular domain D defined on the xy-plane.

Example 5.3: The total mass of a 2D object with non-uniform density

A two-dimensional object has the shape of a parabola and is defined by the parabola $y = x^2$, the line $y = 1$, and the positive y-axis. Calculate the total mass of this object with a non-uniform density $\rho(x,y) = k\,(x+y)$ where k is a constant.

Solution:
As in the previous example, we must evaluate the following double integral over the domain D:

$$M = \int \int_D \rho(x,y)\,dy\,dx = k \int \int_D (x+y)\,dy\,dx \qquad (5.2.10)$$

The plot of the domain D on the xy-plane is shown below in Figure 5.4. Here the region D is non-rectangular and we can choose to do either the x or the y integration first. If we choose to do the y-integration first, we draw a strip parallel to the y-axis, extending between the two functions $y_1(x) = x^2$ and $y_2(x) = 1$.
The y integration will be done with limits defined by $y_1(x)$ and $y_2(x)$, while the integration interval along the x-axis is $[a,b] = [0,1]$.

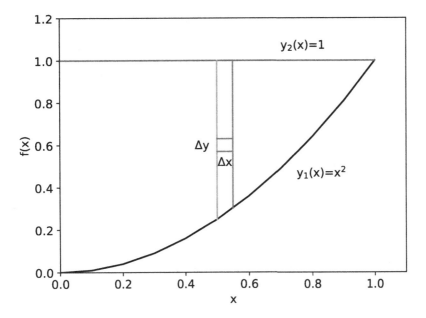

Figure 5.4 Evaluating a double integral over a non-rectangular domain D defined on the xy-plane, as in Example 5.3.

It is now possible to apply (5.2.8):

$$M = k \int_a^b dx \int_{y_1(x)}^{y_2(x)} f(x,y)\, dy = k \int_0^1 dx \left[\int_{x^2}^1 (x+y)\, dy \right] \qquad (5.2.11)$$

The second integral inside the square brackets is calculated by considering x a constant, and the remaining operations consist of applying the basic techniques of integration:

$$M = k \int_0^1 dx \left[x\,y + \frac{y^2}{2} \right]_{x^2}^1 = k \int_0^1 \left(x + \frac{1}{2} - x^3 - \frac{x^4}{2} \right) dx = \frac{13}{20} k \qquad (5.2.12)$$

If we choose to do the x-integration first, we can use a strip parallel to the x-axis extending between the functions $x_1(y) = 0$ and $x_2(y) = \sqrt{y}$, and we apply (5.2.9) to obtain:

$$M = k \int_a^b dy \int_{x_1(y)}^{x_2(y)} f(x,y)\, dx = \int_0^1 dy \int_0^{\sqrt{y}} (x+y)\, dx = \frac{13}{20} k \qquad (5.2.13)$$

to obtain the same numerical value.

In the Python code, we use the `integrate()` command to carry out the integrations using both methods, obtaining the same result.

```
from sympy import symbols, integrate, sqrt
print('-'*28,'CODE OUTPUT','-'*29,'\n')


x, y, k = symbols('x, y, k',real=True)    # define symbols
```

```
f = k*(x+y)                          # density function to integrate

integraly = integrate(f,  (y, x**2, 1), (x,0,1))    # vary y first
integralx = integrate(f,  (x, 0, sqrt(y)), (y,0,1)) # vary x first

print('Integral of f=x+y with y-integration first is =',integraly)
print('Integral of f=x+y with x-integration first is:',integralx)

------------------------- CODE OUTPUT ---------------------------

Integral of f=x+y with y-integration first is = 13*k/20
Integral of f=x+y with x-integration first is: 13*k/20
```

5.3 EVALUATION OF TRIPLE INTEGRALS

Triple integrals can also have rectangular and non-rectangular domains. In this section, we will focus on non-rectangular domains, because the rectangular case is very similar to that of double integrals.

Example 5.4 shows an example of evaluating a triple integral over a three-dimensional domain D defined by the intersection of several planes. Specifically, we evaluate the total charge inside a solid with a uniform volume charge density.

Example 5.4: Total electric charge Q in a non-uniform charge density

Consider the triangular prism shown below enclosed by the planes $z = 1$, $x = y = z = 0$, and $x+y = 1$. Plot the solid in three-dimensions by hand and calculate the total charge $Q = \iiint \rho dV$ inside it, when the volume charge density is given by $\rho = a\,z$, where a is a constant.

Solution:
We must evaluate the triple integral:

$$Q = \iiint \rho \, dV = a \iiint z \, dV = a \iiint_D z \, dx \, dy \, dz \qquad (5.3.1)$$

The plot of the planes $z = 1$, $x = 0$, $y = 0$ and $x + y = 1$ is shown in Figure 5.5. We consider a small volume $\Delta V = \Delta x\, \Delta y\, \Delta z$ as shown in the figure. Since the top surface $z = 1$ does not depend on x or y, the z-integration will be easy to carry out from $z = 0$ to $z = 1$. We can choose for example to do the x integration first, followed by the y-integration and finally the z-integration.

For the x-integration we take a strip parallel to the x-axis shown with black color in the xy-plane. The limits along the x direction are defined by the end points of the strip, in this case $x = 0$ and $x = 1 - y$ (since $x + y = 1$ along the diagonal line on the xy-plane).

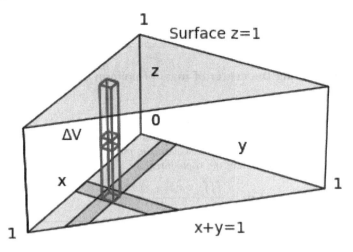

Figure 5.5 Evaluating a triple integral over a three-dimensional domain D, which is defined by the intersection of several planes, from Example 5.4.

$$Q = a \int_0^1 z\, dz \int_{y=0}^{y=1} dy \int_{x=0}^{x=1-y} dx = a \int_0^1 z\, dz \int_{y=0}^{y=1} dy\, [x]_{x=0}^{x=1-y} = a \int_0^1 z\, dz \int_{y=0}^{y=1} (1-y)dy \qquad (5.3.2)$$

$$Q = a \int_0^1 z\, dz \left[y - \frac{y^2}{2} \right]_{y=0}^{y=1} = a \int_0^1 z\, dz \left(\frac{1}{2} \right) = a \left(\frac{1}{2} \right) \left[\frac{z^2}{2} \right]_{z=0}^{z=1} = \frac{a}{4} \qquad (5.3.3)$$

The Python code uses the `integrate()` function to evaluate the triple integral, note that the order of the x and y integrations (x,0,1-y),(y,0,1),(z,0,1) is important in this function. The same result is obtained by evaluating (y,0,1-x),(x,0,1),(z,0,1), which is the order of integration if we used the strip in Figure 5.5 parallel to the y-axis instead.

```
from sympy import symbols, integrate

print('-'*28,'CODE OUTPUT','-'*29,'\n')

x, y, z, a = symbols('x, y, z, a',real=True)  # define symbols

f = a*z    # function to integrate

integralxy = integrate(f,  (x, 0, 1-y), (y,0,1),(z,0,1))  # x-then-y
integralyx = integrate(f,  (y, 0, 1-x), (x,0,1),(z,0,1))  # y-then-x

print('Charge inside the prism (integrate x then y)=',integralxy)
print('Charge inside the prism (integrate y then x)=',integralyx)

---------------------------- CODE OUTPUT ----------------------------

Charge inside the prism (integrate x then y)= a/4
Charge inside the prism (integrate y then x)= a/4
```

Example 5.5 shows the evaluation of the center of mass of a triangular pyramid.

Example 5.5: Evaluating the center of mass of uniform 3D object

Consider the triangular pyramid shown in Figure 5.6, with total mass M and defined by the xz-plane, yz-plane, xy-plane, and $x + y + z = 1$. The uniform mass density of the pyramid is $\rho(x, y, z) = a$, where a is a constant. Calculate the mass M and the z-component of the pyramid's center of mass by evaluating the triple integrals:

$$Z_{CM} = \frac{\iiint_D z\,\rho(x,y,z)\,dx\,dy\,dz}{M} \tag{5.3.4}$$

$$M = \iiint_D \rho(x,y,z)\,dx\,dy\,dz \tag{5.3.5}$$

Solution:
As in the previous example, we choose a small volume $\Delta V = \Delta x\,\Delta y\,\Delta z$ as shown in the figure. We can choose for example to do the z integration first along the direction of the tall box, with the limits of the variable z being $z = 0$ to a point with $z = 1 - x - y$ on the side surface (since $x + y + z = 1$ for this surface).
For the x-integration we take a strip parallel to the x-axis shown with black color in the xy-plane, with the limits $x = 0$ and $x = 1 - y$ (since $x + y = 1$ along the diagonal line). Finally, the limits in the y-direction are $y = 0$ to $y = 1$.

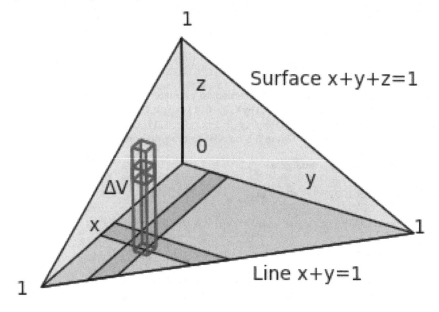

Figure 5.6   Evaluating the center of mass of a triangular pyramid, from Example 5.5.

We need to find the total mass M by evaluating the triple integral:

$$M = \iiint_D \rho(x,y,z)\,dx\,dy\,dz = a \int_{y=0}^{y=1} dy \int_{x=0}^{x=1-y} dx \int_{z=0}^{z=1-x-y} dz \tag{5.3.6}$$

$$M = a \int\limits_{y=0}^{y=1} dy \int\limits_{x=0}^{x=1-y} (1-x-y)\,dx = a \int\limits_{y=0}^{y=1} dy \left[x - \frac{x^2}{2} - xy \right]_{x=0}^{x=1-y} \qquad (5.3.7)$$

$$M = a \int\limits_{y=0}^{y=1} dy \left[(1-y) - \frac{(1-y)^2}{2} - (1-y)\,y \right] = \frac{a}{6} \qquad (5.3.8)$$

Similarly we evaluate the second triple integral Z_{CM} :

$$Z_{CM} = \frac{1}{M} \iiint_D z\,\rho(x,y,z)\,dx\,dy\,dz = \frac{a}{M} \int\limits_{y=0}^{y=1} y\,dy \int\limits_{x=0}^{x=1-y} x\,dx \int\limits_{z=0}^{z=1-x-y} z\,dz \qquad (5.3.9)$$

$$Z_{CM} = \frac{a}{M} \int\limits_{y=0}^{y=1} y\,dy \int\limits_{x=0}^{x=1-y} x\,dx \left[\frac{z^2}{2} \right]_{z=0}^{z=1-x-y} = \frac{a}{2M} \int\limits_{y=0}^{y=1} y\,dy \int\limits_{x=0}^{x=1-y} x\,dx \left[\frac{(1-x-y)^2}{2} \right] = \frac{1}{4}$$
$$(5.3.10)$$

The Python code uses the `integrate()` function to evaluate the two triple integrals. Note that the order of the integrations `(z,0,1-x-y),(x,0,1-y),(z,0,1)` is important in this example.

```
from sympy import symbols, integrate
print('-'*28,'CODE OUTPUT','-'*29,'\n')
x, y, z, a= symbols('x, y, z, a',real=True)  # define symbols

f = a*z   # function to integrate

integral1 = integrate(f,  (z,0,1-x-y),(x, 0, 1-y), (y,0,1)) # find ZCM
M = integrate(a,  (z,0,1-x-y),(x, 0, 1-y), (y,0,1))  # find mass M

print('Total mass M =',M)
print('Z-location of Center of mass of prism =',round(integral1/M,2))

-------------------------- CODE OUTPUT ---------------------------

Total mass M = a/6
Z-location of Center of mass of prism = 0.25
```

5.4   CHANGE OF VARIABLES IN MULTIPLE INTEGRALS

The evaluation of multiple integrals can be simplified in many situations by taking advantage of the symmetry of the problem. In this section we look specifically at how to apply the transformation of variables from Cartesian to polar, or cylindrical, or spherical polar coordinates.

5.4.1   USING POLAR COORDINATES

If the integration domain D has a circular symmetry, and the function has some particular symmetry characteristics, we can transform the variable from Cartesian to polar coordinates. This type of transformation allows one to change the shape of the domain, and to hopefully simplify the operations. In Chapter 4 we saw the transformation equations:

$$x = \rho \cos \phi \qquad (5.4.1)$$

$$y = \rho \sin \phi \tag{5.4.2}$$

$$f(x, y) \to f(\rho \cos \phi, \rho \sin \phi) \tag{5.4.3}$$

We also saw that the differential $dx\, dy$ transformed in the corresponding differential in polar coordinates:

$$dx\, dy = \rho\, d\rho\, d\phi \tag{5.4.4}$$

Example 5.6 shows how to set up and evaluate a double integral over a domain D in polar coordinates.

Example 5.6: The gravitational force exerted by a disk on point mass m

A mass m is located at a distance a above the center of a uniform thin circular disk of radius R and total mass M. Find the gravitational force of attraction exerted by the disk on the mass m, by using Newton's universal law of gravity between two masses m_1 and m_2 located a distance r apart:

$$\boldsymbol{F} = G\frac{m_1 m_2}{r^2}\hat{\boldsymbol{r}} \tag{5.4.5}$$

Here G in the gravitational constant and $\hat{\boldsymbol{r}}$ is the unit vector along the line connecting the two masses m_1 and m_2.

Solution:
We subdivide the circular disk into small areas $dA = dx\, dy$ as shown in Figure 5.7. The area dA contains a mass $dM = \sigma\, dx\, dy$ where σ is the mass area density of the disk defined by $\sigma = M/\left(\pi R^2\right)$, and M is the total mass of the disk.

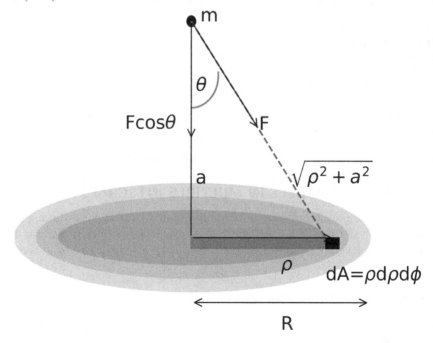

Figure 5.7 Setting up a double integral evaluation over a domain D defined in polar coordinates ρ and ϕ, from Example 5.6.

Because of the symmetry of the problem, the components of the force along the horizontal direction will cancel out. The total gravitational force will be pointing downwards as in the figure, and we need to evaluate only the vertical component $F \cos \theta$ of the force. Using $\cos \theta = a/\sqrt{a^2 + \rho^2}$ we find:

$$dF = G \frac{m \, dM}{a^2 + \rho^2} \cos \theta = G \, a \, \frac{m \, \sigma \, dx \, dy}{(a^2 + \rho^2)^{3/2}} \tag{5.4.6}$$

We transform into polar coordinates using $dx \, dy = \rho \, d\rho \, d\phi$, with the radial variable ρ ranging from $\rho = 0$ to $\rho = R$, and the polar angle from $\phi = 0$ to $\phi = 2\pi$.

$$F = \iint G \, a \frac{m \, \sigma \, dx \, dy}{(a^2 + \rho^2)^{3/2}} = A \, a \int_{\rho=0}^{\rho=R} \frac{\rho}{(a^2 + \rho^2)^{3/2}} \, d\rho \int_{\phi=0}^{\phi=2\pi} d\phi$$

where the constant $A = G \, m \, \sigma$. Integrating we obtain:

$$F = A \, 2\pi \left(1 - \frac{a}{\sqrt{a^2 + R^2}} \right)$$

In the Python code, the `integrate()` command from SymPy is used to evaluate the integral. By transforming from Cartesian to polar coordinates, the integrals over the coordinates ρ, ϕ can be evaluated independently of each other.

```
from sympy import symbols, integrate, simplify, pi
print('-'*28,'CODE OUTPUT','-'*29,'\n')

# define symbols
rho, a, R, phi, A = symbols('rho,  a, R, phi, A',real=True,positive=True)


f = a* rho/((rho**2+a**2)**(3/2))              # function to integrate

res1 = integrate(f, (rho,0,R),(phi,0,2*pi))

print('The gravitational force F =',A*simplify(res1))

--------------------------- CODE OUTPUT ----------------------------

The gravitational force F = A*(-2.0*pi*a/(R**2 + a**2)**0.5 + 2.0*pi)
```

5.4.2   USING CYLINDRICAL COORDINATES

As we saw in Chapter 4, the transformation equations between Cartesian and cylindrical coordinates (ρ, ϕ, z) are:

$$x = \rho \cos \phi \tag{5.4.7}$$

$$y = \rho \sin \phi \tag{5.4.8}$$

$$f(x, y, z) \rightarrow f(\rho \cos \phi, \rho \sin \phi, \, z) \tag{5.4.9}$$

$$dx \, dy \, dz = \rho \, d\rho \, d\phi \, dz \tag{5.4.10}$$

Example 5.7 shows how to set up and evaluate a triple integral using cylindrical coordinates.

Example 5.7: Moment of inertia of a uniform cone

A uniform cone of height H and radius R is rotated around its symmetry axis. Evaluate the moment of inertia I with respect to the rotational axis:

$$I = \iiint r^2 \, dm \qquad (5.4.11)$$

where r is the distance of the infinitesimal mass dm from the axis of rotation.

Solution:
We subdivide the cone into small volumes $dV = dx\, dy\, dz$, these volumes contain a mass $dm = d(x,y,z)\, dx\, dy\, dz$ where $d(x,y,z)$ is the density of the cone defined by $d = M/V$, where M and V are the total mass and volume of the cone, respectively. Because of the symmetry of the problem, we transform into cylindrical coordinates using $dx\, dy\, dz = \rho\, d\rho\, d\phi\, dz$ with the variable z ranging from $z = 0$ to $z = H$, and ϕ from $\phi = 0$ to $\phi = 2\pi$.
The domain is non-rectangular because, as seen in Figure 5.8, the distance ρ of the infinitesimal mass dm from the axis of rotation depends on the value of the z-coordinate, and z ranges from $z = 0$ to $z = H\rho/R$. The triple integral is evaluated over the volume of the cone.

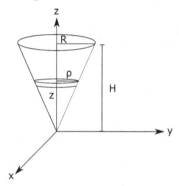

Figure 5.8 Setting up a triple integral evaluation for the cone in Example 5.7, using cylindrical coordinates.

and (5.4.11) gives:

$$I = \frac{M}{V} \iiint r^2 \, dx\, dy\, dz = \frac{M}{V} \iiint \rho^2\, \rho\, d\rho\, d\phi\, dz = \frac{M}{V} \iiint \rho^3\, d\phi\, d\rho\, dz \qquad (5.4.12)$$

$$I = \frac{M}{V} \int_{z=0}^{H} \int_{\theta=0}^{2\pi} \int_{\rho=0}^{Rz/H} \rho^3\, d\phi\, d\rho\, dz \qquad (5.4.13)$$

$$I = \frac{M}{V}\, 2\pi \int_{z=0}^{h} \int_{\rho=0}^{Rz/H} \rho^3\, d\rho\, dz = \frac{M}{V}\, 2\pi \int_{z=0}^{H} \left(\frac{R^4 z^4}{4H^4} \right) dz \qquad (5.4.14)$$

Using the volume of the cone $V = \pi R^2 H/3$ we find:

$$I = \frac{3M}{\pi R^2 H}\, 2\pi \left[\frac{R^4 z^5}{20H^4} \right]_{z=0}^{z=H} = \frac{3M}{\pi R^2 H}\, 2\pi \left[\frac{R^4 H^5}{20H^4} \right] \qquad (5.4.15)$$

$$I = \frac{3}{10} M R^2 \qquad (5.4.16)$$

In the Python code, the `integrate()` command from SymPy is used to evaluate the triple integral.

```
from sympy import symbols, Poly, integrate, pi
print('-'*28,'CODE OUTPUT','-'*29,'\n')

# define symbols
rho, phi, z, H, R, m = symbols('rho, phi, z, H, R, m ',real=True)

I = integrate(rho**3,(z,rho*H/R,H),(rho,0,R),(phi,0,2*pi))

print('Moment of inertia tensor for cone: I= ',I*m/(pi*R**2*H/3))

-------------------------- CODE OUTPUT ----------------------------

Moment of inertia tensor for cone: I=  3*R**2*m/10
```

5.4.3   USING SPHERICAL COORDINATES

As we saw in Chapter 4, the transformation from Cartesian to spherical coordinates is:

$$x = r \cos \phi \sin \theta \tag{5.4.17}$$

$$y = r \sin \phi \sin \theta \tag{5.4.18}$$

$$z = r \cos \theta \tag{5.4.19}$$

$$f(x, y, z) \longrightarrow f(r \cos \phi \sin \theta, r \sin \phi \sin \theta, r \cos \theta) \tag{5.4.20}$$

Note that ϕ can vary between 0 to 2π, while θ can vary between 0 to π.

The differential volume element for spherical coordinates is:

$$dV = dx \, dy \, dz = r^2 dr \, \sin \theta \, d\theta \, d\phi \tag{5.4.21}$$

Example 5.8 shows how to use spherical coordinates to evaluate the gravitational force between a point mass m and a hemisphere with mass M and radius R.

Example 5.8: Gravitational force between a hemisphere and mass m

A mass m is located at the center of a hemisphere with uniform mass distribution and density ρ, a total mass M and radius R. Evaluate the gravitational force of attraction exerted by the hemisphere on the mass m, by using Newton's universal law of gravity between two masses m_1 and m_2 located a distance r apart,

$$\boldsymbol{F} = G\frac{m_1 m_2}{r^2}\hat{r} \tag{5.4.22}$$

Here G in the gravitational constant and $\hat{r}$ is the unit vector along the line connecting the two masses m_1 and m_2.

Solution:
We subdivide the hemisphere into small volumes $dV = dx\,dy\,dz$ as shown in Figure 5.9. This volume contains a mass $dM = \rho\,dx\,dy\,dz$.

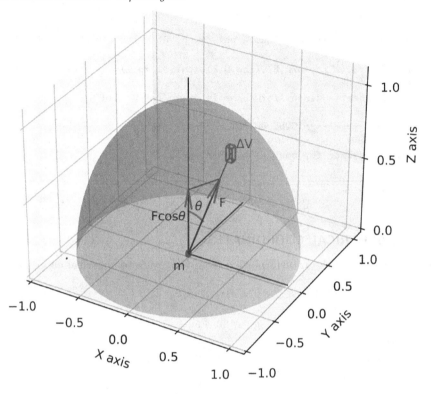

Figure 5.9 Setting up a triple integral evaluation for the hemisphere in Example 5.8, using spherical coordinates.

As in the previous problem, the total gravitational force will point along the z-axis in the figure, due to the symmetry of the problem. We need to evaluate only the vertical (radial) component $F\cos\theta$ of the force:

$$dF = G\frac{m\,dM}{r^2}\cos\theta = G\frac{m\,\rho\,dx\,dy\,dz}{r^2}\cos\theta \tag{5.4.23}$$

Because of the symmetry of the problem, we transform into spherical coordinates using $dx\,dy\,dz = r^2\,dr\,\sin\theta\,d\theta\,d\phi$ with the variable r ranging from $r = 0$ to $r = R$, the polar angle from $\phi = 0$ to $\phi = 2\pi$ and the azimuthal angle $\theta = 0$ to $\theta = \pi/2$.

$$F = G\,m\,\rho\iiint\frac{r^2\,dr\,\sin\theta\,d\theta\,d\phi}{r^2}\cos\theta = G\,m\,\rho\int_{r=0}^{r=R}dr\int_{\phi=0}^{\phi=2\pi}d\phi\int_{\theta=0}^{\theta=\pi/2}\cos\theta\sin\theta\,d\theta \tag{5.4.24}$$

$$F = G\,m\,\rho\,R\,2\pi\int_{\theta=0}^{\theta=\pi/2}\frac{\sin(2\theta)}{2}\,d\theta = G\,m\,\rho\,R\,\pi$$

Substituting the density $\rho = M/V = M/(\frac{2}{3}\pi R^3)$:

$$F = G\,m\,\frac{M}{\frac{2}{3}\pi R^3}\,R\,\pi = 3GmM/(2R^2)$$

In the Python code, the `integrate()` command from SymPy is used to evaluate the triple integral over the hemisphere.

```
from sympy import symbols, integrate, pi, sin, cos
print('-'*28,'CODE OUTPUT','-'*29,'\n')

# define symbols
G, m, M, r, R, theta, phi = symbols('G, m, M, r, R, theta, phi',real=True)

rho = M/((2*pi*R**3)/3)                 # density of hemisphere
f = G*m*rho* cos(theta)*sin(theta)      # function to integrate in spherical

res1 = integrate(f, (r, 0, R), (phi, 0, 2*pi),(theta,0,pi/2))
print('Force on mass m=',res1)

------------------------- CODE OUTPUT ----------------------------

Force on mass m= 3*G*M*m/(2*R**2)
```

5.5 APPLICATION OF MULTIPLE INTEGRALS: MOMENT OF INERTIA TENSOR

In this section, we apply multiple integrals to the description of rotational properties of rigid bodies. *Rigid bodies* are defined as solids in which the particles which make up the solid are at fixed distances from each other, and their rotational properties can be described by the *moment of inertia tensor* **I**. For solids in three dimensions with a continuous mass distribution, the components of the inertia tensor are:

$$\mathbf{I} = \begin{bmatrix} I_{xx} & I_{xy} & I_{xz} \\ I_{yx} & I_{yy} & I_{yz} \\ I_{zx} & I_{zy} & I_{zz} \end{bmatrix} \tag{5.5.1}$$

where the scalar quantities I_{kl} $(k, l = 1, 2, 3$) are the components

$$\left. \begin{aligned} I_{xx} &= \iiint_V \left(y^2 + z^2\right) dm \\ I_{yy} &= \iiint_V \left(x^2 + z^2\right) dm \\ I_{zz} &= \iiint_V \left(x^2 + y^2\right) dm \end{aligned} \right\} \tag{5.5.2}$$

$$I_{xy} = -\iiint_V x\,y\,dm \qquad I_{xz} = -\iiint_V x\,z\,dm \qquad I_{yz} = -\iiint_V y\,z\,dm \tag{5.5.3}$$

where the mass element $dm = \rho\,dx\,dy\,dz$ and $\rho(x, y, z)$ is the mass density of the object. From these equations it is clear that the moment of inertia matrix **I** is symmetric, i.e. $I_{xy} = I_{yx}$, $I_{xz} = I_{zx}$ and $I_{yz} = I_{zy}$. It is important to remember that the elements of the inertia tensor depend on the choice of origin of the coordinate system. The terms I_{xx}, I_{yy}, and I_{zz} are referred to as *moments of inertia* of the rigid body about each axis, and the terms I_{xy}, I_{yz}, I_{xz}, ... are called *products of inertia*.

Example 5.9 shows how to evaluate the elements of the inertia tensor for a square plate with a side a, by using symbolic integration.

Example 5.9: Moment of inertia tensor of a square plate

Consider a uniform square plate of side a of negligible thickness. Evaluate the moment of inertia tensor with respect to a Cartesian coordinate system with the origin at $(0,0)$ located at one of the corners of the plate, and with the xy-axes extending along the sides of the square plate.

Solution:
In order to find the moment of inertia tensor, we first need to choose a coordinate system and identify the limits of integration. In this example we take the z-axis to be perpendicular to the plate, so that $z = 0$, and therefore the elements $I_{xz} = I_{yz} = 0$.

We use Cartesian coordinates for the calculation, with the limits of (x, y) varying from 0 to a. The mass element is $dm = M\, dx\, dy/a^2$ and the moments of inertia elements with $z = 0$ are:

$$I_{xx} = \iint_A \left(y^2 + z^2 \right) dm = \frac{M}{a^2} \int_{x=0}^{a} \int_{y=0}^{a} y^2 dx\, dy$$

$$I_{xx} = \frac{M}{a^2} \int_{x=0}^{a} dx \left[\frac{y^3}{3} \right]_{y=0}^{y=a} = \frac{1}{3} M a^2$$

By symmetry, $I_{xx} = I_{yy} = Ma^2/3$ and we evaluate I_{zz} and I_{xy}:

$$I_{zz} = \iiint_V \left(x^2 + y^2 \right) dm = \frac{M}{a^2} \int_{x=0}^{a} \int_{y=0}^{a} x^2 dx\, dy + \int_{x=0}^{a} \int_{y=0}^{a} y^2 dx\, dy = \frac{2}{3} M a^2$$

$$I_{xy} = -\iint_A x\, y\, dm = -\frac{M}{a^2} \int_{x=0}^{a} x\, dx \int_{y=0}^{a} y\, dy = -\frac{M}{a^2} \int_{x=0}^{a} x\, dx \left[\frac{y^2}{2} \right]_{s0}^{a} = -\frac{1}{4} M a^2$$

$$I = \begin{pmatrix} \frac{1}{3} M a^2 & -\frac{1}{4} M a^2 & 0 \\ -\frac{1}{4} M a^2 & \frac{1}{3} M a^2 & 0 \\ 0 & 0 & \frac{2}{3} M a^2 \end{pmatrix}$$

In the Python code, the `pprint` package is imported and used to format the code output for the printed matrix, and `Matrix` is used to create the matrix with the elements of the inertia tensor.

```
from sympy import symbols, integrate, Matrix
import pprint

print('-'*28,'CODE OUTPUT','-'*29,'\n')

x, y, a, M = symbols('x, y, a, M',real=True)   #  symbols

# evaluate the itegrals using triple itegration in SymPy
Ixx = Iyy  = integrate(x**2, (x,0,a), (y,0,a))*M/a**2

Izz = integrate(x**2+y**2, (x,0,a), (y,0,a))*M/a**2
```

```
Ixy = Iyx =  -integrate(x*y,(x,0,a), (y,0,a))*M/a**2

Ixz = Iyz = 0    # because z=0

I = Matrix([[Ixx,Ixy,Ixz], [Ixy,Iyy,Iyz], [Ixz,Iyz,Izz]])
print('Moment of inertia tensor I =')

pp = pprint.PrettyPrinter(width=41, compact=True)
pp.pprint(I)



--------------------------- CODE OUTPUT ---------------------------

Moment of inertia tensor I =
Matrix([
[ M*a**2/3, -M*a**2/4,           0],
[-M*a**2/4,  M*a**2/3,           0],
[        0,         0, 2*M*a**2/3]])
```

Example 5.10 shows how to evaluate the elements of the inertia tensor for a cube with side a, by using symbolic integration.

Example 5.10: The moment of inertia tensor of a cube

Consider a cube of side a, evaluate the moment of inertia tensor with respect to the Cartesian coordinate system with the origin $(0,0,0)$ located at one of the corners of the cube, and the xyz axes parallel to the sides of the cube.

Solution:
We will use Cartesian coordinates for the calculation, with the limits of (x, y, z) varying from 0 to a. The mass element is $dm = (M/a^3) \, dV = (M/a^3) \, dx \, dy \, dz$ and the moments of inertia elements are:

$$I_{xx} = \iiint_V \left(y^2 + z^2\right) dm = \frac{M}{a^3} \int_{x=0}^{a} \int_{y=0}^{a} \int_{z=0}^{a} \left(y^2 + z^2\right) dx \, dy \, dz$$

$$I_{xx} = \frac{M}{a^3} \int_{x=0}^{a} dx \int_{y=0}^{a} dy \int_{z=0}^{a} \left(y^2 + z^2\right) dz = \frac{M}{a^3} \int_{x=0}^{a} dx \int_{y=0}^{a} dy \left[y^2 z + \frac{z^3}{3}\right]_{z=0}^{z=a}$$

$$I_{xx} = \frac{M}{a^3} \int_{x=0}^{a} dx \left[\frac{y^3}{3}a + \frac{a^3}{3}y\right]_{y=0}^{y=a} = \frac{M}{a^3} \int_{x=0}^{a} dx \left[\frac{a^3}{3}a + \frac{a^3}{3}a\right] = \frac{2}{3}Ma^2$$

The products of inertia are:

$$I_{xy} = -\iiint_V x\,y\,dm = -\frac{M}{a^3} \int_{x=0}^{a} \int_{y=0}^{a} \int_{z=0}^{a} x\,y\,dx\,dy\,dz$$

$$I_{xx} = -\frac{M}{a^3} \int_{x=0}^{a} x\,dx \int_{y=0}^{a} y\,dy \int_{z=0}^{a} dz = -\frac{M}{a^3}\frac{a^2}{2}\frac{a^2}{2}a = -\frac{1}{4}Ma^2$$

By symmetry, $I_{xx} = I_{yy} = I_{zz} = 2Ma^2/3$ and $I_{xy} = I_{xz} = I_{yz} = -Ma^2/4$.

$$I = \begin{pmatrix} \frac{2}{3}Ma^2 & -\frac{1}{4}Ma^2 & -\frac{1}{4}Ma^2 \\ -\frac{1}{4}Ma^2 & \frac{2}{3}Ma^2 & -\frac{1}{4}Ma^2 \\ -\frac{1}{4}Ma^2 & -\frac{1}{4}Ma^2 & \frac{2}{3}Ma^2 \end{pmatrix}$$

Similar to the previous code, the `pprint` package is used to format the code output, and `Matrix` is used to create the 3x3 inertia tensor I.

```python
from sympy import symbols, integrate, Matrix
import pprint

print('-'*28,'CODE OUTPUT','-'*29,'\n')

x, y, z, a, m = symbols('x, y, z, a, m',real=True)  #  symbols

# evaluate the itegrals using triple itegration in SymPy
Ixx = Iyy =Izz = integrate(y**2+x**2, (x,0,a), (y,0,a),( z,0,a))*m/a**3

Ixy = Ixz = Iyz= -integrate(x*y,(x,0,a), (y,0,a), (z,0,a))*m/a**3

I = Matrix([[Ixx,Ixy,Ixz], [Ixy,Iyy,Iyz], [Ixz,Iyz,Izz]])
print('Moment of inertia tensor I =')

pp = pprint.PrettyPrinter(width=41, compact=True)
pp.pprint(I)
```

```
------------------------- CODE OUTPUT ----------------------------

Moment of inertia tensor I =
Matrix([
[2*a**2*m/3,  -a**2*m/4,  -a**2*m/4],
[ -a**2*m/4, 2*a**2*m/3,  -a**2*m/4],
[ -a**2*m/4,  -a**2*m/4, 2*a**2*m/3]])
```

5.5.1 NUMERICAL EVALUATION OF MULTIPLE INTEGRALS

In many cases it is rather tedious to evaluate the elements of the inertia tensor, or it may not possible to obtain analytical expressions for the multiple integrals in two and three-dimensions. In such cases it is usually preferable to use numerical integration, instead of SymPy.

In this section, we show how to carry out a *numerical* integration in three-dimensions by using SciPy. There are several specialized SciPy commands which can evaluate double integrals (e.g. `scipy.integrate.dblquad()`) and triple integrals (e.g.

`scipy.integrate.triplquad()`). Note that the previous examples in this chapter were carried out using *symbolic* integration in SymPy, while in this section we focus on numerical evaluations.

Example 5.11 shows how to evaluate the elements of the inertia tensor for a triangular plate with a variable density, by using two-dimensional numerical integration.

Example 5.11: The moment of inertia tensor of a triangular plate

Consider a right angle uniform triangular plate in the xy-plane with vertices at $(0,0)$, $(1,0)$, and $(0,1)$. The plate has a density $d(x,y,z) = x\,y$ (in SI units). Find the moment of inertia tensor with respect to the Cartesian coordinate system with origin at $(0,0,0)$ and with xy-axes aligned with the vertical sides of the plate.

Solution:
The diagonal of the triangular plate is described by the equation $x + y = 1$. Therefore, the limits for the y integral will range from $y = 0$ to $y = 1 - x$, while the limits for the x integral are from $x = 0$ to $x = 1$.

The mass element is $dm = d(x,y,z)\,dA = x\,y\,dx\,dy$, and the moments of inertia elements are evaluated with $z = 0$ because the triangle is in the xy-plane:

$$I_{xx} = \iint_A \left(y^2 + z^2\right) dm = \iint_A \left(y^2 + z^2\right) x\,y\,dx\,dy = \int_{x=0}^{1}\int_{y=0}^{1-x} y^2\,x\,y\,dx\,dy$$

$$I_{xx} = \int_{x=0}^{1} x\,dx \left[\frac{y^4}{4}\right]_{y=0}^{y=1-x} = \int_{x=0}^{1} x\,dx \left[\frac{(1-x)^4}{4}\right] = \frac{1}{120}$$

By symmetry, $I_{xx} = I_{yy} = M/120$ and we evaluate I_{zz} and I_{xy}:

$$I_{zz} = \iint_A \left(x^2 + y^2\right) dm = \int_{x=0}^{1}\int_{y=0}^{1-x} x^2\,x\,y\,dx\,dy + \int_{x=0}^{1}\int_{y=0}^{1-x} y^2\,x\,y\,dx\,dy = \frac{2}{120}$$

$$I_{xy} = -\iint_A x\,y\,dm = -\int_{x=0}^{1} x^2\,dx \int_{y=0}^{1-x} y^2\,dy = -\int_{x=0}^{1} x^2\,dx \left[\frac{(1-x)^3}{3}\right] = \frac{1}{180}$$

These integrals are rather tedious to evaluate, so we use the numerical integration routines available in SciPy.

The Python code uses the `scipy.integrate.dblquad()` function to evaluate the double integrals. Note that the order of the integrations `(y,0,1-x)`, `(x,0,1)` is important in this example. In order to simplify the code we define a function `f` which evaluates the command `scipy.integrate.dblquad()` for the double integrals and is called for each of the components of the inertia tensor I_{xx}, I_{xy} etc.

The important feature of the double integration here is the implementation of the integration limits in the line of code `dblquad(i, 0, 1, 0, yupper)`. Here `yupper` is the function $1 - x$ representing the upper limit in the y-integration, and `i` is the function to be integrated.

In this example we use a NumPy array to represent the moment of inertia matrix.

```
from scipy.integrate import dblquad
import numpy as np
from sympy import symbols
print('-'*28,'CODE OUTPUT','-'*29,'\n')
```

```
yupper = lambda x: 1-x              #defines the upper limit of integration


# function to integrate i, from y=0 to 1-x, and x from 0 to 1
def f(i):
    return dblquad(i, 0, 1,   0, yupper)[0]


# x*y is the density of the triangular plate
Ixx = Iyy = lambda y, x: (y**2)*x*y
Izz = lambda y, x: (y**2+x**2)*x*y


Ixy = lambda y, x: -(x*y) *x*y
Ixz = Iyz = 0                       #z = 0


I = np.array([[f(Ixx),f(Ixy),Ixz],[f(Ixy),f(Iyy),Iyz],\
[Ixz,Iyz,f(Izz)]])
print('Moment of inertia tensor I=')
I


------------------------- CODE OUTPUT -----------------------------


Moment of inertia tensor I=
array([[ 0.00833333, -0.00555556,  0.        ],
       [-0.00555556,  0.00833333,  0.        ],
       [ 0.        ,  0.        ,  0.01666667]])
```

Example 5.12 shows how to evaluate the elements of the inertia tensor for a triangular pyramid with a variable density, by using numerical integration.

Example 5.12: The moment of inertia tensor of a triangular pyramid

Consider the triangular pyramid with vertices at the origin, $(1,0,0)$, $(0,1,0)$, and $(0,0,1)$, shown previously in Example 5.5. The pyramid has a uniform density $d(x,y,z) = 1$ (in SI units). Find the moment of inertia tensor with respect to the Cartesian coordinate system with the origin at $(0,0,0)$.

Solution:
We already saw in the previous examples how to set up the triple integration over the volume of this triangular pyramid.

The mass element is $dm = d(x,y,z)\, dV = dx\, dy\, dz$, and the moments of inertia elements are:

$$I_{xx} = \iiint_V \left(y^2 + z^2\right) dm = \iiint_V \left(y^2 + z^2\right) dx\, dy\, dz$$

$$I_{xx} = \int_{y=0}^{y=1} dy \int_{x=0}^{x=1-y} dx \left[y^2 z + \frac{z^3}{3}\right]_{z=0}^{z=1-x-y}$$

$$I_{xx} = \int_{y=0}^{y=1} dy \int_{x=0}^{x=1-y} dx \left[y^2\left(1 - x - y\right) + \frac{\left(1 - x - y\right)^3}{3}\right]$$

The integral is rather tedious to evaluate, so we use the alternative numerical integration routines available in SciPy.

The Python code uses the `scipy.integrate.tplquad()` function to evaluate the triple integrals, and we define a function `f(i)` similar to the previous double integration in Example 5.11, to evaluate the elements of the inertia tensor.

Note that the order of the integrations `tplquad(i, 0, 1, 0, x2, 0, z2)` is again important, and we define the upper limits of integration as two functions x2 and z2 .

```
from scipy.integrate import tplquad
import numpy as np
from sympy import symbols
print('-'*28,'CODE OUTPUT','-'*29,'\n')

z2 = lambda y, z: 1-y-z #upper limit for z integration
x2 = lambda z: 1-z      #upper limit for x integration

def f(i):
    return tplquad(i, 0, 1,   0, x2,   0, z2)[0]

Ixx = lambda z, y, x: (y**2+z**2)
Iyy = lambda z, y, x: (x**2+z**2)
Izz = lambda z, y, x: (y**2+x**2)

Ixy = lambda z, y, x: -x*y
Ixz = lambda z, y, x: -x*z
Iyz = lambda z, y, x: -y*z

I = np.array([[f(Ixx),f(Ixy),f(Ixz)],[f(Ixy),f(Iyy),f(Iyz)],\
[f(Ixz),f(Iyz),f(Izz)]])
print('Moment of inertia tensor I=')
I

----------------------- CODE OUTPUT ---------------------------

Moment of inertia tensor I=
array([[ 0.03333333, -0.00833333, -0.00833333],
       [-0.00833333,  0.03333333, -0.00833333],
       [-0.00833333, -0.00833333,  0.03333333]])
```

5.6   END OF CHAPTER PROBLEMS

1. **Evaluation of areas using double integral** – Calculate the area of the ellipse $x^2/9 + y^2 = 1$, in the domain defined by the first quadrant of the Cartesian plane. Do this calculation by hand and symbolically using SymPy.

2. **Evaluation of general double integrals** – A two-dimensional object has the shape of a ellipse $x^2/9 + y^2 = 1$ and is defined by the positive x-axis and the positive y-axis. The object has a non-uniform mass density $\rho(x, y) = x\,y$. Calculate the total mass of this object symbolically using SymPy.

3. **The center of mass of two dimensional objects** – A two-dimensional object has the shape of a ellipse $x^2/9 + y^2 = 1$ and is defined by the positive x-axis and the positive y-axis. The object has a non-uniform mass density $\rho(x, y) = x\,y$. Calculate the center of mass of this object symbolically using SymPy.

4. **The center of mass of a cone** – Calculate the center of mass of a cone of radius a and height h by hand and using SymPy, by evaluating the appropriate triple integrals.

5. **A cone with variable density** – A cone of radius a and height h has a mass density $\rho(x, y) = x^2 + y^2$, with all physical quantities in SI units. Find its the center of mass using cylindrical coordinates, by hand and using SymPy.

6. **The volume of a uniform ellipsoid** – Evaluate the volume of a uniform ellipsoid given by:

$$\frac{x^2}{a^2} + \frac{y^2}{b^2} + \frac{z^2}{c^2} = 1 \tag{5.6.1}$$

by using an appropriate transformation of variables. Here a, b, c are constants. Do this problem using SymPy and by hand.

7. **The mass of an ellipsoid with non-uniform density** – Evaluate the mass of an ellipsoid with non-uniform density $\rho(x, y, z) = 3z^2$ in SI units. The ellipsoid is given by:

$$\frac{x^2}{a^2} + \frac{y^2}{b^2} + \frac{z^2}{c^2} = 1 \tag{5.6.2}$$

Here a, b, c are constants. Do this problem using SymPy and by hand.

8. **Physical properties of a plane sheet** – A plane sheet is defined by the curve $y = \cos x$ from $x = 0$ to $x = \pi/2$. All physical quantities below are in SI units, and the variable surface density is $\rho(x, y) = x + y$. Find

     a. The area under the curve

     b. The mass

     c. The arc length of the curve

     d. The center of mass of the area

     e. The moments of inertia tensor with respect to the x, y, z axes

9. **Electrostatic force between a disk and a point charge** – A point charge q is located at a distance a above the center of a uniform thin circular disk of radius R. The disk carries a total charge Q and has a surface charge density given by $\sigma(\rho) = \rho$ where ρ is the distance from the center of the disk (all physical quantities are in in SI units). Use SymPy to evaluate the electrostatic force of attraction exerted by the disk on the charge q, by using the Coulomb force between two charge q_1 and q_2 located a distance r apart:

$$\boldsymbol{F} = k\frac{q_1 q_2}{r^2}\hat{\boldsymbol{r}} \tag{5.6.3}$$

Here k in the electrostatic constant and $\hat{\boldsymbol{r}}$ is the unit vector along the line connecting the two charges.

10. **Moment of inertia of two-dimensional systems** – Consider a solid thin circular hoop of radius R and mass m, which rotates around the x, y and z axes. Show using Python and by hand that the moments of inertia are $I_z = mR^2/2$ and $I_x = I_y = mR^2/4$.

11. **Moment of inertia of three-dimensional systems** – Find the moments of inertia of the following, using Python and by hand.

    a. A uniform solid cylinder of radius r, height h and mass m rotated around the z-axis.

    b. A uniform solid sphere of radius r and mass m rotated around the z-axis.

12. **Wavefunctions in two-dimensional infinite square well** – The quantum wavefunction for a particle inside an infinite square well of width a in two dimensions is

$$\psi(x, y) = A \sin\left(\frac{\pi x}{a}\right) \sin\left(\frac{2\pi y}{a}\right)$$

where A is a normalization constant, and the well is defined by $0 \leq x \leq a$ and $0 \leq y \leq a$. The wavefunction $\psi(x, y)$ is zero at all points (x, y) outside this square well. Evaluate the following quantities by using double integration and by hand.

    a. Find the normalization constant A, so that the total probability of finding the particle anywhere in space is 1:

$$P_{total} = \int_{\text{All space}} \psi^* \, \psi \, dx \, dy = 1$$

    b. Evaluate the expectation value of the position along the x-axis:

$$\langle x \rangle = \int_{-\infty}^{+\infty} \psi^* \, x \, \psi \, dx \, dy$$

    c. Evaluate the expectation value of the momentum along the y-axis:

$$\langle p_y \rangle = \frac{\hbar}{i} \int_{-\infty}^{+\infty} \psi^* \frac{\partial \psi}{\partial y} dx \, dy$$

    d. Find the expectation value of the kinetic energy:

$$\langle K \rangle = -\frac{\hbar^2}{2m} \int_{-\infty}^{+\infty} \psi^* \left(\frac{\partial^2 \psi}{\partial x^2} + \frac{\partial^2 \psi}{\partial y^2} \right) dx \, dy$$

    e. Find the probability that the particle will be found in the space defined by $0 \leq x \leq a/2$ and $0 \leq y \leq a/2$.

13. **Wavefunction in three-dimensional infinite square well** – The quantum wavefunction for a particle inside an infinite square well of width a in three-dimensions is

$$\psi(x, y, z) = A \sin\left(\frac{\pi x}{a}\right) \sin\left(\frac{2\pi y}{a}\right) \sin\left(\frac{3\pi z}{a}\right)$$

where A is a normalization constant, and the well is defined by $0 \leq x \leq a$, $0 \leq y \leq a$ and $0 \leq z \leq a$. The wavefunction $\psi(x, y, z)$ is zero at all points (x, y, z) outside this square well. Evaluate the following by using triple integration and also by hand.

a. Find the normalization constant A, so that the total probability of finding the particle anywhere in space is 1:

$$P_{total} = \int\limits_{\text{All space}} \psi^* \, \psi \, dx \, dy \, dz = 1$$

b. Evaluate the expectation value of the kinetic energy:

$$\langle K \rangle = -\frac{\hbar^2}{2m} \int\limits_{-\infty}^{+\infty} \psi^* \left(\frac{\partial^2 \psi}{\partial x^2} + \frac{\partial^2 \psi}{\partial y^2} + \frac{\partial^2 \psi}{\partial z^2} \right) dx \, dy \, dz$$

c. Find the probability that the particle will be found in the space defined by $0 \le x \le a/2$, $0 \le y \le a/2$ and $0 \le z \le a/2$.

14. **Center of mass of non-uniform thin quarter-circular plate** – Find the center mass of a non-uniform thin quarter-circular plate of radius $r = 1$ on the first quadrant of the Cartesian xy-plane. The mass density of the quarter-plate is $\rho = x + y$, with all physical quantities in SI units.

15. **Moment of inertia of quarter-circular plate** – Find the moment of inertia I of a uniform quarter-circular plate of radius r located on the first quadrant of the Cartesian xy-plane, when its is rotated around the x, y and z axes.

16. **Center of mass of a triangular plate** – Find the center mass of a uniform triangular plate with right angle sides a, b along the x and y axis respectively.

17. **Moment of inertia of a triangular plate** – Find the moment of inertia I of a uniform triangular plate with right angle sides a, b along the x and y axis respectively, when its is rotated around side a, or rotated around side b.

18. **Normalization of the Maxwell-Boltzmann distribution of speeds** – The distribution of speeds (v_x, v_y, v_z) in an ideal gas follows the Maxwell-Boltzmann distribution:

$$f(v_x, v_y, v_z) = A e^{-m(v_x^2 + v_y^2 + v_z^2)/(2kT)}$$

where T is the temperature of the gas, k is the Boltzmann constant and m is the mass of the atoms. Evaluate the normalization constant A by requiring that the distribution is normalized, i.e.

$$\int\limits_{-\infty}^{\infty} \int\limits_{-\infty}^{\infty} \int\limits_{-\infty}^{\infty} f(v_x, v_y, v_z) \, dv_x \, dv_y \, dv_z = 1$$

19. **Electric energy stored between concentric cylindrical conductors** – The electric field between two concentric cylindrical conductors with radii r_1 and r_2 is given by $\mathbf{E} = (1/r) \, \hat{r}$ V/m where $\hat{r}$ is the unit vector in the radial direction. The energy stored in a region of space where the electric field is $\mathbf{E}$, is given by the three-dimensional integral:

$$W = \frac{1}{2} \int\limits_{all \ space} \epsilon_0 \, |\mathbf{E}|^2 \, dV$$

Using SymPy and also evaluating by hand, evaluate the energy stored in a length L along the z-axis and between the two cylinders.

20. **Moment of inertia tensor for cube with respect to its center of mass** – Evaluate the moment of inertia tensor for a cube of side a and uniform density, with respect to a coordinate system located at the center of mass.

21. **A triangular prism with variable density** – Consider the triangular prism enclosed by the planes $z = 2$, $x = y = z = 0$, and $x + y = 2$. Plot the solid in three-dimensions by hand and calculate the total mass $M = \iiint \rho \, dV$ inside it, when the volume mass density is given by $\rho = x\,y^2$, where all physical quantities are in SI units.

6 Complex Numbers

In this chapter we present an overview of complex numbers and how they are used in Physics. Specifically, we will see how to represent complex numbers, how to perform arithmetic with them, and how to compute functions of complex numbers.

Imaginary and complex numbers arise when learning how to find the roots of quadratic equations such as $x^2 + 1 = 0$, which has the solution $x = \pm\sqrt{-1}$. For simplicity, we use the symbol $i = \sqrt{-1}$ with $i^2 = -1$. An imaginary number is a number like $2i$ which represents $2\sqrt{-1} = \sqrt{-4}$. Imaginary numbers are real numbers multiplied by i, while complex numbers have a real and an imaginary part, e.g. $3 + 2i$. One can think of imaginary and real numbers as subsets of complex numbers.

Before presenting complex numbers, it is important to discuss their applications. We do not obtain complex values by measuring physical quantities. For example, the distance between two points is never $2i$ meters. However, that does not prevent us from using complex numbers to represent physical quantities. For example, it is very convenient to represent oscillations as exponentials with complex arguments, as opposed to sines and cosines. This is done frequently in classical mechanics. In electrical engineering, complex numbers are used routinely to analyze electrical circuits. Complex numbers are also a central component to quantum mechanics, where they are used in the equations for wave functions and operators. In this chapter we will present examples of applications of complex numbers in these diverse areas of science.

6.1 THE COMPLEX PLANE

In general, a complex number takes the form

$$z = x + iy \tag{6.1.1}$$

where x and y are real numbers. We call x the real part of the complex number z and y is called the imaginary part. We often use the notation

$$\left.\begin{aligned} \mathrm{Re}(z) &= x \\ \mathrm{Im}(z) &= y \end{aligned}\right\} \tag{6.1.2}$$

to denote the real and imaginary parts of z.

The complex conjugate z^* of z is found by switching the sign of the imaginary part of z. The complex conjugate z^* of (6.1.1) is

$$z^* = x - iy \tag{6.1.3}$$

For simplicity, one can think of finding the complex conjugate by changing the sign of i in the complex number. For example, if $w = \alpha - i\beta$ (where α and β are real numbers), then $w^* = \alpha + i\beta$.

The magnitude or modulus $|z|$ of a complex number is found by multiplying it by its complex conjugate, and taking the square root of the resulting positive number:

$$|z| = \sqrt{z\,z^*}$$

It is often useful to represent z on the complex plane, which can be thought of as similar to the Cartesian plane. However, the horizontal axis is called the real axis, and the vertical

DOI: 10.1201/9781003294320-6

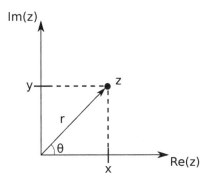

Figure 6.1 The complex plane.

axis is called the imaginary axis. As shown in Figure 6.1, we treat the complex number z as having two Cartesian-like coordinates, one along the real axis and the other along the imaginary.

In Figure 6.1, we see that we can also define a polar-like coordinate system using r and θ with the standard transformations between coordinates

$$x = r\cos\theta \tag{6.1.4}$$
$$y = r\sin\theta \tag{6.1.5}$$
$$r^2 = x^2 + y^2 \tag{6.1.6}$$
$$\theta = \tan^{-1}\left(\frac{y}{x}\right) \tag{6.1.7}$$

Note that $r^2 = |z|^2$. The variable θ is called the phase angle.

Example 6.1 shows how to evaluate the properties of complex numbers, and how to convert them between polar and Cartesian forms.

Example 6.1: Properties of complex numbers

Given the complex number $a = 3 + 4\,i$, use appropriate Python commands to calculate the real and imaginary parts, the complex conjugate, and the magnitude $|z|$. Use the library cmath to evaluate the phase angle θ, and the polar coordinates (r, θ) of z.

Solution:
The code evaluates the properties of the complex number $z = 3 + 4j$, namely its complex conjugate using the method z.conjugate(), the real and imaginary parts using z.real and z.imag respectively, and the magnitude $|\,z\,|$ using the function abs(z). Note that in Python j is used to represent the symbol $i = \sqrt{-1}$.
In addition to these four standard Python commands, we also use the cmath library to evaluate the phase of the complex number using cm.phase(z).
The conversion from Cartesian to polar coordinates is carried out using cm.polar(z), and from polar to Cartesian using cm.rect(z).

```
import cmath as cm

print('-'*28,'CODE OUTPUT','-'*29,'\n')
```

```
z = 3+4j
zconj = z.conjugate()
print('The imaginary part of ',z, 'is ',z.imag)
print('The real part of ',z, 'is ',z.real,'\n')

print('The complex conjugate of ',z, 'is ',z.conjugate())
print('The magnitude or modulus of ',z, 'is ',abs(z))

print('\nUse cmath for angle, polar from, and Cartesian form ')
print('phase = ',round(cm.phase(z),2),' rad','\n')

print('Polar form (r,theta) = ',cm.polar(z))
print('Cartesian form = ',cm.rect(5,0.9272952180016122))


-------------------------- CODE OUTPUT --------------------------

The imaginary part of  (3+4j) is  4.0
The real part of  (3+4j) is  3.0

The complex conjugate of  (3+4j) is  (3-4j)
The magnitude or modulus of  (3+4j) is  5.0

Use cmath for angle, polar form, and Cartesian form
phase =  0.93  rad

Polar form (r,theta) =  (5.0, 0.9272952180016122)
Cartesian form =  (3.0000000000000004+3.9999999999999996j)
```

Example 6.2 shows how to plot complex numbers on the xy-plane.

Example 6.2: Plotting complex numbers in the xy-plane

Plot the complex number $z = 3 + 4j$ and its complex conjugate $z^* = 3 - 4j$ on the xy-plane.

Solution:
In the Python code, the `plt.arrow()` command is used to plot an arrow in Figure 6.2, from the origin $(0,0)$ to the value of the complex number $z = 3 + 4j$, and a second arrow from the origin to the point representing the complex conjugate $z^* = 3 - 4j$.

```
import matplotlib.pyplot as plt
import matplotlib as mpl

mpl.rcParams["font.size"] = 15
print('-'*28,'CODE OUTPUT','-'*29,'\n')

z = 3+4j
zconj = z.conjugate()  # define complex number and its conjugate

# define scale of plot, and plot arrows and text
```

```
plt.scatter([0,5],[-5,5],s=0 )

plt.arrow(0,0,z.real,z.imag, color='red', head_width = 0.1)
plt.arrow(0,0,zconj.real,zconj.imag, color='blue',  head_width = 0.1)

plt.xlabel('x')
plt.ylabel('y')
plt.text(3.5,4,'z=3+4j')
plt.text(3.5,-4,r'z$^{*}$=3-4j')

plt.tight_layout()
plt.show()
```

--------------------------- CODE OUTPUT ----------------------------

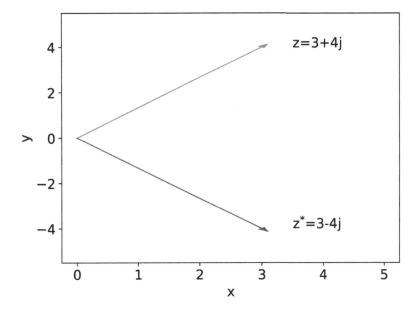

Figure 6.2 The graphical output from Example 6.2, showing a complex number z and its complex conjugate z^* on the xy-plane.

6.2 TRIGONOMETRIC FUNCTIONS AND COMPLEX EXPONENTIALS

Using (6.1.4) and (6.1.5), we can rewrite (6.1.1)

$$z = r\left(\cos\theta + i\sin\theta\right) \tag{6.2.1}$$

Let us compute the Maclaurin series of (6.2.1)

$$z = r \left(\left[1 - \frac{\theta^2}{2!} + \frac{\theta^4}{4!} + \cdots \right] + i \left[\theta - \frac{\theta^3}{3!} + \frac{\theta^5}{5!} + \cdots \right] \right) \tag{6.2.2}$$

$$= r \left(1 + \frac{(i\theta)^2}{2!} + \frac{(i\theta)^3}{3!} + \frac{(i\theta)^4}{4!} + \frac{(i\theta)^5}{5!} + \cdots \right) \tag{6.2.3}$$

Which gives the important polar representation of complex numbers,

$$z = r e^{i\theta} \tag{6.2.4}$$

Note that for (6.2.4) to be valid, θ must be in radians. The value r is the modulus or absolute value of z and is a real number. The value θ is called the phase of z and is also real.

The complex conjugate of z is

$$z^* = r e^{-i\theta} \tag{6.2.5}$$

The Euler formula, one of the most useful formulas involving complex functions, can be obtained by equating (6.2.4) with (6.2.1):

$$e^{i\theta} = \cos\theta + i\sin\theta \tag{6.2.6}$$

It is difficult to overstate the importance of the Euler formula (6.2.6) in physics. For example, (6.2.6) can provide us a formula which connects the trigonometric functions sine and cosine to complex exponentials. Consider the two equations

$$e^{i\theta} = \cos\theta + i\sin\theta \tag{6.2.7}$$

$$e^{-i\theta} = \cos\theta - i\sin\theta \tag{6.2.8}$$

where (6.2.8) is the complex conjugate of (6.2.6). If we add or subtract these equations, we obtain the following relationships

$$\cos\theta = \frac{1}{2} \left(e^{i\theta} + e^{-i\theta} \right) \tag{6.2.9}$$

$$\sin\theta = \frac{1}{2i} \left(e^{i\theta} - e^{-i\theta} \right) \tag{6.2.10}$$

Example 6.3 shows an application of complex numbers in Classical Mechanics, specifically to the verification of the solution of the differential equation for a simple harmonic oscillator.

Example 6.3: The complex exponential and simple harmonic motion

Show that simple harmonic motion can be described using a linear combination of complex exponentials. Specifically, show that the function

$$x(t) = A_1 e^{i\omega_0 t} + A_2 e^{-i\omega_0 t} \tag{6.2.11}$$

is the solution of the following equation

$$m \frac{d^2 x}{dt^2} = -kx \tag{6.2.12}$$

where $\omega_0 = \sqrt{k/m}$. Here A_1, A_2, ω_0 are real constants and t is the time variable.

Solution:
We find the first and second derivatives of $x(t)$:

$$\frac{dx}{dt} = \frac{d}{dt}\left[A_1 e^{i\omega_0 t} + A_2 e^{-i\omega_0 t}\right] = i\,\omega_0\left[A_1 e^{i\omega_0\,t} - A_2 e^{-i\omega_0\,t}\right] \qquad (6.2.13)$$

$$\frac{d^2 x}{dt^2} = (i\,\omega_0)^2\left[A_1 e^{i\omega_0\,t} + A_2 e^{-i\omega_0\,t}\right] \qquad (6.2.14)$$

Substituting $(6.2.14)$ into $(6.2.12)$:

$$m\frac{d^2 x}{dt^2} = m\,(i\,\omega_0)^2\left[A_1 e^{i\omega_0\,t} + A_2 e^{-i\omega_0\,t}\right] = -k\left[A_1 e^{i\omega_0 t} + A_2 e^{-i\omega_0 t}\right]$$

$$m\omega_0^2 = k \;\rightarrow\; \omega_0^2 = k/m$$

The Python code shown here uses SymPy commands `diff(f,x,x)` to evaluate the symbolic second derivative of $x(t)$. These derivatives are then substituted into $(6.2.12)$, and the code uses the logical equal sign `==` to check whether the two sides of the equation are identical. The code returns the statement `True`, verifying that the two sides are identical.

```python
from sympy import symbols, exp, diff, I
print('-'*28,'CODE OUTPUT','-'*29,'\n')

A1, A2, omeg, k, m ,t = symbols('A1, A2, omeg, k, m ,t',real=True)
# define symbols

f = A1*exp(I*omeg*t)+A2*exp(-I*omeg*t)   # define function

secondDeriv = diff( f,t,t)
print('Second Derivative = ',secondDeriv,'\n')

secondDeriv == -omeg**2*f   # check that equation is satisfied

---------------------------- CODE OUTPUT ----------------------------

Second Derivative =  -omeg**2*(A1*exp(I*omeg*t) + A2*exp(-I*omeg*t))

True
```

Example 6.4 shows an application in Quantum Mechanics, demonstrating how to carry out symbolic algebra of complex functions using SymPy.

Example 6.4: Symbolic addition of complex wavefunctions

The wavefunction describing a particle is the linear combination of two functions of the form:

$$\psi(x) = \psi_1(x)\,e^{-i\,E_1\,t} + \psi_2(x)\,e^{-i\,E_2\,t} \qquad (6.2.15)$$

where ψ_1, ψ_2, E_1, E_2 are real quantities. Physically $\psi_1(x)$, $\psi_2(x)$ represent the real-valued wavefunctions for two possible states with energies E_1, E_2. (Note that in the example we use a system of units in which the reduced Planck constant is equal to 1).

Show by hand and by using SymPy that

$$|\psi(x)|^2 = \psi_1(x)^2 + \psi_2(x)^2 + 2\psi_1(x)\,\psi_2(x)\,\cos\left[(E_2 - E_1)\,t\right]$$

Solution:

We evaluate the $|\psi(x)|^2 = \psi(x)\,\psi^*(x)$ with $\psi_1(x)$, $\psi_2(x)$ real functions:

$$|\psi(x)|^2 = \left(\psi_1(x)\,e^{-i\,E_1\,t} + \psi_2(x)\,e^{-i\,E_2\,t}\right)\left(\psi_1(x)\,e^{i\,E_1\,t} + \psi_2(x)\,e^{i\,E_2\,t}\right)$$

$$|\psi(x)|^2 = \psi_1(x)^2 + \psi_2(x)^2 + \psi_1(x)\,\psi_2(x)\left[e^{i\,(E_2-E_1)\,t} + e^{-i\,(E_2-E_1)\,t}\right]$$

$$|\psi(x)|^2 = \psi_1(x)^2 + \psi_2(x)^2 + 2\psi_1(x)\,\psi_2(x)\,\cos\left[(E_2 - E_1)\,t\right]$$

where we used Euler's equation $e^{i\,\theta} + e^{-i\,\theta} = 2\cos\theta$.

Note that the last term in this expression is an oscillatory term, with angular frequency $\omega = E_2 - E_1$. This term is called the interference term, because it describes the interaction of the two wavefunctions $\psi_1(x)$ and $\psi_2(x)$ as a function of time t.

The evaluation using Python is straightforward, we define the symbols using the `symbols` command, and use `Abs(psi**2)` to find the $|\psi(x)|^2$ in SymPy. In this example, the Python output is too long to be shown in one printed line, so we use the text `wrap` library to break up the output into 80-character segments.

In addition, we use the `psisq.rewrite(sin)` method to transform the answer for $|\psi(x)|^2$ from complex exponentials into the more convenient form containing sine and cosine functions.

```python
from sympy import symbols, exp, Abs, sin, I, simplify
import textwrap
print('-'*28,'CODE OUTPUT','-'*29,'\n')

psi1, psi2 ,E1 ,E2, t = symbols('psi1, psi2, E1, E2,t '\
                        ,real=True)
# define symbols

psi = psi1*exp(-I*E1*t)+psi2*exp(-I*E2*t)

psisq = Abs(psi)**2
print('psi^2 complex exponentials = ')

print(textwrap.fill(str(psisq),80))

u = psisq.rewrite(sin)
print('\npsi^2 with trig functions = ')
print(simplify(u))

-------------------------- CODE OUTPUT --------------------------

psi^2 complex exponentials =
psi1**2 + psi1*psi2*exp(I*E1*t)*exp(-I*E2*t) +
psi1*psi2*exp(-I*E1*t)*exp(I*E2*t) + psi2**2

psi^2 with trig functions =
psi1**2 + 2*psi1*psi2*cos(t*(E1 - E2)) + psi2**2
```

Example 6.5 shows how to add two sinusoidal waves described by complex exponential functions $e^{i(k\,x-\omega\,t)}$, resulting in a traveling wave along the x-axis.

Example 6.5: Beat waves

A wave is described by the function

$$f(x,t) = e^{i(k_1\,x-\omega_1\,t)} + e^{i(k_2\,x-\omega_2\,t)} \tag{6.2.16}$$

where x is the position of the wave along the x-axis, t is the elapsed time, and k_1, k_2, ω_1, ω_2 are positive real constants. Evaluate the real part of $f(x,t)$ by hand and by using SymPy, and plot it along the x-axis for two time instants $t = 0$ s and $t = 20$ s. Use the numerical values $\omega_1 = 1$ s^{-1}, $\omega_2 = 1.1$ s^{-1}, $k_1 = 1$ m^{-1}, $k_2 = 1.2$ m^{-1}. Discuss the shape of the resulting wave.

Solution:
We use Euler's identity:

$$f = \cos\left(k_1\,x - \omega_1\,t\right) + i\sin\left(k_1\,x - \omega_1\,t\right) + \cos\left(k_2\,x - \omega_2\,t\right) + i\sin\left(k_2\,x - \omega_2\,t\right)$$

The real part of f is

$$\mathrm{Re}\left[f(x,t)\right] = \cos\left(k_1\,x - \omega_1\,t\right) + \cos\left(k_2\,x - \omega_2\,t\right)$$

Substituting the numerical value ω_1, ω_2, k_1, k_2:

$$\mathrm{Re}\left[f(x,t)\right] = \cos\left(x - t\right) + \cos\left(1.2\,x - 1.1\,t\right) \tag{6.2.17}$$

We verify the result using SymPy, and use `re(f)` to find the real part of the function, and use `f.rewrite(sin)` to convert from complex exponentials to the more familiar trigonometric functions. The result from SymPy is the same as in (6.2.17).
The line `tot=lambdify([t,x],u)` converts the result into a function `tot(t,x)` of the time and position variables, so that it can be plotted at different times and positions along the x-axis.
The resulting plot in Figure 6.3 shows the well known shape of a beat wave, which is produced in this example by adding two cosine waves traveling in opposite directions along the x-axis. The frequencies and the wavenumbers of the two waves are similar to each other, resulting in the characteristic shape of the beat wave.

```
from sympy import symbols, exp, sin, I, re, simplify, lambdify
import numpy as np
import matplotlib.pyplot as plt
print('-'*28,'CODE OUTPUT','-'*29,'\n')
a, b, t, x = symbols('a, b, t, x',real=True)

# f = sum of complex exponentials
f = exp(-I*t)*exp(I*x)+exp(-1.1*I*t)*exp(1.2*I*x)

# transform f from complex exponentials to sine, cosine functions
u = re(f.rewrite(sin))                    # real part of f
print('Real part of f = ',simplify(u))

tot = lambdify([t,x],u)                   # create function for plot

xpos = np.linspace(-40,40,200)            # create x-positions, plot x(t)

# two plots at time t=0 s and t=20 s
```

```
plt.subplot(1,2,1)
plt.plot(xpos,tot(0,xpos),'k')

plt.xlabel('x')
plt.ylabel('re[f(x)]')
plt.title('t = 0s')

plt.subplot(1,2,2)
plt.plot(xpos,tot(20,xpos),'r')

plt.xlabel('x')
plt.ylabel('re[f(x)]')
plt.title('t = 20s')

plt.tight_layout()

plt.show()

------------------------- CODE OUTPUT -----------------------------

Real part of f =   cos(t - x) + cos(1.1*t - 1.2*x)
```

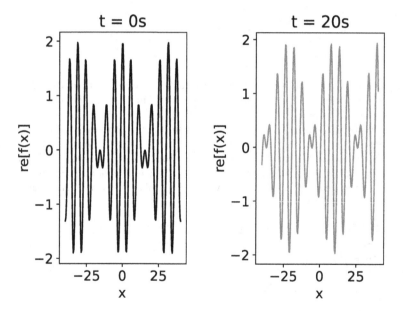

Figure 6.3 The graphical output from Example 6.5, showing the addition of two sinusoidal waves at times $t = 0$ and $t = 20$ s, resulting in a traveling wave along the x-axis.

6.3   ARITHMETIC WITH COMPLEX NUMBERS

Arithmetic with a complex number z is intuitive. One simply treats i as an algebraic symbol while respecting the rule that $i^2 = -1$. In this section, we will examine the basic arithmetic operations, addition, subtraction, multiplication, and division. In what follows, we will consider two complex numbers

$$z_1 = x_1 + iy_1 \tag{6.3.1}$$

$$z_2 = x_2 + iy_2 \tag{6.3.2}$$

Let us begin with addition and subtraction which is performed by adding the real parts and imaginary parts of the two complex numbers separately.

$$z_1 \pm z_2 = (x_1 \pm x_2) + i(y_1 \pm y_2) \tag{6.3.3}$$

Likewise, multiplication is performed in a method similar to expanding formulas like $(x+1)(x-3)$

$$z_1 z_2 = (x_1 + iy_1)(x_2 + iy_2) \tag{6.3.4}$$

$$= x_1 x_2 + x_1 iy_2 + iy_1 x_2 + (iy_1)(iy_2) \tag{6.3.5}$$

Which simplifies to

$$z_1 z_2 = (x_1 x_2 - y_1 y_2) + i(x_1 y_2 + x_2 y_1) \tag{6.3.6}$$

Recall that the modulus of a complex number z is evaluated by multiplying it by its complex conjugate z^*. The square of the absolute value of $z = x + iy$ is found using

$$|z|^2 = z^* z = (x + iy)(x - iy) = x^2 + y^2 = r^2 \tag{6.3.7}$$

In other words, $|z|$ gives the distance between the origin and the point z on the complex plane. The quantity $|z|^2$ is very important in quantum mechanics where it provides the probability of a particle being in a given state, as shown in Example 6.6.

Example 6.6: Probabilities of quantum states

At time $t = 0$ a particle is in the state

$$\psi = \frac{1}{\sqrt{18}}[(2 + 3i)\psi_0 + (1 - 2i)\psi_3]$$

where ψ_0 and ψ_3 are real wavefunctions describing the ground and third excited state of the quantum harmonic oscillator. The probability of the particle being in each state ψ_0 or ψ_3 is found by taking the absolute value of the constant coefficient multiplying the functions ψ_0 or ψ_3. Find the probability of the particle being found in the state ψ_0 and in the state ψ_3.

Solution:
Let P_n be the probability of find the particle in state n, then

$$P_0 = \left[\frac{2 + 3i}{\sqrt{18}}\right]^* \left[\frac{2 + 3i}{\sqrt{18}}\right] = \left[\frac{2 - 3i}{\sqrt{18}}\right]\left[\frac{2 + 3i}{\sqrt{18}}\right] = \frac{13}{18}$$

$$P_3 = \left[\frac{1 - 2i}{\sqrt{18}}\right]^* \left[\frac{1 - 2i}{\sqrt{18}}\right] = \left[\frac{1 + 2i}{\sqrt{18}}\right]\left[\frac{1 - 2i}{\sqrt{18}}\right] = \frac{5}{18}$$

The Python code for calculating P_0 and P_3 is below. We find the probabilities using abs(a) (where a is a complex number) and we round the result to two decimals by using round(abs(a),2). As would be expected, the total probability is $P = P_0 + P_3 = 1$.

```
import numpy as np
print('-'*28,'CODE OUTPUT','-'*29,'\n')


z0 = (2 + 3j)/np.sqrt(18)
z3 = (1 - 2j)/np.sqrt(18)


print('Probability to be in state 0 = ', round(abs(z0)**2,2))
print('Probability to be in state 3 = ', round(abs(z3)**2,2))


-------------------------- CODE OUTPUT ----------------------------


Probability to be in state 0 =  0.72
Probability to be in state 3 =  0.28
```

Next we examine division of two complex numbers. We would like the final answer to be expressed in the traditional form of $z = x + iy$, in other words, the sum of real and imaginary parts. To do this, we multiply the numerator and denominator by the denominator's complex conjugate,

$$\frac{z_1}{z_2} = \frac{z_1}{z_2} \frac{z_2^*}{z_2^*} = \left(\frac{x_1 + iy_1}{x_2 + iy_2} \right) \left(\frac{x_2 - iy_2}{x_2 - iy_2} \right) \tag{6.3.8}$$

Which simplifies to

$$\frac{z_1}{z_2} = \left(\frac{x_1 x_2 + y_1 y_2}{x_2^2 + y_2^2} \right) + i \left(\frac{x_2 y_1 - x_1 y_2}{x_2^2 + y_2^2} \right) \tag{6.3.9}$$

We can also perform the multiplication and division of complex numbers represented by polar coordinates. For example, if $z_1 = r_1 e^{i\theta_1}$ and $z_2 = r_2 e^{i\theta_2}$, then

$$z_1 z_2 = r_1 r_2 e^{i(\theta_1 + \theta_2)} \tag{6.3.10}$$

$$\frac{z_1}{z_2} = \frac{r_1}{r_2} e^{i(\theta_1 - \theta_2)} \tag{6.3.11}$$

When performing arithmetic in polar coordinates, it is important to remember that each phase θ_j is measured in radians, counterclockwise from the positive real axis. You may find it helpful to draw the points on the complex plane to ensure you are working with the correct angle.

Before moving on to the next section, let us return to the concept of a complex conjugate. For example, suppose z is the ratio of two complex numbers

$$z = \frac{x + iy}{\alpha + i\beta} \qquad \text{then} \qquad z^* = \frac{x - iy}{\alpha - i\beta}$$

where x, y, α, β are real quantities. In other words, to find the complex conjugate of z, we need to change the sign of every imaginary part of z. Furthermore, if $z = f + ig$ where f and g are themselves complex numbers, then $z^* = f^* - ig^*$.

The algebra of complex numbers finds an important application in the mathematical description of electronic circuits, as will be described in the next section.

6.4　APPLICATION OF COMPLEX NUMBERS IN AC CIRCUITS

Complex numbers are used in signal analysis and other fields, for a convenient description of periodically varying signals, since periodic signals can often be described as a combination of sine and cosine functions.

In electronics and electrical engineering, the treatment of resistors, capacitors, and inductors can be unified by introducing a single complex number called the *impedance Z*. Electrical impedance Z extends the concept of resistance R to alternating current (AC) circuits. When the circuit is driven with direct current (DC), there is no distinction between impedance Z and resistance R.

In DC circuits, Ohm's law is

$$V = I\,R \tag{6.4.1}$$

where V is the applied DC voltage, I is the resulting current, and R is the resistance. In AC circuits we replace Ohm's law with the more general expression:

$$V = I\,Z \tag{6.4.2}$$

where the impedance Z will be in general a complex number which is a function of the input frequency ω applied to the circuit, with the same units as electrical resistance R, i.e. Ohms (Ω). The voltage and current are commonly represented as sinusoidal complex-valued functions of time in the form:

$$V = |V|e^{i(\omega t + \theta_1)} \tag{6.4.3}$$

$$I = |I|e^{i(\omega t + \theta_2)} \tag{6.4.4}$$

where θ_1 and θ_2 are the phases of the AC voltage and current, respectively. The impedance is evaluated as the ratio of these two complex numbers:

$$Z = \frac{V}{I} = \frac{|V|}{|I|}e^{i(\theta_1 - \theta_2)} \tag{6.4.5}$$

The impedance of an ideal *resistor* R is a purely real number and is referred to as a *resistive impedance*:

$$Z_R = R \tag{6.4.6}$$

Ideal inductors L have a purely imaginary *inductive impedance* Z_L given by:

$$Z_L = i\,\omega\,L \tag{6.4.7}$$

Ideal capacitors C have a purely imaginary *capacitative impedance* Z_C given by:

$$Z_C = \frac{1}{i\,\omega\,C} = \frac{-i}{\omega\,C}$$

Example 6.7 shows how to calculate the current I in an inductor L, when a sinusoidal voltage is applied across it.

Example 6.7: Current and voltage in an inductor AC circuit

Find the current I in an inductor circuit when an AC sinusoidal voltage V of frequency ω and phase θ_1 is applied. Discuss the physical meaning of the result.

Solution:
The impedance is $Z_L = i\,\omega\,L$ and we write the input sinusoidal voltage as a complex exponential:

$$V = |V|e^{i\,(\omega\,t + \theta_1)} = V_0 e^{i\,(\omega\,t + \theta_1)} \tag{6.4.8}$$

where V_0 is the voltage amplitude. The current I is evaluated from Ohm's law for AC circuits:

$$I = \frac{V}{Z_L} = \frac{V_0\,e^{i\,(\omega\,t+\theta_1)}}{i\,\omega\,L} = -i\frac{V_0}{\omega\,L}e^{i\,(\omega\,t+\theta_1)} \tag{6.4.9}$$

We use Euler's identity to replace $-i = e^{-i\,\pi/2}$ to obtain:

$$I = \frac{V_0}{\omega\,L}e^{i\,(\omega\,t+\theta_1-\pi/2)} \tag{6.4.10}$$

The physical meaning of this equation is that the current is a sinusoidal wave with a magnitude $|I| = V_o/(\omega\,L)$. The phase of the current is $\theta_1 - \pi/2$, which is $\pi/2$ radians smaller than the voltage phase θ_1. We usually state that the input voltage is 90° ahead the current.

In the Python code we evaluate symbolically the complex impedance Z, the voltage V and the current I, based on (6.4.8) and (6.4.10) using SymPy. Next we use `lambdify` to create two functions i1 and v1 which will evaluate the values of I, V using the time array `tims`.

The numerical values $V_0=1$ V, $\omega = 60$ s^{-1}, $\theta_1=0$ rad, $L = 0.01$ H, are substituted using the method `.subs`. The real part of the voltage $V(t)$ and of the current $I(t)$ are plotted on the same graph in Figure 6.4, and the phase difference of $\pi/2$ between the two signals is clearly seen in the plot.

```python
import matplotlib.pyplot as plt
import numpy as np
from sympy.utilities.lambdify import lambdify
from sympy import symbols, I, exp, re
print('-'*28,'CODE OUTPUT','-'*29,'\n')


# define symbols
L, V, i, omeg, Vo, theta1, t = symbols('L, V, i, omeg, Vo,\
theta1, t ',positive=True)

Z =  I*omeg*L                 # complex impedance for inductor

V = Vo*exp(I*(omeg*t+theta1))     # input voltage V

i = V/Z                       # Ohm's law for inductor

tims = np.linspace(0,0.3,100)   # sequence of times tims

# substitute numerical values to current and voltage, and take real part
i1 = lambdify(t,re(i.subs({Vo:1,omeg:60,theta1:0,L:0.01})))
v1 = lambdify(t,re(V.subs({Vo:1,omeg:60,theta1:0,L:0.01})))

# plot current i1(t) and voltage v1(t)
plt.plot(tims,i1(tims),'k-',label='I(t)')
plt.plot(tims,v1(tims),'r--',label='V(t)')

plt.xlabel('Time [s]')
plt.ylabel("Current I(t) [A],   Voltage V(t) [V]")
leg = plt.legend()
leg.get_frame().set_linewidth(0.0)
plt.show()
```

------------------------- CODE OUTPUT -----------------------------

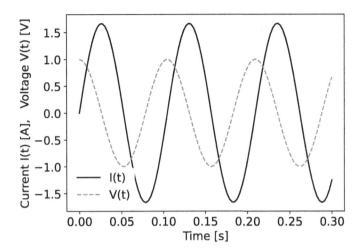

Figure 6.4 The graphical output from Example 6.7 , showing the current I in an inductor L, when a sinusoidal voltage V is applied across it.

The *total impedance* of many simple circuits can be calculated using the rules for combining impedances in series and parallel. These rules are identical to those used for combining resistances in introductory physics classes, except that we are now dealing with complex numbers.

For components $Z_1, Z_2, \cdots Z_n$ connected *in series* as in Figure 6.5, the current through each circuit element is the same and the total impedance Z_{series} is simply the sum of the component impedances:

$$Z_{\text{series}} = Z_1 + Z_2 + \cdots + Z_n \qquad (6.4.11)$$

For components $Z_1, Z_2, \cdots Z_n$ connected *in parallel*, the voltage across each circuit element is the same and the total impedance Z_{parallel} is:

$$\frac{1}{Z_{\text{parallel}}} = \frac{1}{Z_1} + \frac{1}{Z_2} + \cdots + \frac{1}{Z_n} \qquad (6.4.12)$$

Example 6.8 shows how to calculate the total impedance in a series RLC circuit.

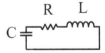

Figure 6.5 Example 6.8: evaluation of the complex impedance of a series RLC circuit.

Example 6.8: Impedance in series RLC circuit

Find the total impedance Z_{eq} in a series RLC circuit, as a function of R, L, C and of the AC frequency ω.

Solution:
In a series circuit we can simply add the impedances:

$$Z_{\text{series}} = Z_R + Z_C + Z_L = R - \frac{i}{\omega C} + i\omega L$$

$$Z_{\text{series}} = R + i\left(\omega L - \frac{1}{\omega C}\right)$$

The magnitude of the total impedance is

$$|Z_{\text{series}}| = \left|R + i\left(\omega L - \frac{1}{\omega C}\right)\right| = \sqrt{R^2 + \left(\omega L - \frac{1}{\omega C}\right)^2}$$

and the phase angle for this complex number is found by dividing the imaginary and real components:

$$\tan\phi = \frac{\text{Im}\left(Z_{\text{series}}\right)}{\text{Re}\left(Z_{\text{series}}\right)} = \frac{\omega L - \frac{1}{\omega C}}{R}$$

The Python code evaluates the real and imaginary parts of the impedance, and stores them in the variables *realz* and *imagz* correspondingly. Notice that in order for Python to evaluate correctly the Z, it needs to be told that the symbols R, L, C, ω are real numbers.
The last two lines of the code evaluate the magnitude and phase of the total impedance Z_{series}.

```
from sympy import symbols, Abs, im, I, re
print('-'*28,'CODE OUTPUT','-'*29,'\n')

R, L, C, omeg = symbols('R, L, C, omeg',real=True)

Z = R + I*(omeg*L-1/(omeg*C))   # complex impedance

print('Magnitude of total impedance Z =',Abs(Z))
print('\ntan(phi) =',im(Z)/re(Z))

-------------------------- CODE OUTPUT ---------------------------

Magnitude of total impedance Z = sqrt(L**2*omeg**2 + R**2 - 2*L/C +
     1/(C**2*omeg**2))

tan(phi) = (L*omeg - 1/(C*omeg))/R
```

6.5   EQUATIONS WITH COMPLEX NUMBERS

It is common in physics to solve algebraic equations that involve complex numbers. Two complex numbers are equal, if and only if their real and imaginary parts are equal. Therefore

$$x + iy = \alpha + i\beta \tag{6.5.1}$$

if and only if $x = \alpha$ *and* $y = \beta$ (here x, y, α, β are assumed to be real). We can extend this to equations of complex variables.

Example 6.9: Solving an equation of complex variables

Solve the following equation for x and y:

$$(x + iy)^2 = 2ix$$

Solution:
One immediate solution is $(x = 0, y = 0)$. However, there are others. To find them, we begin by expanding the left-hand side of the equation.

$$x^2 + 2ixy - y^2 = 2ix$$

Next, we equate the real and imaginary parts on each side.

$$x^2 - y^2 = 0$$

$$2xy = 2x$$

Solving the system of equations yields $(x = 1, y = \pm 1)$.
Using SymPy the possible solutions are obtained using the `solve` command, which determines the pairs of possible solutions (x, y).

```
from sympy import symbols, solve, I
print('-'*28,'CODE OUTPUT','-'*29,'\n')

x, y = symbols('x, y',real=True)      # symbols

sol = solve((x+I*y)**2-2*I*x,[x,y])  # solve equation for (x,y)

print('Possible solutions are the pairs (x,y):\n',sol)

-------------------------- CODE OUTPUT ----------------------------

Possible solutions are the pairs (x,y):
 [(-1, 1), (0, 0), (1, 1)]
```

Example 6.10 shows an application of complex numbers in the solution of the differential equation of the externally driven simple harmonic oscillator (SHO). We will examine this physical system in detail in Chapter 10.

Example 6.10: Solutions of the externally driven damped harmonic oscillator

The differential equation for the externally driven damped harmonic oscillator is:

$$\ddot{x} + 2\gamma \dot{x} + \omega_0^2 x = D \cos(\omega t) \tag{6.5.2}$$

where $x(t)$ is the position at time t, γ is a constant representing the magnitude of the damping force, $\omega_0^2 = k/m$ is the natural frequency of the undamped SHO, and $D \cos(\omega t)$ is the external driving force of frequency ω. Here $\ddot{x}$ and $\dot{x}$ denote the second and first time derivatives of x. One method of solving this differential equation, is by using a trial function $x(t) = A\,e^{i(\omega t - \phi)}$, where A, ϕ are real constants. Show that the constants A, ϕ are the solutions of the following equations:

$$A\left(\omega_0^2 - \omega^2\right) = D\,\cos\phi \tag{6.5.3}$$

$$2\gamma\omega A = D\,\sin\phi \qquad \cdot \tag{6.5.4}$$

Solution:

The algebra in this type of problem is simplified greatly by using complex numbers. Specifically, in this method we replace the $\cos(\omega t)$ term in (6.5.2) with the complex function $e^{i\omega t}$ and carry out the extensive algebra. As we will see in Chapter 10, at the end of this procedure we will take the real part of the solution $x(t)$ in order to obtain the final solution of the differential equation. Replacing $\cos(\omega t)$ with $e^{i\omega t}$ in (6.5.2):

$$\ddot{x} + 2\gamma\,\dot{x} + \omega_0^2\,x = D\,e^{i\,\omega\,t} \tag{6.5.5}$$

We insert the trial solution $x(t) = A\,e^{i(\omega\,t - \phi)}$ into (6.5.5) and evaluate the derivatives to obtain:

$$-A\,\omega^2 e^{i(\omega t - \phi)} + 2\gamma\,\omega\,i\,A e^{i(\omega t - \phi)} + \omega_0^2\,A\,e^{i(\omega t - \phi)} = D\,e^{i\omega t}$$

By canceling out the $e^{i\omega t}$ from all terms, and multiplying both sides by $e^{i\phi} = \cos\phi + i\sin\phi$, we obtain:

$$-A\omega^2 + 2\gamma\,\omega\,i\,A + \omega_0^2\,A = D\,e^{i\,\phi} = D\,(\cos\phi + i\sin\phi)$$

By equating the real parts of the two sides of this equation and separately equating the imaginary parts, we obtain the desired relationships for the constants A, ϕ:

$$A\left(\omega_0^2 - \omega^2\right) = D\,\cos\phi \tag{6.5.6}$$

$$2\gamma\,\omega\,A = D\,\sin\phi \tag{6.5.7}$$

The Python code evaluates the derivatives using `diff` in SymPy, and the $e^{i\omega t}$ term is canceled by dividing and using the `simplify` function. The SymPy functions `re` and `im` are used to obtain the real and imaginary parts of the equation.

The same method using complex numbers can also be used when the driving force is of the form $D\,\sin(\omega t)$.

```python
from sympy import symbols, exp, diff, simplify , I, re, im, solve
print('-'*28,'CODE OUTPUT','-'*29,'\n')


x, t, omeg, omeg0, A, phi , gam, D = symbols(\
'x, t, omeg, omeg0 ,A, phi, gam, D', positive=True)  # symbols


x = A*exp(I*omeg*t-I*phi)  # complex trial function

# ODE for driven SHO
eq = diff(x,t,t)+2*gam*diff(x,t)+omeg0**2*x-D*exp(I*omeg*t)

# divide equation by trial function
ans = simplify(eq/exp(I*omeg*t-I*phi))

print('SHO Equation = ',ans)

print('\nThe first equation for D and A is: ',re(ans),'= 0\n')

print('The second equation for D and A is: ',im(ans),'= 0')
```

```
-------------------------- CODE OUTPUT ----------------------------

SHO Equation =  2*I*A*gam*omeg - A*omeg**2 + A*omeg0**2 - D*exp(I*phi)

The first equation for D and A is:  -A*omeg**2 + A*omeg0**2 - D*cos(phi) = 0

The second equation for D and A is:  2*A*gam*omeg - D*sin(phi) = 0
```

It is often useful to graph them on the complex plane. These graphs can give one insight into what the equation is telling us about a particular system. In order to graph a complex equation, it is often useful to write the complex equation in terms of real and imaginary parts.

For example, consider graphing the quantity $|z - i| = 2$. We rewrite $|z - i| = 2$ in terms of its real and imaginary parts

$$|x + iy - i| = 2 \tag{6.5.8}$$

$$|x + i(y - 1)| = 2 \tag{6.5.9}$$

$$\sqrt{x^2 + (y - 1)^2} = 2 \tag{6.5.10}$$

$$x^2 + (y - 1)^2 = 4 \tag{6.5.11}$$

We see that the quantity $|z - i| = 2$ represents a circle with a radius of 2 centered at the point $(x, y) = (0, 1)$ or $z_0 = i$ in the complex plane.

Sometimes we wish to specify a region of the complex plane. For example, the equation $\text{Im}(z) \geq 3$ represents the complex plane above and including the line $y = 3$. Similarly, $\text{Re}(z) > 0$ is the right half of the complex plane. If we are interested in a line along the complex plane that passes through the origin, polar coordinates are useful. For example, the graph of $\arg(z) = \pi/3$ represents the line passing through the origin at an angle of $\pi/3$ radians counterclockwise from the real axis.

6.6 FUNCTIONS OF A COMPLEX VARIABLE

In this section we will explore how to compute functions when their independent variable is a complex number. Along the way, will we uncover new and useful functions for physics.

6.6.1 EXPONENTIALS, POWERS AND ROOTS OF COMPLEX NUMBERS

Using the Euler formula and what we learned about graphing complex functions, we find that $e^{i\theta}$ describes points on the unit circle centered on the origin of the complex plane. Hence, we find some important expressions using a complex exponential:

$$e^{i\pi/2} = i \tag{6.6.1}$$

$$e^{i\pi} = -1 \tag{6.6.2}$$

$$e^{i3\pi/2} = -i \tag{6.6.3}$$

$$e^{2\pi i} = 1 \tag{6.6.4}$$

Using polar notation, we can compute the powers and roots of complex numbers. Consider the complex number $z = re^{i\theta}$. Then

$$z^n = \left(re^{i\theta}\right)^n = r^n e^{in\theta} \tag{6.6.5}$$

Using DeMoivre's theorem we can compute z^n

$$z^n = r^n \left(\cos\theta + i\sin\theta\right)^n = r^n \left[\cos\left(n\theta\right) + i\sin\left(n\theta\right)\right] \tag{6.6.6}$$

Example 6.11 evaluates the solution of a third order equation with complex numbers, and plots the possible solutions on the complex plane.

Example 6.11: Roots of complex numbers

Find and plot the roots of the equation $z^3 = -8$.

Solution:

The calculations by hand are easiest if one uses Euler's identity to express complex number $z = x + iy$. We write

$$z^3 = -8 = 8e^{i\pi}$$

$$z^{1/3} = 8^{1/3}\left(e^{i\,\pi}\right)^{1/3} = 2e^{i\pi/3} = 2\left(\cos\frac{\pi}{3} + i\sin\frac{\pi}{3}\right) = 1 + i\sqrt{3} \tag{6.6.7}$$

However, $2e^{i\,\pi/3}$ is only one root. For example, we can write $z = -8$ as $8e^{-3\pi\,i}$ by adding 2π to the phase, then

$$z^{1/3} = 8^{1/3}\left(e^{3i\,\pi}\right)^{1/3} = 2e^{i\pi} = -2$$

Furthermore, we can add 2π to the phase again and write $z = -8$ as $8e^{i5\pi}$, then

$$z^{1/3} = 8^{1/3}\left(e^{5i\,\pi}\right)^{1/3} = 2e^{i5\pi/3} = 1 - i\sqrt{3}$$

The Python code uses the function `solve` to obtain the (x,y) pairs which satisfy $z^3 = -8$. The possible roots are contained in the parameter `sol`, and we use the function `float()` to convert them into floating point numbers. As expected, there are 3 possible (x, y) pairs representing the three roots of the equation, each on a circle of radius 2 and centered at the origin. The code uses a function `f(x,y)` which plots the complex roots (x, y) as an arrow on the xy-plane in Figure 6.6.

```
from sympy import symbols, solve,  I
import matplotlib.pyplot as plt
print('-'*28,'CODE OUTPUT','-'*29,'\n')


x, y = symbols('x, y',real=True)    # define symbols


sol = solve((x+I*y)**3+8,[x,y])    # solve equation for (x,y)
print('Possible solutions are the complex numbers (x,y):')


# loop to print possible solutions
for j in range(len(sol)):
    print('x, y = ',sol[j][0], ', ',sol[j][1])


# function for plotting arrows
def f(x1,y1):
    plt.arrow(0,0,x1,y1, color='red', head_width = 0.05,length_includes_head=True)


# set scale of the plot
plt.scatter([-2,2],[-2,2],s=0 )


# loop calling function f to plot arrows
```

```
for j in range(len(sol)):
    f(float(sol[j][0]), float(sol[j][1]))

plt.title('Plotting roots of complex numbers')

# make the size of x and y axes the same on the plot
plt.axis("equal");

#plot circle with radius R=2 centered at (0,0)
circle = plt.Circle((0, 0), 2.0, fill = False)

# use patch library to plot the circle
plt.gca().add_patch(circle)

plt.xlabel('x')
plt.ylabel('y')
plt.show()

------------------------- CODE OUTPUT ---------------------------

Possible solutions are the complex numbers (x,y):
x, y =  -2 ,  0
x, y =  1 ,  -sqrt(3)
x, y =  1 ,  sqrt(3)
```

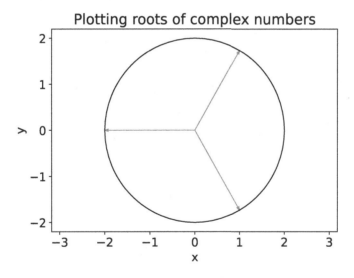

Figure 6.6 The graphical output from Example 6.11, showing the three solutions of a third order equation with complex numbers. The circle is present to illustrate that the roots lie alone a circle of radius 2.

When we plot the roots on the complex plane for Example 6.11, we find that the roots divide the circumference of the circle with radius 2 into three equal parts. In general, computing $z^{1/n}$, results in n points located on a circle with radius $r^{1/n}$. The first point has a phase angle of θ/n, and the other points are spaced $2\pi/n$ apart.

6.6.2 HYPERBOLIC FUNCTIONS

We have already proven the relationships

$$\sin\theta = \frac{1}{2i}\left(e^{i\theta} - e^{-i\theta}\right) \tag{6.6.8}$$

$$\cos\theta = \frac{1}{2}\left(e^{i\theta} + e^{-i\theta}\right) \tag{6.6.9}$$

where we supposed θ is a real number. Now consider the case where $\theta = iy$, where y is a real number. Then we create two new functions, the hyperbolic sine called sinh and the hyperbolic cosine called cosh:

$$\sin(iy) = \frac{i}{2}\left(e^y - e^{-y}\right) = i\sinh y \tag{6.6.10}$$

$$\cos(iy) = \frac{1}{2}\left(e^y + e^{-y}\right) = \cosh y \tag{6.6.11}$$

In addition, we can build other hyperbolic functions based on analogies from the trigonometric functions:

$$\tanh y \quad = \quad \frac{\sinh y}{\cosh y} \tag{6.6.12}$$

$$\operatorname{sech} y \quad = \quad \frac{1}{\cosh y} \tag{6.6.13}$$

$$\operatorname{csch} y \quad = \quad \frac{1}{\sinh y} \tag{6.6.14}$$

$$\coth y \quad = \quad \frac{1}{\tanh y} \tag{6.6.15}$$

The hyperbolic functions appear in many fields of physics. For example, the hyperbolic secant function, (sech) appears in nonlinear wave theory when describing solitons.

There are several useful relationships between the hyperbolic functions. Some of the most useful ones in physics are

$$\cosh^2 z - \sinh^2 z = 1 \tag{6.6.16}$$

$$\frac{d}{dz}\cosh z = \sinh z \tag{6.6.17}$$

$$\frac{d}{dz}\sinh z = \cosh z \tag{6.6.18}$$

The following integral formulas are also useful for the inverse hyperbolic functions (note that we left out the constant of integration).

$$\int \frac{dx}{\sqrt{1+x^2}} = \sinh^{-1} x \tag{6.6.19}$$

$$\int \frac{dx}{\sqrt{x^2-1}} = \cosh^{-1} x \tag{6.6.20}$$

$$\int \frac{dx}{1-x^2} = \tanh^{-1} x \tag{6.6.21}$$

Example 6.12 shows how to prove some of these general properties of hyperbolic functions.

Example 6.12: General properties of hyperbolic functions

Use SymPy to demonstrate the following properties of hyperbolic functions:

$$\cosh^2 z - \sinh^2 z = 1$$

$$\frac{d}{dz}\sinh z = \cosh z$$

$$\int \cosh z \, dz = \sinh z$$

Solution:
We use SymPy the evaluate symbolically the identities. The hyperbolic functions in SymPy are implemented as the functions $\sinh(x)$, $\cosh(x)$. We use diff and integrate for the derivative and integral as usual, and we also use simplify to obtain the results.

```
from sympy import symbols, diff, sinh, cosh, simplify, integrate
print('-'*28,'CODE OUTPUT','-'*29,'\n')

# define symbol x
x = symbols('x ',real=True)

print('cosh(x)**2 - sinh(x)**2 = ',simplify(cosh(x)**2-sinh(x)**2))

print('\nThe derivative of cosh(x) = ', simplify(diff(cosh(x),x)))

print('\nThe indefinite integral of cosh(x) = ', \
simplify(integrate(cosh(x),x)))

--------------------------- CODE OUTPUT ----------------------------

cosh(x)**2 - sinh(x)**2 =  1

The derivative of cosh(x) =  sinh(x)

The indefinite integral of cosh(x) =  sinh(x)
```


6.7   COMPLEX VECTORS

In quantum mechanics, it is often necessary to allow for complex vector components. In this case much of the work presented in Chapter 4 holds true. However, when computing dot products, we need to take the complex conjugate of the first vector in each dot product.

$$\mathbf{v}\cdot\mathbf{u} = \sum_{i=1}^{n} v_i^* u_i \qquad (6.7.1)$$

This will ensure that the magnitude of the vector $\mathbf{v}$, $v^2 = \mathbf{v}^* \cdot \mathbf{v}$ is real.

$$v^2 = \sum_{i=1}^{n} v_i^* v_i \tag{6.7.2}$$

Example 6.13 shows an application in Quantum Mechanics.

Example 6.13: Spin states in quantum mechanics

A particle is in the spin state

$$\chi = (1 + i, 2i) \tag{6.7.3}$$

The state χ, called a *spinor*, is a linear combination of spin up and spin down states in the z-direction. Normalize χ. If the spin of this particle in the z-direction is measured, what is the probability of finding the particle to be in a spin up state.

Solution:
To begin this problem, we first must review some quantum mechanics. The first element in (6.7.3) is the spin up component and the second element is the spin down component. Before measurement, the state of the particle is unknown. We can find the probability of measuring each state by looking at the components of χ. After normalization, the probability of measuring the particle in a spin up state is the first element of the vector times its complex conjugate.
We begin by normalizing χ

$$\chi^2 = \chi^* \cdot \chi = (1 - i, -2i) \cdot (1 + i, 2i) = 2 + 4 = 6 \tag{6.7.4}$$

Hence, the normalized χ is

$$\chi = \left(\frac{1+i}{\sqrt{6}}, \frac{2i}{\sqrt{6}} \right) \tag{6.7.5}$$

The probability of measuring χ to be in a spin up state is

$$\left(\frac{1-i}{\sqrt{6}} \right) \left(\frac{1+i}{\sqrt{6}} \right) = \frac{1}{3} \tag{6.7.6}$$

Below is the Python code used to solve the problem. Note the use of NumPy arrays to represent vectors. Using NumPy arrays can be helpful for numerical evaluations.

```
import numpy as np
from sympy import I,sqrt

chi = np.array([1-I,-2*I])
chi_star = np.conjugate(chi)

# find the norm using the dot product of chi and chi_star
norm_const = sqrt(np.dot(chi,chi_star).simplify())

norm_chi = chi/norm_const
p_up = np.dot(np.conjugate(norm_chi)[0],norm_chi[0])

print('-'*28,'CODE OUTPUT','-'*29,'\n')

print('The normalizing constant is ',norm_const)
print('The probability for spin up is ', p_up.simplify())
```

```
---------------------------- CODE OUTPUT ------------------------------

The normalizing constant is  sqrt(6)
The probability for spin up is  1/3
```

6.8 END OF CHAPTER PROBLEMS

1. **SymPy algebra of complex numbers** – Use Python to convert the following expression in the form $x + iy$, where x and y are real numbers. Find also the polar form of this complex number and plot it in the xy-plane. Do this problem using both SymPy and Numpy.

$$\frac{e^{5\pi i/6} + e^{-\pi i/4}}{(3+i)^2} \qquad (6.8.1)$$

2. **Magnitude, phase angle, polar form and plots of complex numbers** – Find the magnitude and phase angle of the following complex numbers using Python. Plot these complex numbers as arrows in the xy-plane.

   a. $2e^{\pi i/4} + 1$ b. $e^{3\pi i/4}$ c. $\dfrac{e^{-7\pi i/8} + 2}{1+i}$ $(6.8.2)$

3. **Roots of complex numbers** – Find and graph all the roots of the following complex numbers as vectors on a circle in the xy-plane:

   a. $\sqrt{4\dfrac{e^{5\pi i/6}}{3+2i}}$ b. $\sqrt[6]{-4}$ $(6.8.3)$

4. **Euler's equations and trigonometric identities** – Use Euler's formulas to prove the trigonometric identity:

$$2\cos x \, \cos y = \cos(x+y) + \cos(x-y) \qquad (6.8.4)$$

5. **Damped harmonic oscillator and complex numbers** – The differential equation for a RLC circuit oscillator driven by an external voltage $V_o \cos(\omega t)$ is:

$$L\frac{d^2Q}{dt^2} = -\frac{Q}{C} - R\frac{dQ}{dt} + V_0 \, e^{i\omega t} \qquad (6.8.5)$$

   where the driving voltage has an amplitude of V_0, ω is the external frequency, C is the capacitor, inductance L and R is the resistance. $Q(t)$ is the charge in the RLC circuit and and t is the time variable. Find the relations between the constants (A, ϕ) so that $Q(t) = A \, e^{i(\omega t - \phi)}$ is a solution of this ODE. Assume that $L, C, R, \omega, V_0, A, \phi$ are real constants.

6. **Standing waves and complex numbers** – A standing wave in a string is described mathematically by a linear combination of two waves of the same frequency and wavelength, traveling in opposite directions:

$$f(x,t) = e^{i(k\,x - \omega\,t)} + e^{i(k\,x + \omega\,t)} \qquad (6.8.6)$$

   where x is the position of the wave along the x-axis and t is the elapsed time and k, ω are positive constants. Evaluate the real part of $f(x,t)$ by hand and by using SymPy, and plot it along the x-axis for several time instants t. Use reasonable numerical values for k, ω and discuss the shape of the resulting standing wave.

7. **Wavefunctions and complex numbers** – Consider the wavefunction with two possible states described in this chapter, of the form:

$$\psi(x) = A_1\,\psi_1(x)\,e^{-i\,E_1\,t} + A_2\,\psi_2(x)\,e^{-i\,E_2\,t} \qquad (6.8.7)$$

where A_1, A_2, E_1, E_2 are real quantities and $\psi_1(x) = \sin(x)$, $\psi_2(x) = \sin(2x)$, $E_1 = 1$, $E_2 = 4$, $A_1 = 1$, $A_2 = 1$ (all quantities in SI units). Evaluate and plot the $|\psi(x)|^2$ at several time instants t along the x-axis, and discuss the shape of the resulting wave.

8. **A Gaussian function and its Fourier transform: application of complex numbers in Quantum Mechanics** – The wavefunction of a particle as time $t = 0$ is given by the Gaussian function:

$$\psi(x, 0) = A\,\exp\left(-a\,x^2/2\right)$$

a. Normalize this wavefunction so that

$$\int_{-\infty}^{\infty} \psi^*(x, 0)\,\psi(x, 0)\,dx = 1$$

b. Evaluate the Fourier transform $\phi(k)$ of $\psi(x, 0)$ defined by:

$$\phi(k) = \frac{1}{\sqrt{2\pi}} \int_{-\infty}^{\infty} \psi(x, 0)\,e^{-i\,k\,x}\,dx$$

Do this problem by hand and by using SymPy.

9. **The spreading wavepacket in Quantum Mechanics** – A well known application of complex numbers in Quantum Mechanics is the spreading wavepacket, which is described mathematically by the wavefunction:

$$\psi(x, t) = \frac{1}{\sqrt{1 + 2\,i\,t}}\,\exp\left[-\frac{x^2}{1 + 2\,i\,t}\right] \qquad (6.8.8)$$

where t is the time variable, x is the position along the x-axis. Evaluate and plot the $|\psi(x)|^2$ at several time instants t along the x-axis, and discuss the shape of the resulting wave. Do this problem both by hand and using SymPy.

10. **Impedance in parallel RLC circuit** – Calculate the total impedance in a circuit where a resistor R, capacitor C and inductance L are connected in parallel with each other, and the input frequency is ω. Plot the inverse of the total impedance $1/Z_{\text{total}}$ as a function of ω, by using reasonable numerical values of R, L, C. Discuss the physical meaning.

11. **Impedance in series RC circuit** – Calculate the total impedance in a circuit where a resistor R and a capacitor C are in series, while an inductance L is connected in parallel with the series combination of R and C, and the input frequency is ω. Evaluate the total impedance as a function of ω, R, C.

12. **Diffraction phenomena and cosine complex series** – During the mathematical description of diffraction phenomena in optics, we need to evaluate the electric field as a finite sum:

$$E = \sum_{n=1}^{N-1} A\,e^{i(a+n\,b)} \qquad (6.8.9)$$

where A represents the real amplitude of the electric field of the optical wave, and a, b are real constants. Evaluate this sum, both by hand and by using SymPy.

13. **Motion in xy-plane and complex numbers** – A particle moves in the (x, y) plane so that its position as a function of time t is given by the complex number

$$z = x + iy = \frac{i + t}{i - t}$$

Find the magnitude and direction of its position and velocity vectors as a function of time t.

14. **Current and voltage across capacitor** – Calculate the magnitude and the phase of the current I in a circuit consisting of a capacitor $C = 100$ μF. The AC voltage applied across the capacitor is $V = V_0 \sin(\omega t)$ and the frequency $\omega =60$ s^{-1}. Plot the current $I(t)$ and the voltage $V(t)$ on the same graph. How does the phase difference between I and V compare with the corresponding analysis of an inductor L?

15. **Current in series RLC circuit** – Calculate the magnitude and the phase of the current I in a series circuit consisting of a resistor $R =10$ Ohms, capacitor $C =1$ F and an inductor $L = 10$ H. The AC voltage applied to this series RC circuit is $V = 10 \sin(\omega t)$ and the frequency $\omega =60$ s^{-1}. Plot the inverse impedance $1/Z(\omega)$ for the 4 cases where the circuit contains the RC, RL, LC and RLC components, and discuss the physical meaning of these results..

16. **Trig identities with hyperbolic functions** – Use the properties of complex numbers to prove the following identities. Do this problem using SymPy.

    a. $\cosh^2 z + \sinh^2 z = \cosh(2z)$
    b. $\sinh(3z) = 3 \sinh z + 4 \sinh^3 z$
    c. $\cosh(3z) = 4 \cosh^3(z) - 3 \cosh(z)$
    d. $\sinh(2z) = 2 \sinh z \cosh z$

17. **Coupled oscillators** – A mechanical system of coupled oscillators consists of two equal masses m, connected in series with identical springs with spring constants k. The equations of motion for this system are:

$$m\ddot{x}_1 = -kx_1 - k(x_1 - x_2)$$

$$m\ddot{x}_2 = -k(x_2 - x_1) - kx_2$$

where $x_1(t)$ and $x_2(t)$ are the displacements of the masses. Using Python and also by hand, show that the substitution $x_1(t) = A_1 e^{i\omega t}$, $x_2(t) = A_2 e^{i\omega t}$ results in a system of equations for A_1 and A_2 such that:

$$-A_1\omega^2 = -\frac{2k}{m}A_1 + \frac{k}{m}A_2$$

$$-A_2\omega^2 = -\frac{2k}{m}A_2 + \frac{k}{m}A_1$$

Solve this system of equations using Python, and obtain the possible frequencies ω for these coupled oscillators. These frequencies are known as the normal modes of the system.

18. **Spin 1 particle** – In the spin Example of this chapter, the state χ represented a fermion, a spin-1/2 particle which has quantum numbers $s = 1/2$ and $m_s = \pm 1/2$. Hence, before measurement, the particle was in a linear combination of two states. However, bosons have integer values for s. For example, a photon has $s = 1$ and therefore $m_s = -1, 0, 1$. The photon's spin state, before measurement, is in a linear combination of three states. Consider a photon in the state

$$\chi = (i, -2, 1 - i)$$

where the first element represents the $m_s = -1$ component, the second element represents the $m_s = 0$ component, and the last element corresponds to $m_s = 1$. Normalize χ and find the probabilities of finding the photon in each m_s state after measurement.

19. **Infinite square well** – Consider a particle trapped in an infinite square well with sides located at $x = 0$ and $x = a$. The particle is initially in the state

$$\psi = 2i\psi_1 - 3\psi_2 - (2 + 3i)\,\psi_3$$

Normalize ψ. What is the probability of measuring the particle's energy to be E_1?

20. **Infinite series containing complex numbers** –

$$A = \sum_{n=0}^{\infty} \left(\frac{i}{5}\right)^n \tag{6.8.10}$$

$$B = \sum_{n=0}^{\infty} \frac{e^{i n \pi/2}}{3^n} \tag{6.8.11}$$

Evaluate these infinite series, both by hand and by using SymPy. Recall that the following result applies for infinite geometric series :

$$\sum_{n=0}^{\infty} r^n = \frac{1}{1 - r} \qquad \text{for } |r| < 1 \tag{6.8.12}$$

7 Matrices

A matrix is a rectangular grid of numbers, functions, symbols, or other mathematical expressions. In addition to physical quantities, matrices can represent systems of equations and vector transformations. For example, in classical mechanics a matrix is used to represent a rotation (either of a vector or of a coordinate system). In quantum mechanics, matrices can be used to represent operators, which are a mathematical way of representing a physical measurement.

This chapter is organized in two distinct parts. In the first part, we will discuss the basics of matrices and how they are structured. We will revisit vectors and the cross product and see how they can be represented as a matrix and a matrix operation, respectively. We will follow with a discussion of matrix arithmetic and other operations on matrices. In the second part of this chapter, we will demonstrate how to apply matrices to solve systems of linear equations and how to use them to represent linear transformations. We will conclude with a discussion of eigenvectors, eigenvalues, and matrix diagonalization.

7.1 THE STRUCTURE OF A MATRIX

A matrix is a rectangular grid of mathematical expressions. Examples include:

$$A = \begin{pmatrix} 4 & -1 & 3 \\ 5 & 7 & 2 \end{pmatrix} \quad B = \begin{pmatrix} \cos\phi & -\sin\phi \\ \sin\phi & \cos\phi \end{pmatrix} \quad C = \begin{pmatrix} 1 \\ -1 \\ 3 \end{pmatrix} \tag{7.1.1}$$

We normally use unbolded capital letters to represent matrices. Sometimes instead of parentheses, square brackets are used to enclose the elements inside the matrix. There is no difference in meaning between parentheses and square brackets, however parentheses are more commonly used in physics.

The matrix A in (7.1.1) is an example of a 2×3 matrix. It has two rows and three columns. Likewise, B is a 2×2 matrix. Furthermore, B is an example of a *square matrix* which has the same number of rows as columns. Square matrices are important in physics because they represent vector transformations and operators in quantum mechanics. The matrix C in (7.1.1) is a 3×1 matrix can also be used to represent a vector. We will return to matrix representations of vectors later in this section.

It is often useful to identify individual elements in a matrix. To do this, we often use the lower case of the letter used as the matrix's label but with two subscripts. For example, $a_{12} = -1$ because -1 is the element of A in the first row and second column. In general, a_{ij} is the element in the ith row and j^{th} column of the matrix. In physics, it is not unusual to break this convention and use the matrix's label instead (as is common for the moment of inertia tensor). In this case, A_{ij} is the element in the ith row and j^{th} column of the matrix.

Matrices, and their elements, can represent physical quantities. Consider the moment of inertia. In introductory physics, the moment of inertia was a physical quantity associated with an object's size, mass, and rotational axis. For example, the moment of inertia of a thin rod with mass M and length L rotated about its end is $1/3ML^2$. However, the moment of inertia changes when the axis of rotation is changed. In order to capture the complete description of an object's moment of inertia, we need to use a matrix called the moment of inertia tensor which takes the form:

$$I = \begin{pmatrix} I_{xx} & I_{xy} & I_{xz} \\ I_{yx} & I_{yy} & I_{yz} \\ I_{zx} & I_{zy} & I_{zz} \end{pmatrix} \tag{7.1.2}$$

In (7.1.2), the terms I_{xx}, I_{yy}, and I_{zz} are called the moments of inertia about the x, y, and z axes, respectively. In introductory physics classes it is common to identify the rotation axis as the z-axis and then one computes I_{zz}. The terms in (7.1.2) with mismatched subscripts (e.g. I_{xy}, I_{xz} etc.), are called *products of inertia*. In general, an object's angular momentum and angular velocity are not parallel, such as when an automobile tire is not properly balanced. Simply put, the products of inertia measure the misalignment between the object's angular momentum and its angular velocity.

7.1.1　DEFINING MATRICES IN PYTHON

Both the SymPy and NumPy libraries can be used while working with matrices in Python. SymPy is most useful when working with matrices that involve symbols, such as matrix B in (7.1.1). NumPy is most useful when working with matrices (especially large matrices) that contain numbers. In both libraries, a matrix is defined as an array of rows, with each row being a list, as shown in Example 7.1.

Example 7.1: Matrices in Python

Define the matrices A, B and C in (7.1.1) as variables in Python. As output, print the elements A_{23} and B_{12}.

Solution:
The Python code defines the matrix A as a NumPy array, and matrix B is defined using the `Matrix` function in SymPy. By contrast, matrix C is defined in both SymPy and NumPy. In each case, the commands `np.array` and `Matrix` use a list as their input. The first element of the list is the first row of the matrix in the form of an array. The second element is the next row, and so on if there were additional rows (as shown for C). Likewise, the indices for the printed elements correspond to what we discussed above. The first index identifies the row, and the second identifies the column. Note that keeping in the Python tradition, indices begin at 0, hence `A[1,2]` is identifying A_{23} the element of A in the second column and third row. Similarly for B_{12} we use `B[0,1]`.

```
import numpy as np
from sympy import Matrix, sin, cos, symbols

A = np.array([[4, -1, 3],[5, 7, 2]])

p = symbols('p')
B = Matrix([[cos(p),-sin(p)],[sin(p),cos(p)]])

C_numpy = np.array([[1],[-1],[3]])
C_sympy = Matrix([[1],[-1],[3]])

print('-'*28,'CODE OUTPUT','-'*29,'\n')
print('A_23 element of matrix is: ',A[1,2])
print('B_12 element of matrix is: ',B[0,1])
```

```
------------------------------ CODE OUTPUT ------------------------------

A_23 element of matrix is:   2
B_12 element of matrix is:   -sin(p)
```

7.1.2 VECTORS AS MATRICES

In addition to square matrices, matrices with a single column or a single row are also very important in science. They are used to represent vectors and are referred to as *column vectors* or *row vectors*, respectively. The vector C in (7.1.1) is an example of a column vector. If we consider Cartesian coordinates then we can represent the vector C in three equivalent ways

$$1\hat{\mathbf{i}} - 1\hat{\mathbf{j}} + 3\hat{\mathbf{k}} \Leftrightarrow \begin{pmatrix} 1 \\ -1 \\ 3 \end{pmatrix} \Leftrightarrow \begin{pmatrix} 1 & -1 & 3 \end{pmatrix} \qquad (7.1.3)$$

As mentioned in the previous chapter, using a row or column vector representation is useful when working in higher dimensions, or when we do not want to involve unit vectors.

In general, a vector $(A_x\ A_y\ A_z)$ can be normalized by requiring that its magnitude is equal to one. In this matrix notation, the magnitude of $(A_x\ A_y\ A_z)$ is $\sqrt{A_x^2 + A_y^2 + A_z^2}$ and therefore the normalized vector is:

$$\frac{1}{\sqrt{A_x^2 + A_y^2 + A_z^2}}(A_x\ A_y\ A_z) \qquad \text{or} \qquad \frac{1}{\sqrt{A_x^2 + A_y^2 + A_z^2}} \begin{pmatrix} A_x \\ A_y \\ A_z \end{pmatrix} \qquad (7.1.4)$$

In quantum mechanics the spin state of a particle can be represented as a vector called a *spinor*. The dimension of the vector is $2s+1$ where s is the particle's spin quantum number. For example, the spin of an electron is $s = 1/2$, hence its spinor is represented by a 2×1 column vector, or a 1×2 row vector. Similarly, the spin of a graviton is proposed to be $s = 2$. Its spinor is represented by a 5×1 column vector, or a 1×5 row vector.

7.2 MATRIX OPERATIONS

In this section, we discuss arithmetic and other mathematical operations which can be performed on matrices.

7.2.1 MATRIX EQUIVALENCE

There are two conditions required for the matrices A and B to be equal. First, they must have the same size. Suppose A and B are both $m \times n$ matrices. Then the second condition for two matrices to be equal is:

$$A_{ij} = B_{ij} \qquad \text{for all } i = 1, \ldots, m \qquad \text{and} \qquad j = 1, \ldots, n \qquad (7.2.1)$$

7.2.2 MULTIPLICATION BY A SCALAR

Suppose c is a constant scalar (real or complex) and A is an $m \times n$ matrix. Then cA is calculated by multiplying every element of A by the constant c. For example, if

$$A = \begin{pmatrix} x & u \\ y & v \end{pmatrix} \qquad (7.2.2)$$

then

$$cA = \begin{pmatrix} c\,x & c\,u \\ c\,y & c\,z \end{pmatrix} \tag{7.2.3}$$

The same holds for column and row vectors. Hence, multiplying a column or row vector by a scalar is equivalent to multiplying each element by the same scalar.

7.2.3 MATRIX ADDITION

Adding and subtracting matrices is done by simply adding the corresponding matrix elements. If A, B are matrices and $C = A + B$, then

$$C_{ij} = A_{ij} + B_{ij} \tag{7.2.4}$$

In order to add two matrices A and B, then A and B must be the same size. If A and B are $m \times n$ matrices, then C will also be $m \times n$. For example

$$\begin{pmatrix} 1 & 2 & -1 \\ 3 & -2 & 4 \end{pmatrix} + \begin{pmatrix} -2 & -3 & 5 \\ -4 & 5 & 0 \end{pmatrix} = \begin{pmatrix} 1-2 & 2-3 & -1+5 \\ 3-4 & -2+5 & 4+0 \end{pmatrix}$$
$$= \begin{pmatrix} -1 & -1 & 4 \\ -1 & 3 & 4 \end{pmatrix}$$

7.2.4 MATRIX MULTIPLICATION

As we will see in a later section, matrices can represent transformations of vectors. A series of multiple transformations can be combined into one matrix, by multiplying the matrices representing each individual transformation.

Let us consider two matrices A and B. We wish to find the product $C = AB$. If we think of A as consisting of a stack of row vectors and B as a series of column vectors, then C is found by taking the dot product of A's row vectors with B's column vectors. For example,

$$AB = \begin{pmatrix} a & b \\ c & d \end{pmatrix} \begin{pmatrix} x & y \\ u & v \end{pmatrix} = \begin{pmatrix} a\,x + b\,u & a\,y + b\,v \\ c\,x + d\,u & c\,y + d\,v \end{pmatrix} = C \tag{7.2.5}$$

We see in (7.2.5) that C_{11} is the dot product between the first row of A and the first column of B. Likewise, C_{21} is the dot product between the second row of A and the first column of B. In general, C_{ij} is found by taking the dot product of the ith row of A and the jth column of B. In other words,

$$C_{ij} = \sum_k A_{ik} B_{kj} \tag{7.2.6}$$

where k is summed over the number of columns of A, or equivalently, over the number of rows of B.

Notice that the multiplication is only possible if the number of columns of A matches the number of rows of B. For example, we can multiply the two matrices if A is a 2×3 matrix and B is a 3×4 matrix, but not if A is a 2×2 and B is a 3×2. In general if A is an $m \times n$ matrix and B is a $n \times \ell$ matrix, then the product C will be a $m \times \ell$ matrix.

Example 7.2: Rotation of vectors with matrices

The following matrix B represents a counterclockwise rotation of a point in the xy-plane by the angle ϕ about the z-axis.

$$B = \begin{pmatrix} \cos\phi & -\sin\phi \\ \sin\phi & \cos\phi \end{pmatrix}$$

Show that rotating a point by an angle ϕ_1 and then by an angle ϕ_2, is the same as performing a single rotation by the angle $\phi_1 + \phi_2$.

Solution:
We begin by defining two rotation matrices

$$R_1 = \begin{pmatrix} \cos\phi_1 & -\sin\phi_1 \\ \sin\phi_1 & \cos\phi_1 \end{pmatrix} \quad R_2 = \begin{pmatrix} \cos\phi_2 & -\sin\phi_2 \\ \sin\phi_2 & \cos\phi_2 \end{pmatrix}$$

A point (or a vector) in the xy-plane can be represented as a 2×1 column vector $\mathbf{v}$. As we will show later in this chapter, we can rotate the vector $\mathbf{v}$ by an angle ϕ_1, by multiplying $\mathbf{v}$ by R_1. Hence $\mathbf{v}_1 = R_1\mathbf{v}$, where $\mathbf{v}_1$ is the vector $\mathbf{v}$ rotated by an angle ϕ_1. The second rotation (by an angle ϕ_2) is then found using:

$$\mathbf{v}_2 = R_2\mathbf{v}_1 = R_2 R_1 \mathbf{v}$$

Therefore, we can perform the two rotations by a series of multiplications, or we could compute the matrix $R = R_1 R_2$. Let us compute R:

$$R = \begin{pmatrix} \cos\phi_2 & -\sin\phi_2 \\ \sin\phi_2 & \cos\phi_2 \end{pmatrix} \begin{pmatrix} \cos\phi_1 & -\sin\phi_1 \\ \sin\phi_1 & \cos\phi_1 \end{pmatrix}$$

$$= \begin{pmatrix} \cos\phi_2\cos\phi_1 - \sin\phi_2\sin\phi_1 & -\cos\phi_2\sin\phi_1 - \sin\phi_2\cos\phi_1 \\ \sin\phi_2\cos\phi_1 + \cos\phi_2\sin\phi_1 & -\sin\phi_2\sin\phi_1 + \cos\phi_2\cos\phi_1 \end{pmatrix}$$

$$= \begin{pmatrix} \cos(\phi_1 + \phi_2) & -\sin(\phi_1 + \phi_2) \\ \sin(\phi_1 + \phi_2) & \cos(\phi_1 + \phi_2) \end{pmatrix}$$

where we used a trigonometric identity for $\cos(\phi_1 + \phi_2)$ and for $\sin(\phi_1 + \phi_2)$. The Python code below uses SymPy, because we are performing symbolic calculations. We multiply the matrices R_1 and R_2. Note that we use the `simplify()` function to obtain the final answer.

```
from sympy import Matrix, sin, cos, symbols, simplify
import pprint

phi1, phi2 = symbols('phi1, phi2')

R1 = Matrix([[cos(phi1),-sin(phi1)],[sin(phi1),cos(phi1)]])
R2 = Matrix([[cos(phi2),-sin(phi2)],[sin(phi2),cos(phi2)]])
R = R1 * R2

print('-'*28,'CODE OUTPUT','-'*29,'\n')
pp = pprint.PrettyPrinter(width=41, compact=True)

print('The rotational matrix for two rotations is:')
pp.pprint(simplify(R))
```

```
---------------------------- CODE OUTPUT ----------------------------

The rotational matrix for two rotations is:
Matrix([
[cos(phi1 + phi2), -sin(phi1 + phi2)],
[sin(phi1 + phi2),  cos(phi1 + phi2)]])
```

Unlike multiplying numbers, the order of the matrices being multiplied matters. Let A and B be matrices, then the equality $AB = BA$ is not always true. In fact, one might not be able to calculate BA, even if one can find AB. For example, if A is a 1×3 and B is a 3×2 matrix, the AB is a 1×2 matrix. However, the matrix BA cannot be found, because the number of columns in B does not match the number of rows in A.

The difference between the matrices AB and BA is called the *commutator* $[A, B]$ and is defined as:

$$[A, B] = AB - BA \tag{7.2.7}$$

If $[A, B] = 0$, we say that A and B *commute*. The fact that not all pairs of matrices commute requires that we be specific when we discuss multiplication. For example, in the case of the product AB, we say that B is *left multiplied* by A. We could also say that A is *right multiplied* by B.

The commutator plays an important role in quantum mechanics where matrices can represent physical measurements. If the matrices representing two physical measurements commute, then there is no uncertainty principle between them.

Square matrices are common in physics. One important square matrix is the so-called *identity matrix* which is represented by $\mathcal{I}$. The identity matrix has 1's along the diagonal and zeros elsewhere. For example, the 3×3 identity matrix takes the form:

$$\mathcal{I} = \begin{pmatrix} 1 & 0 & 0 \\ 0 & 1 & 0 \\ 0 & 0 & 1 \end{pmatrix} \tag{7.2.8}$$

For any $n \times n$ matrix A, $A\mathcal{I} = \mathcal{I}A = A$ where $\mathcal{I}$ is the $n \times n$ identity matrix. In other words, the identity matrix plays the same role as the number 1 when multiplying scalars.

Furthermore, the *zero matrix* is an $n \times n$ matrix where all of its elements are zero.

7.2.5 TRACE OF A MATRIX

The trace $\text{Tr}(A)$ of the square matrix A, is defined as the sum of the diagonal elements of A:

$$\text{Tr}(A) = \sum_{i=1}^{n} A_{ii} \tag{7.2.9}$$

The trace has some useful properties. Consider a $m \times n$ matrix A, and a $n \times \ell$ matrix B. Then

$$\text{Tr}(AB) = \text{Tr}(BA) \tag{7.2.10}$$

If A and B are two square matrices of the same size, and a, b are scalars then the following identities hold:

$$\text{Tr}(a\,A) = a\text{Tr}(A) \tag{7.2.11}$$

$$\text{Tr}(a\,A + b\,B) = a\text{Tr}(A) + b\text{Tr}(B) \tag{7.2.12}$$

The trace of a matrix plays an important role in quantum mechanics.

7.2.6 TRANSPOSE AND HERMITIAN ADJOINT OF A MATRIX

The *transpose* of the matrix A is denoted as A^{T} and is found by changing its rows into columns (or its columns into rows). For example, consider the 2×3 matrix A and its transpose A^{T}:

$$A = \begin{pmatrix} a & b & c \\ d & e & f \end{pmatrix} \quad A^{\mathrm{T}} = \begin{pmatrix} a & d \\ b & e \\ c & f \end{pmatrix}$$

Notice that while A is a 2×3 matrix, A^{T} is a 3×2 matrix. Furthermore, the rows of A are the columns of A^{T}. Likewise, we could say the columns of A are the rows of A^{T}. If A were an $n \times n$ square matrix, then A^{T} would also be a square matrix of the same size as A. Further note the following general property for two matrices A and B for which the product AB exists:

$$(AB)^{\mathrm{T}} = B^{\mathrm{T}} A^{\mathrm{T}} \tag{7.2.13}$$

In quantum mechanics, we are often interested in finding the *Hermitian adjoint* of the matrix A, which is denoted $A^{\dagger}$ and pronounced A-dagger:

$$A^{\dagger} = \left(A^{\mathrm{T}}\right)^{*} \tag{7.2.14}$$

In other words, to find a matrix's Hermitian adjoint, you must transpose the matrix and compute its complex conjugate. For example

$$A = \begin{pmatrix} 0 & 2i & 1+i \\ -i & 0 & -3 \end{pmatrix} \quad \text{and} \quad A^{\dagger} = \begin{pmatrix} 0 & i \\ -2i & 0 \\ 1-i & -3 \end{pmatrix}$$

A matrix A is called a *Hermitian matrix* if $A = A^{\dagger}$. Hermitian matrices play an important role in quantum mechanics, since operators corresponding to physical measurements can be represented by Hermitian matrices.

In Chapter 6, we discussed taking the dot product of two complex vectors $\mathbf{u}$ and $\mathbf{v}$. Recall that the dot product is done by calculating $\mathbf{u}^{*} \cdot \mathbf{v}$. Now that we have discussed matrix multiplication, we can reformulate the dot product using the Hermitian adjoint of $\mathbf{u}$ when computing the dot product. Hence, the dot product between two complex vectors $\mathbf{u}$ and $\mathbf{v}$ can be expressed as $\mathbf{u}^{\dagger}\mathbf{v}$.

Example 7.3: The Pauli matrices

The *Pauli matrices* (also known as the Pauli spin matrices) are related to angular momentum operators for spin $1/2$ particles in each of three spatial dimensions x, y, z. The Pauli matrices are:

$$\sigma_x = \begin{pmatrix} 0 & 1 \\ 1 & 0 \end{pmatrix} \quad \sigma_y = \begin{pmatrix} 0 & -i \\ i & 0 \end{pmatrix} \quad \sigma_z = \begin{pmatrix} 1 & 0 \\ 0 & -1 \end{pmatrix}$$

(a) Find the Hermitian conjugate of σ_y.
(b) Show that the commutator $[\sigma_x, \sigma_y] = \sigma_x \sigma_y - \sigma_y \sigma_x = 2i\sigma_z$.
(c) Show that the anti-commutator $\{\sigma_x, \sigma_y\} = \sigma_x \sigma_y + \sigma_y \sigma_x = 0$.

Solution:

(a) The Hermitian conjugate of σ_y is

$$\sigma_y^\dagger = \begin{pmatrix} 0 & -i \\ i & 0 \end{pmatrix}$$

and $\sigma_y^\dagger = \sigma_y$. Therefore σ_y is Hermitian. Furthermore, its determinant is $\det(\sigma_y) = 0 + i^2 = -1$ and its trace is $\mathrm{Tr}(\sigma_y) = 0$.

(b) We evaluate the products of the matrices:

$$[\sigma_x, \sigma_y] = \sigma_x \sigma_y - \sigma_y \sigma_x = \begin{pmatrix} 0 & 1 \\ 1 & 0 \end{pmatrix} \begin{pmatrix} 0 & -i \\ i & 0 \end{pmatrix} - \begin{pmatrix} 0 & -i \\ i & 0 \end{pmatrix} \begin{pmatrix} 0 & 1 \\ 1 & 0 \end{pmatrix}$$

$$[\sigma_x, \sigma_y] = \begin{pmatrix} 0 \cdot 0 + 1 \cdot i & 0 \cdot (-i) + 1 \cdot 0 \\ 1 \cdot 0 + 0 \cdot i & 1 \cdot (-i) + 0 \cdot 0 \end{pmatrix} - \begin{pmatrix} 0 \cdot 0 - i \cdot 1 & 0 \cdot 1 - i \cdot 0 \\ i \cdot 0 + 0 \cdot 1 & i \cdot 1 + 0 \cdot 0 \end{pmatrix}$$

$$[\sigma_x, \sigma_y] = \begin{pmatrix} i & 0 \\ 0 & -i \end{pmatrix} - \begin{pmatrix} -i & 0 \\ 0 & i \end{pmatrix} = \begin{pmatrix} 2i & 0 \\ 0 & -2i \end{pmatrix} = 2i\sigma_z$$

(c) Similarly

$$\sigma_x \sigma_y + \sigma_y \sigma_x = \begin{pmatrix} 0 & 1 \\ 1 & 0 \end{pmatrix} \begin{pmatrix} 0 & -i \\ i & 0 \end{pmatrix} + \begin{pmatrix} 0 & -i \\ i & 0 \end{pmatrix} \begin{pmatrix} 0 & 1 \\ 1 & 0 \end{pmatrix}$$

$$\sigma_x \sigma_y + \sigma_y \sigma_x = \begin{pmatrix} i & 0 \\ 0 & -i \end{pmatrix} + \begin{pmatrix} -i & 0 \\ 0 & i \end{pmatrix} = \begin{pmatrix} 0 & 0 \\ 0 & 0 \end{pmatrix}$$

In this example, it is helpful to use NumPy arrays for the Pauli matrices because they contain numbers and not symbols. Here the matrices are defined using `np.array`, and we use the NumPy functions `np.conjugate()` and `np.transpose()` to evaluate the Hermitian adjoint of the matrices. We also use the function `np.array_equal` to test whether the two arrays are equal.

```
import numpy as np
from sympy import simplify
print('-'*28,'CODE OUTPUT','-'*29,'\n')


s_x = np.array([[0,1],[1,0]])
s_y = np.array([[0,-1j],[1j,0]])
s_z = np.array([[1,0],[0,-1]])


s_x_dagger = np.conjugate(np.transpose(s_x))
s_y_dagger = np.conjugate(np.transpose(s_y))
s_z_dagger = np.conjugate(np.transpose(s_z))

print('\nIs s_x Hermitian? ', np.array_equal(s_x, s_x_dagger))
print('Is s_y Hermitian? ', np.array_equal(s_y, s_y_dagger))
print('Is s_z Hermitian? ', np.array_equal(s_z, s_z_dagger))


commut = simplify(s_x @ s_y - s_y @ s_x )
print('\nThe commutator [s_x,s_y] =', commut)


print( '\nThe equation [s_x, s_y] = 2i*s_z is ',\
  np.array_equal(s_x @ s_y - s_y @ s_x, 2j* s_z ))
```

```
------------------------- CODE OUTPUT ----------------------------


Is s_x Hermitian?  True
Is s_y Hermitian?  True
Is s_z Hermitian?  True


The commutator [s_x,s_y] = [[2.0*I, 0], [0, -2.0*I]]


The equation [s_x, s_y] = 2i*s_z is  True
```

Table 7.1 shows a partial list of matrix-related functions and methods from both NumPy and SymPy. Note that the @ operator is the recommended method for computing the matrix product AB between two-dimensional arrays A and B in NumPy, and requires Python version 3.5 or later.

Table 7.1

Partial list of matrix/arrays functions and methods from NumPy and SymPy.

	NumPy	SymPy
Determinant of matrix A	linalg.det(A)	A.det()
Trace of matrix A	trace(A)	A.trace()
Inverse of a general matrix A	inv(A)	A.inv()
Eigenvalues of a general matrix a	linalg.eigvals(a)	A.eigenvals
Transpose of matrix A	transpose(A)	A.T
Complex conjugate of matrix A	conjugate(A)	
Norm of Matrix or vector a	linalg.norm(a)	A.eigenvects
Matrix product of 2D arrays A, B	A @ B (operator)	A*B
Dot product of arrays A, B	dot(a,b)	a.dot(b)
Cross product of arrays A, B	cross(a,b)	a.cross(b)
Solve system of linear scalar/matrix equations	linalg.solve(a, b)	solve(a,b)

7.3 THE DETERMINANT

The *determinant* of a matrix A, denoted by $\det(A)$, is a number associated with a square matrix. Computers can easily compute this value for you. However, it is helpful to understand how to perform the calculation by hand for 2×2 and 3×3 matrices. In this section, we will present how to calculate the determinant (both by hand and with Python) and some of its properties. We will also revisit the cross product, and show its connection with the determinant.

7.3.1 CALCULATING DETERMINANTS

Consider the general 2×2 matrix A, its determinant $\det(A)$ can be found using:

$$A = \begin{pmatrix} a_{11} & a_{12} \\ a_{21} & a_{22} \end{pmatrix} \qquad \det(A) = \begin{vmatrix} a_{11} & a_{12} \\ a_{21} & a_{22} \end{vmatrix} = a_{11}a_{22} - a_{12}a_{21} \qquad (7.3.1)$$

The vertical lines in (7.3.1) denote the determinant of the matrix.

The method of finding the determinant of a 2×2 matrix is central for finding the determinants of larger square matrices. Let us now consider the determinant of a 3×3 matrix A:

$$\det(A) = \begin{vmatrix} a_{11} & a_{12} & a_{13} \\ a_{21} & a_{22} & a_{23} \\ a_{31} & a_{32} & a_{33} \end{vmatrix} \qquad (7.3.2)$$

To calculate this determinant, we will use the following formula:

$$\det(A) = a_{11} \begin{vmatrix} a_{22} & a_{23} \\ a_{32} & a_{33} \end{vmatrix} - a_{12} \begin{vmatrix} a_{21} & a_{23} \\ a_{31} & a_{33} \end{vmatrix} + a_{13} \begin{vmatrix} a_{21} & a_{22} \\ a_{31} & a_{32} \end{vmatrix} \qquad (7.3.3)$$

Notice that each term has a similar structure. Consider the first term in (7.3.3). We multiply a_{11} by the determinant of the matrix that remains when we eliminate the first row and first column of A. The next term multiplies the element a_{12} by the determinant of the matrix remaining when the first row and second column are removed. Finally, the third term is a_{13} multiplied by the determinant of the matrix remaining when the first row and third column of A are removed. The determinant of the matrix that remains when we remove the element containing a_{ij} is called the *minor* M_{ij} of a_{ij}.

This pattern would persist if we had a 4×4 matrix. However, the minors would now consist of the determinant of the remaining 3×3 matrix. The pattern of alternating signs would also persist, i.e.:

$$\det(A) = a_{11}M_{11} - a_{12}M_{12} + a_{13}M_{13} - a_{14}M_{14}$$

More generally, we can write the determinant by expansion along any row as:

$$\det(A) = \sum_{j=1}^{n} (-1)^{i+j} a_{ij} |M_{ij}|$$

As you can see, these calculations can become cumbersome to do by hand.

Strictly speaking, there is nothing special about using the minors of the first row of A. We could use any row or column. However, in physics that is rarely done, so we will stick with the procedure outlined above.

Example 7.4: The determinant of a 3×3 matrix

Consider the matrix

$$C = \begin{pmatrix} 2 & 3 & -1 \\ 1 & 4 & -2 \\ 5 & 9 & 0 \end{pmatrix}$$

(a) Calculate the determinant of C by hand and using Python.
(b) Use SymPy to define two *symbolic* 3×3 matrices A and B and demonstrate the general identity:

$$\det (AB) = \det (A) \det (B)$$

Solution:
(a) We begin by using the method of minors.

$$\det(C) = 2 \begin{vmatrix} 4 & -2 \\ 9 & 0 \end{vmatrix} - 3 \begin{vmatrix} 1 & -2 \\ 5 & 0 \end{vmatrix} + (-1) \begin{vmatrix} 1 & 4 \\ 5 & 9 \end{vmatrix}$$
$$= 2 (0 + 18) - 3 (0 + 10) - 1 (9 - 20) = 17$$

In this part we use the function `array()` in NumPy to define the given matrix C, and the function `linalg.det(A)` to evaluate the determinant, within the linear algebra submodule `linalg`.

(b) In this part we define two symbolic 3×3 matrices A and B using `MatrixSymbol` from SymPy. Next we use the `A.as_mutable()` method to change these symbolic matrices into mutable matrices with explicit elements, so that we can evaluate their determinants.
The NumPy function `array_equal()` is used to test the validity of the equation $\det (AB) = \det (A) \det (B)$.

```
from sympy import MatrixSymbol
from numpy import array, array_equal, linalg

print('-'*28,'CODE OUTPUT','-'*29,'\n')

# PART (a): use NumPy to find determinant
C = array([[2,3,-1], [1,4,-2], [5,9,0]])

print('Using NumPy in part (a), the det(C)=',linalg.det(C))

# PART (b): Create (3x3) symbolic matrices with symbols A, B
A = MatrixSymbol('A', 3, 3)
B = MatrixSymbol('B', 3, 3)

# make symbolic matrices into mutable matrices with explicit elements
Am = A.as_mutable()
Bm = B.as_mutable()
print('\n A =')
Am
print('\n B =')
Bm
print('\n The identity det(AB)=det(A)det(B) is: ',\
      array_equal(Am.det()*Bm.det(),Bm.det()*Am.det()))

------------------------- CODE OUTPUT -----------------------------

Using NumPy in part (a), the det(C)= 17.0


 A =
Matrix([
[A[0, 0], A[0, 1], A[0, 2]],
[A[1, 0], A[1, 1], A[1, 2]],
[A[2, 0], A[2, 1], A[2, 2]]])
```

```
B =
Matrix([
[B[0, 0], B[0, 1], B[0, 2]],
[B[1, 0], B[1, 1], B[1, 2]],
[B[2, 0], B[2, 1], B[2, 2]]])

The identity det(AB)=det(A)det(B) is:   True
```

While you will often find yourself using a computer to calculate determinants for matrices larger than 3×3, there are some helpful rules that could help you calculate a determinant faster than inputting the matrix into the computer.

1. If each element of a row or a column of the matrix A is multiplied by the scalar k, then the value of the determinant is also multiplied by k.

2. The determinant is zero if

   a. all elements of one row (or column) are zero.

   b. two rows or two columns are identical.

   c. two rows or two columns are proportional to each other.

3. If two rows or two columns are interchanged, then the determinant changes sign.

4. The determinant of the transpose of a matrix is the same as that of the original matrix: $\det(A) = \det(A^{\mathrm{T}})$

5. We can add to each element of a row, a scalar k times the corresponding element of another row, without changing the value of the determinant. The same is true for columns.

Below are some useful identities for the calculation of determinants, where A and B are $n \times n$ matrices, $\mathcal{I}$ is the identity matrix and c is a constant.

$$\det \mathcal{I} = 1 \tag{7.3.4}$$
$$\det(AB) = \det A \det B \tag{7.3.5}$$
$$\det(cA) = c^n \det A \tag{7.3.6}$$
$$\det\left(A^{\mathrm{T}}\right) = \det A \tag{7.3.7}$$
$$\det\left(A^{-1}\right) = [\det A]^{-1} \tag{7.3.8}$$

where A^{-1} is the inverse of matrix A. Inverse matrices are discussed later in this chapter.

7.3.2 DETERMINANTS AS REPRESENTATIONS OF AREA AND VOLUME

Now that we know how to calculate the determinant, it is useful to connect it to a geometrical concept.

Here we will demonstrate the connection between the determinant and area in two-dimensions. As we saw in Chapter 4, the cross product $\mathbf{A} \times \mathbf{B}$ of any two vectors $\mathbf{A} = A_x\hat{\mathbf{i}} + A_y\hat{\mathbf{j}} + A_z\hat{\mathbf{k}}$ and $\mathbf{B} = B_x\hat{\mathbf{i}} + B_y\hat{\mathbf{j}} + B_z\hat{\mathbf{k}}$, represents the area of the parallelogram defined by the two vectors $\mathbf{A}, \mathbf{B}$:

$$\text{Area} = |\mathbf{A} \times \mathbf{B}| \tag{7.3.9}$$

We also saw that:

$$\mathbf{A} \times \mathbf{B} = (A_y B_z - A_z B_y)\,\hat{\mathbf{i}} + (A_z B_x - A_x B_z)\,\hat{\mathbf{j}} + (A_x B_y - A_y B_x)\,\hat{\mathbf{k}} \qquad (7.3.10)$$

Which can be rewritten as the determinant:

$$\mathbf{A} \times \mathbf{B} = \begin{vmatrix} \hat{\mathbf{i}} & \hat{\mathbf{j}} & \hat{\mathbf{k}} \\ A_x & A_y & A_z \\ B_x & B_y & B_z \end{vmatrix} \qquad (7.3.11)$$

Notice that in (7.3.11), the first vector $\mathbf{A}$ in the cross product $\mathbf{A} \times \mathbf{B}$ is the second row of the determinant. Likewise, the second vector $\mathbf{B}$ in the cross product $\mathbf{A} \times \mathbf{B}$ is the third row. It is important to preserve this order, because as we know from our discussion of the cross product and of the determinant rules, the cross product is anti-commutative, i.e. the determinant changes sign when the order of the vectors in the cross product $\mathbf{A} \times \mathbf{B}$ is reversed.

If the two vectors $\mathbf{A}, \mathbf{B}$ are located in the xy-plane, then $A_z = B_z = 0$ and the cross product becomes:

$$\mathbf{A} \times \mathbf{B} = (A_x B_y - A_y B_x)\,\hat{\mathbf{k}} \qquad (7.3.12)$$

By combining (7.3.9) and (7.3.12), we obtain:

$$\text{Area} = |\mathbf{A} \times \mathbf{B}| = |A_x B_y - A_y B_x| = \begin{vmatrix} A_x & A_y \\ B_x & B_y \end{vmatrix} \qquad (7.3.13)$$

Therefore, in two-dimensions the absolute value of the determinant gives the area of the parallelogram defined by the two vectors $\mathbf{A}, \mathbf{B}$. If we include the sign of the determinant, then we obtain the *oriented area* of the parallelogram. The oriented area is the same as the area. However, the oriented area is negative when the angle from the first to the second vector defining the parallelogram is in a clockwise direction.

Similarly, it can be shown that the volume of the parallelopiped spanned by any three vectors $\mathbf{A}, \mathbf{B}$ and $\mathbf{C}$ is equal to the absolute value of the determinant of the matrix with columns (or rows) the components of the vectors:

$$\text{Volume} = |\mathbf{C} \cdot (\mathbf{A} \times \mathbf{B})| = \begin{vmatrix} A_x & A_y & A_z \\ B_x & B_y & B_z \\ C_x & C_y & C_z \end{vmatrix} \qquad (7.3.14)$$

During the evaluation of a determinant, we can treat the unit vectors as constants, as shown in Example 7.5.

Example 7.5: The Lorenz force

Consider a particle with a charge q which is moving at a time t with a velocity $\mathbf{v} = (v_x, v_y, v_z)$ through a magnetic field $\mathbf{B} = (B_x, B_y, B_z)$. Find the components of the Lorenz force $\mathbf{F} = q\,(\mathbf{v} \times \mathbf{B})$ acting on the particle.

Solution:

The Lorenz force $\mathbf{F} = q\,(\mathbf{v} \times \mathbf{B})$ can be found using the determinant equation for the cross product $\mathbf{v} \times \mathbf{B}$:

$$\begin{vmatrix} \hat{\mathbf{i}} & \hat{\mathbf{j}} & \hat{\mathbf{k}} \\ v_x & v_y & v_z \\ B_x & B_y & B_z \end{vmatrix} = \hat{\mathbf{i}}\begin{vmatrix} v_y & v_z \\ B_y & B_z \end{vmatrix} - \hat{\mathbf{j}}\begin{vmatrix} v_x & v_z \\ B_x & B_z \end{vmatrix} + \hat{\mathbf{k}}\begin{vmatrix} v_x & v_y \\ B_x & B_y \end{vmatrix}$$

$$= \hat{\mathbf{i}}\,(v_y B_z - v_z B_y) - \hat{\mathbf{j}}\,(v_x B_z - v_z B_x) + \hat{\mathbf{k}}\,(v_x B_y - v_y B_x)$$

To obtain the force, we multiply by the charge q:

$$\mathbf{F} = \hat{\mathbf{i}}q\,(v_y B_z - v_z B_y) - \hat{\mathbf{j}}q\,(v_x B_z - v_z B_x) + \hat{\mathbf{k}}q\,(v_x B_y - v_y B_x)$$

In the Python code we define the symbols for the component vectors using `symbols` from SymPy. We next use `array` to define the matrices in NumPy, and the NumPy function `cross(v,B)` to evaluate the cross product. We can find the specific components of the force $\mathbf{F}$ as the elements `F[0]`, `F[1]` and `F[2]`.

```python
from numpy import array, cross
from sympy import symbols

print('-'*28,'CODE OUTPUT','-'*29,'\n')

Bx, By, Bz, vx, vy, vz, q = symbols('Bx, By, Bz, vx, vy, vz, q')

B = array([Bx, By, Bz])
v = array([vx, vy, vz])

F = q * cross(v,B)
print('The Lorenz force is F=')
print(F)

print('\nFx = ',F[0])
print('Fy = ',F[1])
print('Fz = ',F[2])


-------------------------- CODE OUTPUT ----------------------------


The Lorenz force is F=
[q*(-By*vz + Bz*vy) q*(Bx*vz - Bz*vx) q*(-Bx*vy + By*vx)]

Fx =  q*(-By*vz + Bz*vy)
Fy =  q*(Bx*vz - Bz*vx)
Fz =  q*(-Bx*vy + By*vx)
```

7.3.3 THE JACOBIAN DETERMINANT AND VOLUME TRANSFORMATIONS

The Jacobian determinant J (or simply Jacobian) is used when making a change of variables from one coordinate system to another. For example, the Jacobian can be very useful when evaluating the multiple integral of a function.

Let us consider a general situation, where the variables $(y_1, \ldots, y_n)$ are known function of n other variables $(x_1, \ldots, x_n)$. The Jacobian of $(y_1, \ldots, y_n)$ with respect to $(x_1, \ldots, x_n)$

is constructed using the partial derivatives of these functions:

$$J\left(\frac{y_1,\ldots,y_n}{x_1,\ldots,x_n}\right) = \begin{vmatrix} \dfrac{\partial y_1}{\partial x_1} & \cdots & \dfrac{\partial y_1}{\partial x_n} \\ \vdots & \ddots & \vdots \\ \dfrac{\partial y_m}{\partial x_1} & \cdots & \dfrac{\partial y_m}{\partial x_n} \end{vmatrix} \qquad (7.3.15)$$

A simpler notation for this Jacobian is also $J(y_1,\ldots,y_n)$. An important general theorem in vector analysis states that:

$$dy_1\, dy_2\ldots dy_n = |J|\, dx_1\, dx_2\ldots dx_n \qquad (7.3.16)$$

This result can be very useful when evaluating multiple integrals of a function f over a domain D:

$$\iiint_D f(y_1\, y_2\ldots y_n)\, dy_1\, dy_2\ldots dy_n = \iiint_D f(x_1\, x_2\ldots x_n)\, |J|\, dx_1\, dx_2\ldots dx_n$$

For instance, the Jacobian determinant of the transformation from Cartesian to Polar coordinates is found from:

$$J(\rho,\phi) = \begin{vmatrix} \dfrac{\partial x}{\partial \rho} & \dfrac{\partial x}{\partial \phi} \\ \dfrac{\partial y}{\partial \rho} & \dfrac{\partial y}{\partial \phi} \end{vmatrix} \qquad (7.3.17)$$

By substituting the x- and y-coordinates in polar coordinates, i.e. $x = \rho\cos\phi$ and $y = \rho\sin\phi$, we find:

$$J(\rho,\phi) = \begin{vmatrix} \dfrac{\partial(\rho\cos\phi)}{\partial\rho} & \dfrac{\partial(\rho\cos\phi)}{\partial\phi} \\ \dfrac{\partial(\rho\sin\phi)}{\partial\rho} & \dfrac{\partial(\rho\sin\phi)}{\partial\phi} \end{vmatrix} = \begin{vmatrix} \cos\phi & -\rho\sin\phi \\ \sin\phi & \rho\cos\phi \end{vmatrix} = \rho\cos^2\phi + \rho\sin^2\phi = \rho \quad (7.3.18)$$

The Jacobian determinant of this transformation is then is equal to ρ. According to (7.3.16), an infinitesimal volume in the Cartesian coordinate system is transformed into an infinitesimal volume in the polar coordinate system thus:

$$dx\, dy = |J|\, d\rho\, d\phi = \rho\, d\rho\, d\phi \qquad (7.3.19)$$

This is of course the same result we obtained in Chapter 4, during our discussion of the polar coordinates.

Once the function is transformed and the domain evaluated, it is possible to calculate double integrals using polar coordinates:

$$\iint_D f(x,y)\, dx\, dy = \iint_T f(\rho\cos\phi, \rho\sin\phi)\, \rho\, d\rho\, d\phi \qquad (7.3.20)$$

The Jacobian for the transformation from Cartesian to cylindrical coordinates is the same as in polar coordinates, i.e. equal to ρ, and therefore the volume element $dx\, dy\, dz$ becomes $\rho\, d\rho\, d\phi\, dz$ according to:

$$dV = dx\, dy\, dz = \rho\, d\rho\, d\phi\, dz \qquad (7.3.21)$$

It is then possible to calculate a triple integral in cylindrical coordinates, by using the equation:

$$\iiint_D f(x,y,z)\,dx\,dy\,dz = \iiint_D f(\rho\cos\phi, \rho\sin\phi, z)\,\rho\,d\rho\,d\phi\,dz. \qquad (7.3.22)$$

The Jacobian determinant of the transformation from Cartesian to spherical polar coordinates is the following:

$$J(r,\theta,\phi) = \begin{vmatrix} \dfrac{\partial x}{\partial r} & \dfrac{\partial x}{\partial \theta} & \dfrac{\partial x}{\partial \phi} \\[2mm] \dfrac{\partial y}{\partial r} & \dfrac{\partial y}{\partial \theta} & \dfrac{\partial y}{\partial \phi} \\[2mm] \dfrac{\partial z}{\partial r} & \dfrac{\partial z}{\partial \theta} & \dfrac{\partial z}{\partial \phi} \end{vmatrix} = \begin{vmatrix} \sin\theta\cos\phi & r\cos\theta\cos\phi & -r\sin\theta\sin\phi \\ \sin\theta\sin\phi & r\cos\theta\sin\phi & r\sin\theta\cos\phi \\ \cos\theta & -r\sin\theta & 0 \end{vmatrix} \qquad (7.3.23)$$

$$J(r,\theta,\phi) = r^2\sin\theta \qquad (7.3.24)$$

The differential volume element dV is:

$$dV = dx\,dy\,dz = r^2 dr\,\sin\theta\,d\theta\,d\phi \qquad (7.3.25)$$

and we can calculate triple integrals $I = \iiint f(x,y,z)\,dx\,dy\,dz$ in spherical coordinates by using the integration formula:

$$I = \iiint f(r\sin\theta\cos\phi, r\sin\theta\sin\phi, r\cos\theta)\,r^2\sin\theta\,dr\,d\theta\,d\phi \qquad (7.3.26)$$

Example 7.6 shows how to evaluate the area of an ellipse by transforming the Cartesian coordinates (x,y) into a new coordinate system (x',y'). The example demonstrates the use of the Jacobian to evaluate multiple integrals.

Example 7.6: Example of using the Jacobian: The area of an ellipse

Evaluate the area of an ellipse

$$\frac{x^2}{a^2} + \frac{y^2}{b^2} = 1 \qquad (7.3.27)$$

by using an appropriate transformation of variables. Here a, b are the semi-major and semi-minor axis of the ellipse.

Solution:
We use two new variables $x' = x/a$ and $y' = y/b$, so that (7.3.27) becomes:

$$\left(x'\right)^2 + \left(y'\right)^2 = 1 \qquad (7.3.28)$$

The ellipse has now been effectively transformed into a circle of radius $R = 1$ in the x'-y' coordinate system. We evaluate the Jacobian of (x,y) with respect to (x',y'):

$$J(x,y) = \begin{vmatrix} \dfrac{\partial x}{\partial x'} & \dfrac{\partial x}{\partial y'} \\[2mm] \dfrac{\partial y}{\partial x'} & \dfrac{\partial y}{\partial y'} \end{vmatrix} = \begin{vmatrix} a & 0 \\ 0 & b \end{vmatrix} = a\,b \qquad (7.3.29)$$

and the differential area element is:

$$dx\,dy = |J|\,dx'\,dy' = a\,b\,dx'\,dy' \qquad (7.3.30)$$

The area is given by:

$$A = \iint_D dx\,dy = \iint_{D'} |J|\,dx'\,dy' = a\,b \iint_{D'} dx'\,dy'$$

The integral over the circular domain D' of radius $R = 1$ is of course the area of a circle $A = \pi R^2 = \pi$, so that the area of the ellipse is:

$$A = \pi a\,b$$

In the Python code, `diff()` from SymPy is used to evaluate the Jacobian, and `integrate()` is used to evaluate the double integral over the over the circular domain D' of radius $R = 1$:

$$A = 4 \int_{x'=0}^{x'=1} \int_{y'=0}^{y'=\sqrt{1-(x')^2}} dx\,dy = \pi a\,b$$

```
from sympy import symbols, integrate, diff, pi, simplify, sqrt
print('-'*28,'CODE OUTPUT','-'*29,'\n')

x, y, a, b, xprime, yprime = symbols('x, y, a, b, xprime, yprime',\
real=True)  # symbols

x = a*xprime
y = b*yprime

J = diff(x,xprime)*diff(y,yprime)-diff(x,yprime)*diff(y,yprime)
print('Jacobian determinant J(x,y) = ',simplify(J),'\n')

A = integrate(4*J,(yprime,0,sqrt(1-xprime**2)),(xprime,0,1),)
print('Area of ellipse A = ',A)

-------------------------- CODE OUTPUT ----------------------------

Jacobian determinant J(x,y) =  a*b

Area of ellipse A =  pi*a*b
```

7.4   MATRICES AND SYSTEMS OF LINEAR EQUATIONS

Systems of linear equations arise frequently in science and engineering. Solving a large system of linear equations with the methods of elementary algebra can be cumbersome at best. Matrices provide an easier method.

In this section, we will discuss how to represent and solve systems of linear equations. We can represent these equations as a matrix in two different ways. The first method involves writing the system of equations as a matrix equation and computing the inverse of one of the matrices. The second method uses an *augmented matrix*. Operations on the rows of the augmented matrix will allow us to solve the system of equations.

We will demonstrate both methods symbolically and using Python, by solving the general system of equations:

$$\left.\begin{array}{l} a\,x + b\,y = e \\ c\,x + d\,y = f \end{array}\right\} \tag{7.4.1}$$

The methods demonstrated in this section are easily extended to higher dimensional systems of equations (i.e. systems with more than two equations and two unknowns).

7.4.1   REPRESENTING SYSTEMS OF LINEAR EQUATIONS AS A MATRIX EQUATION

Let us begin by rewriting (7.4.1) as a matrix equation

$$\left(\begin{array}{cc} a & b \\ c & d \end{array} \right) \left(\begin{array}{c} x \\ y \end{array} \right) = \left(\begin{array}{c} e \\ f \end{array} \right) \tag{7.4.2}$$

where a, b, c, d are constants and we solve (7.4.2) for the variables x and y. Notice that (7.4.2) is in the form $A\mathbf{x} = \mathbf{b}$ where in (7.4.2)

$$A = \left(\begin{array}{cc} a & b \\ c & d \end{array} \right) \quad \mathbf{x} = \left(\begin{array}{c} x \\ y \end{array} \right) \quad \mathbf{b} = \left(\begin{array}{c} e \\ f \end{array} \right) \tag{7.4.3}$$

The matrix A is called the *matrix of coefficients*, $\mathbf{x}$ is called the *solution vector*, and $\mathbf{b}$ is called the constant vector. Note that in this case the coefficient b is not the magnitude of the vector $\mathbf{b}$.

If we ignore the matrix and vector natures of A, $\mathbf{x}$, and $\mathbf{b}$, we would use basic algebra to solve $A\mathbf{x} = \mathbf{b}$ by dividing both sides by A. Of course, we cannot divide matrices. However, we can sometimes find matrices A^{-1} such that

$$A^{-1}A = \mathcal{I} \tag{7.4.4}$$

where A^{-1} is called the *left inverse of the matrix* A. Note that, although not commonly used in physics, there is a right inverse of the matrix as well, such that $AA^{-1} = \mathcal{I}$. In general, the left and right inverses of a matrix are not equal. It should also be noted that not all square matrices have inverses. We will discuss the conditions of when a matrix has an inverse later in this chapter. For now, let us assume that the matrix A has an inverse.

If we can find the matrix A^{-1}, then we could find $\mathbf{x}$ by multiplying both sides of (7.4.2) by A^{-1} on the left:

$$\left.\begin{array}{r} A\mathbf{x} = \mathbf{b} \\ A^{-1}A\mathbf{x} = A^{-1}\mathbf{b} \\ \mathbf{x} = A^{-1}\mathbf{b} \end{array}\right\} \tag{7.4.5}$$

How do we find A^{-1}? In practice, we use a computer. However, it is useful to know how to also do the calculation by hand. The formula for the inverse A^{-1} of a matrix is:

$$A^{-1} = \frac{1}{\det(A)} C^T \tag{7.4.6}$$

where C is called the *cofactor matrix* and its elements are defined as:

$$c_{ij} = (-1)^{i+j} m_{ij} \tag{7.4.7}$$

where m_{ij} is the determinant of the minor M_{ij} of the matrix A. Note that in the case of a 2×2 matrix, the minor is a scalar and the determinant of a scalar is simply the value of the scalar. Note that we have uncovered one condition necessary for a matrix to have an inverse: its determinant cannot be zero.

Example 7.7 allows us to solve (7.4.1) using $\mathbf{x} = A^{-1}\mathbf{b}$.

Example 7.7: The inverse of a Matrix

(a) Calculate the inverse of the matrix

$$A = \begin{pmatrix} a & b \\ c & d \end{pmatrix}$$

(b) Use the inverse matrix from (a) to solve the system of equations

$$\left. \begin{array}{l} a\,x + b\,y = e \\ c\,x + d\,y = f \end{array} \right\}$$

Solution:

We begin by computing the determinant of A

$$\det(A) = a\,d - c\,b$$

If $ad \neq cb$, then the matrix A will have an inverse. Assuming that is true, let us proceed with finding the cofactor matrix:

$$\begin{array}{ll} c_{11} = (-1)^2\,(d) = d & c_{12} = (-1)^3\,(c) = -c \\ c_{21} = (-1)^3\,(b) = -b & c_{22} = (-1)^4\,(a) = a \end{array}$$

Hence, the cofactor matrix is

$$C = \begin{pmatrix} d & -c \\ -b & a \end{pmatrix}$$

and the inverse is

$$A^{-1} = \frac{1}{a\,d - b\,c} \begin{pmatrix} d & -b \\ -c & a \end{pmatrix}$$

The solution of $A\mathbf{x} = \mathbf{b}$ is found using $\mathbf{x} = A^{-1}\mathbf{b}$:

$$\begin{pmatrix} x \\ y \end{pmatrix} = \frac{1}{a\,d - b\,c} \begin{pmatrix} d & -b \\ -c & a \end{pmatrix} \begin{pmatrix} e \\ f \end{pmatrix} = \frac{1}{a\,d - b\,c} \begin{pmatrix} d\,e - b\,f \\ -c\,e + a\,f \end{pmatrix} \quad (7.4.8)$$

Therefore

$$x = \frac{d\,e - b\,f}{a\,d - b\,c}$$

$$y = \frac{-c\,e + a\,f}{a\,d - b\,c}$$

In the Python code we define the coefficient matrix A and the symbols for a, b, c, d using `Matrix` and `symbols` in SymPy.

We evaluate the determinant and the inverse of the matrix A by using the methods `,det` and `.inv`. We also check that the product $A^{-1}A$ is equal to the identity matrix $\mathcal{I}$.

```
from sympy import symbols, Matrix, simplify, solve
import pprint

print('-'*28,'CODE OUTPUT','-'*29,'\n')

pp = pprint.PrettyPrinter(width=41, compact=True)

a, b, c, d, e, f, x, y = symbols('a, b, c, d, e, f, x, y')
A = Matrix([[a,b],[c,d]])

det = A.det()
inverse = A.inv()

print('The determinant of A =', det)

print('\n The inverse of A is: ')
pp.pprint(inverse)

print('\n The product inverseA*A is')
print(simplify(inverse*A))

sol=solve([a*x+b*y-e,c*x+d*y-f],(x,y))
print('\n The solution of the system is:')
print('x =',sol[x])
print('y =',sol[y])

--------------------------- CODE OUTPUT ----------------------------

The determinant of A = a*d - b*c

 The inverse of A is:
Matrix([
[ d/(a*d - b*c), -b/(a*d - b*c)],
[-c/(a*d - b*c),  a/(a*d - b*c)]])

 The product inverseA*A is
Matrix([[1, 0], [0, 1]])

 The solution of the system is:
x = (-b*f + d*e)/(a*d - b*c)
y = (a*f - c*e)/(a*d - b*c)
```

7.4.2 REPRESENTING SYSTEMS OF LINEAR EQUATIONS AS AN AUGMENTED MATRIX

The second method of solving the system $A\mathbf{x} = \mathbf{b}$, involves using an augmented matrix. An augmented matrix is made by appending a system of equation's constant vector $\mathbf{b}$ to the end of its matrix of coefficients A. For example, the augmented matrix of

$$\left.\begin{aligned} -2x + y &= 1 \\ x + y &= 4 \end{aligned}\right\} \tag{7.4.9}$$

is

$$\begin{pmatrix} -2 & 1 & 1 \\ 1 & 1 & 4 \end{pmatrix} \qquad (7.4.10)$$

We then perform a series of operations on the rows of the augmented matrix, in order to get it into *reduced row echelon form* . A matrix is in reduced row echelon form (also known as reduced echelon form) if all rows consisting of only zeros are at the bottom, and the leading coefficient of a nonzero row is always strictly to the right of the leading coefficient above it. Although some texts do not require it, we will require that the leading coefficient must be one.

For example, we will show that the reduced row echelon form of the augmented matrix (7.4.10) is:

$$\begin{pmatrix} 1 & 0 & 1 \\ 0 & 1 & 3 \end{pmatrix} \qquad (7.4.11)$$

Notice that the reduced row echelon form has the identity matrix as its first two columns. The solution found in (7.4.8) appears as the rightmost column of (7.4.11). Recall that we wrote the matrix of coefficients such that its first column contains the coefficients of x and its second column contains the coefficients of y. Further recall, that the solution vector $\mathbf{x}$ has x as its first element and y as its second. Therefore, the row with the 1 in the first column corresponds to the solution for x, i.e. we immediately have the solution $x = 1$. Similarly, the row with the 1 in the second column corresponds to the solution for $y = 3$.

How do we obtain the reduced row echelon form of a matrix (augmented or otherwise)? There are three *elementary row operations* we can perform. They are:

1. Interchange two rows

2. Multiply (or divide) a row by a nonzero constant

3. Add a multiple of one row to another.

Note for operation 3, subtraction would involve multiplying one of the rows by -1 then adding.

Here are the steps we used to get (7.4.11) from (7.4.10).

$$\begin{pmatrix} -2 & 1 & 1 \\ 1 & 1 & 4 \end{pmatrix} \xrightarrow{a} \begin{pmatrix} 1 & -1/2 & -1/2 \\ 1 & 1 & 4 \end{pmatrix} \xrightarrow{b} \begin{pmatrix} 1 & -1/2 & -1/2 \\ 0 & 3/2 & 9/2 \end{pmatrix} \xrightarrow{c} \qquad (7.4.12)$$

$$\begin{pmatrix} 1 & -1/2 & -1/2 \\ 0 & 1 & 3 \end{pmatrix} \xrightarrow{d} \begin{pmatrix} 1 & 0 & 1 \\ 0 & 1 & 3 \end{pmatrix} \qquad (7.4.13)$$

In step a, we divided the first row by -2. In step b, we made the second row by taking the matrix from step a and subtracting the first row from the second. In step c, we multiplied the second row by $2/3$. In step d, we divided the second row by 2 and added it to the first.

The augmented matrix can tell us when a solution exists for system of equations. The *rank* of a matrix is the number of nonzero rows remaining when a matrix has been row reduced. Let A be the matrix of coefficients for a system of n equations, and M be the corresponding augmented matrix for the system of equations.

1. If rank$(A) <$ rank(M), then the equations have no solution.

2. If rank$(A) =$ rank$(M) = n$, where n is the number of unknowns, then there is a unique solution to the system of equations.

3. If rank$(A) =$ rank$(M) = R < n$, then we can find the value of R of the unknowns in terms of the remaining $n - R$ unknowns.

The rank of a matrix can be found using the **rank** method in SymPy.

The process outlined above in steps $a - d$ is tedious, and rarely beneficial to do by hand. Fortunately, the SymPy library has the method **rref** which works well. The first element of the output in the method **rref** is the reduced matrix.

Example 7.8 illustrates the value of treating systems of equations as a matrix equation, and illustrates how to use the reduced matrix method.

Example 7.8: Kirchhoff's laws for electronics

Calculate the current in each resistor in the circuit shown in Figure 7.1, by using the .rref method in SymPy.

Solution:

We begin with the top loop of the circuit. Starting to the left of the battery with voltage V_1 and proceeding through the loop clockwise, Kirchhoff's loop rule gives:

$$10 - i_1 - 2i_2 - 5 = 0$$

where we have dropped the units for voltage and resistance. If we begin to the left of the battery V_2 and proceed counter clockwise through the bottom loop, the loop rule gives:

$$5 + 2i_2 - 3i_3 = 0$$

Finally, using the junction rule at the junction to the right of R_2 we obtain:

$$i_1 = i_2 + i_3$$

The resulting system of equations is:

$$i_1 + 2i_2 = 5$$
$$2i_2 - 3i_3 = -5$$
$$i_1 - i_2 - i_3 = 0$$

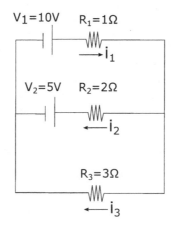

Figure 7.1　The circuit for Example 7.8. The arrows denote the assumed direction of current.

The augmented matrix for this system is

$$\begin{pmatrix} 1 & 2 & 0 & 5 \\ 0 & 2 & -3 & -5 \\ 1 & -1 & -1 & 0 \end{pmatrix}$$

where the first column is the coefficient if i_1, the second column is the coefficient if i_2, and the third column is the coefficient if i_3.

The Python code uses SymPy to reduce the augmented matrix to reduced row echelon form, with the result:

$$\begin{pmatrix} 1 & 0 & 0 & 35/11 \\ 0 & 1 & 0 & 10/11 \\ 10 & 0 & 1 & 25/11 \end{pmatrix}$$

Hence, we see that

$$i_1 = \frac{35}{11}A \quad i_2 = \frac{10}{11}A \quad i_3 = \frac{25}{11}A$$

In Problem 1, you will solve this problem using the inverse of the coefficient matrix.

```python
from sympy import Matrix

A = Matrix([[1,2,0,5],[0,2,-3,-5],[1,-1,-1,0]])

print('-'*28,'CODE OUTPUT','-'*29,'\n')
Ared=A.rref()[0]

print('The reduced matrix is:')
print(Ared)

print('\nCurrent i1 =', Ared[0,3])
print('Current i2 =', Ared[1,3])
print('Current i3 =', Ared[2,3])

---------------------------- CODE OUTPUT ----------------------------

The reduced matrix is:
Matrix([[1, 0, 0, 35/11], [0, 1, 0, 10/11], [0, 0, 1, 25/11]])

Current i1 = 35/11
Current i2 = 10/11
Current i3 = 25/11
```

7.4.3 HOMOGENEOUS EQUATIONS

We conclude this section with a special case of equations where $\mathbf{b} = 0$. In other words, the constants on the right-hand side of the system of equations are all zero. These are called *homogeneous equations*. For example, consider the system of equations and the reduced row echelon form of its corresponding augmented matrix.

$$\begin{aligned} x - 2y = 0 \\ x + y = 0 \end{aligned} \qquad \begin{pmatrix} 1 & 0 & 0 \\ 0 & 1 & 0 \end{pmatrix} \tag{7.4.14}$$

The rank of the augmented matrix (i.e. the number of non-zero rows) for (7.4.14) is equal to 2, the same as the number of unknowns. Therefore, there is a solution to (7.4.14), and it is $x = y = 0$. This is called the *trivial solution* of the system (7.4.14).

Now let's consider another system of equations and the associated row reduced matrix:

$$\begin{matrix} x + y = 0 \\ -3x - 3y = 0 \end{matrix} \quad \begin{pmatrix} 1 & 1 & 0 \\ 0 & 0 & 0 \end{pmatrix} \tag{7.4.15}$$

The rank of the augmented matrix for (7.4.15) is 1. However, there are two unknowns. Hence, in addition to the trivial solution, there are many values of x and y that satisfy (7.4.15) (i.e. $y = -x$). This is due to the fact that we have only one independent equation in (7.4.15), $x + y = 0$.

Homogeneous equations always have the trivial solution. If the number of independent equations is equal to the number of unknowns (i.e. the rank of the augmented matrix equals the number of unknowns), then the trivial solution is the only solution to the system. If there are less independent equations than unknowns (i.e. the rank of the augmented matrix is less than that of the number of unknowns), then there are an infinite number of solutions.

In physics, it is often the case that we are interested in systems where there are n homogeneous equations and n unknowns. In this case, the system of homogeneous equations takes the form

$$A\mathbf{x} = \mathbf{0} \tag{7.4.16}$$

where A is an $n \times n$ matrix of coefficients, and $\mathbf{x}$ is a $n \times 1$ vector of unknowns. If A is invertible, then the rank$(A) = n$, and the trivial solution $\mathbf{x} = \mathbf{0}$ is the only solution.

However, consider the case where rank$(A) < n$ (i.e. there are fewer independent equations than unknowns). Then at least one row of the row reduced echelon form of A is all zeros. In that case the det$(A) = 0$, the matrix is not invertible, and there are an infinite number of solutions.

We can summarize the above in the following way. A system of n homogeneous equations with n unknowns has solutions other than the trivial solution, if and only if, the determinant of its coefficient matrix is equal to zero.

Example 7.9 illustrates how to use Python to solve a system of homogeneous linear equations. We find the natural oscillating frequencies of the coupled system of two masses m_1 and m_2 attached to three springs with spring constants k_1, k_2, and k_3 and to two fixed walls.

Example 7.9: The normal modes of a two-mass three-spring system

Consider the coupled system of two equal masses m attached to three springs with equal spring constants k, which are attached to two fixed walls, as shown in Figure 7.2. Each spring in the figure obeys Hooke's law.

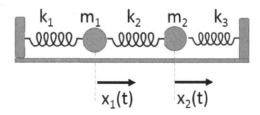

Figure 7.2 A system of two coupled harmonic oscillators, consisting of two masses, connected with three springs and attached to two walls.

(a) Write Newton's second law $F = m\,a$ for each of the two masses, by denoting with $x_1(t)$ and $x_2(t)$ the horizontal displacements of the two masses from their respective equilibrium points.
(b) Substitute $x_1(t) = A_1 e^{i\omega t}$ and $x_2(t) = A_2 e^{i\omega t}$ and solve the system of linear homogeneous equations, to find the natural frequencies of the system.

Solution:

The force on the first mass due to the first spring is $-kx_1$. The middle spring will be stretched by a distance $(x_1 - x_2)$, and the force on the first mass due to this middle spring will be $-k(x_1 - x_2)$. Newton's second law $F = ma$ for the two masses gives:

$$\left. \begin{array}{l} m\dfrac{d^2x_1}{dt^2} = -kx_1 - k(x_1 - x_2) \\[4mm] m\dfrac{d^2x_2}{dt^2} = -k(x_2 - x_1) - kx_2 \end{array} \right\} \tag{7.4.17}$$

Since we expect oscillatory motion, we try solutions of the form:

$$x_1(t) = A_1 e^{i\omega t} \qquad x_2(t) = A_2 e^{i\omega t} \tag{7.4.18}$$

where A_1 and A_2 are the unknown amplitudes of oscillation for the two masses, and ω is the unknown frequency of oscillation. Substituting these into (7.4.17) and canceling the $e^{i\omega t}$ terms, gives:

$$\left. \begin{array}{l} -m\,\omega^2 A_1 = -kA_1 - k(A_1 - A_2) \\[2mm] -m\,\omega^2 A_2 = -k(A_2 - A_1) - kA_2 \end{array} \right\} \tag{7.4.19}$$

This can be written in matrix form:

$$\begin{pmatrix} -\omega^2 m + 2k & -k \\ -k & -\omega^2 m + 2k \end{pmatrix} \begin{pmatrix} A_1 \\ A_2 \end{pmatrix} = \begin{pmatrix} 0 \\ 0 \end{pmatrix} \tag{7.4.20}$$

From the previous discussion, this system of linear homogeneous equations is of the form $A\mathbf{x} = \mathbf{0}$ and will have a non-zero solution only if the determinant of the coefficient matrix A is zero, i.e.:

$$\begin{vmatrix} -\omega^2 m + 2k & -k \\ -k & -\omega^2 m + 2k \end{vmatrix} = 0 \tag{7.4.21}$$

$$\left(\omega^2 m - 2k\right)\left(\omega^2 m - 2k\right) - k^2 = 0 \tag{7.4.22}$$

Solving for ω, we obtain four possible solutions, only two of which are positive and are the natural frequencies of the system:

$$\omega_1 = \sqrt{\frac{k}{m}}, \quad \omega_2 = \sqrt{\frac{3k}{m}} \tag{7.4.23}$$

In the Python code, we enter the differential equations for the variables x1,x2 to obtain the two symbolic equations eq1, eq2. Note the use of the expand() function to get rid of the parentheses in the evaluation of eq1, eq2.
The matrix of the coefficients A is obtained using the .coeff() method on eq1 and eq2 in Python.
Finally, the natural frequencies are obtained using the solve() function in SymPy, to solve the equation $\det(A) = 0$. The final result is the same as the answer obtained analytically above.

```
from sympy import symbols, exp, I, diff, solve, expand, symbols,  Matrix
import pprint
pp = pprint.PrettyPrinter(width=41, compact=True)


k, m, omega, x1, x2, A1, A2, t = \
symbols("k, m, omega, x1, x2, A1, A2, t")
```

```
#  define all symbols for variables

print('-'*28,'CODE OUTPUT','-'*29,'\n')

# define x1 and x2 as complex exponentials
x1 = A1*exp(I*omega*t)
x2 = A2*exp(I*omega*t)

# differential equations for x1, x2 divided by exp(I*omega*t)
# use expand to get rid of parentheses
eq1 = expand((m*diff(x1,t,t)+k*x1+k*(x1-x2))/exp(I*omega*t))
eq2 = expand((m*diff(x2,t,t)+k*x2+k*(x2-x1))/exp(I*omega*t))

# Matrix A has coefficients of A1, A2 in the system of equations
A = Matrix([[eq1.coeff(A1),eq1.coeff(A2)], [eq2.coeff(A1),eq2.coeff(A2)]])

print('The matrix of coefficients for A1, A2 is:\n')
pp.pprint(A)

# set determinant det(A)=0 and solve for omega
sol = solve(A.det(),omega)

# print natural frequencies omega1, omega2, omega3, omega4
print('\nNatural frequency omega1 = ',sol[0])
print('Natural frequency omega2 = ',sol[1])
print('\nNatural frequency omega3 = ',sol[2])
print('Natural frequency omega4 = ',sol[3])

-------------------------- CODE OUTPUT ----------------------------

The matrix of coefficients for A1, A2 is:

Matrix([
[2*k - m*omega**2,              -k],
[              -k, 2*k - m*omega**2]])

Natural frequency omega1 =  -sqrt(3)*sqrt(k/m)
Natural frequency omega2 =  sqrt(3)*sqrt(k/m)

Natural frequency omega3 =  -sqrt(k/m)
Natural frequency omega4 =  sqrt(k/m)
```

7.5   MATRICES AS REPRESENTATIONS OF LINEAR OPERATORS

In addition to representing systems of equations, matrices can also be used to represent linear transformations of vectors. In this context, a transformation is a function that maps one vector into another. Before we discuss matrices as linear transformations, we must first review linear functions and operators.

A linear vector function f must satisfy two rules

1. $f(\mathbf{v}_1 + \mathbf{v}_2) = f(\mathbf{v}_1) + f(\mathbf{v}_2)$ where $\mathbf{v}_1$ and $\mathbf{v}_2$ are vectors.

2. $f(c\,\mathbf{v}) = c\,f(\mathbf{v})$ where c is a scalar and $\mathbf{v}$ is a vector

The dot product is an example of a linear function of a vector. Consider a given vector $\mathbf{u}$ and a constant c, and define the function f such that $f(\mathbf{v}) = \mathbf{u} \cdot \mathbf{v}$. Then

$$
\begin{aligned}
f(\mathbf{v}_1 + \mathbf{v}_2) &= \mathbf{u} \cdot (\mathbf{v}_1 + \mathbf{v}_2) = \mathbf{u} \cdot \mathbf{v}_1 + \mathbf{u} \cdot \mathbf{v}_2 = f(\mathbf{v}_1) + f(\mathbf{v}_2) \\
f(c\,\mathbf{v}) &= \mathbf{u} \cdot (c\,\mathbf{v}) = c(\mathbf{u} \cdot \mathbf{v}) = c\,f(\mathbf{v})
\end{aligned}
\tag{7.5.1}
$$

Hence, the dot product satisfies the requirements of a linear function. However, the function $f(\mathbf{v}) = |v|$ that returns the magnitude of a vector, does not. For example, consider two vectors $\mathbf{u}$ and $\mathbf{v}$:

$$
f(\mathbf{u} + \mathbf{v}) = |\mathbf{u} + \mathbf{v}| \neq |\mathbf{v}| + |\mathbf{u}|
\tag{7.5.2}
$$

and therefore

$$
f(\mathbf{u} + \mathbf{v}) \neq f(\mathbf{u}) + f(\mathbf{v})
\tag{7.5.3}
$$

An *operator* is a mapping of a function from one space to another (possibly the same space). Consider the derivative d/dx:

$$
\frac{d}{dx} x^2 = 2x
\tag{7.5.4}
$$

In this case, the derivative operator d/dx transformed the function x^2 to the function $2x$. In other words, the operator d/dx mapped one function to another. In fact, the derivative is a *linear operator*. Consider the functions $f(x)$ and $g(x)$ and the constant c:

$$
\begin{aligned}
\frac{d}{dx}(f(x) + g(x)) &= \frac{d}{dx}f(x) + \frac{d}{dx}g(x) \\
\frac{d}{dx}(c\,f(x)) &= c\frac{d}{dx}f(x)
\end{aligned}
\tag{7.5.5}
$$

Another way of describing an operator is that an operator is a rule which provides instructions on what to do with its argument. In the case of d/dx, that rule is to take a derivative of the function that follows the operator.

In general, we say that the operator O is a linear operator if it satisfies these two conditions:

$$
O(X + Y) = O(X) + O(Y)
\tag{7.5.6}
$$

$$
O(c\,X) = c\,O(X)
\tag{7.5.7}
$$

where c is a scalar and X and Y are numbers, functions, vectors, or whatever is appropriate for the operator O to act upon.

For example, consider the operator $O(x) = x^2$. This is not a linear operator because

$$
O(x + y) = (x + y)^2 \neq x^2 + y^2
\tag{7.5.8}
$$

and therefore

$$
O(x + y) \neq O(x) + O(y)
\tag{7.5.9}
$$

Now let us return to matrices. For concreteness, consider a 2×2 matrix A multiplied by the 2×1 vector $\mathbf{v}$:

$$
A\mathbf{v} = \begin{pmatrix} a_{11} & a_{12} \\ a_{21} & a_{22} \end{pmatrix} \begin{pmatrix} v_1 \\ v_2 \end{pmatrix} = \begin{pmatrix} a_{11}v_1 + a_{12}v_2 \\ a_{21}v_1 + a_{22}v_2 \end{pmatrix} = \begin{pmatrix} u_1 \\ u_2 \end{pmatrix} = \mathbf{u}
\tag{7.5.10}
$$

Earlier this chapter, we saw the equation $A\mathbf{v} = \mathbf{u}$ as an example of matrix multiplication and as a representation of a system of equations. However, there is another way we can think about $A\mathbf{v} = \mathbf{u}$. We can think of the matrix A as transforming the vector $\mathbf{v}$ into the vector $\mathbf{u}$. In this context, the matrix A is a transformation from the vector space $\mathbb{R}^2$ to itself. In general, an $n \times n$ matrix is a transformation from the vector space $\mathbb{R}^n$ to itself.

Matrices are linear operators. We know using the basic rules of matrix operation that for an $n \times n$ matrix A, $n \times 1$ vectors $\mathbf{u}$ and $\mathbf{v}$, and scalar c:

$$\begin{aligned} A\left(\mathbf{u} + \mathbf{v}\right) &= A\mathbf{u} + A\mathbf{v} \\ A\left(c\mathbf{v}\right) &= c\left(A\mathbf{v}\right) \end{aligned} \tag{7.5.11}$$

We will now look at a special type of transformation, the *orthogonal transformations*. A matrix R representing an orthogonal transformation is called an *orthogonal matrix*. A matrix R is orthogonal, if its inverse is equal to its transpose:

$$R^{-1} = R^{\mathrm{T}} \tag{7.5.12}$$

It can be shown (see Problem 4) that for an orthogonal matrix R

$$\det(R) = \pm 1 \tag{7.5.13}$$

Orthogonal transformations preserve the length of a vector. In other words if R is the matrix representing an orthogonal transformation and $\mathbf{v}' = R\mathbf{v}$ then $|\mathbf{v}'| = |\mathbf{v}|$.

Not only do orthogonal matrices preserve the length of a vector, they also preserve areas (see Problem 5), making them an important class of transformations in physics. Example 7.10 shows how rotations in the xy-plane are represented by an orthogonal rotation matrix.

Example 7.10: The rotation matrix

Consider the vector $\mathbf{v}$ shown in Figure 7.3. The vector is rotated counterclockwise by an angle ϕ to create the new vector $\mathbf{v}'$. Note that the rotation does not change the length of the vector.
(a) Compute the components of $\mathbf{v}'$ in terms of the components of $\mathbf{v}$.
(b) Show $R^{-1} = R^{\mathrm{T}}$
(c) Show that the magnitude of the vector is preserved.

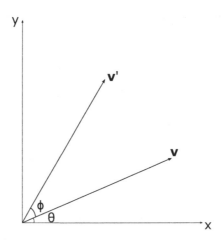

Figure 7.3 Example 7.10: The vector $\mathbf{v}$ is rotated by the angle ϕ.

Solution:

(a) We begin by using basic trigonometry for the components of $\mathbf{v}'$ and noting that $\mathbf{v}'$ and $\mathbf{v}$ have the same magnitude ($v' = v$):

$$v'_x = v \cos(\phi + \theta) = v[\cos\phi\cos\theta - \sin\phi\sin\theta] = v_x \cos\phi - v_y \sin\phi \qquad (7.5.14)$$

$$v'_y = v \sin(\phi + \theta) = v[\sin\phi\cos\theta + \cos\phi\cos\theta] = v_x \sin\phi + v_y \cos\phi \qquad (7.5.15)$$

which can be rewritten as

$$\begin{pmatrix} v'_x \\ v'_y \end{pmatrix} = \begin{pmatrix} \cos\phi & -\sin\phi \\ \sin\phi & \cos\phi \end{pmatrix} \begin{pmatrix} v_x \\ v_y \end{pmatrix} \qquad (7.5.16)$$

The 2×2 matrix in (7.5.16) is called the *rotation matrix*. We see that this matrix can be used to represent the rotation of a vector or a coordinate system.

Notice that the determinant of the rotation matrix is 1, satisfying the requirement for an orthogonal matrix.

(b) To show $R^{-1} = R^{\mathsf{T}}$, we must show that $R\,R^{\mathsf{T}} = \mathcal{I}$ where $\mathcal{I}$ is the identity matrix.

$$R\,R^{\mathsf{T}} = \begin{pmatrix} \cos\phi & -\sin\phi \\ \sin\phi & \cos\phi \end{pmatrix} \begin{pmatrix} \cos\phi & \sin\phi \\ -\sin\phi & \cos\phi \end{pmatrix}$$

$$R\,R^{\mathsf{T}} = \begin{pmatrix} \cos^2\phi + \sin^2\phi & 0 \\ 0 & \cos^2\phi + \sin^2\phi \end{pmatrix} = \begin{pmatrix} 1 & 0 \\ 0 & 1 \end{pmatrix}$$

(c) We need to show that $\left(v'_x\right)^2 + \left(v'_y\right)^2 = v_x^2 + v_y^2$.

Using (7.5.14) and (7.5.15):

$$\left(v'_x\right)^2 + \left(v'_y\right)^2 = (v_x \cos\phi - v_y \sin\phi)^2 + (v_x \sin\phi + v_y \cos\phi)^2$$

$$= v_x^2 \left(\cos^2\phi + \sin^2\phi\right) + v_y^2 \left(\cos^2\phi + \sin^2\phi\right) = v_x^2 + v_y^2$$

```
from sympy import Matrix, simplify, cos, sin, symbols
import pprint
pp = pprint.PrettyPrinter(width=41, compact=True)

print('-'*28,'CODE OUTPUT','-'*29,'\n')

p, vx, vy = symbols('p, vx, vy')

# define matrices R, v
R = Matrix([[cos(p),-sin(p)],[sin(p),cos(p)]])
v = Matrix([vx,vy])

vprime = R @ v
print('The rotated vector v =')
pp.pprint(vprime)

print('\n The determinant of R=',simplify(R.det()))

Rinv = R.inv()
print('\n The inverse of R =')
pp.pprint(simplify(Rinv))

print('\n The product of Rinverse @ R =')
```

```
pp.pprint(simplify(Rinv @ R))

print("\n The magnitude of v' =",simplify(vprime[0]**2+vprime[1]**2))

--------------------------- CODE OUTPUT ---------------------------

The rotated vector v =
Matrix([
[vx*cos(p) - vy*sin(p)],
[vx*sin(p) + vy*cos(p)]])

 The determinant of R= 1

 The inverse of R =
Matrix([
[ cos(p), sin(p)],
[-sin(p), cos(p)]])

 The product of Rinverse @ R =
Matrix([
[1, 0],
[0, 1]])

 The magnitude of v' = vx**2 + vy**2
```

7.6 EIGENVALUES AND EIGENVECTORS

As we saw earlier, we can think of the equation $A\mathbf{v} = \mathbf{u}$ as the matrix A transforming the vector $\mathbf{v}$ into the vector $\mathbf{u}$. Let us assume that A is an $n \times n$ matrix and $\mathbf{v}$ and $\mathbf{u}$ are $n \times 1$ vectors. We can ask a specific question. Are there vectors $\mathbf{v}$ such that $\mathbf{u} = \lambda\mathbf{v}$, where λ is a scalar? In other words we want to find the vectors $\mathbf{v}$ that satisfy:

$$A\mathbf{v} = \lambda\mathbf{v} \qquad (7.6.1)$$

A vector $\mathbf{v}$ that satisfies (7.6.1) for a given matrix A is called an *eigenvector* of A, and λ is called an *eigenvalue* of A associated with the eigenvector $\mathbf{v}$. Eigenvectors and eigenvalues play important roles in science. In classical mechanics, the eigenvectors of a rigid body's inertia tensor correspond to the body's axes of symmetry, and the eigenvalues are the moments of inertia about those axes. In quantum mechanics, eigenvectors correspond to potential states a particle can occupy after a measurement is made, while eigenvalues are potential results for a measurement.

Another way of interpreting (7.6.1) is to ask: Are there any vectors $\mathbf{v}$ whose line of direction are not changed by the matrix A? In other words, if $\lambda > 1$, the vector $\mathbf{v}$ is stretched (likewise, it would be compressed if $0 < \lambda < 1$), but its direction is not changed since multiplying a vector by a positive scalar does not change the vector's direction. If $\lambda < 0$, then the vector may be stretched or compressed (depending on the value of $|\lambda|$) but it also now points in the opposite direction. However, the vector would still lie along the line of its original direction. Eigenvectors are then special, because their line of direction is not changed by the matrix A.

How do we find the eigenvectors and eigenvalues of the matrix A? We can rewrite (7.6.1) in the following way

$$(A - \lambda\mathcal{I})\,\mathbf{v} = 0 \qquad (7.6.2)$$

Note that $\mathcal{I}$ is the $n \times n$ identity matrix, and multiplying by $\mathcal{I}$ does not change the vector equation.

The matrix equation (7.6.2) is a system of n homogeneous equations whose solution is **v**. The system (7.6.2) has a nontrivial solution if and only if

$$\det(A - \lambda\mathcal{I}) = 0 \tag{7.6.3}$$

Although it is not obvious, (7.6.3) will provide a nth-degree polynomial in λ, called the *characteristic polynomial* of the matrix A. We can solve the characteristic polynomial for λ. The $n \times n$ matrix A will have n eigenvalues. Once you have obtained the eigenvalues of A from (7.6.3), you can find the eigenvector **v** associated with each eigenvalue by using (7.6.2).

In general, an $n \times n$ matrix A will have n eigenvectors, one for each eigenvalue. However, a *degeneracy* can occur. For example, an eigenvalue is *degenerate* if it has more than one linearly independent eigenvector (see Chapter 9 for a discussion of linear independence). Degeneracies are common in problems in classical and quantum mechanics, and they often illustrate a symmetry in the problem.

In practice, you will often use a computer to find the eigenvalues and eigenvectors of a matrix. However, it is important to know how to do the calculation by hand, especially for the 2×2 and 3×3 cases. The next two examples show how to find eigenvectors and eigenvalues by hand and using Python.

Example 7.11: The eigenvalues and eigenvectors of a 2×2 matrix

Find the eigenvalues and eigenvectors of the matrix

$$A = \begin{pmatrix} 2 & 1 \\ 1 & 2 \end{pmatrix}$$

Solution:
To create the characteristic polynomial, let us write out the matrix $A - \lambda\mathcal{I}$

$$A - \lambda\mathcal{I} = \begin{pmatrix} 2 & 1 \\ 1 & 2 \end{pmatrix} - \begin{pmatrix} \lambda & 0 \\ 0 & \lambda \end{pmatrix} = \begin{pmatrix} 2-\lambda & 1 \\ 1 & 2-\lambda \end{pmatrix}$$

Next, we insert into (7.6.3) to get the characteristic polynomial

$$\begin{vmatrix} 2-\lambda & 1 \\ 1 & 2-\lambda \end{vmatrix} = (2-\lambda)^2 - 1 = 0$$

The roots of the characteristic polynomial are $\lambda = 1, 3$. Now that we have the eigenvalues, we find each eigenvector using (7.6.2). Let us look at the process in detail using $\lambda = 1$ by writing out the matrix $A - \lambda\mathcal{I}$:

$$A - \lambda\mathcal{I} = \begin{pmatrix} 2 & 1 \\ 1 & 2 \end{pmatrix} - \begin{pmatrix} 1 & 0 \\ 0 & 1 \end{pmatrix} = \begin{pmatrix} 1 & 1 \\ 1 & 1 \end{pmatrix}$$

We then insert into (7.6.2):

$$\begin{pmatrix} 1 & 1 \\ 1 & 1 \end{pmatrix}\begin{pmatrix} x \\ y \end{pmatrix} = \begin{pmatrix} 0 \\ 0 \end{pmatrix}$$

where we used x and y as place holders for the components of the vector $\mathbf{v}$. In this case, we find the solution requires that $y = -x$ but we don't have a specific value. We will choose $x = 1/\sqrt{2}$ to keep the eigenvector normalized (i.e. has a length of 1). Hence, the eigenvector for $\lambda = 1$ is $1/\sqrt{2}(1, -1)$.

Let us use SymPy to find the eigenvalues and eigenvectors using the .eigenvalues and .eigenvects methods.

The eigenvalues method returns a dictionary of the eigenvalues and their multiplicity (following the colon). The eigenvects returns a list of tuples. Each tuple contains the eigenvalue, the eigenvalue's multiplicity, and the eigenvalue's unnormalized eigenvector.

Specifically, the eigenvalue $\lambda = 3$ has a multiplicity of 1, and the corresponding unnormalized eigenvector is $(1, 1)$. The eigenvalue $\lambda = 1$ has a multiplicity of 1, with the corresponding unnormalized eigenvector being $(-1, 1)$.

```
from sympy import Matrix

A = Matrix([[1,2],[2,1]])

print('-'*28,'CODE OUTPUT','-'*29,'\n')

print('The eigenvalues are: ')
print(A.eigenvals())

print('\nThe eigenvectors are:\n',A.eigenvects()[0][2])
print('\n',A.eigenvects()[1][2])

---------------------------- CODE OUTPUT ----------------------------

The eigenvalues are:
{3: 1, -1: 1}

The eigenvectors are:
 [Matrix([
[-1],
[ 1]])]

 [Matrix([
[1],
[1]])]
```

In a previous chapter we saw how to calculate the elements of the 3×3 moment of inertia tensor for a solid. In Example 7.12 we show how to evaluate the eigenvalues and eigenvectors of a given inertia tensor.

Example 7.12: The inertia tensor of a cube

The inertia tensor for a cube of mass m and side b rotating about a corner is

$$A = \frac{mb^2}{12} \begin{pmatrix} 8 & -3 & -3 \\ -3 & 8 & -3 \\ -3 & -3 & 8 \end{pmatrix}$$

Using $m = 1$ kg and $b = 0.5$ m, find the eigenvalues and eigenvectors of the cube.

Solution:

Again, we will find the eigenvalues and one eigenvector by hand. For simplicity we will define a new variable $c = mb^2/12$. The eigenvalues are then found by solving (7.6.3)

$$\begin{vmatrix} 8c - \lambda & -3c & -3c \\ -3c & 8c - \lambda & -3c \\ -3c & -3c & 8c - \lambda \end{vmatrix} = 0$$

Computing this determinant provides the following characteristic polynomial.

$$(2c - \lambda)(11c - \lambda)^2 = 0$$

Hence, we have two distinct eigenvalues, $2c$ and $11c$ (with a multiplicity of 2).
Let us next find the eigenvector associated with $\lambda = 2c$ using (7.6.2)

$$(A - \lambda \mathcal{I})\,\mathbf{v} = c \begin{pmatrix} 6 & -3 & -3 \\ -3 & 6 & -3 \\ -3 & -3 & 6 \end{pmatrix} \begin{pmatrix} x \\ y \\ z \end{pmatrix} = \begin{pmatrix} 0 \\ 0 \\ 0 \end{pmatrix}$$

which produces the homogeneous system of equations

$$2x - y - z = 0$$
$$-x + 2y - z = 0$$
$$-x - y + 2z = 0$$

which has the solution $(1, 1, 1)$. After normalization, the eigenvector is

$$\frac{1}{\sqrt{3}}(1, 1, 1)$$

Notice that this eigenvector points along the cube's diagonal, an axis of symmetry.

Next, we use SymPy to find the other two eigenvectors. The eigenvalue $\lambda = 2c$ has a multiplicity of 1, and the corresponding unnormalized eigenvector is $(1, 1, 1)$. The eigenvalue $\lambda = 11c$ has a multiplicity of 2, with the corresponding unnormalized eigenvectors being $(-1, 1, 0)$ and $(-1, 0, 1)$.

```
from sympy import Matrix, symbols

c = symbols('c')
A = Matrix([[8*c,-3*c,-3*c],[-3*c,8*c,-3*c],[-3*c,-3*c,8*c]])

print('-'*28,'CODE OUTPUT','-'*29,'\n')

print('\nThe eigenvalues are: ')
print(A.eigenvals())

print('\n The eigenvectors are:')
print(A.eigenvects()[0][2][0])
print(A.eigenvects()[1][2][0])
print(A.eigenvects()[1][2][1])



-------------------------- CODE OUTPUT ---------------------------


The eigenvalues are:
{2*c: 1, 11*c: 2}

 The eigenvectors are:
Matrix([[1], [1], [1]])
Matrix([[-1], [1], [0]])
Matrix([[-1], [0], [1]])
```

7.7 DIAGONALIZATION OF A MATRIX

It is sometimes easier to describe a system by using a carefully chosen coordinate system. For example, when dealing with problems that involve spherical symmetry, using spherical coordinates can simplify the mathematical description of the problem. With this in mind, let us examine a specific problem in classical mechanics, angular momentum.

Recall from introductory physics the formula $\ell = I\omega$ where ℓ is the angular momentum of an object, I is the object's moment of inertia, and ω is the object's angular velocity. This formula holds in simple situations where an axis of rotation is along an object's axis of symmetry. In this case, the angular momentum and the angular velocities are along the same direction (the axis of rotation). However, such an alignment is not always the case. In more advanced classical mechanics courses, you learn the formula

$$\mathbf{L} = I\,\boldsymbol{\omega} \tag{7.7.1}$$

where $\mathbf{L}$ is the object's angular momentum vector, $\boldsymbol{\omega}$ is the object's angular velocity vector, and I is the object's moment of inertia tensor, which is a 3×3 symmetric matrix (i.e. $I^{\mathrm{T}} = I$).

Cartesian coordinates are typically used for (7.7.1). Let us now consider a coordinate transformation which is described using the 3×3 matrix S

$$\boldsymbol{\omega} = S\,\boldsymbol{\omega}' \qquad \mathbf{L} = S\mathbf{L}' \tag{7.7.2}$$

where primes denote the quantity in the new (primed) coordinate system. Let us further suppose that the matrix S is invertible. Then we can rewrite (7.7.1) as

$$\mathbf{L}' = S^{-1}\mathbf{L} = S^{-1}\,I\,S\,\boldsymbol{\omega}' = D\,\boldsymbol{\omega}' \tag{7.7.3}$$

where D is a 3×3 matrix defined by

$$D = S^{-1}\,I\,S \tag{7.7.4}$$

We say that the matrices D and I are *similar* matrices. In general, two matrices A and B are similar if there exists an invertible matrix S such that

$$A = S^{-1}\,B\,S \tag{7.7.5}$$

Similar matrices have the same rank, trace, determinant, and eigenvalues.

The matrix D in (7.7.4) serves as the moment of inertia tensor in the primed coordinate system. Now let us suppose we want our new coordinate system to be such that

$$L_i{}' = d_{ii}\,\omega_i' \tag{7.7.6}$$

where the subscript $i = 1, 2, 3$ denotes the component of the vectors $\mathbf{L}'$ and $\boldsymbol{\omega}'$ and d_{ii} is a scalar (and an element of the matrix D).

Equation (7.7.6) is close to the introductory physics case where the angular momentum and angular velocity vectors are parallel. However, because each d_{ii} could be different, the two vectors $\mathbf{L}'$ and $\boldsymbol{\omega}'$ will not be parallel. Regardless, the condition in (7.7.6) means that D is a *diagonal matrix* (i.e. the only nonzero terms in D are on the diagonal).

Now let us rewrite (7.7.4):

$$I\,S = S\,D \tag{7.7.7}$$

While (7.7.7) has the product of two 3×3 matrices on each side, we could instead focus on one column of S at a time. The ith column of IS in (7.7.7) becomes:

$$I\,S_i = d_{ii}\,S_i \tag{7.7.8}$$

In other words, the columns of the matrix S are the eigenvectors of I, and the elements along the diagonal of the matrix D are the eigenvalues of I. The coordinate system that puts I in diagonal form is the one whose axes are the eigenvectors of I. In classical mechanics, we call the eigenvectors of I *the principal axes*, and the eigenvalues of I are called *the principal moments of inertia*.

In physics, diagonalization can often be used to provide a coordinate system which decouples equations. For example, notice that according to (7.7.6), L_1' would depend only on ω_1' and not the other components of ω'. We say that (7.7.6) is an example of a decoupled equation, because only one component appears in the equation.

The process that we have outlined using the inertia tensor, is a specific example of what is known as *diagonalizing* a matrix. Matrix diagonalization is a specific similarity transformation which uses the eigenvectors of the matrix to create a similar matrix that is diagonal.

The above discussion suggests the following procedure for diagonalizing an inertia tensor:
(a) Construct a matrix S whose columns are the eigenvectors of I.
(b) Find the inverse matrix S^{-1}.
(c) Evaluate the product $D = S^{-1} I S$, the matrix D will be the diagonal form of the inertia matrix, with the principal moments of inertia along the diagonal.

An example of diagonalizing the Pauli spin matrix σ_y is found in Problem 19.

7.8   END OF CHAPTER PROBLEMS

1. **Electronics: Kirchhoff's laws** – Solve the system of equations in Example 7.8 by hand and using Python:

$$i_1 + 2i_2 = 5$$
$$2i_2 - 3i_3 = -5$$
$$i_1 - i_2 - i_3 = 0$$

(a) Using the `solve()` in SymPy
(b) Using the inverse of the coefficient matrix.

2. **Tests for linearity (functions)** – Are the following functions linear? Prove analytically your answers by showing that the function either does or does not satisfy the conditions required for a function to be linear. Also demonstrate each using Python's symbolic capabilities. (a) $f(x) = mx + b$, (b) $f(x) = x^2$, (c) $f(x) = \sin x$, (d) $f(\mathbf{v}) = \mathbf{a} \times \mathbf{v}$, where $\mathbf{a}$ is a given vector, (e) $f(\mathbf{v}) = \mathbf{a} + \mathbf{v}$, where $\mathbf{a}$ is a given vector.

3. **Tests for linearity (functions)** – Are the following functions linear? Prove analytically your answers by showing that the function either does or does not satisfy the conditions required for a function to be linear. Also demonstrate each using Python's symbolic capabilities. (a) The definite integral of a function $f(x)$ with respect to x from -1 to 1 , (b) $D = \frac{d}{dx}$ where if $f = f(x)$ then $Df = df/dx$, (c) $D^2 + 2D + 1$, where $D^2 = d^2/dx^2$ and D is defined as in part b above, (d) The transpose of a square matrix. For Python, use a generic 3×3 matrix, (e) The determinant of a square matrix. For Python, use a generic 3×3 matrix.

4. **Orthogonal matrices** – A matrix R is orthogonal if $R^{-1} = R^{\mathrm{T}}$.

   a. Is the matrix

   $$A = \begin{pmatrix} 0 & 1 \\ 1 & 0 \end{pmatrix}$$

   orthogonal?

b. Using the rules for determinants, prove $\det(R) = \pm 1$ for any orthogonal matrix R.

c. Show that the eigenvalues of A are ± 1, and that its eigenvectors are real and orthogonal to each other. This is a general property of any orthogonal matrix.

5. **Determinants and areas, revisited** – Consider the parallelogram with an area A and spanned by the two-dimensional vectors $\mathbf{v}$ and $\mathbf{u}$. The area spanned by the parallelogram is equal to the determinant of the matrix whose columns are $\mathbf{v}$ and $\mathbf{u}$. Now consider a linear transformation M, represented by a 2×2 matrix, which is applied to both $\mathbf{v}$ and $\mathbf{u}$, creating new vectors $M\mathbf{v}$ and $M\mathbf{u}$. The vectors $M\mathbf{v}$ and $M\mathbf{u}$ span a new parallelogram. Using Python to handle the algebra, show that the area of the new parallelogram (spanned by $M\mathbf{v}$ and $M\mathbf{u}$) is $\det(M) \times A$.

6. **(2 × 2) Matrix operations** – Let $c = 2$ and

$$A = \begin{pmatrix} -1 & 3 \\ 0 & 4 \end{pmatrix} \quad B = \begin{pmatrix} 2 & -1 \\ -2 & 5 \end{pmatrix}$$

Calculate by hand and using Python: (a) $cA + B$, (b) AB, (c) BA, (d) A^{-1}, (e) $\det(AB)$. Does it equal $\det(A)\det(B)$?

7. **(3 × 3) Matrix operations** – Let $c = 2$ and

$$A = \begin{pmatrix} 0 & -1 & 3 \\ 1 & 2 & -2 \\ 4 & 0 & -4 \end{pmatrix} \quad B = \begin{pmatrix} 1 & -1 & 0 \\ -3 & 0 & 1 \\ 5 & 2 & 1 \end{pmatrix}$$

Calculate by hand and using Python: (a) $cA + B$, (b) AB, (c) BA, (d) A^{-1}, (e) $\det(AB)$. Does it equal $\det(A)\det(B)$?

8. **Real symmetric matrix** – Real symmetric matrices are common in physics. The inertia tensor is one such example. Consider the real symmetric matrix

$$H = \begin{pmatrix} a & h \\ h & a \end{pmatrix}$$

where a and h are real values. Using Python, show that the eigenvalues of H are real and that its eigenvectors are perpendicular. Find the matrix that diagonalizes H.

9. **Eigenvalues and Eigenvectors** – Find the eigenvalues and eigenvectors of the following matrices by hand and using Python

$$A = \begin{pmatrix} 1 & -2 \\ -2 & 1 \end{pmatrix} \quad B = \begin{pmatrix} -19 & 24 \\ -12 & 17 \end{pmatrix}$$

$$C = \begin{pmatrix} -3 & 0 & -4 \\ -6 & 0 & -6 \\ 8 & 1 & 9 \end{pmatrix} \quad D = \begin{pmatrix} -3 & -2 & -4 \\ -12 & 3 & 0 \\ 10 & 2 & 7 \end{pmatrix}$$

10. **4 × 4 Matrix inverse and determinant** – Calculate the determinant of the 4×4 matrix:

$$D = \begin{pmatrix} 3 & 2 & 4 & 1 \\ 0 & 3 & 0 & 1 \\ 1 & 2 & 7 & 0 \\ 1 & 2 & 3 & 0 \end{pmatrix}$$

by hand and using Python. Use SymPy to evaluate the inverse D^{-1} of the matrix D, and verify the result by showing that the product of D and D^{-1} is the identity matrix.

11. **Rotation matrices** – The rotation of a vector on the xy-plane by an angle ϕ can be described by the 2×2 matrix:

$$R(\phi) = \begin{pmatrix} \cos\phi & \sin\phi \\ -\sin\phi & \cos\phi \end{pmatrix}$$

a. Find the inverse of the rotation matrix $R(\phi)$. Explain your result.

b. The above result can be generalized for rotational matrices in three dimensions:

$$S(\phi) = \begin{pmatrix} \cos\phi & \sin\phi & 0 \\ -\sin\phi & \cos\phi & 0 \\ 0 & 0 & 1 \end{pmatrix}$$

Find the inverse of the rotation matrix $S(\phi)$.

c. Show that the product of two matrices $S(\phi)\,S(\theta)$ describes a rotation by an angle $\phi + \theta$ in three dimensions.

12. **Hermitian matrices** – In quantum mechanics, measurements are represented using Hermitian matrices. Show that the matrices below are Hermitian, calculate their eigenvalues and eigenvectors, and write the unitary matrix U that diagonalizes each matrix by a similarity transformation. Perform the calculations by hand and using Python.

$$A = \begin{pmatrix} 1 & -i \\ i & 2 \end{pmatrix} \quad B = \begin{pmatrix} 2 & 1-i \\ 1+i & 2 \end{pmatrix}$$

13. **The triple scalar product revisited** – The triple scalar product can be found using the determinant:

$$\mathbf{A} \cdot (\mathbf{B} \times \mathbf{C}) = \begin{vmatrix} A_x & A_y & A_z \\ B_x & B_y & B_z \\ C_x & C_y & C_z \end{vmatrix}$$

a. If $\mathbf{A} = 2\hat{\mathbf{i}} - \hat{\mathbf{j}} - \hat{\mathbf{k}}$, $\mathbf{B} = 2\hat{\mathbf{i}} + 3\hat{\mathbf{j}} - \hat{\mathbf{k}}$, and $\mathbf{C} = \mathbf{A} = \hat{\mathbf{i}} + \hat{\mathbf{j}} + \hat{\mathbf{k}}$, calculate $\mathbf{A} \cdot (\mathbf{B} \times \mathbf{C})$ by hand and using Python

b. Using SymPy, prove that $\mathbf{A} \cdot (\mathbf{B} \times \mathbf{C}) = \mathbf{B} \cdot (\mathbf{C} \times \mathbf{A}) = \mathbf{C} \cdot (\mathbf{A} \times \mathbf{B})$ for *any* three vectors $\mathbf{A}, \mathbf{B}, \mathbf{C}$.

14. **Polarizability tensor** – If a weak uniform electric field $\mathbf{E}$ is applied to an atom, the atom's nucleus will shift slightly in the direction of the electric field. The atom's electron will shift slightly in the opposite direction. The separation of charge has polarized the atom, and thus a dipole moment has been induced on the atom. The strength of the atom's dipole moment $\mathbf{p}$ depends on the atom's atomic polarizabiltiy α such that

$$\mathbf{p} = \alpha \mathbf{E} \tag{7.8.1}$$

Molecules are more complicated. For example, in a carbon dioxide molecule, the oxygen atoms are on opposite sides of the carbon atom. The molecule's geometry affects is polarizabiliy. Carbon dioxide's polarizability for fields along the molecule's axis is more than twice that for fields that are perpendicular. To capture the added complexity, we use a 3×3 matrix instead of a constant for the polarizability of a molecule α.

    c. Using Cartesian coordinates, interpret the meaning of each element of α. For example, what does the term α_{xx} represent? How about α_{yz}? You may want to create symbolic vectors and matrices for $\mathbf{p}$, α, and $\mathbf{E}$ in Python and multiply them to help illustrate your answer.

    d. Using what you know about the inertia tensor, how do we interpret the eigenvectors and eigenvalues of α?

15. **Cramer's Rule** –

    a. Consider the following system of equations

$$a_{11}x + a_{12}y = b_1$$
$$a_{21}x + a_{22}y = b_2$$

    or in matrix form:

$$A \begin{pmatrix} x \\ y \end{pmatrix} = b$$

    Use SymPy to show that each solution (x, y) can be written as the ratio of determinants. This method of solving a system of linear equations is called *Cramer's rule*.

    b. Cramer's rule can be generalized to the solution to n linear equations with n unknowns. For example, use SymPy to show that each solution (x, y, z) of the system of equations

$$a_{11}x + a_{12}y + a_{13}z = b_1$$
$$a_{21}x + a_{22}y + a_{23}z = b_2$$
$$a_{31}x + a_{32}y + a_{33}z = b_3$$

can also be written as the ratio of determinants.

16. **Systems of linear equations** – Solve the following systems of linear equations by hand and using the following four methods discussed in this book:

$$
\begin{array}{ll}
2x - 3y = -5 & x - y + z = 0 \\
x + 4y = 2 & x + y + 2z = 1 \\
 & -x + 3y - 3z = -1
\end{array}
$$

    a. The inverse matrix method

    b. Cramer's rule method (see Problem 15)

    c. The `solve()` command in SymPy

    d. The reduced matrix method.

17. **Rank of matrices and solutions to systems of linear equations** – Without solving for the unknowns, use the `.rank()` method for matrices in SymPy, to identify how many variables can be evaluated in each of the following system of equations

$$
\begin{aligned}
x - y &= 0 \\
x + y &= 2
\end{aligned}
\qquad
\begin{aligned}
x - y + z &= 1 \\
3x - y + 2z &= 6 \\
-x + 3y - 2z &= 2
\end{aligned}
$$

18. **Eigenvalues and Eigenvectors of the two-mass and three-spring system** – Consider the system of two masses m_1 and m_2 and three different springs with constants $k_1, k_2,$ and k_3 (from left to right) connected as shown in Figure 7.2.

    a. Show that the natural frequencies of the system can be found from the equation:

    $$
    \det \begin{bmatrix} \frac{k_1+k_2}{m_1} - \omega^2 & \frac{-k_2}{m_1} \\ \frac{-k_2}{m_2} & \frac{k_2+k_3}{m_2} - \omega^2 \end{bmatrix} = 0
    $$

    b. For the special case of $k_1 = k_2 = k_3 = k$ show that:

    $$
    \omega_1 = \sqrt{\frac{k\,(m_1 + m_2 - z)}{m_1 m_2}}
    \qquad
    \begin{pmatrix} A_1 \\ A_2 \end{pmatrix} = \begin{pmatrix} \frac{m_1 - m_2 + z}{m_1} \\ 1 \end{pmatrix}
    $$

    $$
    \omega_2 = \sqrt{\frac{k\,(m_1 + m_2 + z)}{m_1 m_2}}
    \qquad
    \begin{pmatrix} A_1 \\ A_2 \end{pmatrix} = \begin{pmatrix} \frac{m_1 - m_2 - z}{m_1} \\ 1 \end{pmatrix}
    $$

    where $z = \sqrt{m_1^2 - m_1 m_2 + m_2^2}$.

19. **Eigenvalues and eigenvectors of a Pauli matrix** – Find the eigenvalues and normalized eigenvectors of the Pauli matrix

    $$
    \sigma_y = \begin{pmatrix} 0 & -i \\ i & 0 \end{pmatrix}
    $$

    by hand. Write down the matrix that diagonalizes σ_x. Using Python, check the diagonalization using a similarity transformation.

20. **Quantum Mechanics: The angular momentum matrices** – The matrix representing a physical operator in quantum mechanics must be Hermitian. For example, the angular momentum for a system with orbital quantum number $l = 1$ can be represented by the following matrices.

    $$
    L_x = \frac{\hbar}{\sqrt{2}} \begin{pmatrix} 0 & 1 & 0 \\ 1 & 0 & 1 \\ 0 & 1 & 0 \end{pmatrix}
    \qquad
    L_y = \frac{\hbar}{i\sqrt{2}} \begin{pmatrix} 0 & 1 & 0 \\ -1 & 0 & 1 \\ 0 & -1 & 0 \end{pmatrix}
    \qquad
    L_z = \hbar \begin{pmatrix} 1 & 0 & 0 \\ 0 & 1 & 0 \\ 0 & 0 & 1 \end{pmatrix}
    $$

    a. Show that these three matrices are Hermitian.

    b. The eigenvalues of L_x are the possible values obtained in a measurement of the particle's angular momentum. Find the eigenvalues and eigenvectors of L_x by hand and by using Python.

21. **Quantum Mechanics: Commutators of angular momentum matrices** – Consider again the three matrices L_x, L_y and L_z from the previous problem:

$$L_x = \frac{\hbar}{\sqrt{2}} \begin{pmatrix} 0 & 1 & 0 \\ 1 & 0 & 1 \\ 0 & 1 & 0 \end{pmatrix} \qquad L_y = \frac{\hbar}{i\sqrt{2}} \begin{pmatrix} 0 & 1 & 0 \\ -1 & 0 & 1 \\ 0 & -1 & 0 \end{pmatrix} \qquad L_z = \hbar \begin{pmatrix} 1 & 0 & 0 \\ 0 & 0 & 0 \\ 0 & 0 & -1 \end{pmatrix}$$

   a. Evaluate the matrix $L^2 = L_x^2 + L_y^2 + L_z^2$ which represents the total angular momentum

   b. Find the eigenvalues of L^2 using Python.

   c. Show that the following identity holds:

$$L_x L_y - L_y L_x = i\hbar L_z$$

   The quantity $[L_x, L_y] = L_x L_y - L_y L_x$ is called the commutator of L_x and L_y.

   d. Show that the following identity holds:

$$[L^2, L_z] = L^2 L_z - L_z L^2 = 0$$

22. **The Jacobian and coordinate transformations** – An ellipsoid with uniform density is described by:

$$\frac{x^2}{a^2} + \frac{y^2}{b^2} + \frac{z^2}{c^2} = 1$$

   where a, b, c are constants. Do this problem using SymPy and by hand.

   a. Evaluate the volume of this ellipsoid, assuming a uniform density. Use an appropriate transformation of variables, and evaluate the corresponding Jacobian of the transformation.

   b. Evaluate the volume of an ellipsoid with non-uniform density $\rho(x, y, z) = 3z^2$ in SI units.

23. **Unitary matrices** – A unitary matrix is a matrix whose Hermitian adjoint is also its inverse, i.e. $A^{-1} = A^\dagger$. Under what condition(s) is the matrix

$$A = \begin{pmatrix} a & ih \\ ih & a \end{pmatrix}$$

unitary? Assume that a and h are real positive constants.

24. **Quantum mechanics: Functions of matrices** – Sometimes in quantum mechanics, we need to compute the functions of matrices beyond simple powers. This is done using power series. For example

$$e^A = I + A + \frac{A^2}{2!} + \cdots$$

   where I is the identity matrix. Consider the matrix

$$A = \begin{pmatrix} 0 & -1 & 5 \\ 0 & 0 & 3 \\ 0 & 0 & 0 \end{pmatrix}$$

   a. Show that A^3 is the 3×3 matrix with zero elements.

   b. Calculate the matrices for $\sin(A)$, $\cos(A)$, and $\exp(A)$.

8 Vector Analysis

Up to this point, we have focused on vector arithmetic and transformations. We also examined how vectors change as a function of time. However, in fields like electromagnetism, we need to know more about how vectors change than time derivatives alone.

In this chapter, we will introduce the concept of a *vector field*, which is a rule that assigns a vector to each point in space. We can then ask questions such as, how does the vector field change from one point in space to another. This will involve taking derivatives with respect to spatial coordinates, however, the direction in which the derivative is taken will matter. Likewise, we can integrate vector fields along curves. We will conclude this chapter with a presentation of three important theorems which describe the integrals of vector field derivatives.

8.1 SCALAR AND VECTOR FIELDS

Both scalars and vectors can be used to represent physical quantities. Scalars are used in the case when a magnitude alone is enough to describe a quantity. Vectors are used when a quantity requires both a magnitude and a direction for a complete description. Often in introductory physics, we think of scalars and vectors as having a particular value at a particular time. For example, the mass of an object may be 3.0 kg or a projectile's velocity may be $\left(3\hat{\mathbf{i}} - 2\hat{\mathbf{j}}\right)$ m/s. However, these quantities can depend on location.

A *scalar field* is a function that assigns a single number to every point in space. For example, the temperature of a metal plate could be described by the function

$$T(x, y) = T_0 e^{-\sigma\left(x^2 + y^2\right)} \tag{8.1.1}$$

where T_0 and σ are real positive constants and x and y are spatial coordinates on the plate.

Likewise, a *vector field* is an assignment of a vector to each point in space. For example, consider the velocity of the water on a river's surface. The water may move in different directions and different rates as we move along the river. For example, the velocity could be described using

$$\mathbf{v} = x\hat{\mathbf{i}} + y\hat{\mathbf{j}} \tag{8.1.2}$$

where x and y are coordinates on the river's surface and the units of $\mathbf{v}$ are meters per second.

The next two examples shows how to use Python to plot both scalar and vector fields.

Example 8.1: Plotting a scalar field

The temperature of a metal plate is described by (8.1.1) where $T_0 = 273$ K and $\sigma = 0.4\ m^{-2}$. Plot the temperature as a function of x and y.

Solution:

We use the command `pcolormesh` to create the two-dimensional density plot of T. The command `colorbar` creates the scale that appears to the right of the plot in Figure 8.1.

The code line `contour(x,y,T,5)` creates 5 contour lines within the plot, at which the temperature $T = $ constant. In this example the function $T(x, y)$ is symmetric with respect to x, y and the curves of constant T (called level sets, or contours of T) are circles. The command `clabel` places the values of the temperature at the corresponding contour lines.

```python
import numpy as np
import matplotlib.pyplot as plt
import matplotlib.cm as cm


coord_range = np.linspace(-1,1,100)


x, y = np.meshgrid(coord_range,coord_range)


T0 , sigma = 273.0, 0.4
T = T0 * np.exp(-sigma * (x**2 + y**2))


# use contour() to plot the contours and
# use clabel() to label the contours


plt.pcolormesh(x,y,T, cmap = cm.hot)
plt.colorbar();


plt.clabel(plt.contour(x,y,T,5))


plt.xlabel('x coordinate')
plt.ylabel('y coordinate')
plt.axis('equal');
plt.title('Plot of T(x,y) and contours')
plt.show()
```

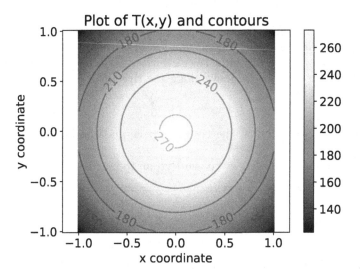

Figure 8.1 Graphical output from Example 8.1, showing plot of a scalar field and the respective contours.

Notice that in Example 8.1, the value of the scalar field changes with location. In this chapter, we will learn how to answer questions such as how quickly does the scalar field change as the location changes and, at a given location, in what direction does the field change most rapidly.

Example 8.2: Plotting a vector field

Suppose the velocity of the water on the surface of a river is described by (8.1.2). Plot the velocity vector field as a function of x and y.

Solution:

We will use the command `quiver(x,y,vx,vy)` to create the two-dimensional vector field, as shown in Figure 8.2. The parameters `vx`, `vy` determine the length of the arrows at points (x, y).

```python
import numpy as np
import matplotlib.pyplot as plt

coord_range = np.linspace(-1,1,10)

x, y = np.meshgrid(coord_range,coord_range)

# vector field components
vx =  x
vy =  y

# plot arrows at the points (x,y) with length (vx, vy)
plt.quiver(x,y,vx,vy, color = 'r')

plt.xlabel('x coordinate')

plt.axis('equal');
plt.title('Plot of 2D vector field F = x.i +y.j')
plt.show()
```

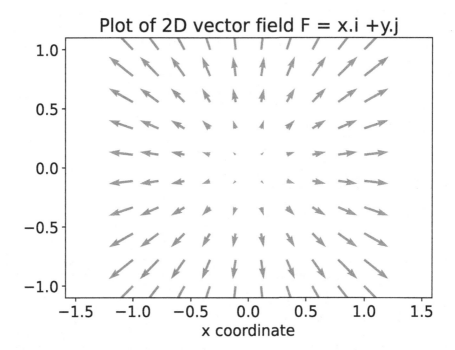

Figure 8.2 Graphical output from Example 8.2, showing plot of a vector field.

In Example 8.2, we see that the direction and magnitude (as represented by an arrow's length) changes with location. In the next several sections, we will examine how scalar fields and vector fields change as a function of spatial coordinates. We will see that there are three derivatives we can find.

1. Gradient: In which direction is a scalar field most rapidly changing? What is that rate of change of the scalar field?

2. Divergence of the vector field: What is the net out-flow rate of the vector field?

3. Curl of the vector field: What is the rotational rate of a vector field?

The next several sections will be devoted to each of the above derivatives.

8.2 THE GRADIENT OF A SCALAR FIELD

Consider the temperature inside an old house on a cold day. Near the windows, the temperature might be cool whereas, on the second floor, it might be warmer. The temperature inside the house is a scalar field because the temperature (a scalar quantity) depends on location. We could ask questions like, how does the temperature of the house change as we go from one floor to the next. We could also ask, at a given location, in which direction does the temperature change most rapidly? Both of these questions can be answered using what is called the *gradient*.

Consider a one-dimensional scalar field $f(x)$. In this case, dimension refers to the number of independent variables. From Chapter 2, we know that the differential

$df = f(x + dx) - f(x)$ can be written as

$$df = \frac{\partial f}{\partial x} dx \qquad (8.2.1)$$

In other words, the amount f changes (df) can be found by taking f's rate of change (df/dx) and multiplying it by the change of the independent variable (dx). However, just like the house's temperature, many real-world scalar fields have more than one dimension (i.e. more than one independent variable). In such a case we define

$$df = f(x + dx, y + dy, z + dz) - f(x, y, z) \qquad (8.2.2)$$

The differential df tells us how much f changes when all of its independent variables change. We can Taylor expand the first term in the right-hand side of (8.2.2)

$$f(x + dx, y + dy, z + dz) = f(x, y, z) + \frac{\partial f}{\partial x} dx + \frac{\partial f}{\partial y} dy + \frac{\partial f}{\partial z} dz + \cdots \qquad (8.2.3)$$

Then (8.2.2) becomes

$$df = \frac{\partial f}{\partial x} dx + \frac{\partial f}{\partial y} dy + \frac{\partial f}{\partial z} dz \qquad (8.2.4)$$

$$= \left(\frac{\partial f}{\partial x} \hat{\mathbf{i}} + \frac{\partial f}{\partial y} \hat{\mathbf{j}} + \frac{\partial f}{\partial z} \hat{\mathbf{k}} \right) \cdot \left(dx\hat{\mathbf{i}} + dy\hat{\mathbf{j}} + dz\hat{\mathbf{k}} \right) \qquad (8.2.5)$$

The operator appearing in the right-hand side of (8.2.5) is very important in physics and is called the *Del Operator* denoted by the symbol *nabla* ∇:

$$\nabla = \left(\hat{\mathbf{i}} \frac{\partial}{\partial x} + \hat{\mathbf{j}} \frac{\partial}{\partial y} + \hat{\mathbf{k}} \frac{\partial}{\partial z} \right) \qquad (8.2.6)$$

The del operator takes different forms in other coordinates systems (see below). When the del operator is applied to a scalar field f, the resulting vector ∇f is called the *gradient* of f and is a vector function:

$$\nabla f = \left(\frac{\partial f}{\partial x} \hat{\mathbf{i}} + \frac{\partial f}{\partial y} \hat{\mathbf{j}} + \frac{\partial f}{\partial z} \hat{\mathbf{k}} \right) \qquad (8.2.7)$$

If we define $d\mathbf{r} = dx\,\hat{\mathbf{i}} + dy\,\hat{\mathbf{j}} + dz\,\hat{\mathbf{k}}$, then (8.2.5) can be rewritten as

$$df = \nabla f \cdot d\mathbf{r} \qquad (8.2.8)$$

Notice that (8.2.8) is in a form similar to (8.2.1), where a change df of the function f is proportional to the change $d\mathbf{r}$ of the independent variables. Hence, we can interpret ∇f as the rate of change of the function f, as we change its independent variables. Since we are in higher dimensions, the rate of change depends on the direction. You may have experienced this if you went hiking on a trail, where your elevation is a function of your location x and y. For example, there may be a steep decline to your right, while the trail ahead of you is relatively flat.

To help us better understand the gradient, we will use the concept of *level sets or contours of the function f*. The level set of a scalar function f is a curve in two-dimensions, or a surface in three-dimensions, on which the value of f is constant. Hence, along these contour curves or surfaces we have $df = 0$.

If $d\mathbf{r}$ is the displacement along a level set of the function f, then:

$$df = \nabla f \cdot d\mathbf{r} = 0 \qquad (8.2.9)$$

Thus, the two vectors ∇f and $d\mathbf{r}$ are perpendicular to each other, i.e. $\nabla f \perp d\mathbf{r}$. We conclude that the gradient of f is perpendicular to the level sets of f.

Let us consider the gravitational potential energy $V = m\,g\,z$ of an object of mass m, at a height z above the Earth's surface. The level sets of $V(z)$ are lines parallel to the Earth's surface. The value of the level sets increases with height. According to (8.2.9), ∇V in this case points vertically, along the same axis as the force of gravity. In fact, the force of gravity $\mathbf{F} = -\nabla V = -\left(dV/dz\right)\hat{\mathbf{k}} = -m\,g\,\hat{\mathbf{k}}$. In general, the relationship between a force $\mathbf{F}$ and the gradient of its associated potential energy V is:

$$\mathbf{F} = -\nabla V \tag{8.2.10}$$

which shows us that forces point in the direction of most rapidly decreasing potential energy.

The usual SymPy functions can still be applied to scalar and vector fields when using Python to compute the Del operator. However, care must be taken when also using `Coordsys3D` to establish a coordinate system. Consider the case where we establish a coordinate system using `R = Coordsys3D('R')`. To compute partial derivatives with respect to the x-coordinate in the R coordinate system, we would need to use `diff(B, R.x)` instead of the usual `diff(B, x)`. The latter will cause an error in the code.

Example 8.3 shows how to evaluate and plot the gravitational force between two masses, together with the contours of the associated scalar gravitational potential energy $V(x, y, z)$.

===

Example 8.3: Gravitational potential energy

In general, the gravitational potential energy of a mass m near the Earth can be found using

$$V = -\frac{G\,m\,M}{\sqrt{x^2 + y^2 + z^2}} \tag{8.2.11}$$

where M is the mass of the Earth and the coordinate system has the Earth's center at the origin. Calculate the force of gravity $\mathbf{F} = -\nabla V$ in Cartesian coordinates. Plot the level sets of V and the vector field $\mathbf{F}$ using $G = 1$, $M = 1$, and $m = 1$.

Solution:

We begin by calculating the x-component of $\mathbf{F}$:

$$\frac{\partial V}{\partial x} = \frac{\partial}{\partial x}\left(-\frac{G\,m\,M}{\sqrt{x^2 + y^2 + z^2}}\right) = \frac{G\,m\,M\,x}{\left(x^2 + y^2 + z^2\right)^{3/2}}$$

Similarly, for the other components we have

$$\mathbf{F} = -\nabla V$$

$$\mathbf{F} = -\left(\frac{\partial V}{\partial x}\hat{\mathbf{i}} + \frac{\partial V}{\partial y}\hat{\mathbf{j}} + \frac{\partial V}{\partial z}\hat{\mathbf{k}}\right) = -G\,m\,M\frac{\left(x\hat{\mathbf{i}} + y\hat{\mathbf{j}} + z\hat{\mathbf{k}}\right)}{\left(x^2 + y^2 + z^2\right)^{3/2}}$$

The Python code evaluates the gradient using the `Del()` operator. This operator is applied to the given potential energy V using `-delop(V)` to evaluate the gravitational force F. A simpler alternative to using `Del()` and `delop(V)` is the abbreviated form `gradient(V)`. We will use this abbreviated form for all three operators discussed in this chapter, the gradient, divergence and curl.

The line `F = delop(V).doit()` evaluates the gradient and stores it in the parameter F. Finally, to extract the x-component of the force, we use the `F.coeff(R.i)` method, where `R.i` represents the unit vector $\hat{i}$ along the x-axis.

The code also plots the level sets and vector field for $z = 0$, and for simplicity we set the constants $m = M = G = 1$. We use `np.mgrid` to evaluate the points (x,y) where the arrows are plotted using the `quiver` function.

Notice that the vectors in Figure 8.3 representing the force are locally perpendicular to the level sets (contours) of V. As expected, the length of the arrows increases toward the center where the force of gravity is stronger.

```python
from sympy.vector import CoordSys3D, gradient, Del
from sympy import symbols, sqrt
import numpy as np
import matplotlib.pyplot as plt

print('-'*28,'CODE OUTPUT','-'*29,'\n')

R = CoordSys3D('R')
delop = Del()

G, m, Me = symbols('G, m, Me')

V = -G*m*Me/sqrt(R.x**2 + R.y**2 + R.z**2)
F =-delop(V).doit() # evaluate all 3 components of gradient vector F

# use .coef method to get components Fx, Fy, Fz
print('x-component of gradient=',F.coeff(R.i))
print('y-component of gradient=',F.coeff(R.j))
print('z-component of gradient=',F.coeff(R.k))

# Create grid for radius r and polar angle theta
# r=0.4 to r=1, and theta=0,2*pi
r, theta = np.mgrid[0.4:1:5j, 0.0:2*np.pi:50j]

# calculate values of x,y where arrows will be drawn
x = r*np.cos(theta)
y = r*np.sin(theta)

# Gravitational potential V with m=M=G=1
V = 1/(x**2 + y**2)**(1/2)

# creating plot of Force and contours where V=constant
fig, ax = plt.subplots()
ax.set_aspect('equal')

# Components of Force along x and y axis
Fx = -x/(x**2 + y**2)**(3/2)
Fy = -y/(x**2 + y**2)**(3/2)

plt.quiver(x,y,Fx,Fy)
plt.contour(x,y,V,10)

plt.xlabel('x')
plt.ylabel('y')
plt.title('Gravitational field and contours')
```

```
plt.show()

--------------------------- CODE OUTPUT ----------------------------

x-component of gradient= -R.x*G*Me*m/(R.x**2 + R.y**2 + R.z**2)**(3/2)
y-component of gradient= -R.y*G*Me*m/(R.x**2 + R.y**2 + R.z**2)**(3/2)
z-component of gradient= -R.z*G*Me*m/(R.x**2 + R.y**2 + R.z**2)**(3/2)
```

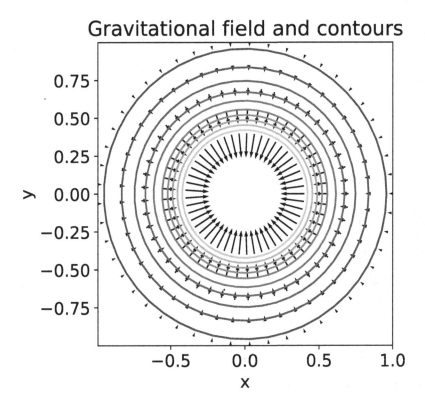

Figure 8.3 Graphical output from Example 8.3, showing a plot of the gravitational force between two masses.

We can define a directional derivative df/dn of a scalar field f which gives the rate of change of $f(x, y, z)$ in the direction of the unit vector $\hat{\mathbf{n}} = n_x\hat{\mathbf{i}} + n_y\hat{\mathbf{j}} + n_z\hat{\mathbf{k}}$:

$$\frac{df}{dn} = \hat{\mathbf{n}} \cdot \nabla f = n_x\frac{\partial f}{\partial x} + n_y\frac{\partial f}{\partial y} + n_z\frac{\partial f}{\partial z} \qquad (8.2.12)$$

Now we ask the question, in what direction does the scalar field f change most rapidly? We can compute the dot product as:

$$\frac{df}{dn} = \hat{\mathbf{n}} \cdot \nabla f = |\hat{\mathbf{n}}|\,|\nabla f|\cos\theta \qquad (8.2.13)$$

where θ is the angle between $\hat{\mathbf{n}}$ and ∇f. The direction in which the directional derivative is greatest would occur when $\hat{\mathbf{n}} \cdot \nabla f$ is maximal, i.e. when the angle θ equals zero. Therefore,

∇f points in the direction of the most rapid increase of f. Likewise, the direction of greatest *decrease* in the scalar field is $-\nabla f$.

Let us return to the hiking example. Consider a surface whose height above sea level at a point (x, y) is $H(x, y)$. The gradient of H at a point (x, y) is a vector pointing in the direction of the steepest slope, or steepest grade at that point. The steepness of the slope at that point is given by the magnitude of the gradient vector.

Example 8.4 shows how to evaluate the directional derivative of a scalar function.

=====

Example 8.4: The directional derivative

Evaluate the directional derivative of the scalar function $f(x, y, z) = x\,y\,z$ along the direction of the vector $\mathbf{v} = \hat{\mathbf{i}} + \hat{\mathbf{j}} + \hat{\mathbf{k}}$.

Solution:

We evaluate the gradient of the function

$$\nabla f = \frac{\partial f}{\partial x}\hat{\mathbf{i}} + \frac{\partial f}{\partial y}\hat{\mathbf{j}} + \frac{\partial f}{\partial z}\hat{\mathbf{k}} = y\,z\,\hat{\mathbf{i}} + x\,z\,\hat{\mathbf{j}} + x\,y\,\hat{\mathbf{k}}$$

and the normalized vector $\hat{\mathbf{n}}$ along the given direction $\mathbf{v} = \hat{\mathbf{i}} + \hat{\mathbf{j}} + \hat{\mathbf{k}}$ is:

$$\hat{\mathbf{n}} = \frac{\hat{\mathbf{i}} + \hat{\mathbf{j}} + \hat{\mathbf{k}}}{|\hat{\mathbf{i}} + \hat{\mathbf{j}} + \hat{\mathbf{k}}|} = \frac{\hat{\mathbf{i}} + \hat{\mathbf{j}} + \hat{\mathbf{k}}}{\sqrt{3}}$$

The directional derivative is the dot product of $\hat{\mathbf{n}}$ and ∇f:

$$\frac{df}{dn} = \hat{\mathbf{n}} \cdot \nabla f = n_x\frac{\partial f}{\partial x} + n_y\frac{\partial f}{\partial y} + n_z\frac{\partial f}{\partial z} = \frac{1}{\sqrt{3}}\,(y\,z + x\,z + x\,y)$$

In the Python code we first evaluate the normalized vector $\hat{\mathbf{n}}$ by using the method `v.normalize()` in SymPy. Next, we show two methods of evaluating the directional derivative using `sympy.vector`. In the first method we evaluate the gradient using `gradient(f)`, and then find the dot product of $\hat{\mathbf{n}}$ and ∇f using the method `n.dot(gradient(f))`.
In the second method we use the dedicated function `directional_derivative(f)`. The result from the two methods is the same.

```
from sympy import simplify
from sympy.vector import CoordSys3D, directional_derivative, gradient

print('-'*28,'CODE OUTPUT','-'*29,'\n')
C = CoordSys3D('C')        # Cartesian system named C
                           # has unit vectors C.i , C.j , C.k


f = C.x * C.y * C.z        # scalar field f(x, y, z)=xyz


v = C.i + C.j + C.k        # vector v defines desired direction


n = v.normalize()          # n= normalized vector v
print('Normalized vector n =',n)


# directional deriv=dot product of n and gradient(f)
```

```
print('\nDirectional derivative using gradient(f) is:')
print(simplify(n.dot(gradient(f))))

# Shortcut method for directional deriv,
# using the dedicated function directional_derivative
print('\nThe alternative method using directional_derivativ gives:')
print(simplify(directional_derivative( C.x*C.y*C.z, n)))

------------------------- CODE OUTPUT -------------------------

Normalized vector n = (sqrt(3)/3)*C.i + (sqrt(3)/3)*C.j + (sqrt(3)/3)*C.k

Directional derivative using gradient(f) is:
sqrt(3)*(C.x*C.y + C.x*C.z + C.y*C.z)/3

The alternative method using directional_derivativ gives:
sqrt(3)*(C.x*C.y + C.x*C.z + C.y*C.z)/3
```

8.2.1 THE GRADIENT IN OTHER COORDINATE SYSTEMS

Next, we will demonstrate how to obtain the gradient in cylindrical coordinates.

To begin, the total differential df of a scalar function in cylindrical coordinates is given by:

$$df = f\left(\rho + d\rho, \phi + d\phi, z + dz\right) - f\left(\rho, \phi, z\right) \tag{8.2.14}$$

which can be rewritten as

$$df = \frac{\partial f}{\partial \rho}d\rho + \frac{\partial f}{\partial \phi}d\phi + \frac{\partial f}{\partial z}dz = \nabla f \cdot d\mathbf{r} \tag{8.2.15}$$

In general, the gradient of any scalar function f in cylindrical coordinates $(\rho,\ \phi,\ z)$ will be of the form:

$$\nabla f = (\nabla f)_\rho\, \hat{\boldsymbol{\rho}} + (\nabla f)_\phi\, \hat{\boldsymbol{\phi}} + (\nabla f)_z\, \hat{\mathbf{z}} \tag{8.2.16}$$

where $(\nabla f)_\rho$, $(\nabla f)_\phi$, $(\nabla f)_z$ are the components of the gradient along the unit vectors $\hat{\boldsymbol{\rho}}$, $\hat{\boldsymbol{\phi}}$, $\hat{\mathbf{z}}$, respectively. The vector $d\mathbf{r}$ in cylindrical coordinates is given by:

$$d\mathbf{r} = d\rho\,\hat{\boldsymbol{\rho}} + \rho\,d\phi\,\hat{\boldsymbol{\phi}} + dz\,\hat{\mathbf{z}} \tag{8.2.17}$$

Taking the dot product of (8.2.16) and (8.2.17), and substituting into (8.2.15), we find:

$$\frac{\partial f}{\partial \rho}d\rho + \frac{\partial f}{\partial \phi}d\phi + \frac{\partial f}{\partial z}dz = (\nabla f)_\rho\, d\rho + (\nabla f)_\phi\, \rho\, d\phi + (\nabla f)_z\, dz \tag{8.2.18}$$

Therefore:

$$(\nabla f)_\rho = \frac{\partial f}{\partial \rho}, \quad (\nabla f)_\phi = \frac{1}{\rho}\frac{\partial f}{\partial \phi}, \quad (\nabla f)_z = \frac{\partial f}{\partial z}$$

Using (8.2.16), the expression of the gradient in cylindrical coordinates is:

$$\nabla f(\rho, \phi, z) = \frac{\partial f}{\partial \rho}\hat{\boldsymbol{\rho}} + \frac{1}{\rho}\frac{\partial f}{\partial \phi}\hat{\boldsymbol{\phi}} + \frac{\partial f}{\partial z}\hat{\mathbf{z}} \tag{8.2.19}$$

The same method can be used to obtain the del operator in spherical coordinates.

In summary, the del operator in various coordinate systems is :

$$\nabla = \hat{\mathbf{i}}\frac{\partial}{\partial x} + \hat{\mathbf{j}}\frac{\partial}{\partial y} + \hat{\mathbf{k}}\frac{\partial}{\partial z} \qquad \text{Cartesian Coordinates} \qquad (8.2.20)$$

$$\nabla = \hat{\rho}\frac{\partial}{\partial \rho} + \hat{\phi}\frac{1}{\rho}\frac{\partial}{\partial \phi} + \hat{\mathbf{z}}\frac{\partial}{\partial z} \qquad \text{Cylindrical Coordinates} \qquad (8.2.21)$$

$$\nabla = \hat{\mathbf{r}}\frac{\partial}{\partial r} + \hat{\theta}\frac{1}{r}\frac{\partial}{\partial \theta} + \hat{\phi}\frac{1}{r\sin\theta}\frac{\partial}{\partial \phi} \qquad \text{Spherical Coordiantes} \qquad (8.2.22)$$

The `sympy.vector` library supports the evaluation of the gradient, divergence and curl in orthogonal curvilinear coordinate systems.

Example 8.5 shows how to evaluate the gradient of a scalar function in cylindrical and spherical coordinates.

Example 8.5: The gradient in cylindrical coordinates

(a) Evaluate the gradient of the scalar function $f(\rho, \phi, z) = \rho\sin\phi\cos z$ in cylindrical coordinates (ρ, ϕ, z), both analytically and using the functions available in `sympy.vector`.
(b) Evaluate the gradient of the scalar function $f(r, \phi, z) = r\sin\theta\cos\phi$ in spherical coordinates (r, θ, ϕ), both analytically and using the functions available in `sympy.vector`.

Solution:

(a) Using (8.2.21) for cylindrical coordinates:

$$\nabla f = \hat{\rho}\frac{\partial f}{\partial \rho} + \hat{\phi}\frac{1}{\rho}\frac{\partial f}{\partial \phi} + \hat{\mathbf{z}}\frac{\partial f}{\partial z} = \hat{\rho}\sin\phi\cos z + \hat{\phi}\cos\phi\cos z + \hat{\mathbf{z}}\left(-\rho\sin\phi\sin z\right)$$

(b) Using (8.2.22) for spherical coordinates, the gradient is:

$$\nabla f = \hat{\mathbf{r}}\frac{\partial f}{\partial r} + \hat{\theta}\frac{1}{r}\frac{\partial f}{\partial \theta} + \hat{\phi}\frac{1}{r\sin\theta}\frac{\partial f}{\partial \phi}$$

In the Python code we define the coordinate system C and transform it into cylindrical coordinates using `transformation='cylindrical'`. This transformed system has unit vectors c.R, C.P, C.Z and the cylindrical coordinate parameters (ρ, ϕ, z) are represented as c.r, c.p, c.z. The gradient is evaluated using `gr=gradient(f)` and we use `gr.coeff()` to find the coefficients of the unit vectors in the cylindrical system.
Similarly with the spherical coordinate system N, we use `transformation='spherical'` and proceed in the same manner.

```
# Gradient of a vector in cylindrical coordinates
from sympy import sin, cos, flatten
from sympy.vector import CoordSys3D, gradient, curl, divergence

print('-'*28,'CODE OUTPUT','-'*29,'\n')

# part (a)
# define Cartesian system named c, with unit vectors c.i, c.j, c.k
# Transform it to system with cylindrical coordinates (rho, phi, z)

# variables are c.r, c.t, c.z for rho, phi, z
# unit vectors c.R, c.P, c.Z
```

```
c = CoordSys3D('c', transformation='cylindrical',\
variable_names = list("rpz"),vector_names=list("RPZ") )

scalar = c.r*sin(c.p)*cos(c.z)

print('scalar =',scalar)

gr = gradient(scalar)
print('\nThe gradient of this scalar in cylindrical is:')
print(gr)

print('\nThe r-component of the gradient is:', gr.coeff(c.R))
print('The theta-component of the gradient is:', gr.coeff(c.P))
print('The z-component of the gradient is:', gr.coeff(c.Z))

# part (b) spherical coordinates
# variables are c.r, c.theta, c.phi
N = CoordSys3D('N', transformation='spherical',\
variable_names=list("rtp"),vector_names=list("RTP") )

scalar = N.r*sin(N.t)*cos(N.p)
print('\n-------Gradient in spherical coordinates-------')
print('scalar =',scalar)

gr=gradient(scalar)
print('\nThe gradient of this scalar in spherical coordinates is:')
print(gr)

print('\nThe r-component of the gradient is:', gr.coeff(N.R))
print('The theta-component of the gradient is:', gr.coeff(N.T))
print('The phi-component of the gradient is:', gr.coeff(N.P))
```

`-------------------------- CODE OUTPUT --------------------------`

```
scalar = c.r*sin(c.p)*cos(c.z)

The gradient of this scalar in cylindrical is:
(sin(c.p)*cos(c.z))*c.R + (cos(c.p)*cos(c.z))*c.P + (-c.r*sin(c.p)*sin(c.z))*c.Z

The r-component of the gradient is: sin(c.p)*cos(c.z)
The theta-component of the gradient is: cos(c.p)*cos(c.z)
The z-component of the gradient is: -c.r*sin(c.p)*sin(c.z)

-------Gradient in spherical coordinates-------
scalar = N.r*sin(N.t)*cos(N.p)

The gradient of this scalar in spherical coordinates is:
(sin(N.t)*cos(N.p))*N.R + (cos(N.p)*cos(N.t))*N.T + (-sin(N.p))*N.P

The r-component of the gradient is: sin(N.t)*cos(N.p)
The theta-component of the gradient is: cos(N.p)*cos(N.t)
The phi-component of the gradient is: -sin(N.p)
```

8.3   PROPERTIES OF THE GRADIENT

The gradient operator follows many of the same derivative rules that one would expect. The following are true for scalar fields f and g and constant c:

$$\nabla\left(cf\right) = c\nabla f \tag{8.3.1}$$

$$\nabla\left(fg\right) = f\left(\nabla g\right) + g\left(\nabla f\right) \tag{8.3.2}$$

Example 8.6 shows how to prove the second of these equations.

Example 8.6: The derivative rule for gradient

Demonstrate the general property of the gradient, for any scalar functions $f(x, y, z)$ and $g(x, y, z)$, by hand and demonstrate using SymPy.

$$\nabla\left(fg\right) = f\left(\nabla g\right) + g\left(\nabla f\right)$$

Solution:

$$
\begin{aligned}
\nabla\left(f\,g\right) &= \frac{\partial\left(f\,g\right)}{\partial x}\hat{\mathbf{i}} + \frac{\partial\left(f\,g\right)}{\partial y}\hat{\mathbf{j}} + \frac{\partial\left(f\,g\right)}{\partial z}\hat{\mathbf{k}} \\
&= f\frac{\partial g}{\partial x}\hat{\mathbf{i}} + g\frac{\partial f}{\partial x}\hat{\mathbf{i}} + f\frac{\partial g}{\partial y}\hat{\mathbf{j}} + g\frac{\partial f}{\partial y}\hat{\mathbf{j}} + f\frac{\partial g}{\partial z}\hat{\mathbf{k}} + g\frac{\partial f}{\partial z}\hat{\mathbf{k}} \\
&= f\left(\frac{\partial g}{\partial x}\hat{\mathbf{i}} + \frac{\partial g}{\partial y}\hat{\mathbf{j}} + \frac{\partial g}{\partial z}\hat{\mathbf{k}}\right) + g\left(\frac{\partial f}{\partial x}\hat{\mathbf{i}} + \frac{\partial f}{\partial y}\hat{\mathbf{j}} + \frac{\partial f}{\partial z}\hat{\mathbf{k}}\right) \\
&= f\left(\nabla g\right) + g\left(\nabla f\right)
\end{aligned}
$$

In the Python code we define the coordinate system C within `sympy.vector`, with unit vectors `C.i`, `C.j`, `C.k`. We define the symbols f, g to be functions of the coordinate parameters `C.x`, `C.y`, `C.k`, and we evaluate the left-hand side (LHS) and right-hand side (RHS) of this equation using the `gradient()` function.
We verify that the two sides of the equation are identical by using the logical statement LHS==RHS, which returns a True value for the equation.

```
from sympy.vector import CoordSys3D, gradient

print('-'*28,'CODE OUTPUT','-'*29,'\n')

C = CoordSys3D('C')     # define Cartesian system, named here C

# The scalar field f(x,y,z) is a function of the Cartesian
# coordinate variables C.x, C.y, C.z

from sympy import symbols, Function
f, g = symbols('f, g', cls=Function)

# Define the scalar fields as fscalar(x,y,z), gscalar(x,y,z),
fscalar = f(C.x, C.y, C.z)
gscalar = g(C.x, C.y, C.z)
```

```
# Construct the expression for the LHS: grad(f.g)
lhs = gradient(fscalar*gscalar)

# Construct the expression for the rhs: grad(f).g+grad(g).f
rhs = gradient(fscalar)*gscalar+gradient(gscalar)*fscalar

# Compare the two sides
print('The equation grad(f.g)=grad(f).g+grad(g).f is ',lhs==rhs)



------------------------- CODE OUTPUT -----------------------------


The equation grad(f.g)=grad(f).g+grad(g).f is  True
```

8.4 DIVERGENCE

The divergence of a vector field $\mathbf{v} = v_x\hat{\mathbf{i}} + v_y\hat{\mathbf{j}} + v_z\hat{\mathbf{k}}$ is a *scalar* function found by taking a dot product between the del operator and a vector field:

$$\nabla \cdot \mathbf{v} = \frac{\partial v_x}{\partial x} + \frac{\partial v_y}{\partial y} + \frac{\partial v_z}{\partial z} \tag{8.4.1}$$

The divergence of a vector field appears often in fields such as electromagnetism and fluid dynamics.

Roughly speaking the divergence is a measure of the net outward flow rate of a vector field. For example, consider air as it is heated or cooled. The relevant vector field for this example is the velocity of the air. If air is heated in a region, it will expand in all directions such that the velocity field points outward from that region and toward regions with lower temperature. In this situation, the divergence of the velocity field in that region would have a positive value, and the region is called a *source*. If the air cools and contracts, the divergence is negative and the region is called a *sink*. The divergence is a vector operator that measures the magnitude of a vector field's source or sink at a given point, in terms of a signed scalar.

To illustrate that (8.4.1) is a measure of outflow rate, let us consider the following situation. A wire frame box is submerged into a river as shown in Figure 8.4. The box contains sensors to measure the mass flow rate of the water $\mathbf{A} = \rho\mathbf{v}$ where ρ is mass density (measured in kg/m^3 and assumed constant for simplicity) and $\mathbf{v} = \mathbf{v}(x, y)$ is the fluid velocity (measured in m/s).

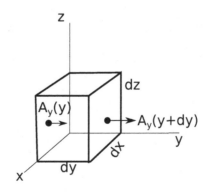

Figure 8.4 A wire fame box submerged in a fluid.

Suppose we want to know the flow rate (as measured in $\text{kg/m}^2\text{/s}$) through the box. We could do that by adding up the net flow rate flowing along each direction. For example, in the y-direction we would measure

$$[A_y(y + dy) - A_y(y)]\, dx dz \tag{8.4.2}$$

which has units of kg/s. In the y-direction, the area of the box is $dx dz$. Hence, the flow rate is $\mathbf{A}$ times area.

Next, we Taylor expand $A_y(y + dy)$:

$$\left[A_y(y) + \frac{\partial A_y}{\partial y} dy + \cdots - A_y(y)\right] dx dz = \frac{\partial A_y}{\partial y} dx dy dz \tag{8.4.3}$$

Repeat this procedure for the x- and z-directions and add the results to get the net outflow rate

$$\left[\frac{\partial A_x}{\partial x} + \frac{\partial A_y}{\partial y} + \frac{\partial A_z}{\partial z}\right] dx dy dz = (\nabla \cdot \mathbf{A})\, dx dy dz \tag{8.4.4}$$

We see that the divergence is measuring a net outflow rate per unit volume.

We can extend this concept further. If the net outflow rate of the fluid is positive, then the fluid density inside the box must be decreasing (unless there is a source inside the box). In this case we could write

$$\frac{\partial \rho}{\partial t} + \nabla \cdot \mathbf{A} = 0 \tag{8.4.5}$$

which is known as the continuity equation. In this example, the continuity equation is describing the conservation of mass.

If $\mathbf{u}$, $\mathbf{v}$ are vector functions and f is a scalar function, then the divergence obeys the following product rule:

$$\nabla \cdot (f\mathbf{v}) = f(\nabla \cdot \mathbf{v}) + \mathbf{v} \cdot \nabla f \tag{8.4.6}$$

However, the divergence of a cross vector product is less intuitive:

$$\nabla \cdot (\mathbf{u} \times \mathbf{v}) = \mathbf{v} \cdot (\nabla \times \mathbf{u}) - \mathbf{u} \cdot (\nabla \times \mathbf{v}) \tag{8.4.7}$$

Example 8.7: Solenoidal vector fields

A vector field $\mathbf{E}$ is called *solenoidal* if its divergence is zero, i.e. $\nabla \cdot \mathbf{E} = 0$.

(a) The electric field due to a charge density $\rho(x, y, z)$ is given by

$$\mathbf{E} = 3x^2 y^2 z\, \hat{\mathbf{i}} + 2x^3 y z\, \hat{\mathbf{j}} + x^3 y^2\, \hat{\mathbf{k}}$$

(in SI units) and satisfies Gauss's law in its differential form:

$$\nabla \cdot \mathbf{E} = \frac{\rho}{\epsilon_0}$$

where ϵ_0 is a constant known as the permittivity of free space. Using Gauss's law, find the charge density ρ.

(b) Test whether this vector field is solenoidal, by using the function `is_solenoidal()` in the library `vector.sympy`.

Solution:

(a) We begin by finding $\nabla \cdot \mathbf{E}$

$$
\begin{aligned}
\nabla \cdot \mathbf{E} &= \frac{\partial E_x}{\partial x} + \frac{\partial E_y}{\partial y} + \frac{\partial E_z}{\partial z} \\
&= \frac{\partial}{\partial x}\left(3x^2 y^2 z\right) + \frac{\partial}{\partial y}\left(2x^3 y z\right) + \frac{\partial}{\partial z}\left(x^3 y^2\right) \\
&= 6x y^2 z + 2x^3 z
\end{aligned}
$$

Therefore

$$
\rho = 2\,x\,z\,\epsilon_0\left(3y^2 + x^2\right)
$$

(b) The Python code finds $\rho(x, y, z)$ and tests if $\mathbf{E}$ is solenoidal.

```python
from sympy.vector import CoordSys3D, divergence, is_solenoidal
from sympy import symbols


R = CoordSys3D('R')
E0 = symbols('E0')     # permitivity constant


# The Electric field E
E = 3* R.x**2 * R.y**2 * R.z * R.i \
    + 2 * R.x**3 * R.y * R.z * R.j \
    + R.x**3 * R.y**2 * R.k

print('-'*28,'CODE OUTPUT','-'*29,'\n')

print('The charge density rho = ')
print(E0*divergence(E))

print('\nThe statement   E is a solenoidal field   is :',is_solenoidal(E))

--------------------------- CODE OUTPUT ----------------------------

The charge density rho =
E0*(2*R.x**3*R.z + 6*R.x*R.y**2*R.z)

The statement   E is a solenoidal field   is : False
```

Like the gradient, the divergence takes on different forms depending on the coordinate system being used:

$$
\nabla \cdot \mathbf{v} = \frac{\partial v_x}{\partial x} + \frac{\partial v_y}{\partial y} + \frac{\partial v_z}{\partial z} \tag{8.4.8}
$$

$$
\nabla \cdot \mathbf{v} = \frac{1}{r}\frac{\partial\left(r\,v_r\right)}{\partial r} + \frac{1}{r}\frac{\partial\left(v_\phi\right)}{\partial \phi} + \frac{\partial v_z}{\partial z} \tag{8.4.9}
$$

$$
\nabla \cdot \mathbf{v} = \frac{1}{r^2}\frac{\partial\left(r^2\,v_r\right)}{\partial r} + \frac{1}{r\sin\theta}\frac{\partial}{\partial \theta}\left(v_\theta \sin\theta\right) + \frac{1}{r\sin\theta}\frac{\partial\left(v_\phi\right)}{\partial \phi} \tag{8.4.10}
$$

8.5 CURL

We can also use the del operator to measure the rotational properties of a vector field, called the *curl*. The curl of a vector field $\mathbf{v}(x, y, z) = v_x\hat{\mathbf{i}} + v_y\hat{\mathbf{j}} + v_z\hat{\mathbf{k}}$ is a vector function defined as:

$$\nabla \times \mathbf{v} = \left(\frac{\partial v_z}{\partial y} - \frac{\partial v_y}{\partial z}\right)\hat{\mathbf{i}} + \left(\frac{\partial v_x}{\partial z} - \frac{\partial v_z}{\partial x}\right)\hat{\mathbf{j}} + \left(\frac{\partial v_y}{\partial x} - \frac{\partial v_x}{\partial y}\right)\hat{\mathbf{k}} \qquad (8.5.1)$$

From a physical point of view, the curl at a point (x, y, z) is proportional to the on-axis torque to which a tiny pinwheel would be subjected, if it were centered at that point. We will illustrate this later by examining the curl of a velocity vector $\mathbf{v}$. The curl is commonly found in electromagnetism and classical mechanics.

The cross product operation can also be found as a determinant in Cartesian coordinates:

$$\nabla \times \mathbf{v} = \begin{vmatrix} \hat{\mathbf{i}} & \hat{\mathbf{j}} & \hat{\mathbf{k}} \\ \frac{\partial}{\partial x} & \frac{\partial}{\partial y} & \frac{\partial}{\partial z} \\ v_x & v_y & v_z \end{vmatrix} \qquad (8.5.2)$$

If $\mathbf{u}$ and $\mathbf{v}$ are any vector functions and f is any scalar function, the curl also follows a product rule:

$$\nabla \times (f\mathbf{v}) = (\nabla f) \times \mathbf{v} + f (\nabla \times \mathbf{v}) \qquad (8.5.3)$$

However, the curl of a cross product is more complex:

$$\nabla \times (\mathbf{u} \times \mathbf{v}) = (\mathbf{v} \cdot \nabla)\mathbf{u} - (\mathbf{u} \cdot \nabla)\mathbf{v} + \mathbf{u}(\nabla \cdot \mathbf{v}) - \mathbf{v}(\nabla \cdot \mathbf{u}) \qquad (8.5.4)$$

The gradient of the dot product of two vector fields is also complex

$$\nabla(\mathbf{u} \cdot \mathbf{v}) = \mathbf{u} \times (\nabla \times \mathbf{v}) + \mathbf{v} \times (\nabla \times \mathbf{u}) + (\mathbf{u} \cdot \nabla)\mathbf{v} + (\mathbf{v} \cdot \nabla)\mathbf{u} \qquad (8.5.5)$$

We can use (8.5.4) to motivate the concept of the curl measuring a rotation. Consider a point on a rotating rigid body. The point has a tangential velocity which can be found using

$$\mathbf{v} = \boldsymbol{\omega} \times \mathbf{r} \qquad (8.5.6)$$

where $\boldsymbol{\omega}$ is the angular velocity of the object (and therefore also of the point we are considering), and $\mathbf{r}$ denotes the location of the point on the body relative to the origin. We will assume $\boldsymbol{\omega}$ is constant and passes through the origin.

Next, we calculate $\nabla \times \mathbf{v}$:

$$\nabla \times \mathbf{v} = \nabla \times (\boldsymbol{\omega} \times \mathbf{r}) = (\nabla \cdot \mathbf{r})\boldsymbol{\omega} - (\boldsymbol{\omega} \cdot \nabla)\mathbf{r} \qquad (8.5.7)$$

Using $\boldsymbol{\omega} = \omega_x\hat{\mathbf{i}} + \omega_y\hat{\mathbf{j}} + \omega_z\hat{\mathbf{k}}$ and $\mathbf{r} = x\hat{\mathbf{i}} + y\hat{\mathbf{j}} + z\hat{\mathbf{k}}$, we can show

$$\nabla \times \mathbf{v} = 2\boldsymbol{\omega} \qquad (8.5.8)$$

In other words, the curl of the velocity is proportional to the angular velocity $\boldsymbol{\omega}$ (which describes a rotation about an axis).

The curl takes a different form in cylindrical and spherical coordinates. To compute the curl in spherical and cylindrical coordinates, choose the appropriate form of the Del Operator from (8.2.20), (8.2.21), and (8.2.22) and evaluate the cross product between Del and the vector of interest. The results are:

$$\nabla \times \mathbf{v} = \left(\frac{\partial v_z}{\partial y} - \frac{\partial v_y}{\partial z} \right) \hat{\mathbf{i}} + \left(\frac{\partial v_x}{\partial z} - \frac{\partial v_z}{\partial x} \right) \hat{\mathbf{j}} + \left(\frac{\partial v_y}{\partial x} - \frac{\partial v_x}{\partial y} \right) \hat{\mathbf{k}} \qquad (8.5.9)$$

$$\nabla \times \mathbf{v} = \left(\frac{1}{\rho} \frac{\partial v_z}{\partial \phi} - \frac{\partial v_\phi}{\partial z} \right) \hat{\boldsymbol{\rho}} + \left(\frac{\partial v_\rho}{\partial z} - \frac{\partial v_z}{\partial \rho} \right) \hat{\boldsymbol{\phi}} + \frac{1}{\rho} \left(\frac{\partial (\rho v_\phi)}{\partial \rho} - \frac{\partial v_\rho}{\partial \phi} \right) \hat{\mathbf{z}} \qquad (8.5.10)$$

$$\nabla \times \mathbf{v} = \frac{1}{r \sin \theta} \left(\frac{\partial}{\partial \theta} (v_\phi \sin \theta) - \frac{\partial v_\theta}{\partial \phi} \right) \hat{\mathbf{r}} + \frac{1}{r} \left(\frac{1}{\sin \theta} \frac{\partial v_r}{\partial \phi} - \frac{\partial}{\partial r} (r v_\phi) \right) \hat{\boldsymbol{\theta}}$$
$$+ \frac{1}{r} \left(\frac{\partial (r v_\theta)}{\partial r} - \frac{\partial v_r}{\partial \theta} \right) \hat{\boldsymbol{\phi}} \qquad (8.5.11)$$

Example 8.8 shows how to evaluate the curl of a vector in spherical coordinates.

Example 8.8: Evaluation of the curl and divergence in cylindrical coordinates

Evaluate the curl and the divergence of the vector function $\mathbf{v} = r \sin \theta \, \hat{\mathbf{r}} + \cos \phi \, \hat{\boldsymbol{\theta}}$ in spherical coordinates (r, θ, ϕ), both analytically and using the functions available in `sympy.vector`.

Solution:

Using the expression for the curl in spherical coordinates:

$$\nabla \times \mathbf{v} = \frac{1}{r \sin \theta} \left(\frac{\partial}{\partial \theta} (v_\phi \sin \theta) - \frac{\partial v_\theta}{\partial \phi} \right) \hat{\mathbf{r}} + \frac{1}{r} \left(\frac{1}{\sin \theta} \frac{\partial v_r}{\partial \phi} - \frac{\partial}{\partial r} (r v_\phi) \right) \hat{\boldsymbol{\theta}}$$
$$+ \frac{1}{r} \left(\frac{\partial (r v_\theta)}{\partial r} - \frac{\partial v_r}{\partial \theta} \right) \hat{\boldsymbol{\phi}}$$

with $v_r = r \sin \theta$, $v_\phi = 0$, $v_\theta = \cos \phi$:

$$\nabla \times \mathbf{v} = \frac{1}{r \sin \theta} \left(0 - \frac{\partial v_\theta}{\partial \phi} \right) \hat{\mathbf{r}} + \frac{1}{r} (0 - 0) \hat{\boldsymbol{\theta}} + \frac{1}{r} \left(\frac{\partial (r v_\theta)}{\partial r} - \frac{\partial v_r}{\partial \theta} \right) \hat{\boldsymbol{\phi}}$$

$$\nabla \times \mathbf{v} = \frac{1}{r \sin \theta} (\sin \phi) \, \hat{\mathbf{r}} + \frac{1}{r} (\cos \phi - r \cos \theta) \, \hat{\boldsymbol{\phi}}$$

Using the expression for the divergence in spherical coordinates with $v_r = r \sin \theta$, $v_\phi = 0$, $v_\theta = \cos \phi$:

$$\nabla \cdot \mathbf{v} = \frac{1}{r^2} \frac{\partial (r^2 v_r)}{\partial r} + \frac{1}{r \sin \theta} \frac{\partial}{\partial \theta} (v_\theta \sin \theta) + \frac{1}{r \sin \theta} \frac{\partial (v_\phi)}{\partial \phi}$$

$$\nabla \cdot \mathbf{v} = \frac{1}{r^2} 3 r^2 \sin \theta + \frac{1}{r \sin \theta} \cos \phi \cos \theta + 0 = 3 \sin \theta + \frac{1}{r \sin \theta} \cos \phi \cos \theta$$

In the Python code we define the spherical coordinate system c and transform it into cylindrical coordinates using `transformation='spherical'`, and the curl and divergence are evaluated using `curl(v)` and `divergence(v)`, respectively.

```
from sympy import sin, cos, flatten
from sympy.vector import CoordSys3D, gradient, curl, divergence

print('-'*28,'CODE OUTPUT','-'*29,'\n')
```

```
# define Cartesian system named c, with unit vectors c.i, c.j, c.k
# Transform it to system with spherical coordinates (r, theta, phi)

c = CoordSys3D('c', transformation='spherical',\
variable_names=list("rtp"),vector_names=list("RTP") )

# variables are c.r, c.t, c.p, for r, theta, phi
# unit vectors are c.R, c.T, c.P

v = c.r*sin(c.t)*c.R + cos(c.p)*c.T

curl_v = curl(v)
div_v = divergence(v)

print('\nThe vector v =',v)

print('\nThe curl of vector in spherical is:')
print(curl_v)

print('\nThe r-component of the curl is:', curl_v.coeff(c.R))
print('The theta-component of the curl is:', curl_v.coeff(c.T))
print('The pi-component of the curl is:', curl_v.coeff(c.P))

print('\nThe divergence of vector in spherical is:')
print(div_v)

---------------------------- CODE OUTPUT ----------------------------


The vector v = (c.r*sin(c.t))*c.R + (cos(c.p))*c.T

The curl of vector in spherical is:
(sin(c.p)/(c.r*sin(c.t)))*c.R + ((-c.r*cos(c.t) + cos(c.p))/c.r)*c.P

The r-component of the curl is: sin(c.p)/(c.r*sin(c.t))
The theta-component of the curl is: 0
The pi-component of the curl is: (-c.r*cos(c.t) + cos(c.p))/c.r

The divergence of vector in spherical is:
3*sin(c.t) + cos(c.p)*cos(c.t)/(c.r*sin(c.t))
```

One important property of the curl is that for any scalar function $V(x, y, z)$

$$\nabla \times (\nabla V) = 0 \qquad (8.5.12)$$

In words: the curl of the gradient of any scalar functions is zero. Example 8.9 how to prove this identity for any scalar function $V(x, y, z)$.

Example 8.9: Evaluation of the curl of a gradient

Consider a scalar function $V = V(x, y, z)$. Prove that $\nabla \times (\nabla V) = 0$ analytically and demonstrate it using the `sympy.vector` package.

Solution:

If $V = V(x, y, z)$ then

$$\nabla V = \frac{\partial V}{\partial x}\hat{\mathbf{i}} + \frac{\partial V}{\partial y}\hat{\mathbf{j}} + \frac{\partial V}{\partial z}\hat{\mathbf{k}}$$

and

$$\nabla \times (\nabla V) = \begin{vmatrix} \hat{\mathbf{i}} & \hat{\mathbf{j}} & \hat{\mathbf{k}} \\ \frac{\partial}{\partial x} & \frac{\partial}{\partial y} & \frac{\partial}{\partial z} \\ \frac{\partial V}{\partial x} & \frac{\partial V}{\partial y} & \frac{\partial V}{\partial z} \end{vmatrix}$$

Expanding, we find:

$$\nabla \times (\nabla V) = \left(\frac{\partial}{\partial y}\frac{\partial V}{\partial z} - \frac{\partial}{\partial z}\frac{\partial V}{\partial y}\right)\hat{\mathbf{i}} + \left(\frac{\partial}{\partial z}\frac{\partial V}{\partial x} - \frac{\partial}{\partial x}\frac{\partial V}{\partial z}\right)\hat{\mathbf{j}} + \left(\frac{\partial}{\partial x}\frac{\partial V}{\partial y} - \frac{\partial}{\partial y}\frac{\partial V}{\partial x}\right)$$

which is zero due to the properties of partial derivatives presented in Chapter 2. In the Python code, we define the symbol V as a function using `V = symbols('V', cls=Function)`, and use `V(C.x, C.y, C.z)` as a symbol of the scalar function $V(x, y, z)$. The curl of the gradient of V is then evaluated to be zero using `curl(gradient(V))`.

```
# Proof of curl of the gradient of f = 0
from sympy.vector import CoordSys3D, curl, gradient
from sympy import symbols, Function

V = symbols('V', cls=Function)

print('-'*28,'CODE OUTPUT','-'*29,'\n')

C = CoordSys3D('C')     # define Cartesian system, named here C

# The scalar field V(x,y,z) is a function of the Cartesian
# coordinate variables C.x, C.y, C.z

# Define the scalar field as ffield(x,y,z)
ffield = V(C.x, C.y, C.z)

# Construct the expression for the (curl of the gradient)
print('The curl of the gradient is always ',curl(gradient(ffield)))


---------------------- CODE OUTPUT ----------------------------

The curl of the gradient is always  0
```

Conservative forces can be written as the gradient of a scalar function (called the potential energy). The curl of any conservative force is zero. Example 8.10 shows how to test whether a force is conservative or not, using two different methods.

Example 8.10: Conservative forces

(a) Use an analytical method to test whether the force

$$\mathbf{F} = c_1 yz\hat{\mathbf{i}} - c_2 zy\hat{\mathbf{j}} + c_3(x+y)\hat{\mathbf{k}}$$

is conservative.
(b) Repeat using the function `is_conservative()` in `sympy.vector`.

Solution:

(a) We compute $\nabla \times \mathbf{F}$

$$\nabla \times \mathbf{F} = \left(\frac{\partial F_z}{\partial y} - \frac{\partial F_y}{\partial z}\right)\hat{\mathbf{i}} + \left(\frac{\partial F_x}{\partial z} - \frac{\partial F_z}{\partial x}\right)\hat{\mathbf{j}} + \left(\frac{\partial F_y}{\partial x} - \frac{\partial F_x}{\partial y}\right)\hat{\mathbf{k}}$$
$$= (c_3 + c_2 y)\,\hat{\mathbf{i}} + (c_1 y - c_3)\,\hat{\mathbf{j}} - (c_1 z)\,\hat{\mathbf{k}}$$

Since $\nabla \times \mathbf{F} \neq 0$, the force is not conservative.
(b) The Python code evaluates the curl symbolically, and the function `is_conservative()` is used to test whether the force $\mathbf{F}$ is conservative, yielding a `False` statement.

```
from sympy.vector import CoordSys3D, curl, is_conservative
from sympy import symbols

print('-'*28,'CODE OUTPUT','-'*29,'\n')
R = CoordSys3D('R')
c1, c2, c3 = symbols('c1,c2,c3')

F = c1*R.y*R.z*R.i - c2*R.z*R.y*R.j + c3*(R.x+R.y)*R.k

print('The curl(F) is:')
print(curl(F))

print('The statement F is conservative is ',is_conservative(F))


---------------------------- CODE OUTPUT ----------------------------

The curl(F) is:
(R.y*c2 + c3)*R.i + (R.y*c1 - c3)*R.j + (-R.z*c1)*R.k
The statement F is conservative is  False
```

8.6 SECOND DERIVATIVES USING ∇

We have already seen one second derivative, $\nabla \times (\nabla V) = 0$ for any scalar function V. Another important second derivative is

$$\nabla \cdot (\nabla \times \mathbf{v}) = 0 \tag{8.6.1}$$

for any vector field $\mathbf{v}$. In words: the divergence of the curl of any vector functions is zero. Example 8.11 shows how to prove this identity for any vector $\mathbf{v}$.

Example 8.11: Evaluation of the divergence of a curl

(a) Prove that the divergence of the curl of any vector functions is zero, $\nabla \cdot (\nabla \times \mathbf{v}) = 0$.
(b) Demonstrate this also using the functions in sympy.vector.

Solution:

(a)

$$\nabla \cdot (\nabla \times \mathbf{v}) = \nabla \cdot \left[\left(\frac{\partial F_z}{\partial y} - \frac{\partial F_y}{\partial z} \right) \hat{\mathbf{i}} + \left(\frac{\partial F_x}{\partial z} - \frac{\partial F_z}{\partial x} \right) \hat{\mathbf{j}} + \left(\frac{\partial F_y}{\partial x} - \frac{\partial F_x}{\partial y} \right) \hat{\mathbf{k}} \right]$$

$$\nabla \cdot (\nabla \times \mathbf{v}) = \frac{\partial}{\partial x} \left(\frac{\partial F_z}{\partial y} - \frac{\partial F_y}{\partial z} \right) + \frac{\partial}{\partial y} \left(\frac{\partial F_x}{\partial z} - \frac{\partial F_z}{\partial x} \right) + \frac{\partial}{\partial z} \left(\frac{\partial F_y}{\partial x} - \frac{\partial F_x}{\partial y} \right)$$

In general for any scalar function g we have:

$$\frac{\partial^2 g}{\partial x \, \partial y} = \frac{\partial^2 g}{\partial y \, \partial x}$$

$$\nabla \cdot (\nabla \times \mathbf{v}) = \left(\frac{\partial^2 F_z}{\partial x \, \partial y} - \frac{\partial^2 F_y}{\partial x \, \partial z} \right) + \left(\frac{\partial^2 F_x}{\partial y \, \partial z} - \frac{\partial^2 F_z}{\partial y \, \partial x} \right) + \left(\frac{\partial^2 F_y}{\partial z \, \partial x} - \frac{\partial^2 F_x}{\partial z \, \partial y} \right) = 0$$

(b) In the Python code, we define the symbol A_x as a function using Ax = symbols('Ax', cls=Function), and use Ax(C.x, C.y, C.z) as a symbol of the x-component $A_x(x, y, z)$. The divergence of the curl of $\mathbf{A}$ is then evaluated to be zero, by using divergence(curl(A)).

```
from sympy.vector import CoordSys3D, divergence, curl

C = CoordSys3D('C')      # define Cartesian system, named here C

print('-'*28,'CODE OUTPUT','-'*29,'\n')

# The vector field A(x,y,z) is a function of the Cartesian
# coordinate variables C.x, C.y, C.z
# the unit vectors in C are C.i, C.j, C.k

from sympy import symbols, Function
Ax, Ay, Az, A = symbols('Ax, Ay, Az, A', cls=Function)

# Define the vector field as A(x,y,z)
A = Ax(C.x, C.y, C.z)*C.i+ Ay(C.x, C.y, C.z)*C.j+ Az(C.x, C.y, C.z)*C.k

# Construct the divergence of the curl
print('The  divergence of the curl of any vector is: ',\
divergence(curl(A)))
```

```
-------------------------- CODE OUTPUT ----------------------------

The   divergence of the curl of any vector is:   0
```

Another important (and often nonzero) second derivative, is the Laplace operator or *Laplacian* ∇^2. The Laplacian is a *scalar* operator that can be applied to either vector or scalar fields. The Laplacian is defined as:

$$\nabla^2 = \nabla \cdot \nabla = \frac{\partial^2}{\partial x^2} + \frac{\partial^2}{\partial y^2} + \frac{\partial^2}{\partial z^2} \tag{8.6.2}$$

The Laplacian appears in important partial differential equations including the Laplace equation and the Schrodinger equation, to name a few. We will discuss these in Chapter 11.

There are other second derivatives, but they are related to the ones we already saw. For example, it can be shown that

$$\nabla \times (\nabla \times \mathbf{v}) = \nabla (\nabla \cdot \mathbf{v}) - \nabla^2 \mathbf{v} \tag{8.6.3}$$

Example 8.12: The Laplace equation

Use Python to show that the electric potential

$$V = \frac{q}{4\pi\epsilon_0} \left(\frac{1}{\sqrt{x^2 + y^2 + z^2}} \right)$$

satisfies Laplace's equation $\nabla^2 V = 0$.

Solution:

The analytical evaluation of the Laplacian here by hand is complex and long. However, we can easily calculate the second partial derivatives using `diff(V,x,2)`, and use `simplify()` to check the answer.

```
from sympy import diff, symbols, pi, simplify

q, e0, x, y, z = symbols('q, e0,x,y,z')

V = q/(4*pi*e0)*((x**2+y**2+z**2)**(-1/2))

laplacian = diff(V,x,2) + diff(V,y,2) + diff(V,z,2)

print('-'*28,'CODE OUTPUT','-'*29,'\n')
print('The Laplacian of the potential V = ',simplify(laplacian))

-------------------------- CODE OUTPUT ----------------------------

The Laplacian of the potential V =  0
```

8.7   LINE INTEGRALS

We will take a break from differentiating vectors, to examine the integration of vector fields. Consider a force $\mathbf{F}(x, y, z)$ which moves a particle of mass m from the point $\mathbf{r}_1$ to the point $\mathbf{r}_2$. The force $\mathbf{F}$ does work on the particle which can be found using

$$W = \int_{\mathbf{r}_1}^{\mathbf{r}_2} \mathbf{F} \cdot d\mathbf{r} \qquad (8.7.1)$$

For non-conservative forces, the amount of work $\mathbf{F}$ does on the particle depends on the path C taken between the points $\mathbf{r}_1$ and $\mathbf{r}_2$. In general we write

$$W = \int_C \mathbf{F} \cdot d\mathbf{r} \qquad (8.7.2)$$

The integral in (8.7.2) is an example of a *line integral*. The value of a line integral depends on the vector field $\mathbf{F}$ and the path C along which $\mathbf{F}$ is integrated. In order to calculate integrals, one can break them up into component integrals, by writing out the dot product of the components of the force vector $\mathbf{F} = F_x\hat{\mathbf{i}} + F_y\hat{\mathbf{j}} + F_z\hat{\mathbf{k}}$ and the differential $d\mathbf{r} = dx\hat{\mathbf{i}} + dy\hat{\mathbf{j}} + dz\hat{\mathbf{k}}$, as follows:

$$W = \int_C F_x dx + \int_C F_y dy + \int_C F_z dz \qquad (8.7.3)$$

One would then evaluate the individual integrals along the path C. The next several examples demonstrate how to calculate line integrals along different types of integration paths.

Example 8.13 illustrates two different methods of evaluating a line integral over a straight line integration path on the xy-plane.

===

Example 8.13: Work done along a straight-line path

(a) Find the work done by the force $\mathbf{F} = xy\hat{\mathbf{i}} - y^2\hat{\mathbf{j}}$ along the straight line path from the origin to the point $(2, 1)$ as depicted in Figure 8.5.
(b) Find the work along this line using the `ParametricRegion` and the `ParametricIntegral` functions in `sympy.vector`.

Solution:

(a) This problem is most easily done in Cartesian coordinates. We write for the work

$$W = \int_C F_x dx + \int_C F_y dy = \int_C xy dx - \int_C y^2 dy \qquad (8.7.4)$$

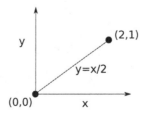

Figure 8.5   The path of integration for Example 8.13.

The straight line path shown in the figure can be written as $y = x/2$ and therefore the differentials are related by $dy = dx/2$. The x-coordinate the straight line integration path changes from $x = 0$ to $x = 2$. By substituting these into the above expression, we end up with simple integrals over the x-axis only:

$$W = \int_{x=0}^{x=2} x \frac{x}{2} dx - \int_{x=0}^{x=2} \left(\frac{x}{2}\right)^2 \frac{dx}{2} = \int_0^2 \frac{3}{8} x^2 dx = 1 \qquad (8.7.5)$$

The Python code for Method 1 uses the `integrate` function in SymPy to evaluate the line integral.

We can also evaluate the work by introducing the parameter $x = t$ such that t varies from $t = 0$ to $t = 2$. Along the line connecting the origin to the point $(2, 1)$, we then have $y = t/2$, and also $dx = dt$ and $dy = dt/2$. The work becomes:

$$W = \int_C F_x dx + \int_C F_y dy = \int_{t=0}^{t=2} \left(t^2/2\right) dt - \int_{t=0}^{t=2} (t/2)^2 dt/2 = 1$$

(b) In Method 2 of the Python code we define the line from the origin to the point $(2, 1)$ by using `ParametricRegion((t,t/2),(t,0,2)`, and store this as the variable `diagonal`. The work is calculated using `ParametricIntegral(F,diagonal)` where F represents the force vector $\mathbf{F} = xy\hat{\mathbf{i}} - y^2\hat{\mathbf{j}}$.

```
from sympy import integrate, symbols
from sympy.vector import  CoordSys3D, ParametricIntegral,\
ParametricRegion

print('-'*28,'CODE OUTPUT','-'*29,'\n')

x, y, t, x1, y1 = symbols('x, y, t, x1, y1')

C = CoordSys3D('C')     # define Cartesian system, named here C

# Method 1: using two line integrals

intx = integrate(x1*(x1/2), (x1,0,2))

inty = integrate(-y1**2, (y1,0,1))

print('The line integral using two line integrals =',intx + inty)

# Method 2: using ParametricRegion and ParametricIntegral

x = t
y = t/2

F = x*y*C.i - y**2*C.j   # Force vector

diagonal = ParametricRegion((t, t/2), (t, 0, 2))

W = ParametricIntegral(F, diagonal)
print('The line integral using ParametricIntegral =',W)

---------------------------- CODE OUTPUT ----------------------------

The line integral using two line integrals = 1
The line integral using ParametricIntegral = 1
```

In Example 8.14 we evaluate the same integral by choosing a different integration path between the origin and the point $(2, 1)$.

Example 8.14: Work done along a path consisting of multiple line segments

(a) Find the work done by the force in Example 8.13 along the line path $(0,0) \to (2,0) \to (2,1)$, indicated in Figure 8.6. Does the work done by the force depend on the path taken between points $(0,0)$ and $(2,1)$?

(b) Find the work along this line using the `ParametricIntegral` function in `sympy.vector`.

Solution:

(a) The work in this case is calculated in two segments, as the sum of two segments $(0,0) \to (2,0)$, added to the work in segment $(2,0) \to (2,1)$.

In the first segment we have $y = 0$ and therefore $dy = 0$, while x varies from $x = 0$ to $x = 2$.

$$W_1 = \int_{(0,0)\to(2,0)} F_x dx + \int_{(0,0)\to(2,0)} F_y dy \tag{8.7.6}$$

$$= \int_{(0,0)\to(2,0)} x\,y\big|_{y=0}\,dx - \int_{(0,0)\to(2,0)} y^2\big|_{y=0}\,dy = 0 + 0 = 0 \tag{8.7.7}$$

The first integral in (8.7.7) is zero because $y = 0$ along the path from $(0,0) \to (2,0)$. The second integral in (8.7.7) is zero because $dy = 0$ along the path.

Along the second segment from $(2,0) \to (2,1)$, we have $x = 2$ and $dx = 0$, while y varies from $y = 0$ to $y = 1$.

$$W_2 = \int_{(2,0)\to(2,1)} F_x dx + \int_{(2,0)\to(2,1)} F_y dy \tag{8.7.8}$$

$$= \int_{y=0}^{y=1} xy\,dx - \int_{y=0}^{y=1} y^2\,dy \tag{8.7.9}$$

$$= 0 - \frac{1}{3}y^3\Big|_0^1 = -\frac{1}{3} \tag{8.7.10}$$

The first integral in (8.7.9) is zero because $dx = 0$ along this segment. The work done along the entire path is $W = W_1 + W_2 = -1/3$.

(b) In the Python code we use the same method as in the previous example, with `ParametricRegion((t,0),(t,0,2)` defining the horizontal line of the integration path with $x = t$ and $y = 0$. Similarly, `ParametricRegion((2,t),(t,0,1)` defines the vertical line of the integration path with $x = 2$ and $y = t$. The work is calculated using `ParametricIntegral(F,line1)` plus `ParametricIntegral(F,line2)`, where F represents again the force vector $\mathbf{F} = xy\hat{\mathbf{i}} - y^2\hat{\mathbf{j}}$.

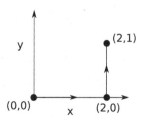

Figure 8.6 The path of integration for Example 8.14.

```
from sympy import symbols
from sympy.vector import  CoordSys3D, ParametricIntegral,\
ParametricRegion

print('-'*28,'CODE OUTPUT','-'*29,'\n')

C = CoordSys3D('C')      # define Cartesian system, named here C

t = symbols('t')

# using ParametricRegion and ParametricIntegral

# Force vector
F = C.x*C.y*C.i - C.y**2*C.j

line1 = ParametricRegion((t, 0), (t, 0, 2))

integr1 = ParametricIntegral(F, line1)
print('integral along line 1 =', integr1)

line2 = ParametricRegion((2, t), (t, 0, 1))
integr2 = ParametricIntegral(F, line2)
print('integral along line 1 =',integr2)

print('\nWork done is :', integr1+integr2)

-------------------------- CODE OUTPUT ----------------------------

integral along line 1 = 0
integral along line 1 = -1/3

Work done is : -1/3
```

Examples 8.13 and 8.14 show clearly that the work done by this force depends on the path taken between the two points $(0,0)$ and $(2,1)$. Such a force is called a *nonconservative force*. We will see later on that nonconservative forces cannot be associated with a potential energy function.

In Example 8.15 we evaluate the line integral around a quarter circular path.

Example 8.15: Work done along a curved path

(a) Find the work done by the force $\mathbf{F} = xy\hat{\mathbf{i}} - y^2\hat{\mathbf{j}}$ in Example 8.13, along the unit quarter circle from the point $(1,0)$ to the point $(0,1)$ as indicated in the Figure 8.7.
(b) Find the work along this line using the `ParametricIntegral` function in `sympy.vector`.

Solution:

(a) This is a problem that is best done in polar coordinates. Let ϕ be the angle measured counter clockwise with respect to the x-axis, denoted in the figure above. A particle moving along the quarter circle would have a constant radial coordinate $\rho = 1$. Therefore, the conversion from

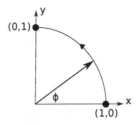

Figure 8.7 The path of integration for Example 8.15.

Cartesian to polar coordinates would be: $x = \cos\phi$ and $y = \sin\phi$. In order to rewrite the force vector, we need to convert the unit vectors from Cartesian to polar coordinates. In Chapter 4, we showed

$$\hat{\mathbf{i}} = \cos\phi\,\hat{\boldsymbol{\rho}} - \sin\phi\,\hat{\boldsymbol{\phi}}$$

$$\hat{\mathbf{j}} = \sin\phi\,\hat{\boldsymbol{\rho}} + \cos\phi\,\hat{\boldsymbol{\phi}}$$

By inserting the coordinate transformations for x, y, $\hat{\mathbf{i}}$, and $\hat{\mathbf{j}}$ into the force we obtain:

$$\mathbf{F} = \sin\phi\cos(2\phi)\,\hat{\boldsymbol{\rho}} - 2\cos\phi\sin^2\phi\,\hat{\boldsymbol{\phi}}$$

where we used the trigonometric identity $\cos(2\phi) = \cos^2\phi - \sin^2\phi$. We can use the infinitesimal displacement in polar coordinates, $d\mathbf{r} = \hat{\boldsymbol{\rho}}\,d\rho + \hat{\boldsymbol{\phi}}\,\rho\,d\phi$, but this simplifies to $d\mathbf{r} = \rho\,d\phi\,\hat{\boldsymbol{\phi}}$, because the particle is not moving radially away from the origin (*i.e.* $\rho = 1$ and $d\rho = 0$). After inserting $\mathbf{F}$ and $d\mathbf{r}$ into (8.7.2) and setting $\rho = 1$, we obtain:

$$W = \int_0^{\pi/2} (-2\cos\phi\sin^2\phi)\,d\phi = -2/3$$

(b) In the Python code we use the same method as in the previous two examples, with

path = ParametricRegion((cos(t),sin(t)),(t,0,pi/2)

defining the quarter circular integration path, with the angle varying from $t = 0$ to $t = \pi/2$. The work is calculated once more using ParametricIntegral(F,path)), where F represents again the force vector $\mathbf{F} = xy\hat{\mathbf{i}} - y^2\hat{\mathbf{j}}$.

```
from sympy import  symbols, sin ,cos, pi
from sympy.vector import  CoordSys3D, ParametricIntegral,\
ParametricRegion

print('-'*28,'CODE OUTPUT','-'*29,'\n')

C = CoordSys3D('C')      # define Cartesian system, named here C

t = symbols('t')

# Force vector
F= C.x*C.y*C.i - C.y**2*C.j

path = ParametricRegion((cos(t), sin(t)), (t, 0, pi/2))

integr = ParametricIntegral(F, path)
print('integral using ParametricIntegral =', integr)
```

```
----------------------------- CODE OUTPUT -----------------------------

integral using ParametricIntegral = -2/3
```

8.8 CONSERVATIVE FIELDS

As we saw in Example 8.13, the value of a line integral can depend on the path taken. When the line integral of a vector field $\mathbf{F}$ depends only on the end points, and not on the path taken between them, we say that $\mathbf{F}$ is a *conservative field*.

Conservative fields are important in many applications. Let us consider a conservative force $\mathbf{F}$. The work done by $\mathbf{F}$ between the points $\mathbf{r}_0$ and $\mathbf{r}$ is independent of the path taken between $\mathbf{r}_0$ and $\mathbf{r}$. We can define a function called the *potential energy*

$$V(\mathbf{r}) - V(\mathbf{r}_0) = -\int_{\mathbf{r}_0}^{\mathbf{r}} \mathbf{F} \cdot d\mathbf{r}' \tag{8.8.1}$$

$$V(\mathbf{r}) = -\int_{\mathbf{r}_0}^{\mathbf{r}} \mathbf{F} \cdot d\mathbf{r}' \tag{8.8.2}$$

where we have set as a reference point $V(\mathbf{r}_0) = 0$. The definition (8.8.1) works because of the path-independence of $\mathbf{F}$. Without path independence, V would be a multivalued function for any point $\mathbf{r}$, because different paths would produce different values for the same value of $\mathbf{r}$.

Next we consider a small change in potential energy along the path.

$$dV = -\mathbf{F} \cdot d\mathbf{r} \tag{8.8.3}$$

Using the chain rule, we can write

$$dV = \frac{\partial V}{\partial x}dx + \frac{\partial V}{\partial y}dy + \frac{\partial V}{\partial z}dz \tag{8.8.4}$$

which can be rewritten as

$$dV = \nabla V \cdot d\mathbf{r} \tag{8.8.5}$$

Equating (8.8.3) to (8.8.5), we find

$$\mathbf{F} = -\nabla V \tag{8.8.6}$$

Hence, a conservative field can be found from the gradient of a scalar potential function. In classical mechanics V is called the potential energy. However, in electromagnetism, we can write an electric field as $\mathbf{E} = -\nabla V$ and V is called the *electric potential*.

If we work in Cartesian coordinates then

$$F_x = -\frac{\partial V}{\partial x} \quad F_y = -\frac{\partial V}{\partial y} \quad F_z = -\frac{\partial V}{\partial z} \tag{8.8.7}$$

Example 8.16: Finding the potential energy from a force

(a) Show that the force $\mathbf{F} = -\left(3c_1x^2y^2z + c_2y\right)\hat{\mathbf{i}} - \left(2c_1x^3yz + c_2x\right)\hat{\mathbf{j}} - c_1x^3y^2\hat{\mathbf{k}}$ is conservative, where c_1 and c_2 are constants, and find the potential energy using an analytical method.
(b) Verify the result in (a) by finding the potential energy with the `scalar_potential` function in `sympy.vector`.

Solution:

We evaluate first the $\nabla \times \mathbf{F}$:

$$\nabla \times \mathbf{F} = \left(\frac{\partial F_z}{\partial y} - \frac{\partial F_y}{\partial z}\right)\hat{\mathbf{i}} + \left(\frac{\partial F_x}{\partial z} - \frac{\partial F_z}{\partial x}\right)\hat{\mathbf{j}} + \left(\frac{\partial F_y}{\partial x} - \frac{\partial F_x}{\partial y}\right)\hat{\mathbf{k}} = 0$$

The curl is zero, therefore $\mathbf{F}$ is conservative.
Next, we need to calculate the line integral:

$$V = -\int \mathbf{F} \cdot d\mathbf{r}$$

Since the force is conservative, we can find the potential by integrating over *any* path between the two integration points, We choose a path that starts at the origin and integrate

$$\int \left[\left(3c_1x^2y^2z + c_2y\right)dx + \left(2c_1x^3yz + c_2x\right)dy + \left(c_1x^3y^2\right)dz\right]$$

to the point (x, y, z) along the path

$$(0,0,0) \rightarrow (x,0,0) \rightarrow (x,y,0) \rightarrow (x,y,z)$$

We are breaking this path into three segments and then we will sum the result of the integral along each segment.
Along the first segment, $y = z = 0$ and $dy = dz = 0$ and the integral is zero. Along the second segment, x is constant, $z = 0$, and $dx = dz = 0$. The integral is

$$\int_0^y c_2x\,dy = c_2xy$$

The final leg holds x and y constant and $dx = dy = 0$. The integral is

$$\int_0^z c_1x^3y^2\,dz = c_1x^3y^2z$$

Hence, we find

$$V = -c_1x^3y^2z - c_2xy$$

(b) We verify the result from (a), using the `scalar_potential` function in `sympy.vector`.

```
from sympy.vector import CoordSys3D,curl, scalar_potential
from sympy import symbols

R = CoordSys3D('R')
c1, c2 = symbols('c1, c2')

x = R.x
y = R.y
```

```
z = R.z

F = -(3*c1*x**2*y**2*z + c2*y)*R.i - \
    (2*c1*x**3*y*z+ c2*x)*R.j - c1*x**3*y**2*R.k

print('-'*28,'CODE OUTPUT','-'*29,'\n')
print('The curl of the force F is: ',curl(F))

print('\nThe potential V for this force is: \n')
print(scalar_potential(F, R))

------------------------- CODE OUTPUT ----------------------------

The curl of the force F is:  0

The potential V for this force is:

-R.x**3*R.y**2*R.z*c1 - R.x*R.y*c2
```

8.9   AREA INTEGRALS AND FLUX

In physics, it is common to integrate a vector or scalar field over an area. Consider a surface S. At each point on the surface of S, we can define a unit vector $\hat{n}$ normal to S. If S is a closed surface, like a sphere, then it is common for $\hat{n}$ to point outward from the surface. In this case $\hat{n}$ is called the *unit outward normal vector*.

Integrating over areas usually requires that we break the area up into infinitesimal area elements. Consider a surface differential dS as shown in Figure 8.8 which is the area of an infinitesimal segment of S (not shown in the figure). We define the vector $d\mathbf{S} = \hat{n}dS$ (Figure 8.8) as the vector which is perpendicular to the surface element dS and has the same direction as $\hat{n}$.

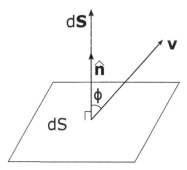

Figure 8.8   A vector field **v** with a flux through an area element $\hat{n}dS$.

The integral

$$\iint\limits_{S} \mathbf{A} \cdot d\mathbf{S} = \iint\limits_{S} A \cdot \hat{n}dS \qquad (8.9.1)$$

is called the surface integral of the vector **A** over the surface S. For example, the integral

$$\Phi = \iint\limits_{S} \mathbf{E} \cdot d\mathbf{S} \qquad (8.9.2)$$

of the electric field vector $\mathbf{E}$ is called the *flux Φ of the electric field* through the surface S. These types of integrals appear in Maxwell's equations, which are the fundamental blocks of electromagnetism.

In addition to surface integrals of the type found in (8.9.1), we also come across various surface integrals involving scalar functions ϕ or vector functions, such as:

$$\iint_S \phi \, dS \qquad \iint_S \phi \, d\mathbf{S} \qquad \iint_S \mathbf{A} \times d\mathbf{S} \qquad\qquad (8.9.3)$$

These types of integrals can be defined as the limits of sums, similar to what is done for integrals over a single variable. The notation $\oint \mathbf{A} \cdot d\mathbf{S}$ is used to indicate integration over the *closed* surface S.

To evaluate surface integrals, it is convenient to express them as double integrals taken over the projected area of the surface S on one of the coordinate planes. We now discuss these surface integrals in the context of electric and magnetic fields.

In electromagnetism and fluid dynamics, we often need to know the flux of a vector field $\mathbf{v}$ through a surface, such as that shown in Figure 8.8. The flux can be thought of, loosely, as the flow of a vector field through an area. To find the flux of a vector field, we need to know the component of $\mathbf{v}$ perpendicular to the area. In Figure 8.8, the vector $\hat{\mathbf{n}}$ points in a direction that is normal to the surface area element. The unit vector $\hat{\mathbf{n}}$ is often referred to as the unit outward normal because if the area is a closed surface (such as a sphere), then $\hat{\mathbf{n}}$ points outward from the enclosed volume. If $\mathbf{v}$ were perpendicular to $\hat{\mathbf{n}}$, then $\mathbf{v}$ cannot be considered to be flowing through the area, but instead, along it. To find the flux $d\Phi$ through the area element dS, we need to know the component of $\mathbf{v}$ along $\hat{\mathbf{n}}$.

$$d\Phi = \mathbf{v} \cdot \hat{\mathbf{n}} dS \qquad\qquad (8.9.4)$$

However, if dS is only a small section of a greater surface area, then we must integrate to obtain the total flux

$$\Phi = \oint_S \mathbf{v} \cdot \hat{\mathbf{n}} dS \qquad\qquad (8.9.5)$$

where the integration is done over the entire area of the surface S.

In electromagnetism, the integration is often done over a *Gaussian surface* which encloses a charge distribution. The integral form of Gauss's law relates the flux of the electric field $\mathbf{E}$ through the Gaussian surface to the charge enclosed by the Gaussian surface q_{enc}.

$$\epsilon_0 \oint_S \mathbf{E} \cdot \hat{\mathbf{n}} dS = q_{\text{enc}} \qquad\qquad (8.9.6)$$

where ϵ_0 is the permittivity permittivity of free space constant. Example 8.17 shows how to evaluate the charge q_{enc} when we know the electric field, in spherical coordinates.

Example 8.17: Gauss's law

(a) The electric field in a region is $\mathbf{E} = k\,\hat{\mathbf{r}}$ in spherical coordinates. Using (8.9.6), find the total charge contained in the sphere of radius R centered at the origin.

(b) Find the total charge using the `ParametricRegion` function to define the surface of the sphere, and using the `vector_integrate` functions in `sympy.vector` to integrate the electric field over this surface.

Solution:

(a) Consider a small surface dS on the sphere, and note that the unit vector on the surface points along the radial direction, so $\hat{\mathbf{n}} = \hat{\mathbf{r}}$ and

$$\mathbf{E} \cdot \hat{\mathbf{n}}\, dS = (k\,\hat{\mathbf{r}}) \cdot \hat{\mathbf{r}}\, dS$$

Using $\hat{\mathbf{r}}\hat{\mathbf{r}} = 1$:

$$\mathbf{E} \cdot \hat{\mathbf{n}}\, dS = k\, dS$$

and (8.9.6), becomes

$$q_{\text{enc}} = k \int dS \epsilon_0 = 4\pi\, k\, R^2 \epsilon_0$$

(b) In the Python code the surface of the sphere is easily defined using the `ParametricRegion()` function with the appropriate range of the spherical coordinates (r, θ, ϕ). We write the $\hat{\mathbf{r}}$ as $\mathbf{r}/r$ in Cartesian coordinates;

$$\hat{\mathbf{r}} = \frac{\mathbf{r}}{r} = \frac{x\,\hat{\mathbf{i}} + y\,\hat{\mathbf{j}} + c\,\hat{\mathbf{k}}}{\sqrt{x^2 + y^2 + z^2}}$$

and store it in variable `runitvector`. The `vector_integrate()` function is used to integrate the electric field over the sphere's surface in the line of code `vector_integrate(runitvector, sphere)` E0. The result from the code is the same as the analytical result from (a). This is of course the electric field $\mathbf{E}$ of a point charge q_{enc} located at the origin, which according to Coulomb's law produces a radial electric field equal to:

$$\mathbf{E} = \frac{q_{\text{enc}}}{4\pi\,\epsilon_0 R^2}\,\hat{\mathbf{r}}$$

```
from sympy import sin, cos, pi, symbols, sqrt
from sympy.vector import CoordSys3D, ParametricRegion, vector_integrate
from sympy.abc import  theta, phi

E0, R, k, r = symbols('E0, R, k, r', positive=True)

# define the spherical coordinate system with (r, theta, phi)
C = CoordSys3D('C')

# Define the surface of the sphere with angles phi and theta, and r = R
sphere = ParametricRegion((R*sin(theta)*cos(phi), \
   R*sin(theta)*sin(phi), R*cos(theta)),(theta, 0, pi), (phi, 0, 2*pi))

runitvector = (C.x*C.i+C.y*C.j+C.z*C.k)/(sqrt(C.x**2+C.y**2+C.z**2))

charge = vector_integrate(runitvector, sphere) E0

print('-'*28,'CODE OUTPUT','-'*29,'\n')
print('The total charge inside the sphere is: ',charge)


-------------------------- CODE OUTPUT ----------------------------

The total charge inside the sphere is:  4*pi*R**2*E0
```

In the next example the electric field **E** is given in Cartesian coordinates, and we integrate over a hemisphere.

Example 8.18: Electric flux

The electric field in a region is given by $\mathbf{E} = A\,\hat{\mathbf{k}}$ in Cartesian coordinates, where A is constant. Using SymPy, calculate the flux through a hemisphere of radius R centered at the origin above the xy-plane.

Solution:

(a) As in the previous problem, the unit vector on the surface of a sphere points along the radial direction, so $\hat{\mathbf{n}} = \hat{\mathbf{r}}$ and

$$\mathbf{E} \cdot \hat{\mathbf{n}}\,dS = \left(A\,\hat{\mathbf{k}}\right) \cdot \hat{\mathbf{r}}\,dS$$

The dot product $\hat{\mathbf{k}} \cdot \hat{\mathbf{r}} = \cos\theta$, where θ is the polar angle in spherical coordinates. Note that on the sphere, θ represents the angle between the radial direction and the z-axis.
Therefore

$$\mathbf{E} \cdot \hat{\mathbf{n}}\,dS = (A\,\cos\theta)\,r^2 \sin\theta\,d\theta\,d\phi$$

Using $r = R$ on the surface of the hemisphere, (8.9.6) becomes:

$$q_{\text{enc}} = A\,\epsilon_0\,R^2 \int_0^{2\pi} \int_0^{\pi/2} \cos\theta \sin\theta\,d\theta\,d\phi = \pi\,R^2\,A\,\epsilon_0$$

The function `ParametricRegion()` can be used to define the Gaussian surface. The `vector_integrate()` function can be used to integrate the vector **E** over the hemisphere.

```
from sympy import sin, cos, pi, symbols
from sympy.vector import CoordSys3D, ParametricRegion, vector_integrate
from sympy.abc import R, theta, phi, A

E0 = symbols('E0')

# Cartesian coordinate system C
C = CoordSys3D('C')

# define parametric surface of hemisphere
hemisphere = ParametricRegion((R*sin(theta)*cos(phi), \
                    R*sin(theta)*sin(phi), R*cos(theta)),\
                    (theta, 0, pi/2), (phi, 0, 2*pi))
flux = E0* vector_integrate(A*C.k, hemisphere)

print('-'*28,'CODE OUTPUT','-'*29,'\n')
print('The flux through the hemisphere is: ', flux)

----------------------- CODE OUTPUT ----------------------------

The flux through the hemisphere is:  pi*A*E0*R**2
```

8.10   GREEN'S THEOREM IN THE PLANE

There are two theorems in the field of vector calculus which are particularly useful in physics, the divergence theorem and Stokes's theorem. Here, we will develop a foundational theorem, called Green's theorem, from which the divergence and Stokes's theorems can be derived.

Consider the area A bounded by the contour C in Figure 8.9.

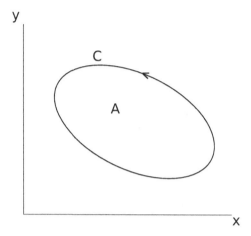

Figure 8.9   A contour C in the xy-plane bounds an area A.

Suppose that $L(x, y)$ and $M(x, y)$ are continuous functions with continuous first derivatives. Green's theorem states

$$\int\int_A \left(\frac{\partial L}{\partial x} - \frac{\partial M}{\partial y} \right) dx dy = \oint_C (M dx + L dy) \tag{8.10.1}$$

where $\oint_C$ represents a contour integral counterclockwise around the curve C that bounds the area A.

Green's theorem allows us to choose between computing an area integral or a contour integral, whichever is easier. To see the usefulness of Green's theorem, consider Example 8.19.

Example 8.19: Green's theorem and work

Using Green's theorem, calculate the work done by the force $\mathbf{F} = c_1 x y^2 \hat{\mathbf{i}} + c_2 x y^3 \hat{\mathbf{j}}$ around the unit square, with one corner at the origin as shown in Figure 8.10.

Solution:

The path of integration is shown in Figure 8.10. The work done around this contour is

$$W = \oint \left(x y^2 dx + y^3 x dy \right)$$

However, instead of doing the contour integral, which would need four integrations (one for each side), we can use (8.10.1) to calculate a much simpler area integral.

$$M = x y^2 \qquad L = y^3 x$$
$$\frac{\partial M}{\partial y} = 2 x y \qquad \frac{\partial L}{\partial x} = y^3$$

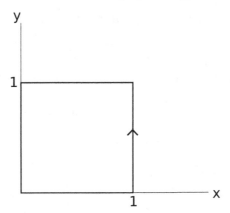

Figure 8.10 The contour for Example 8.19.

Therefore

$$W = \int_0^1 \int_0^1 \left(y^3 - 2\,x\,y\right) dx\,dy$$

In the Python code we use the **integrate** function in SymPy to evaluate this integral over the square.

```
from sympy import integrate, symbols, diff

x, y = symbols('x, y')

dMdy = diff(x*y**2, y)
dLdx = diff(y**3*x, x)
print('-'*28,'CODE OUTPUT','-'*29,'\n')

print('The area integral is : ',integrate(dMdy - dLdx, (x,0,1),(y,0,1)))

----------------------- CODE OUTPUT ----------------------------

The area integral is :  1/4
```

Consider the work done by a force $\mathbf{F} = F_x\hat{\mathbf{i}} + F_y\hat{\mathbf{j}}$ moving a particle of mass m along a closed path C on the xy-plane. As we saw earlier,

$$W = \oint_C \left(F_x dx + F_y dy\right) \tag{8.10.2}$$

If $\mathbf{F}$ is a conservative force, then $W = 0$, and Green's theorem says:

$$W = \int\int_A \left(\frac{\partial F_y}{\partial x} - \frac{\partial F_x}{\partial y}\right) dx dy = 0 \tag{8.10.3}$$

where A is the area bounded by C. Therefore

$$\frac{\partial F_y}{\partial x} = \frac{\partial F_x}{\partial y} \tag{8.10.4}$$

In other words, $\nabla \times \mathbf{F} = 0$. since $F_z = 0$ in this example. This is another illustration of the fact that the curl of a conservative force is zero. The curl is a useful way of determining whether or not a force is conservative.

8.11 THE DIVERGENCE THEOREM

Consider the area A shown in Figure 8.11, which is bounded by the contour C. The vector $d\mathbf{r}$ points in the direction of a displacement tangent to C, while $\hat{\mathbf{n}} ds$ is a vector that is normal to C. The vector $\hat{\mathbf{n}}$ is called the unit outward normal vector. Note that $d\mathbf{r} \cdot \hat{\mathbf{n}} \, ds = 0$. Hence, we can write

$$d\mathbf{r} = dx\,\hat{\mathbf{i}} + dy\,\hat{\mathbf{j}} \tag{8.11.1}$$

$$\hat{\mathbf{n}}\,ds = dy\,\hat{\mathbf{i}} - dx\,\hat{\mathbf{j}} \tag{8.11.2}$$

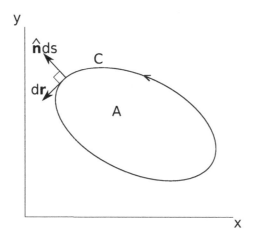

Figure 8.11 A contour C bounding a closed area A. The vector $d\mathbf{r}$ illustrates a displacement around C. The vector $\hat{\mathbf{n}} ds$ points in a direction that is outward from A and normal to C.

Next, consider a vector field defined on the xy-plane,

$$\mathbf{v} = v_x(x, y)\,\hat{\mathbf{i}} + v_y(x, y)\,\hat{\mathbf{j}} \tag{8.11.3}$$

According to Green's theorem, we can write

$$\int\int_A \left(\frac{\partial v_x}{\partial x} + \frac{\partial v_y}{\partial y} \right) dx\,dy = \oint_C (-v_y\,dx + v_x\,dy) \tag{8.11.4}$$

The right-hand side can be rewritten as:

$$\oint_C (-v_y\,dx + v_x\,dy) = \oint_C \left(v_x\,\hat{\mathbf{i}} + v_y\,\hat{\mathbf{j}} \right) \cdot \left(\hat{\mathbf{i}}\,dy - \hat{\mathbf{j}}\,dx \right) = \oint_C \mathbf{v} \cdot \hat{\mathbf{n}}\,ds \tag{8.11.5}$$

Therefore

$$\int\int_A \left(\frac{\partial v_x}{\partial x} + \frac{\partial v_y}{\partial y} \right) dx\,dy = \oint_C \mathbf{v} \cdot \hat{\mathbf{n}}\,ds \tag{8.11.6}$$

Generalizing to three dimensions, we have

$$\int\int\int_\tau \nabla \cdot \mathbf{v}\,d\tau = \oint_\sigma \mathbf{v} \cdot \hat{\mathbf{n}}\,d\sigma \tag{8.11.7}$$

where the volume τ is bounded by the surface area σ. Here we changed notation for the volume from V to τ to avoid confusion with the vector field $\mathbf{v}$. Equation (8.11.7) is called the *divergence theorem* and is of central importance in electromagnetism and fluid mechanics.

The left-hand side of (8.11.7) can be thought of as the net outward flow of the vector field $\mathbf{v}$ from the volume τ. The divergence theorem says that the net outward flow is equal to the flow through the volume's surface, the right-hand side of (8.11.7). Furthermore, the divergence theorem allows us to replace the volume integral of a divergence with an area integral.

In Example 8.20 we apply the divergence theorem to Gauss's law for electric fields, and we show how to evaluate the charge density when one knows the corresponding electric field.

Example 8.20: Gauss's law for electric fields

(a) Consider the integral form of Gauss's law for electric fields

$$\epsilon_0 \oint_A \mathbf{E} \cdot \hat{n}\, dS = Q_{enc}$$

where the integral is calculated over an area A that encloses the charge Q_{enc}. Derive the differential form of Gauss's law:

$$\nabla \cdot \mathbf{E} = \rho/\epsilon_0 \qquad (8.11.8)$$

where the charge density is defined by $Q_{enc} = \int \rho\, d\tau$. Provide a physical interpretation.

(b) Use the result from part (a) to find the charge density ρ creating the electric field given by $\mathbf{E} = k\, r\, \hat{r}$, where $\hat{r}$ is the unit vector along the radial direction and k is a constant.

Solution:

(a) We can use the divergence theorem to rewrite the left-hand side of Gauss's law

$$\epsilon_0 \int_\tau \nabla \cdot \mathbf{E}\, d\tau = Q_{enc}$$

where τ is the volume bounded by the surface area A. Note that the single integral sign is actually denoting a triple integral. Next, we write $Q_{enc} = \int \rho\, d\tau$ where ρ is the density of the charge enclosed in the area A.

$$\epsilon_0 \int_\tau \nabla \cdot \mathbf{E}\, d\tau = \int_\tau \rho\, d\tau$$

Therefore

$$\nabla \cdot \mathbf{E} = \rho/\epsilon_0 \qquad (8.11.9)$$

Gauss's law encapsulates most of the rules of electric field lines that you learned in your introductory physics course. It says that the net outflow of the electric field $\mathbf{E}$ depends on the magnitude and sign of the charge density enclosed by the surface A, often called a *Gaussian surface*. For example, an isolated positive charge has field lines that point radially away from it, while a negative charge has field lines that point radially toward it.

(b) We evaluate the divergence of the given electric field. Because of the spherical symmetry of the electric field, we use the expression (8.4.10) for the divergence of a vector field in spherical coordinates:

$$\nabla \cdot \mathbf{E} = \frac{1}{r^2} \frac{\partial \left(r^2\, E_r\right)}{\partial r} + \frac{1}{r \sin\theta} \frac{\partial}{\partial \theta} \left(E_\theta \sin\theta\right) + \frac{1}{r \sin\theta} \frac{\partial \left(E_\phi\right)}{\partial \phi}$$

In this example we have $E_\theta = E_\phi = 0$ and $E_r = k\,r$ and we obtain:

$$\rho = \epsilon_0\,\nabla\cdot\mathbf{E} = \epsilon_0\,\frac{1}{r^2}\,\frac{\partial\left(r^2\,E_r\right)}{\partial r} = \frac{1}{r^2}\,\frac{\partial\left(r^2\,k\,r\right)}{\partial r} = 3\,\epsilon_0\,k$$

In the Python code we evaluate the divergence in spherical coordinates, with the spherical variables denoted by c.r, c.theta, c.phi , and the corresponding unit vectors by c.R, c.Theta, c.Phi . In this notation the given electric field is written as E = k * c.r * c.R, and the divergence() function evaluates the charge density.

```
from sympy.vector import CoordSys3D, divergence
from sympy import symbols

k, E0 = symbols('k, E0')

print('-'*28,'CODE OUTPUT','-'*29,'\n')

# define Cartesian system named c, with unit vectors c.i, c.j, c.k
# Transform it to system with cylindrical coordinates (r, theta, z)

c = CoordSys3D('c', transformation='spherical',\
variable_names = ("r", "theta", "phi"),vector_names=("R","Theta","Phi"))

# spherical variables are c.r, c.theta, c.phi
# unit vectors are c.R, c.Theta, c.Phi

# Electric field E = k *r *(radial unit vector)
E = k * c.r * c.R

rho = E0  * divergence(E,c)
print('The charge density rho = ',rho)


---------------------------- CODE OUTPUT ----------------------------

The charge density rho =   3*E0*k
```

8.12   STOKES'S THEOREM

Now let us return to Green's theorem. We will again consider the contour shown in Figure 8.11 and the vector field $\mathbf{v} = v_x(x,y)\,\hat{\mathbf{i}} + v_y(x,y)\,\hat{\mathbf{j}}$. Note that $\mathbf{v}$ is not necessarily a velocity. According to Green's theorem

$$\int\int_A \left(\frac{\partial v_y}{\partial x} - \frac{\partial v_x}{\partial y}\right)dx\,dy = \oint_C \left(v_x dx + v_y dy\right) \tag{8.12.1}$$

Note that the left-hand side of (8.12.1) is the z-component of $\nabla\times\mathbf{v}$ and the right-hand side is $\mathbf{v}\cdot d\mathbf{r}$. We can generalize (8.12.1) to three dimensions

$$\int\int_S (\nabla\times\mathbf{v})\cdot\hat{\mathbf{n}}dS = \oint_C \mathbf{v}\cdot d\mathbf{r} \tag{8.12.2}$$

where S is the area of an open surface (such as a hemisphere) bounded by the contour C, dS is an infinitesimal area element of S, and $d\mathbf{r}$ is an infinitesimal displacement around C in a *counterclockwise* direction. The vector $\hat{\mathbf{n}}$ is the unit outward normal vector to the surface A. Equation (8.12.2) is called *Stokes's theorem* and is of central importance in vector analysis.

The left-hand side of (8.12.2) calculates the circulation of a vector field across the surface A, while the right-hand side calculates the circulation of the vector field around the contour bounding the surface. Stokes's theorem says those two quantities are the same. Hence, when convenient, we can replace an integration of a curl on a surface, with a contour integral around the surface.

Example 8.21: Demonstration of Stokes theorem

Test the Stokes theorem for $\mathbf{v} = xy\hat{\mathbf{i}} + 2yz\hat{\mathbf{j}} + 3xz\hat{\mathbf{k}}$ using the triangle shown in Figure 8.12, located on the yz-plane.

Solution:

We will begin by calculating the left-hand side of (8.12.2):

$$\nabla \times \mathbf{v} = \begin{vmatrix} \hat{\mathbf{i}} & \hat{\mathbf{j}} & \hat{\mathbf{k}} \\ \frac{\partial}{\partial x} & \frac{\partial}{\partial y} & \frac{\partial}{\partial z} \\ xy & 2yz & 3xz \end{vmatrix} = -2y\hat{\mathbf{i}} - 3z\hat{\mathbf{j}} - x\hat{\mathbf{k}}$$

Note that the triangle is in the yz-plane, and therefore, $\hat{\mathbf{n}} = \hat{\mathbf{i}}$. This can be found by the right-hand rule. Curl the fingers of your right hand in a counterclockwise direction around the triangle's perimeter and the thumb will point in the direction of the x-axis. Hence,

$$\int\int_A (\nabla \times \mathbf{v}) \cdot \hat{\mathbf{n}} da = \int_{z=0}^{2} \int_{y=0}^{2-z} (-2y)\, dy dz = -8/3 \qquad (8.12.3)$$

Next, we evaluate the right-hand side of (8.12.2). Using $\mathbf{v} \cdot d\mathbf{r}$ on each side of the triangle, the integral will have three terms.

Along the bottom side of the triangle, $d\mathbf{r} = \hat{\mathbf{j}} dy$ and $x = z = 0$. Hence

$$\int_0^2 2yz dy = 0$$

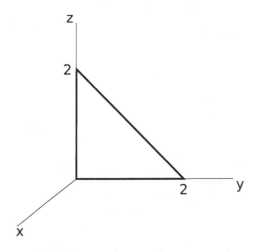

Figure 8.12 The triangle used for Example 8.21 to demonstrate the Stokes theorem.

You should convince yourself that the integral along the leg of the triangle on the z-axis is also zero. The integral along the hypotenuse is found by calculating

$$\int \left(xy\hat{\mathbf{i}} + 2yz\hat{\mathbf{j}} + 3xz\hat{\mathbf{k}} \right) \cdot \left(dy\hat{\mathbf{j}} + dz\hat{\mathbf{k}} \right)$$

Along the hypotenuse, $dz = -dy$, $x = 0$, and $z = 2 - y$. Therefore, the integral becomes

$$\int_{2}^{0} 2y \left(2 - y \right) dy = -8/3$$

The right-hand side is equal to the left-hand side. Therefore, Stokes's theorem has been demonstrated.

In the Python code we use `ParametricRegion` to define the surface of the triangle, and store the triangular parametric region of integration in the parameter `triangle`, before using `vector_integrate` to carry out the integration in (8.12.3).

The line integrals in the right-hand side of Stokes's theorem are evaluated using the `integrate()` function in SymPy.

```
# parametricRegion integration triangle
from sympy import integrate
from sympy.vector import curl, CoordSys3D,vector_integrate,\
ParametricRegion

print('-'*28,'CODE OUTPUT','-'*29,'\n')
from sympy.abc import x,y,z

C = CoordSys3D('C')      # define Cartesian system, named here C

v = C.x*C.y*C.i + 2*C.y*C.z*C.j + 3*C.x*C.z*C.k

print('The curl is =', curl(v))

u = curl(v).coeff(C.i)
print('\nThe x-coefficient of the curl is =', u)

triangle = ParametricRegion((y, z), (z, 0, 2), (y, 0, 2 - z))
print('\nUsing vector_integrate, the surface integral = ',\
vector_integrate(-2*C.y, triangle))

int1 = integrate(x*y,(x,0,2)).subs(y,0)
int2 = integrate(2*y*(2-y),(y,2,0))
int3 = integrate(3*x*z,(x,0,0))
print('\nUsing the integrate function, the line integral =',\
int1+int2+int3)

# another method, using the integrate function in sympy
print('\nUsing  integrate, the surface integral =',\
integrate(-2*y , (y,0,2-z) , (z,0,2)))

--------------------------- CODE OUTPUT ----------------------------

The curl is = (-2*C.y)*C.i + (-3*C.z)*C.j + (-C.x)*C.k

The x-coefficient of the curl is = -2*C.y
```

```
Using vector_integrate, the surface integral =  -8/3

Using the integrate function, the line integral = -8/3

Using  integrate, the surface integral = -8/3
```

Recall from introductory physics, that Ampere's law states:

$$\oint_C \mathbf{B} \cdot d\boldsymbol{\ell} = \mu_0 I_{\text{enc}} \tag{8.12.4}$$

which finds the magnetic field $\mathbf{B}$ due to a current I_{enc} by integrating along a closed contour C (called the Amperian loop) which encloses the current. We can rewrite the left-hand side of Ampere's law using Stokes's theorem

$$\oint_C \mathbf{B} \cdot d\boldsymbol{\ell} = \int_S (\nabla \times \mathbf{B}) \cdot \hat{\mathbf{n}} \, dS \tag{8.12.5}$$

where S is the area enclosed by the Amperian loop, and $\hat{\mathbf{n}}$ is normal to the area. Again, we use a single integrand to represent the double integral over the area.

Next, we can rewrite the current using the current density $\mathbf{J}$ piercing through the Amperian loop :

$$I_{\text{enc}} = \int_S \mathbf{J} \cdot \hat{\mathbf{n}} \, dS \tag{8.12.6}$$

Therefore, Ampere's law becomes

$$\int_S (\nabla \times \mathbf{B}) \cdot \hat{\mathbf{n}} \, dS = \mu_0 \int_A \mathbf{J} \cdot \hat{\mathbf{n}} \, dS \tag{8.12.7}$$

or

$$\nabla \times \mathbf{B} = \mu_0 \mathbf{J} \tag{8.12.8}$$

Equation (8.12.8) tells us that current densities $\mathbf{J}$ create rotational magnetic fields. For example, the magnetic field lines around a current carrying wire take the shape of concentric circles. Furthermore, if you curl the fingers of your right hand around the direction of the field, your thumb will point in the direction of the current. Notice that (8.12.8) states the curl of $\mathbf{B}$ points in the same direction as $\mathbf{J}$.

8.13   END OF CHAPTER PROBLEMS

1. **Scalar field and vector field plots** – Consider the temperature scalar field from Example 8.1. Using Python, plot the scalar field again using the same values of T_0 and σ in the Example. This time, include a vector representing ∇T at the point (1,2). Create another plot which superimposes the gradient field on top of the scalar field.

2. **Divergence identities** – Prove $\nabla \cdot (\mathbf{u} \times \mathbf{v}) = \mathbf{v} \cdot (\nabla \times \mathbf{u}) - \mathbf{u} \cdot (\nabla \times \mathbf{v})$ by hand and demonstrate using Python.

3. **The divergence theorem for a cube** – Test the divergence theorem for the vector field $\mathbf{v} = 3xy\hat{\mathbf{i}} - 2yz\hat{\mathbf{j}} - xz\hat{\mathbf{k}}$. Use as your volume, the unit cube with one corner at the origin in the first octant of three dimensional space. Set up the integrals by hand, but use Python to evaluate them.

4. **The divergence theorem for a sphere** – Test the divergence theorem for the vector field $\mathbf{v} = 3xy\hat{\mathbf{i}} - 2yz\hat{\mathbf{j}} - xz\hat{\mathbf{k}}$. Use as your volume, a sphere with unit radius centered at the origin. Set up the integrals by hand, but use Python to evaluate them.

5. **Testing Stokes theorem** – Test Stokes theorem for the vector $\mathbf{v} = -\left(z^2 + y^2\right)\hat{\mathbf{i}} + xy\hat{\mathbf{j}} - \left(z^2 + 4x\right)\hat{\mathbf{k}}$ along a unit square with one corner in the origin of the xy-plane's first quadrant. Set up the integrals by hand, but use Python to evaluate them.

6. **Testing Stokes's theorem** – Test Stokes's theorem for the vector $\mathbf{v} = -2xy^2\hat{\mathbf{i}} + x^2yz\hat{\mathbf{j}} - 4xy\hat{\mathbf{k}}$ for the quarter disk $x^2 + y^2 \leq 1$ in the first quadrant of the xy-plane. Set up the integrals by hand, but use Python to evaluate them.

7. **The gradient** – Suppose the temperature of a rectangular plate is given by $T = 3x^2y - y^2x$. Plot the scalar field. Find ∇T and plot the resulting vector field over top the scalar field. Calculate the directional derivative in the direction of $\mathbf{n} = 2\hat{\mathbf{i}} - \hat{\mathbf{j}}$ at the point $(-1, 3)$.

8. **The gradient in Cartesian coordinates** – Using Cartesian coordinates, calculate (by hand and using Python) the gradient of (a) $f = \ln r$, (b) $f = 1/r$.
   Note that r is the radial spherical coordinate. Do each problem by hand and using Python. Plot the gradient vector field in the xy-plane.

9. **Intersecting surfaces** – Find the angle between the surfaces $x^2 + y^2 + z^2 = 9$ and $z = x^2 + y^2 - 3$ at the point (2,-1,2). Plot each surface and its gradient vector at the point (2,-1,2) on the same graph.

10. **Divergence** –

    a. The velocity of a fluid is given by
    $$\mathbf{v} = -x\left(z^2 + y^2\right)\hat{\mathbf{i}} + (xy - z)\hat{\mathbf{j}} - \left(z^2 - 2y\right)\hat{\mathbf{k}}$$
    Calculate $\nabla \cdot \mathbf{v}$ at the point (-1,1,2).

    b. Calculate he the divergence of the following velocity vector field in spherical coordinates:
    $$\mathbf{v} = \frac{2}{r}\hat{\mathbf{r}}$$
    Do this problem by hand and using SymPy.

11. **Divergence identities** – Using Python's symbolic capabilities, demonstrate the following identities:

    c. $\nabla \cdot (\mathbf{A} + \mathbf{B}) = \nabla \cdot \mathbf{A} + \nabla \cdot \mathbf{B}$
    d. $\nabla \cdot (f\mathbf{v}) = \mathbf{v} \cdot \nabla f + f(\nabla \cdot \mathbf{v})$ where f is a scalar field.

12. **Curl** – Find the curl of the vector $\mathbf{v}$:
    $$\mathbf{v} = -x\left(z^2 + y^2\right)\hat{\mathbf{i}} + (xy - z)\hat{\mathbf{j}} - \left(z^2 - 2y\right)\hat{\mathbf{k}}$$
    at the point (0,1,-1) by hand and using Python. Calculate $\nabla \times (\nabla \times \mathbf{v})$ at the same point, and compute the angle between the vectors $\nabla \times (\nabla \times \mathbf{v})$ and $\nabla \times \mathbf{v}$.

13. **Curl identities** – Using Python's symbolic capabilities, demonstrate the following identities:
   (a) $\nabla \times (\mathbf{A} + \mathbf{B}) = \nabla \times \mathbf{A} + \nabla \times \mathbf{B}$
   (b) $\nabla \times (f\mathbf{A}) = (\nabla f) \times \mathbf{A} + f(\nabla \times \mathbf{A})$
   where f is a scalar field.

14. **Work done by a force** – Calculate the work done by the force $\mathbf{F} = xy\hat{\mathbf{i}} - y^2x\hat{\mathbf{j}}$ between the origin and the point (1,1) along these integration paths:

   (a) The x-axis to the point (1,0) and then parallel to the y-axis to the point (1,1).
   (b) A straight line path joining the origin to the point (1,1).
   (c) The quarter circle centered at the point (1,0).
   (d) Is this force conservative? How do you know?

15. **Work done by a force** – Calculate the work done by the force $\mathbf{F} = k\rho\hat{\boldsymbol{\rho}}$ (where ρ is in polar coordinates) between the origin and the point (1,1) along along these integration paths:

   (a) The x-axis to the point (1,0) and then parallel to the y-axis to the point (1,1).
   (b) A straight line path joining the origin to the point (1,1).
   (c) The quarter circle centered at the point (1,0).

16. **Work done by a force along a parameterized curve** – Calculate the work done by the force $\mathbf{F} = x^2y\hat{\mathbf{i}} - y^2x\hat{\mathbf{j}} - xyz\hat{\mathbf{k}}$ along the curve $x = t^2$, $y = 3t$, and $z = t^4$ from $t = 0$ to $t = 2$. Plot the curve and the force vector at the point (1,3,1).

17. **Scalar fields and cross products** – If f and g are differentiable scalar fields, prove that $\nabla \cdot (\nabla f \times \nabla g) = 0$.

18. **Potential energy** – Find the potential energy associated with the conservative force $\mathbf{F} = \left(2xy + z^3\right)\hat{\mathbf{i}} + x^2\hat{\mathbf{j}} + 3xz^2\hat{\mathbf{k}}$. Use Python to show that you found the correct potential energy.

19. **Flux through a plane** – Calculate the flux of $\mathbf{v} = 5z\hat{\mathbf{i}} - 4\hat{\mathbf{j}} + y\hat{\mathbf{k}}$ through the part of the surface $2x + 3y + 6z = 12$ located in the first octant. Plot the plane and $\mathbf{v}$ and the vector perpendicular to the plane at the point on the plane located at $x = 0.5$, $y = 0.5$.

20. **Flux of a curl** – Suppose $\mathbf{F} = (y + 3)\hat{\mathbf{i}} + (x - 2xz)\mathbf{j} + xy\hat{\mathbf{k}}$. Evaluate the flux of $\nabla \times \mathbf{F}$ through the surface of the sphere $x^2 + y^2 + z^2 = a^2$ above the xy-plane. Using Python, plot the hemisphere and the vectors $\hat{\mathbf{r}}$ and $\nabla \times \mathbf{F}$ at the coordinates:
   (a) $x = 0$, $y = 0$
   (b) $x = 0.5$, $y = 0.5$
   (c) $x = -0.5$, $y = -0.25$

21. **Flux through a cube** – Consider the vector $\mathbf{F} = (xy + 3)\hat{\mathbf{i}} + \left(y^2 - z^2\right)\mathbf{j} + xy\hat{\mathbf{k}}$. Calculate the flux of $\mathbf{F}$ through the unit cube with one corner at the origin bounded by $x = 1$, $y = 1$, and $z = 1$. Using Python, plot the cube and $\hat{\mathbf{n}}$ for each face (in the middle of the face) and $\mathbf{F}$ at the middle of each face.

22. **Green's theorem** – Using Green's theorem, evaluate the integral

$$\oint (x + \cos y)\, dx + x \sin y\, dy$$

along the triangle in the xy-plane with corners at the origin, $(\pi/2, 0)$, and $(\pi/2, 1)$. Set the integral up by hand and use SymPy's `ParametricRegion` to solve the integral.

23. **Areas and Green's theorem** – Show that the area bounded by a closed curve C is given by

$$\frac{1}{2} \oint_C x\, dy - y\, dx$$

Consider the closed contour defined by the x-axis from $x = 0$ to $x = 4$, the y-axis from $y = 0$ to $y = 4$, and the curve $y^{1/2} + x^{1/2} = 4$. Using Python, plot the curve $y^{1/2} + x^{1/2} = 4$ and find the area enclosed by the contour.

24. **The electric field due to a dipole** – Consider an electric dipole at the origin with dipole moment $\mathbf{p} = p\hat{\mathbf{k}}$, where $p = qd$, q is the charge, and d is the dipole distance. The electric potential due to the dipole can be shown to be in spherical coordinates:

$$V(r, \theta) = \frac{p \cos \theta}{4\pi \epsilon_0 r^2}$$

where ϵ_0 is a constant called the permittivity of free space. By hand, and using Python, compute the electric field using $\mathbf{E} = -\nabla V$. Using Python, plot the electric field and the potential scalar field both in the yz-plane using $q = 1.6 \times 10^{-19}$C, $d = 1$pm, and $\epsilon_0 = 8.885 \times 10^{-12}$ F/m. Note, you may consider normalizing $\mathbf{E}$ for plotting purposes.

9 Vector Spaces

In this chapter, we present the useful concept of vector spaces. Linear algebra is the branch of mathematics dealing with vector spaces and linear transformations (which we examined in Chapter 7). Although abstract in nature, it is difficult to understate the importance of linear algebra in a diverse set of fields, ranging from engineering and physics to computer science. We will present only a small sample of the field of linear algebra, however linear vector spaces are widely used in classical and quantum physics.

9.1 DEFINITION OF A VECTOR SPACE

We begin with a technical definition of a *linear vector space*, which we will refer to simply a *vector space*. Although we have shied away from such definitions so far in this book, it is important to do so for vector spaces, because understanding what qualifies as a vector space can provide insight into the types of systems we study in physics.

A vector space V is a set of elements called vectors with the following properties:

1. The set V is *closed under addition*. In other words, if the vectors $\mathbf{u}$ and $\mathbf{v}$ are elements of the vector space V, then so is the vector $\mathbf{u} + \mathbf{v}$.

2. The set V is *closed under multiplication by a scalar*. Consider a scalar a, which in physics is most often either a real or complex number, and the vector $\mathbf{v}$, which is an element of V. The set V is closed under multiplication by a scalar, if $a\mathbf{v}$ is also an element of V.

In addition to the properties above, the following conditions must hold for all elements $\mathbf{u}$, $\mathbf{v}$, $\mathbf{w}$ of the vector space and scalars a and b.

1. Commutativity - $\mathbf{u} + \mathbf{v} = \mathbf{v} + \mathbf{u}$

2. Associativity of vector addition - $(\mathbf{u} + \mathbf{v}) + \mathbf{w} = \mathbf{u} + (\mathbf{v} + \mathbf{w})$

3. Additive identity - There exists a vector $\mathbf{0}$ in V such that $\mathbf{v} + \mathbf{0} = \mathbf{0} + \mathbf{v} = \mathbf{v}$

4. Associativity of scalar multiplication - $a(b\mathbf{v}) = (ab)\mathbf{v}$

5. Distributivity of scalar addition - $(a + b)\mathbf{v} = a\mathbf{v} + b\mathbf{v}$

6. Multiplicative identity - There exists a scalar 1 such that $1\mathbf{v} = \mathbf{v}$

Admittedly, properties 3 and 6 are obvious if we consider the scalars to be real or complex numbers. However, the definition of vector fields can be generalized to any set of scalars which in mathematics is called a *field*. This generalization, while interesting, is beyond the scope of the text.

Before moving on, it is important to motivate the above definition with concrete examples. First, consider the vector space $\mathbb{R}^3$ which has elements of the familiar form

$$\mathbf{v} = a\hat{\mathbf{i}} + b\hat{\mathbf{j}} + c\hat{\mathbf{k}} \tag{9.1.1}$$

where a, b and c are real numbers. From what we learned in Chapter 4, the set of all vectors of the form (9.1.1) are a vector space. We know that adding two vectors of the form (9.1.1) creates another vector of the same form. In addition, multiplying $\mathbf{v}$ by a scalar only changes the length of $\mathbf{v}$ (and/or flips its direction), but the result is still a vector in $\mathbb{R}^3$.

DOI: 10.1201/9781003294320-9

A less obvious example of a vector space is the set of all polynomials of the form

$$f(x) = c_0 + c_1 x + c_2 x^2 \tag{9.1.2}$$

Note that all of the above requirements are met. For example, the sum of two second degree polynomials is also a second-degree polynomial. Vector spaces consisting of functions are very important in physics. We will examine them later in this chapter. For now, we focus on finite-dimensional linear vector spaces.

9.2 FINITE DIMENSIONAL VECTOR SPACES

In this section, we will examine the properties of vectors spaces like $\mathbb{R}^2$ and $\mathbb{R}^3$, where the scalars can be real or complex. What we learn in this section will be generalized to more abstract function spaces in the next section.

9.2.1 LINEAR INDEPENDENCE

Consider the vector space V and two of its elements $\mathbf{v}_1$ and $\mathbf{v}_2$. We say that a third vector $\mathbf{v}_3$ (also an element of V) is a *linear combination* of the vectors $\mathbf{v}_1$ and $\mathbf{v}_2$, if

$$\mathbf{v}_3 = a_1 \mathbf{v}_1 + a_2 \mathbf{v}_2 \tag{9.2.1}$$

where a_1 and a_2 are scalars. The term linear combination means that we are adding the scalar multiples of two or more vectors. Recall, that for V to be a vector space, the result of a linear combination of its elements must also be in V.

Note that if the vector $\mathbf{v}$ is an element of the vector space, then so must be the linear combination $\mathbf{v} - \mathbf{v} = \mathbf{0}$, which is why the vector $\mathbf{0}$ is needed for the vector space. For example, any plane in $\mathbb{R}^3$ not passing through the origin is not a vector space. However, the Cartesian plane is a vector space and is often denoted as $\mathbb{R}^2$.

Let us return to the concept of vector addition. Consider the set of vectors $\{\mathbf{v}_1, \mathbf{v}_2, \ldots, \mathbf{v}_n\}$ which for simplicity we represent using $\{\mathbf{v}_i\}$. The set of vectors $\{\mathbf{v}_i\}$ is said to be linearly independent if there is no set of scalars $\{c_1, c_2, \ldots, c_n\}$ not all zero such that

$$\sum_{i=1}^{n} c_i \mathbf{v}_i = 0 \tag{9.2.2}$$

In other words, if (9.2.2) is true, then at least one of the vectors is a linear combination of the others. Each vector in a linearly independent set is independent of each other. The simplest example of a linearly independent set in $\mathbb{R}^3$ (Cartesian space) are $\left\{\hat{\mathbf{i}}, \hat{\mathbf{j}}, \hat{\mathbf{k}}\right\}$, as there are no linear combinations of any two elements that would produce the third. However, the set $\left\{\mathbf{v}_1 = 3\hat{\mathbf{i}} - 2\hat{\mathbf{j}}, \mathbf{v}_2 = -\hat{\mathbf{i}} + \hat{\mathbf{j}}, \mathbf{v}_3 = \hat{\mathbf{i}}\right\}$ is not linearly independent, because $\mathbf{v}_3 = \mathbf{v}_1 + 2\mathbf{v}_2$.

Example 9.1: Linear independence of vectors

(a) Are the three vectors $\mathbf{v}_1 = (1, 0, -1)$, $\mathbf{v}_2 = (1, 1, 1)$, and $\mathbf{v}_3 = (-1, 1, 0)$ linearly independent?
(b) Repeat part (a) for the three vectors $\mathbf{v}_1 = (1, 1, 1)$, $\mathbf{v}_2 = (14, -1, -7)$, and $\mathbf{v}_3 = (-3, 2, 4)$. Note that we wrote the vectors as lists of components for compactness.

Solution:

(a) We want to find scalars c_1, c_2, and c_3 such that

$$c_1\mathbf{v}_1 + c_2\mathbf{v}_2 + c_3\mathbf{v}_3 = 0 \tag{9.2.3}$$

We can rewrite (9.2.3) as the system of equations

$$\left.\begin{array}{r} c_1 + c_2 - c_3 = 0 \\ c_2 + c_3 = 0 \\ -c_1 + c_2 = 0 \end{array}\right\} \tag{9.2.4}$$

or as the matrix equation

$$\begin{pmatrix} 1 & 1 & -1 \\ 0 & 1 & 1 \\ -1 & 1 & 0 \end{pmatrix} \begin{pmatrix} c_1 \\ c_2 \\ c_3 \end{pmatrix} = \begin{pmatrix} 0 \\ 0 \\ 0 \end{pmatrix} \tag{9.2.5}$$

If the determinant of the coefficient matrix is non-zero, then only the trivial solution exists and the set of vectors is linearly independent. Using Python we find that the determinant is not zero. Therefore, the vectors $\mathbf{v}_1$, $\mathbf{v}_2$, and $\mathbf{v}_3$ are linearly independent.

(b) We use the system of equations

$$\left.\begin{array}{r} c_1 + c_2 + c_3 = 0 \\ 14c_1 - c_2 - 7c_3 = 0 \\ -3c_1 + 2c_2 + 4c_3 = 0 \end{array}\right\} \tag{9.2.6}$$

$$\begin{pmatrix} 1 & 1 & 1 \\ 14 & -1 & -7 \\ -3 & 2 & 4 \end{pmatrix} \begin{pmatrix} c_1 \\ c_2 \\ c_3 \end{pmatrix} = \begin{pmatrix} 0 \\ 0 \\ 0 \end{pmatrix} \tag{9.2.7}$$

We use Python to find that the determinant is zero, and therefore the vectors $\mathbf{v}_1$, $\mathbf{v}_2$, and $\mathbf{v}_3$ are linearly dependent. Furthermore, we can find the coefficients c_i by using solve() in SymPy:

$$c_1 = \frac{2}{5}c_3 \qquad c_2 = \frac{-7}{5}c_3$$

```python
from sympy import Matrix, solve, symbols

c1, c2, c3 = symbols('c1, c2, c3')

print('-'*28,'CODE OUTPUT','-'*29,'\n')

# part (a)
A = Matrix([[1,1,-1],[0,1,1],[-1,1,0]])

print('The determinant of A in part (a) is = ', A.det())

# part (b)
A = Matrix([[1,1,1],[14,-1,-7],[-3,2,4]])

coeffs = solve([c1 + c2 + c3,  14*c1 - c2 -7*c3,\
    -3*c1 + 2*c2 + 4*c3 ],[c1,c2,c3])

print('\nThe determinant of A in part (b) is = ', A.det())
print('\nThe coefficients for part (b) are: ', coeffs)
```

```
----------------------------- CODE OUTPUT -----------------------------

The determinant of A in part (a) is = -3

The determinant of A in part (b) is = 0

The coefficients for part (b) are:  {c1: 2*c3/5, c2: -7*c3/5}
```

9.2.2 BASIS VECTORS

A set of vectors $\mathbf{u} = \{\mathbf{u}_i\}$ in the vector space V *span* V, if all vectors in V can be written as a linear combination of the vectors in the set $\mathbf{u}$. For example, the set $\left\{\hat{\mathbf{i}}, \hat{\mathbf{j}}\right\}$ span the vector space $\mathbb{R}^2$ because any vector in the Cartesian plane can be written as a linear combination of $\hat{\mathbf{i}}$ and $\hat{\mathbf{j}}$.

We can often create multiple sets of vectors which span the space V. However, one of the most important of such sets are a *basis* set. A basis is a set of vectors that are linearly independent and span a vector space. While the set $\left\{\hat{\mathbf{i}}, \hat{\mathbf{j}}\right\}$ is an obvious basis for $\mathbb{R}^2$, it is not the only one. The set $\left\{\hat{\mathbf{i}} + \hat{\mathbf{j}}, \hat{\mathbf{i}} - \hat{\mathbf{j}}\right\}$ is also a basis for $\mathbb{R}^2$. Basis vectors are important in quantum mechanics. For example, we can represent the spin state of a spin-1/2 particle by a vector called a spinor (a, b), where a and b are complex numbers. Any spin state for a given spin-1/2 particle is a linear combination of the spin states $(1,0)$ and $(0,1)$, or spin-up and spin-down, respectively.

The *dimension* of a vector space is equal to the number of basis vectors.

9.2.3 BEYOND CARTESIAN COORDINATES

Next, we will generalize the work above beyond the Cartesian coordinate system. Let us consider an n-dimensional vector space, $\mathbb{V}$. Any vector $\mathbf{v}$ in the vector space $\mathbb{V}$ can be written as

$$\mathbf{v} = \sum_{i=1}^{n} v_i \hat{\mathbf{e}}_i \tag{9.2.8}$$

where v_i is a scalar and $\hat{\mathbf{e}}_i$ belongs to a set of mutually orthogonal unit vectors that form a basis in $\mathbb{V}$. The same definitions for linear independence, span, linear combinations, and basis apply to $\mathbb{V}$.

As we saw in Chapter 6, the dot or inner product of two *complex* vectors $\mathbf{v}$ and $\mathbf{u} = \sum u_i \hat{\mathbf{e}}_i$ is found using:

$$\mathbf{v} \cdot \mathbf{u} = \sum_{i=1}^{n} v_i^* u_i \tag{9.2.9}$$

Note the complex conjugate components v_i^* appearing in this inner product. The magnitude of $\mathbf{v}$ is $v = \sqrt{\mathbf{v}^* \cdot \mathbf{v}}$, and the coefficient v_i in (9.2.8) is found using the dot product:

$$v_i = \mathbf{v} \cdot \hat{\mathbf{e}}_i \tag{9.2.10}$$

Two vectors $\mathbf{v}$ and $\mathbf{u}$ are orthogonal if $\mathbf{v} \cdot \mathbf{u} = 0$. If $v = u = 1$, and $\mathbf{v} \cdot \mathbf{u} = 0$, then we say $\mathbf{v}$ and $\mathbf{u}$ are *orthonormal*.

The Schwarz inequality relates $\mathbf{v} \cdot \mathbf{u}$ to the magnitude of each vector

$$\left| \sum_{i=1}^{n} v_i^* u_i \right| \leq \sqrt{\sum_{i=1}^{n} v_i^* v_i} \sqrt{\sum_{i=1}^{n} u_i^* u_i} \tag{9.2.11}$$

and is often useful in quantum mechanics, for example when deriving uncertainty relationships, like the famous Heisenberg uncertainty principle $\Delta x \Delta p \geq \hbar/2$.

Vector spaces can contain other vector spaces of smaller dimensions, called subspaces.

9.2.4 THE GRAM-SCHMIDT METHOD

Sometimes we need to take a basis set and turn them into an orthonormal basis. The Gram-Schmidt method gives us a procedure of doing exactly that. Working through the method helps develops one's intuition for vector components and dot products. Consider a set of n basis vectors $\{\mathbf{A}_i\}$. We can transform $\{\mathbf{A}_i\}$ into a mutually orthonormal set $\{\hat{\mathbf{e}}_i\}$ using the following procedure:

1. Normalize $\mathbf{A}_1$ such that $\hat{\mathbf{e}}_1 = \mathbf{A}/A$.

2. Subtract from $\mathbf{A}_2$ its component along the direction of $\hat{\mathbf{e}}_1$ and normalize the result.

$$\hat{\mathbf{e}}_2 = \frac{\mathbf{A}_2 - \hat{\mathbf{e}}_1 \left(\hat{\mathbf{e}}_1 \cdot \mathbf{A}_2 \right)}{\left| \mathbf{A}_2 - \hat{\mathbf{e}}_1 \left(\hat{\mathbf{e}}_1 \cdot \mathbf{A}_2 \right) \right|} \tag{9.2.12}$$

3. Subtract from $\mathbf{A}_3$ its components along the directions of $\hat{\mathbf{e}}_1$ and $\hat{\mathbf{e}}_2$ and normalize the result.

$$\hat{\mathbf{e}}_3 = \frac{\mathbf{A}_3 - \hat{\mathbf{e}}_1 \left(\hat{\mathbf{e}}_1 \cdot \mathbf{A}_3 \right) - \hat{\mathbf{e}}_2 \left(\hat{\mathbf{e}}_2 \cdot \mathbf{A}_3 \right)}{\left| \mathbf{A}_3 - \hat{\mathbf{e}}_1 \left(\hat{\mathbf{e}}_1 \cdot \mathbf{A}_3 \right) - \hat{\mathbf{e}}_2 \left(\hat{\mathbf{e}}_2 \cdot \mathbf{A}_3 \right) \right|} \tag{9.2.13}$$

4. Repeat for the remaining vectors, continually subtracting the components of $\mathbf{A}_i$ that are parallel to previously calculated unit vectors.

The resulting set of vectors is built by making them independent of the previous vectors in the new set. Example 9.2 illustrates the Gram-Schmidt procedure by hand and using Python.

========

Example 9.2: The Gram-Schmidt method

Consider the three vectors $\mathbf{A}_1 = (0,0,0,3)$, $\mathbf{A}_2 = (2,0,0,4)$, and $\mathbf{A}_3 = (1,3,4,2)$, where the vectors are written as lists of components for compactness. Use the Gram-Schmidt method to find a mutually orthonormal basis.

Solution:

We begin with normalizing $\mathbf{A}_1$. It should be easy to see that $\hat{\mathbf{e}}_1 = (0,0,0,1)$. Next we find $\hat{\mathbf{e}}_2$ using (9.2.12)

$$\hat{\mathbf{e}}_2 = \frac{(2,0,0,4) - \hat{\mathbf{e}}_1 \, (4)}{\left| (2,0,0,4) - \hat{\mathbf{e}}_1 \, (4) \right|} = \frac{(2,0,0,0)}{\left| (2,0,0,0) \right|} = (1,0,0,0) \tag{9.2.14}$$

Finally, we calculate $\hat{\mathbf{e}}_3$ using (9.2.13)

$$\hat{\mathbf{e}}_3 = \frac{(1,3,4,2) - \hat{\mathbf{e}}_1 \, (2) - \hat{\mathbf{e}}_2 \, (1)}{\left| (1,3,4,2) - \hat{\mathbf{e}}_1 \, (2) - \hat{\mathbf{e}}_2 \, (1) \right|} = \frac{(0,3,4,0)}{\left| (0,3,4,0) \right|} = \left(0, \frac{3}{5}, \frac{4}{5}, 0 \right) \tag{9.2.15}$$

Both the NumPy and SymPy libraries provide ways for implementing the Gram-Schmidt algorithm. In NumPy, we can find the basis vectors using a method similar to what is done by hand. In SymPy, the command GramSchmidt will perform the calculation on a list of vectors, each stored using the command Matrix. The option True in the GramSchmidt command normalizes the output.

```
import numpy as np
from sympy.matrices import Matrix, GramSchmidt

#NumPy method
a1 = np.array([0,0,0,3])
a2 = np.array([2,0,0,4])
a3 = np.array([1,3,4,2])

e1 = a1/np.linalg.norm(a1)

e2_numerator = a2 - e1*np.dot(e1,a2)
e2 = e2_numerator/np.linalg.norm(e2_numerator)

e3_numerator = a3 - e1*np.dot(e1,a3) - e2*np.dot(e2,a3)
e3 = e3_numerator/np.linalg.norm(e3_numerator)

#SymPy Method

a = [Matrix([0,0,0,3]), Matrix([2,0,0,4]), Matrix([1,3,4,2])]
GS = GramSchmidt(a,True)

print('-'*28,'CODE OUTPUT','-'*29,'\n')
print('The SymPy command GramSchmidt produces:')
for i in GS:
    print(i)

print('\n The NumPy method produces:')
print(e1)
print(e2)
print(e3)

-------------------------- CODE OUTPUT ----------------------------

The SymPy command GramSchmidt produces:
Matrix([[0], [0], [0], [1]])
Matrix([[1], [0], [0], [0]])
Matrix([[0], [3/5], [4/5], [0]])

 The NumPy method produces:
[0. 0. 0. 1.]
[1. 0. 0. 0.]
[0. 0.6 0.8 0. ]
```

9.3 VECTOR SPACES OF FUNCTIONS

In the previous section, we focused on familiar vectors like $\mathbf{v} = 2\hat{\mathbf{i}} - \hat{\mathbf{j}}$ as elements of a Euclidean vector space. However, vector spaces can also be spaces of functions. We still use the term vector, even when functions are the elements of interest. It is sometimes necessary to distinguish between vectors of the type $\mathbf{v}$ from functions. In those situations, we will use the term *Euclidean vectors* for vectors of the type $\mathbf{v}$. In this section, we will see how the definitions and concepts of vector spaces need to change when working with functions as vectors.

9.3.1 DEFINITION

The basic definition of a vector space as presented in Section 9.1 does not change when functions are used as vectors.

For example, the set of all quadratic polynomials, $f(x) = c_0 + c_1 x + c_2 x^2$ is a vector space. It satisfies all of the conditions in Section 9.1. The sum of two quadratic polynomials is a quadratic polynomial. Multiplying a quadratic polynomial by a scalar does not change the degree of the polynomial. The function $g(x) = 0$ serves as the additive identity and the scalar $a = 1$ serves as the multiplicative identity. The associativity and distributivity requirements are also met by the basic rules of arithmetic for polynomials. As we will see later, the dimension of this space is 3.

Vector spaces of functions can be infinite dimensional. The mathematics of infinite dimensional vector spaces are beyond the scope of this text. However, we will conclude this chapter with a few comments on such spaces. Next, we will address linear independence, norms, orthogonality, and basis vectors for vector spaces of functions.

9.3.2 FUNCTIONS AND LINEAR INDEPENDENCE

Functions can be linear combinations of other functions. For example, the cosine function is a linear combination of complex exponentials:

$$\cos \theta = \frac{1}{2} e^{i\theta} + \frac{1}{2} e^{-i\theta} \tag{9.3.1}$$

However, it is difficult to apply a linear independence condition like $\sum c_i \mathbf{v}_i = 0$ to functions. Instead, we use the *Wronskian*, W. Consider the set of functions $\{f_i(x)\}$ for $i = 1 \ldots n$. The Wronskian for the set of functions $\{f_i(x)\}$ is the determinant of the $n \times n$ matrix

$$W = \begin{vmatrix} f_1(x) & f_2(x) & f_3(x) & \cdots & f_n(x) \\ f_1'(x) & f_2'(x) & f_3'(x) & \cdots & f_n'(x) \\ f_1''(x) & f_2''(x) & f_3''(x) & \cdots & f_n''(x) \\ \vdots & \vdots & \vdots & \ddots & \vdots \\ f_1^{(n)}(x) & f_2^{(n)}(x) & f_3^{(n)}(x) & \cdots & f_n^{(n)}(x) \end{vmatrix} \tag{9.3.2}$$

If, for a given set of functions $\{f_i(x)\}$, $W = 0$ for all values of x, then the functions are linearly dependent.

Example 9.3: The Wronskian

Consider the two sets of functions below. Using the Wronskian, determine whether the sets are linearly independent or linearly dependent.

(a) $\{1, \sin x, \sin 2x\}$
(b) $\{e^{ix}, e^{-ix}, \sin x, \cos x\}$

Solution: We will do part (a) by hand.

$$W = \begin{vmatrix} 1 & \sin x & \sin 2x \\ 0 & \cos x & 2\cos 2x \\ 0 & -\sin x & -4\sin 2x \end{vmatrix} = 2\sin x \cos 2x - 4\sin 2x \cos x$$

which is not zero for all values of x. Therefore, $\{1, \sin x, \sin 2x\}$ are linearly independent. The

SymPy command `wronskian()` can be used to compute the Wronskian. The first argument is a list of the function while the second argument is the independent variable.

```
from sympy import symbols, exp, sin, cos, I
from sympy.matrices.dense import wronskian

x = symbols('x')

W = wronskian([exp(I*x),exp(-I*x),sin(x),cos(x)],x)

print('-'*28,'CODE OUTPUT','-'*29,'\n')
print('The Wronskian = ', W)

-------------------------- CODE OUTPUT ----------------------------

The Wronskian =  0
```

9.3.3 INNER PRODUCTS AND FUNCTIONS

The dot product can be generalized to an *inner product*. The inner product is a way to multiply two vectors in a vector space with the result being a scalar. We already saw the definition of the dot product between two n-dimensional complex Euclidean vectors $\mathbf{u}$ and $\mathbf{v}$

$$\mathbf{v} \cdot \mathbf{u} = \sum_{i=1}^{n} v_i^* u_i \tag{9.3.3}$$

Functions do not have components in the same sense as a Euclidean vector, hence we cannot apply (9.3.3) directly. However, the central mathematical operations in (9.3.3), complex conjugation, multiplication, and summation, can be applied to functions. The inner product between two functions $f(x)$ and $g(x)$ on the interval $a \le x \le b$ is defined as

$$\langle f|g \rangle = \int_a^b f^*(x)g(x)dx \tag{9.3.4}$$

Note that we used the bracket notation from quantum mechanics, where the function in the *bra* $\langle f|$ is complex conjugated and is also the first function written after the integral sign. The function in the *ket* $|g\rangle$ is written second. The order of the functions typically matters, especially in fields such as quantum mechanics.

We can use (9.3.4) to define the norm $|\langle f|f \rangle|$ of a function f on the interval $a \le x \le b$

$$|\langle f|f \rangle| = \sqrt{\int_a^b f^*(x)f(x)dx} \tag{9.3.5}$$

A function can be normalized by dividing f by $|\langle f|f \rangle|$. In quantum mechanics, we are often interested in working with normalized functions.

Finally, two functions are orthogonal if

$$\langle f|g\rangle = \int_a^b f^*(x)g(x)dx = 0 \qquad (9.3.6)$$

The inner product has many of the same properties of the dot product, as shown in Table 9.1.

Table 9.1

Inner product relationships in a vector space. Here f, g, and h are functions and a, b are complex scalars.

Property	Formula
Symmetry	$\langle f|g\rangle^* = \langle g|f\rangle$
Zero norm	$\langle f|f\rangle = 0$ if and only if $f = 0$
Linearity of the bra	$\langle a\,f + b\,g\,|h\rangle = a^*\langle f|h\rangle + b^*\langle g|h\rangle$
Linearity of the ket	$\langle f|a\,g + b\,h\rangle = a\langle f|g\rangle + b\langle f|h\rangle$
Factoring of scalars	$\langle a\,f|b\,g\rangle = a^*\,b\,\langle f|g\rangle$
Schwarz Inequality	$|\langle f|g\rangle|^2 \leq \langle f|f\rangle\langle g|g\rangle$

Example 9.4 demonstrates an orthonormal set of functions which form a vector space.

======

Example 9.4: Hermite polynomials

The *Hermite equation*

$$\frac{d^2y}{dx^2} - 2x\frac{dy}{dx} + 2ny = 0 \qquad (9.3.7)$$

is an ordinary differential equation associated with the Schrödinger equation for the quantum harmonic oscillator. We will examine ordinary differential equations and their solutions, in a later chapter. Here we present the solution to (9.3.7) which are called *Hermite polynomials* $y = H_n(x)$.

$$H_n(x) = (-1)^n\, e^{x^2}\,\frac{d^n}{dx^n}e^{-x^2} \qquad (9.3.8)$$

We first saw Hermite polynomials in Chapter 3. Here, we will focus on their general orthonormality condition.
(a) Compute the first three Hermite polynomials, $H_0(x)$, $H_1(x)$, and $H_2(x)$ using (9.3.8).
(b) The Hermite polynomials have a weighted orthogonality relationship

$$\int_{-\infty}^{\infty} e^{-x^2} H_n(x)H_m(x)dx = \begin{cases} 0 & n \neq m \\ 2^n n!\sqrt{\pi} & n = m \end{cases} \qquad (9.3.9)$$

Using Python, demonstrate (9.3.9) for $H_0(x)$, $H_1(x)$, and $H_2(x)$.

Solution: We will do part (a) by hand.

$$H_0(x) = (-1)^0 \, e^{x^2} \frac{d^0}{dx^0} e^{-x^2} = 1$$

$$H_1(x) = (-1) \, e^{x^2} \frac{d}{dx} e^{-x^2} = -e^{x^2} \left(-2x e^{-x^2} \right) = 2x$$

$$H_2(x) = (-1)^2 \, e^{x^2} \frac{d^2}{dx^2} e^{-x^2} = e^{x^2} \left(-2e^{-x^2} + 4x^2 e^{-x^2} \right) = 4x^2 - 2$$

The Python code below shows how to generate the Hermite polynomials and also solves part (b). Note the command `hermite_poly(n,x)` has two arguments. The first argument n is the order of the Hermite polynomial. and the second argument x is the polynomial's independent variable.

```python
from sympy import symbols, integrate, oo, exp
from sympy.polys.orthopolys import hermite_poly

x = symbols('x')

def ortho(n,m):
    return integrate(exp(-x**2)*hermite_poly(n,x)*hermite_poly(m,x),(x,-oo,oo))

print('-'*28,'CODE OUTPUT','-'*29,'\n')
print('H_0(x) = ', hermite_poly(0,x))
print('H_1(x) = ', hermite_poly(1,x))
print('H_2(x) = ', hermite_poly(2,x))

print('\nThe ortogonality relationships are:')
print('H0-H0 = ', ortho(0,0))
print('H0-H1 = ', ortho(0,1))
print('H0-H2 = ', ortho(0,2))
print('H1-H1 = ', ortho(1,1))
print('H1-H2 = ', ortho(1,2))
print('H2-H2 = ', ortho(2,2))


---------------------------- CODE OUTPUT ----------------------------

H_0(x) =  1
H_1(x) =  2*x
H_2(x) =  4*x**2 - 2

The ortogonality relationships are:
H0-H0 =  sqrt(pi)
H0-H1 =  0
H0-H2 =  0
H1-H1 =  2*sqrt(pi)
H1-H2 =  0
H2-H2 =  8*sqrt(pi)
```

9.3.4 BASIS FUNCTIONS

A set of functions $\{f_i\}$ for $i = 1 \ldots n$ are a basis set in the vector space V, if the elements in the set $\{f_i\}$ are linearly independent and span the vector space. Basis sets are very important in quantum mechanics, as they can represent possible states of a system. Example 9.5 provides an example of a basis set for a vector space of functions.

Example 9.5: Basis functions

Consider the set of all polynomials $f(x) = c_0 + c_1 x + c_2 x^2$. Show that the set of functions $\{1, x, x^2\}$ is a basis set for the vector space of all quadratic polynomials.

Solution: We have already established that the set of all quadratic polynomials is a vector space.

It is straightforward to show that the set $\{1, x, x^2\}$ spans the vector space. Next, we need to show that the set $\{1, x, x^2\}$ is linearly independent. We can do that using Python. The non-zero Wronskian shows that the set is linearly independent.

The basis set has a length of three, therefore the dimension of the space of all quadratic polynomials is 3. In an abstract sense, we can think of a quadratic polynomial $f(x) = c_0 + c_1 x + c_2 x^2$ as the point (c_0, c_1, c_2) in this vector space.

```
from sympy import symbols
from sympy.matrices.dense import wronskian

x = symbols('x')

W = wronskian([1,x,x**2],x)

print('-'*28,'CODE OUTPUT','-'*29,'\n')
print('The Wronskian = ', W)

------------------------- CODE OUTPUT ----------------------------

The Wronskian =  2
```

9.3.5   THE GRAM-SCHMIDT METHOD FOR FUNCTIONS

The Gram-Schmidt method can be extended to functions. The method is similar to that used for Euclidean vectors. Let $\{f_i\}$ for $i = 1 \ldots n$ be a set of basis functions for a vector space, and $\{e_i\}$ be the corresponding set of orthonormal basis functions generated from $\{f_i\}$. We can find the set $\{e_i\}$ using

$$e_i = \frac{f_i - \sum_{j<i}\langle e_j|f\rangle\, e_j}{\left|f_i - \sum_{j<i}\langle e_j|f\rangle\, e_j\right|} \qquad (9.3.10)$$

where the denominator is the norm of the numerator.

As with Euclidean vectors, we are creating e_i by making each f_i independent of the previous elements in the set. In Example 9.6, we demonstrate the Gram-Schmidt procedure on the set of quadratic polynomials.

Example 9.6: The Gram-Schmidt method for functions

Starting with the set $\{1, x, x^2\}$, use Python to create an orthonormal basis on the interval $-1 \leq x \leq 1$ for the vector space of quadratic polynomials.

Solution: We use a Python code similar to the code we used for Euclidean vectors. The orthonormal basis of the vector space of quadratic polynomials is:

$$e_0 = \frac{\sqrt{2}}{2}$$
$$e_1 = \frac{\sqrt{6}}{2}x$$
$$e_2 = \frac{3\sqrt{10}}{4}\left(x^2 - \frac{1}{3}\right)$$

```python
from sympy import symbols, integrate, sqrt

x = symbols('x')

f0, f1, f2 = 1, x, x**2

# function to integrate two functions
def inner_prod(g1,g2):
    return integrate(g1*g2,(x,-1,1))

e0 = f0/sqrt(inner_prod(f0,f0))

# numerator of e1
e1_num = f1 - e0*inner_prod(e0,f1)
e1 = e1_num/sqrt(inner_prod(e1_num,e1_num))

e2_num = f2 - e0*inner_prod(e0,f2) - e1*inner_prod(e1,f2)
e2 = e2_num/sqrt(inner_prod(e2_num,e2_num))

print('-'*28,'CODE OUTPUT','-'*29,'\n')

print('The orthonormal basis of the vector space is:')
print('e0 = ', e0)
print('e1 = ', e1)
print('e2 = ', e2)
```

```
-------------------------- CODE OUTPUT ----------------------------

The orthonormal basis of the vector space is:
e0 =   sqrt(2)/2
e1 =   sqrt(6)*x/2
e2 =   3*sqrt(10)*(x**2 - 1/3)/4
```

9.4 INFINITE DIMENSIONAL VECTOR SPACES

Although the mathematics of infinite-dimensional vector spaces are beyond the scope of this book, infinite-dimensional vector spaces are important in physics. For example, consider the set of periodic functions $f(t)$ with period τ. Fourier's theorem tells us that we can write:

$$f(t) = \frac{a_0}{2} + \sum_{n=1}^{\infty} a_n \cos(n\omega t) + b_n \sin(n\omega t) \tag{9.4.1}$$

and we showed in Chapter 2 how to find the coefficients a_n and b_n. Notice that using the language above, periodic functions are a linear combination of the infinite dimensional basis set

$$\{\cos(n\omega t), \sin(n\omega t)\} \tag{9.4.2}$$

where $n = 0, 1, 2 \cdots$.

As we will see in a later chapter, the solutions to partial differential equations often involve infinite sums, which can be thought of as a linear combination of an infinite-dimensional basis set. Partial differential equations are of central importance in fields like thermodynamics, electromagnetism, and quantum mechanics. In particular, in quantum mechanics particle states are often linear combinations of an infinite-dimensional set of basis functions. In the next section we will provide a brief outline of vector spaces in quantum mechanics.

9.4.1 QUANTUM MECHANICS AND HILBERT SPACE

The linear algebra concepts discussed in this section are central to the field of quantum mechanics. To illustrate this, we will give a brief overview of the infinite square well, one of the simplest problems in quantum mechanics.

Consider a particle of mass m trapped in an infinite square well of width a. Physically, this represents a free particle between the points $x = 0$ and $x = a$. However, at $x = 0$ and $x = a$, the particle experiences an infinite force directed toward the center of the well. A classical equivalent of this problem is a toy car on a frictionless track bouncing between two walls, where the collision with each wall is elastic.

In classical mechanics, we would want to know the particle's position as a function of time. In quantum mechanics, we cannot find that. Instead, we find the particle's wave function which, among other things, can tell us the probability of finding the particle between any two points.

The probability of finding a particle described by a wavefunction $\psi(x)$ between two points $x = c_1$ and $x = c_2$ is

$$P = \int_{c_1}^{c_2} \psi^*(x)\psi(x)dx \tag{9.4.3}$$

The time independent wave function of a particle interacting with a potential energy $V(x)$ and with a given energy E is found by solving the Schrödinger equation

$$-\frac{\hbar^2}{2m}\frac{\partial^2 \psi}{\partial x^2} + V(x)\psi = E\psi \tag{9.4.4}$$

In a later chapter, we will discuss how to solve partial differential equations like the Schrödinger equation. For now, it is sufficient to know that for physical situations, solutions ψ exist.

In order for the wave function to represent a probability, we also require an additional restriction on ψ, that it must be normalized, i.e.:

$$|\langle\psi|\psi\rangle| = \int_{-\infty}^{\infty} \psi^*(x)\psi(x)dx = 1 \tag{9.4.5}$$

Normalization means that the total probability of finding the particle anywhere on the interval $-\infty \le x \le \infty$ must be one. In other words, the particle must exist somewhere.

In the case of the infinite square well, we solve (9.4.4) for $V = 0$ between $x = 0$ and $x = a$ and that $\psi = 0$ outside that interval. The time-independent wavefunction ψ for a particle in an infinite square well is

$$\psi(x) = \sum_{i=1}^{\infty} c_n \sqrt{\frac{2}{a}} \sin\left(\frac{n\pi}{a}x\right) \tag{9.4.6}$$

where c_n is a constant.

Notice the infinite sum in (9.4.6), which can be interpreted as a linear combination of vectors of the form

$$\psi_n = \sqrt{\frac{2}{a}} \sin\left(\frac{n\pi}{a}x\right) \tag{9.4.7}$$

We say that the state $\psi(x)$ is a linear superposition of the functions ψ_n, which are linearly independent and span the space of all possible ψ's on the interval $x \in [0, a]$.

Furthermore, the functions ψ_n are orthonormal

$$\langle\psi_m|\psi_n\rangle = \int_0^a \psi_m^*(x)\psi_n(x)dx = \delta_{mn} = \begin{cases} 0 & m \ne n \\ 1 & m = n \end{cases} \tag{9.4.8}$$

and hence form an orthonormal basis for the space of infinite square well wavefunctions ψ.

Just like we can compute $\hat{\mathbf{i}} \cdot \mathbf{v}$ to obtain the x-component of the Euclidean vector $\mathbf{v}$, we can use the inner product (9.4.8) to calculate c_j coefficients:

$$\langle\psi_j|\psi\rangle = \int_0^a \psi_j^*\psi dx = \sum_{n=1}^{\infty}\int_0^a \psi_j^*(c_n\psi_n)dx = c_j \tag{9.4.9}$$

The functions ψ_n are also complete. In other words, we can write any function f on the interval $x \in [0, a]$ as a linear combination of the $\psi_n(x)$:

$$f(x) = \sum_{i=1}^{n} a_n \psi_n(x) \tag{9.4.10}$$

Hence, we have closure under addition and multiplication by a scalar. Furthermore, the inner product continues to hold for these functions.

$$\langle f|g\rangle = \int_0^a f^*(x)g(x)dx \tag{9.4.11}$$

In quantum mechanics, it is physically necessary to require $\langle f|f\rangle < \infty$ and if f is a wave function, then $\langle f|f\rangle = 1$. Physicists call the set of all functions f satisfying (9.4.10) and (9.4.11) (with $\langle f|f\rangle < \infty$) a *Hilbert space*. A Hilbert space is a vector space of functions.

In our example of the infinite square scribes well, each function ψ_n represents a particle state with an energy

$$E_n = \frac{n^2\hbar^2}{2\pi^2 m a^2} \tag{9.4.12}$$

where m is the mass of the particle. This expression for the energy is found by solving (9.4.4).

When the particle is observed experimentally, its wavefunction collapses onto one of the ψ_n states with a probability $|c_n|^2$, given by:

$$|c_n|^2 = |\langle \psi_n | \psi \rangle|^2 \tag{9.4.13}$$

We can think of c_n as the ψ_n component of (9.4.6) in the Hilbert space. Large values of c_n mean that we are more likely to measure the value E_n when measuring the particle's energy. To better understand this, consider the vector $\mathbf{v} = 10\hat{\mathbf{i}} + 2\hat{\mathbf{j}}$. The vector $\mathbf{v}$ points mostly along the x-axis, its x component is much larger than its y component. Likewise, if a term c_j in (9.4.6) is very large, then the corresponding ψ_j term makes up a significant portion of the sum, and the particle is more likely to be found in the state ψ_j with an energy E_j when measured.

This subsection only scratches the surface when it comes to discussing the relationship between linear algebra and quantum mechanics. For example, we have yet to mention the fact that ψ_n is often referred to as an eigenstate of the Schrödinger equation for infinite square well with an eigenvalue E_n, thus allowing us to describe quantum mechanics in the language of matrices. In a quantum mechanics course, you will see how linear algebra can be used to simplify calculations in quantum mechanics, as well as provide an elegant interpretation.

9.5　END OF CHAPTER PROBLEMS

1. **Linear combinations** – Use Python to write the following sets of vectors as a linear combination of the vectors $\mathbf{v} = (-1, 5, 8)$ and $\mathbf{u} = (4, 2, -3)$.

   a. $(-11, 11, 30)$

   b. $(28/3, -52/3, -36)$

   c. $(-45, -17, 41)$

   d. $(-7, 24, 41)$

2. **Linear independence of Euclidean vectors** – Determine if the following sets of vectors of linearly independent or linearly dependent. If they are linearly dependent, identify the linearly independent subset and write each vector as a linear combination of the independent vectors.

   a. $(1, 1, 1)$, $(-1, 0, 1)$ $(0, 1, 0)$

   a. $(0, -9, 3, 11)$, $(1, 3, -4, 2)$, $(-4, -30, 22, 14)$, $(3, 18, -15, -5)$

   b. $(1, 2, 3)$, $(1, 1, 1)$, $(-1, 0, 1)$, $(-3, 4, 3)$

3. **The Gram-Schmidt method for Euclidean vectors** – For each given set of vectors, use the Gram-Schmidt method to find an orthonormal set

   a. $(0, 0, 2, 0)$, $(1, -4, 0, 0)$, $(1, 1, 1, 1)$

   b. $(1, 0, 1, 0)$, $(0, 0, 3, 0)$, $(1, 1, 0, 1)$

   c. $(a, 0, 0, 0)$, $(b, c, d, 0)$, (u, v, x, y), $(0, m, 0, n)$

4. **Schwarz Inequality** – By hand and using Python, show that the Schwarz inequality is satisfied with the vectors $\mathbf{v} = (1 + i, 2, 4 - 2i, 3i)$ and $\mathbf{u} = (1, i - 1, 3i, 2)$.

5. **Orthonormal basis for a subspace** – Using the Gram-Schmidt method, find a basis for a four dimensional subspace of $\mathbb{R}^5$ using the vectors $\mathbf{A}_1 = (0, 2, 0, 4, 0)$, $\mathbf{A}_2 = (1, 0, 0, 3, 1)$, $\mathbf{A}_3 = (1, 3, 4, 2, -1)$, and $\mathbf{A}_4 = (-2, 2, 0, 3, 0)$.

6. **Linear independence of functions** – Use the Wronskian to determine if each set of functions are linearly independent.

   a. $\sin x$, $\cos x$, $x \sin x$

   b. e^x, e^{-x}, $\sinh x$, $\cosh x$

   c. x, x^3, x^5, x^7

   d. $\sinh x$, $\tanh x$, $\cosh x$

   e. $\sin x$, $\cos x$, $\sinh x$, $\cosh x$

7. **Hermite polynomial identities** – Demonstrate using Python and prove using (9.3.8) the following recursion relationships between Hermite polynomials:

   a. $H_n'(x) = 2nH_{n-1}(x)$, (where the prime denotes differentiation with respect to x)

   b. $H_{n+1}(x) = 2xH_n(x) - 2nH_{n-1}(x)$

8. **Laguerre polynomials** – Another important set of special functions are known as *Laguerre polynomials* which solve the differential equation $xy'' + (1 - x)y' + ny = 0$. It can be shown that $y = L_n(x)$ where

$$L_n(x) = \frac{1}{n!}e^x \frac{d^n}{dx^n}\left(x^n e^{-x}\right) \tag{9.5.1}$$

   a. Using (9.5.1) and the Python command `laguerre_poly` (from the `sympy.polys.orthopolys` library) write out $L_0(x)$, $L_1(x)$, and $L_2(x)$.

   b. Using the functions from part (a), show

$$\int_0^\infty e^{-x} L_n(x) L_m(x) dx = \begin{cases} 0 & n \neq m \\ 1 & n = m \end{cases} \tag{9.5.2}$$

9. **Associated Laguerre polynomials** – The differential equation $xy'' + (k + 1 - x)y' + ny = 0$ appears when solving the Schrödinger equation for the Hydrogen atom. The solution $y = L_n^k(x)$ are called associated Laguerre polynomials and are defined by

$$L_n^k(x) = \frac{x^{-k}e^x}{n!}\frac{d^n}{dx^n}\left(x^{n+k}e^{-x}\right) \tag{9.5.3}$$

   a. Using (9.5.3) and the Python command `laguerre_poly` (from the `sympy.polys.orthopolys` library) write out $L_0^0(x)$, $L_1^0(x)$, and $L_1^1(x)$.

   b. Using the functions from part (a), show

$$\int_0^\infty x^k e^{-x} L_n^k(x) L_m^k(x) dx = \begin{cases} 0 & n \neq m \\ \frac{(n+k)!}{n!} & n = m \end{cases} \tag{9.5.4}$$

10. **Gram-Schmidt method for functions** – Are the following sets of functions vector spaces? If so, find a basis and the dimension of the space.

  a. The set of polynomials of degree less than or equal to 3.

  b. The set of polynomials of degree less than or equal to 4, such that $f(x) = c_0 + c_2x^2 + c_3x^3 + c_4x^4$

  c. Linear combinations of the set of functions $\{x, \sin x, x \sin x, e^x \sin x, (3 - 2e^x) \sin x, x(1 + 5\sin x)\}$

11. **Fourier basis functions** – Prove that $\{\cos(n\omega t), \sin(n\omega t)\}$ for $n = 0, 1, \ldots$ is a set of mutually orthonormal functions.

12. **Fourier series and complex exponentials** – Consider a periodic function $f(t)$ with a period τ. In addition to writing $f(t)$ as a sum of sines and cosines, we can also write it as a sum of complex exponentials

$$f(t) = \sum_{n=-\infty}^{\infty} c_n e^{in\omega t} \tag{9.5.5}$$

where $\omega = 2\pi/\tau$ and c_n is a complex constant.

  a. Show that the functions $\{\exp(in\omega t)\}$ are orthonormal.

  b. Find a formula for c_n.

13. **Infinite square well** – Consider a particle trapped in an infinite square well with sides located at $x = 0$ and $x = a$. The particle is initially in the state

$$\psi = 2i\psi_1 - 3\psi_2 - (2 + 3i)\psi_3$$

Normalize ψ. What is the probability of measuring the particle's energy to be E_1? *Hint: Use (9.4.8) for the dot product.*

14. **Completeness of ψ_n** – Consider a particle in the infinite square well (with width a) in the state
$$\psi(x) = \begin{cases} A & 0 \le x \le 3a/4 \\ 0 & 3a/4 \le x \le a \end{cases}$$

  a. Normalize this wavefunction.

  b. Write $\psi(x)$ as a linear combination of ψ_n as in (9.4.10), using (9.4.9) to obtain the a_i's.

  c. What is the probability of finding this particle in the ground state with energy E_1?

15. **The quantum harmonic oscillator** – A particle with mass m and potential energy $V = 1/2m\omega^2x^2$ (where ω is an angular frequency) is called the quantum harmonic oscillator. Solving the Schrödinger equation with the potential V results in the time-independent wave function for the quantum harmonic oscillator:

$$\psi_n(x) = \left(\frac{m\omega}{\pi\hbar}\right)^{1/4} \frac{1}{\sqrt{2^n n!}} H_n\left(\sqrt{\frac{m\omega}{\hbar}}x\right) \exp\left(-\frac{m\omega}{2\hbar}x^2\right) \tag{9.5.6}$$

where n is a positive integer (or zero) corresponding to the oscillator's energy state $E_n = (n + 1/2)\hbar\omega$. In classical mechanics, the particle could not exist beyond $x = a$

where $E_n = 1/2m\omega^2 a^2$. In other words, the total energy of the particle is equal to the maximum value of its potential energy. However, in quantum mechanics, there is a nonzero probability for the particle to be in the so called *classically forbidden region*. Using Python, plot the probability the particle is in the classically forbidden region (i.e. $-\infty \leq x \leq a$ and $a \leq x \leq \infty$) as a function of n for $0 \leq n \leq 5$. Explain the shape of the graph.

10 Ordinary Differential Equations

In this chapter we introduce different types of ordinary differential equations (ODEs), which you will encounter during your undergraduate education. Differential equations appear in all branches of physics, from classical mechanics to quantum physics. We start by looking at several examples of ODEs, some of which have analytical solutions and can be solved exactly, while other examples will require numerical integration methods. We will learn how to use SymPy to obtain the analytical solutions of the ODEs and will study the ODEs for oscillating systems, which are of fundamental importance in all branches of science. For such systems, we will discuss both natural oscillations, and forced or driven oscillations under an external force, and the important physical concept of resonance. The chapter continues with a discussion of the analogy between an oscillating mechanical system and an electrical RLC circuit. We then introduce the solution of systems of differential equations, and the important concept of phase space trajectories. We conclude the chapter with two sections covering Legendre and Bessel equations, two differential equations important in physics. In those sections, we demonstrate how to solve differential equations by the method of Frobenius series solutions.

10.1 DEFINITIONS AND EXAMPLES OF DIFFERENTIAL EQUATIONS

Simply put, a *differential equation* is any equation which contains the derivative of a function. Differential equations are an essential tool for scientists, and they are used in many scientific areas to describe the change of a physical quantity as a function of a variable characterizing the system.

If the equation contains partial derivatives, it is called a *partial differential equation* (PDE), otherwise it is an *ordinary differential equation* (ODE). For example, Newton's second law is an ODE and can be written in several forms:

$$F = m\frac{d^2x}{dt^2} = m\frac{dv}{dt} \tag{10.1.1}$$

The *order* of a differential equation is the order of the highest derivative in the equation. For example, the order of Newton's law when written as $F(x, t) = m \, dv/dt$ is one, because the highest derivative in this equation is a first derivative of the dependent variable v, with respect to the independent variable t. Similarly, the order of the following ODE is 3:

$$\frac{d^3x}{dt^3} + \frac{dx}{dt} + 5 = 0 \tag{10.1.2}$$

A *linear* differential equation is of the form:

$$a_0 y(t) + a_1 y'(t) + a_2 y''(t) + a_3 y(t) + \cdots = b \tag{10.1.3}$$

where $a_0, a_1...$ are constants or functions of the independent variable t, and primes denote differentiation with respect to t. The ODE $y''(t) + t \, y(t) + t^2 = 0$ is linear, while the ODE $y''(t) + t \, y^2(t) = 0$ is not linear, because of the term $y^2(t)$. This chapter focuses on linear differential equations. We will examine the techniques for studying nonlinear differential equations in Chapter 12.

DOI: 10.1201/9781003294320-10

An ODE is called *homogeneous*, when every term contains the dependent variable y, or a derivative of y. For example, $y''(t) + t\,y^2(t) = 0$ is a homogeneous ODE, while $y''(t) + t\,y^2(t) + 3 = 0$ is an inhomogeneous ODE. The following is an example of a homogeneous, second-order, linear ODE:

$$\frac{d^2u}{dx^2} - x\frac{du}{dx} + u = 0 \tag{10.1.4}$$

Solving differential equations is of central importance in the field of physics and engineering, and many other sciences. Examples of each type of equation described above abound in the field. For example, the following homogeneous, linear, second-order ODE, with constant coefficients is the equation which describes a simple harmonic oscillator with no frictional forces:

$$\frac{d^2y}{dt^2} + \omega^2 y = 0 \tag{10.1.5}$$

where y is the oscillator's displacement from equilibrium and ω is the oscillator's natural frequency.

Another example is the second-order, nonlinear ODE, which describes the motion of a simple pendulum of length L and angular displacement from the vertical θ:

$$L\frac{d^2\theta}{dt^2} + g\sin\theta = 0 \tag{10.1.6}$$

Partial differential equations are also important in physics. For example, the *Laplace equation* :

$$\frac{\partial^2 u}{\partial x^2} + \frac{\partial^2 u}{\partial y^2} = 0 \tag{10.1.7}$$

is a homogeneous, second-order, linear partial differential equation with constant coefficients and is a fundamental equation in electromagnetism. We will study the Laplace equation in Chapter 11.

A *solution* of an ODE such as (10.1.5), is any function $y(t)$ which when substituted into the ODE, gives an identity of the two sides of the ODE. There are generally two types of solutions to ODEs, *general solutions* and *particular solutions*. The general solution of an ODE includes arbitrary constants, which are obtained by evaluating indefinite integrals during the solution of the ODE (see the next section). There are an infinite number of solutions to an ODE, and the general solution describes them all.

However, in applications, we usually want a *particular solution*, that is, one which satisfies the differential equation and some other requirements as well. Usually we solve ODEs using *initial conditions* which the system must satisfy at time $t = 0$. Essentially, we are integrating the ODE using definite integrals (see the next section). For example, to find the particular solution to (10.1.5), we need to specify the position y and the speed v of the oscillator at time $t = 0$. In other cases, we may solve ODEs using *boundary conditions*; for example, we may specify that the solution of (10.1.5) must have specific values at two times t_1, t_2.

In this chapter we will discuss how to solve ODEs using three different methods: the analytical method (by hand), the symbolic methods based on SymPy, and the Python numerical integration methods based on SciPy. It is important that you learn how to solve ODEs using all three methods. Solving an ODE by hand is very important, however in many cases this is not possible, and one must use the computer to solve the ODEs numerically. However, it is good practice to compare the solutions obtained by hand with the symbolic or numerical solutions from the computer, so that you understand the limitations of these approaches. No matter which method you use to solve the ODE, the computer is an indispensable tool for the graphical presentation of the results.

10.2   SEPARABLE DIFFERENTIAL EQUATIONS

As a first simple example, we use a problem which is familiar from your introductory physics classes, the motion of a particle moving under the influence of a constant acceleration a. In this situation we can write Newton's law as the simple differential equation $dv/dt = v'(t) = a$, or :

$$\frac{dv}{dt} = a \qquad (10.2.1)$$

This is a differential equation which says that $v(t)$, the solution to the differential equation, is a function whose first derivative is equal to a. Of course we know that $v(t) = a\,t$ is a solution that works. However, there are an infinite number of other solutions as well, since we can add a constant to $v(t)$ and still have a solution to our ODE. Hence, the general solution is

$$v(t) = a\,t + c \qquad (10.2.2)$$

where c is a constant. This differential equation is simple enough to solve. However, most ODEs are not that simple, and many cannot be solved at all. Before discussing how to solve an ODE, let's first point out a few things about our example.

1. Equation (10.2.1) is a *first-order ODE*, because the highest derivative in the equation is a first derivative. As mentioned previously, an nth-order ODE is an ODE whose highest derivative is an nth derivative.

2. The ordinary derivative implies that v is a function of only one variable t, which is the variable of differentiation.

3. The number of arbitrary constants in the general solution of (10.2.1) is equal to the order of the ODE. This is an important general property of linear ODEs: all *linear* differential equations of order n have a solution containing n independent arbitrary constants.

To formally solve (10.2.1), we need to separate the variables of the equation. Colloquially speaking, this means getting all the terms with v on one side of the equation, and all the terms that are either constant or depend on t, on the other. This process is called *separation of variables* and is performed by treating the derivative as a fraction, and multiplying both sides of (10.2.1) by dt,

$$dv = a\,dt \qquad (10.2.3)$$

$$\int dv = \int a\,dt \qquad (10.2.4)$$

$$v = a\,t + c_1 \qquad (10.2.5)$$

To solve (10.2.1), we carried out an indefinite integral after separating out variables. Note that both integrals would produce constants of integration, but since both are constants, we can combine them into one arbitrary constant c_1. You can double check the solution by computing the derivative of $a\,t + c_1$, to check that it satisfies (10.2.1).

However, the infinite number of solutions $v = a\,t + c_1$ is not very helpful. Which solution describes the actual path taken by the particle? In order to specify the particular solution for an ODE, we need to include initial conditions, the value of our function at a particular time (normally at $t = 0$). Suppose we know that at $t = 0$, the particle has an initial speed, $v(0) = 3$ m/s. Then we can solve for the arbitrary constant by inserting the initial condition into our general solution,

$$v(0) = (a)\,(0) + c_1 = 3 \qquad (10.2.6)$$

which gives $c_1 = 3$ m/s. Our particular solution is then, $v(t) = a\,t + 3$ (in SI units). A different initial condition will give a different particular solution.

What happens in the case of second order ODEs? Second order ODEs are more common in physics. There are many techniques to solve them, but here we will demonstrate only one. Consider the same ODE as above, written in terms of the position variable $x(t)$:

$$\frac{d^2 x}{dt^2} = a \tag{10.2.7}$$

Separation of variables does not make sense here, because we cannot integrate terms like $d^2 x$. However, we saw that the solution for the speed was

$$v(t) = dx/dt = a\,t + c_1 \tag{10.2.8}$$

and we can rewrite this equation as:

$$\frac{dx}{dt} = a\,t + c_1 \tag{10.2.9}$$

We can use separation of variable once more:

$$\int dx = \int (a\,t + c_1)\,dt \tag{10.2.10}$$

$$x(t) = \frac{1}{2}a\,t^2 + c_1 t + c_2 \tag{10.2.11}$$

where c_2 is the constant of integration obtained by performing the above integral. Hence, we pick up an additional constant of integration in our solution, giving us two arbitrary constants for the solution of the second order ODE (10.2.7). Loosely speaking, we see that the number of arbitrary constants in the solution of an nth-order ODE is equal to n, because we need to do n integrations in order to solve the equation, and each integration produces a new arbitrary constant.

In order to find the position $x(t)$ of the particle as a function of time, it is clear that one initial condition will not be enough to determine both constants c_1 and c_2. Hence, we will need to specify both $x(t)$ and dx/dt at a particular time (usually $t = 0$). Suppose that $x(0) = 3$ and $v(0) = 1$. Then we have:

$$x(0) = \frac{1}{2}a\,(0)^2 + c_1\,(0) + c_2 = 3 \tag{10.2.12}$$

$$v(0) = a\,(0) + c_1 = 1 \tag{10.2.13}$$

where (10.2.13) is the derivative of the general solution evaluated at $t = 0$. The result here is that $c_1 = 1$ and $c_2 = 3$, and the particular solution is $x(t) = a\,t^2/2 + t + 3$.

In summary, in order to solve for the particular solution of an nth-order ODE, we need n initial conditions. We can also specify the value of x at two different times, as opposed to knowing the initial values of x and its first derivative. In classical mechanics, it is most common to know the initial conditions of the position and velocity. However, in other fields, such as electromagnetism and thermodynamics, it is often more common to know the value of a function, say the temperature, at two different locations. In this case, we have what is known as a *boundary value problem*, and the conditions that provide the constants of integration are known as boundary conditions.

In the next section we will see how to solve the same differential equations using the symbolic capabilities of Python.

10.3 SYMBOLIC INTEGRATION OF DIFFERENTIAL EQUATIONS USING SYMPY

In Example 10.1 we show how we can use SymPy to solve the same differential equation as in the previous section.

Example 10.1: Symbolic solution of ODE $dv/dt = a$ **using SymPy**

Write a SymPy code to solve the ODE:

$$\frac{dv}{dt} = a \qquad (10.3.1)$$

with the initial condition that the velocity of the particle at time $t = 0$ is v_0, i.e. $v(0) = v_0$.

Solution:

In the Python code we first define which variables are to be treated as symbols, and which ones as functions. The next line solves the differential equation using dsolve in SymPy, and the result is assigned to the function variable soln. As we saw previously in this book, the first time derivative $v'(t)$ is evaluated symbolically using the command Derivative(v(t),t).
The differential equation $v'(t) - a = 0$ is defined inside the dsolve command as Derivative(v(t),t)-a. Note that the differential equation in dsolve must have a zero right-hand side. As a short-hand notation, we import the Derivative symbol as D. Alternatively, we can also use diff instead of D in the code.
The first part of the code prints the general solution soln of our ODE, using the .rhs method. As expected, the general solution is C1 + a*t , and it contains the constant of integration C1. In the second part of the code we find the integration constant C1, by specifying the initial condition that the solution must satisfy.
The initial condition $v(0) = v0$ is specified in the code line initCondits={v(0):v0}.
The dsolve command now contains the additional argument ics=initCondits, which is used to find the constant C1. The code prints the final result as a*t+v0.
This is of course the same kinematic equation $v(t) = v0 + a\,t$, which we found previously by hand.

```
from sympy import Function, symbols, dsolve, Derivative as D

print('-'*28,'CODE OUTPUT','-'*29,'\n')

v = Function('v')
t, v0, a = symbols('t, v0, a',real=True)

# use dsolve to obtain the general solution
soln = dsolve(D(v(t), t)-a, v(t),simplify=True).rhs

print('General symbolic solution v(t) =',soln)

# solve again using dsolve, with initial conditions
initCondits = {v(0): v0}
soln = dsolve(D(v(t), t)-a, v(t),simplify=True,ics=initCondits).rhs

print('\nUsing initial conditions, the solution v(t) =',soln)

---------------------------- CODE OUTPUT ----------------------------
```

General symbolic solution v(t) = C1 + a*t

Using initial conditions, the solution v(t) = a*t + v0

Example 10.2 shows how we would integrate symbolically the second order differential equation:

$$\frac{d^2 x}{dt^2} = a \tag{10.3.2}$$

Example 10.2: Symbolic solution of ODE $x''(t) = a$ using SymPy

(a) Solve the ODE both by hand and using SymPy:

$$\frac{d^2 x}{dt^2} = a \tag{10.3.3}$$

with the initial conditions $x(0) = x0$ and $x'(0) = v0$.
(b) Plot the solution from part (a), with the numerical values $x(0) = 0$ m, $x'(0) = v(0) = 1$ m/s and $a = -9.8$ m/s^2.

Solution:

Once more, we need to tell Python that the acceleration a and the time t are real variables, and that the position symbol x is a function. As we saw previously in this book, the second time derivative $x''(t)$ is evaluated symbolically using the command `Derivative(x(t),t,t)`, with two t's in the argument meaning the second derivative with respect to time.
The next line solves the differential equation using `dsolve`, and the result is assigned to the function variable *soln*. The differential equation $x''(t) - a = 0$ is defined inside the `dsolve` command as `Derivative(x(t),t,t)-a`.
Finally, the code prints the general solution of our differential equation using `soln.rhs`. As expected, the general solution contains two constants of integration C1 , C2 in the expression C1 + C2*t + a*t**2/2. This is of course the same result as we found by hand in (10.2.11).

Part (b) finds the constants C1,C2 by specifying the initial conditions, as specified in the line
`initCondits={x(0):x0, Derivative(x(t),t).subs(t, 0):v0}`
The additional argument ics=initCondits is once more used for finding C1,C2. The code prints the final result as a*t**2/2 +t*v0 +x0 .
This is the same as the well known kinematic equation for objects moving with a constant acceleration:

$$x(t) = x_0 + v_0\, t + \frac{1}{2} a\, t^2 \tag{10.3.4}$$

In many situations we want to plot the symbolic solution of the ODE. In this example we change the symbolic result `soln`, into a computable function named `xsoln()` by using `lambdify` in the line of code `xsoln = lambdify(t, soln,'numpy')`.
Once the function `xsoln` is defined in the code, it can be called and evaluated for different values of the time t. The graphical output is shown in Figure 10.1.

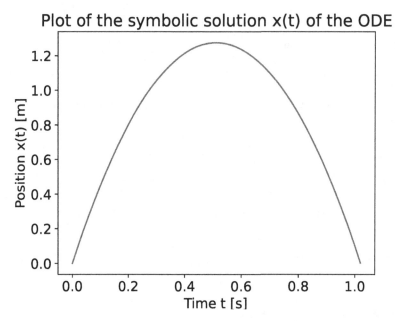

Figure 10.1 Graphical output from Example 10.2, which shows how to integrate symbolically the second order differential equation.

```python
from sympy import symbols, Function, lambdify, Derivative as D, dsolve
import numpy as np

import matplotlib.pyplot as plt

print('-'*28,'CODE OUTPUT','-'*29,'\n')

# define function x and various symbols
x = Function('x')
a, t, x0, v0 = symbols('a, t, x0, v0',real=True)

# initial conditions for ode are x(0)=x0 and v(0)=v0
initCondits = {x(0): x0, D(x(t),t).subs(t, 0): v0}

# use dsolve to find the solution with given initial conditions
soln = dsolve(D(x(t), t, t) -a, x(t),simplify=True,ics=initCondits).rhs
print('The symbolic solution is x(t)=',soln)

# make a numpy-ready function x(t) from the SymPy results
# using lamdify and substitute numerical values of a, x0, v0
xsoln = lambdify(t, soln.subs({a:-9.8, v0:5, x0:0}),'numpy')

# use the lambda function above to evaluate x(t) values from 0 to 1.03 s
xvals = np.arange(0,1.03,.02)
yvals = xsoln(xvals)

# plot values x(t)
plt.plot(xvals,yvals)
```

```
plt.xlabel('Time t [s]')
plt.ylabel('Position x(t) [m]')
plt.title('Plot of the symbolic solution x(t) of the ODE')
plt.show()

-------------------------- CODE OUTPUT ----------------------------

The symbolic solution is x(t)= a*t**2/2 + t*v0 + x0
```

10.4 GENERAL SOLUTION OF FIRST ORDER LINEAR ODES: THE INTEGRATING FACTOR METHOD

A linear ODE of first order with variable coefficients has the general form:

$$y'(x) + f(x)\, y(x) = g(x), \tag{10.4.1}$$

where the functions $f(x)$ and $g(x)$ may or may not vary with the independent variable x. In this section we obtain the general solution for this equation.

In general, equations of this form can be solved by multiplying throughout with the *integrating factor*

$$\exp\left(\int f(x)\, dx\right) \tag{10.4.2}$$

to obtain:

$$y'(x) \exp\left(\int f(x)\, dx\right) + f(x)\, y(x) \exp\left(\int f(x)\, dx\right) = g(x) \exp\left(\int f(x)\, dx\right) \tag{10.4.3}$$

This can be simplified by using the product rule:

$$\frac{d}{dx}\left[y(x) \exp\left(\int f(x)\, dx\right)\right] = g(x) \exp\left(\int f(x)\, dx\right) \tag{10.4.4}$$

By integrating both sides, we obtain:

$$y(x) \exp\left(\int f(x)\, dx\right) = \int g(x) \exp\left(\int f(x)\, dx\right)\, dx + C_1 \tag{10.4.5}$$

and solving for $y(x)$ we obtain the desired solution:

$$y(x) = \frac{\int g(x) \exp\left(\int f(x)\, dx\right)\, dx + C_1}{\exp\left(\int f(x)\, dx\right)} \tag{10.4.6}$$

In summary, the solution of the first-order linear ODE of the form:

$$y'(x) + f(x)\, y(x) = g(x) \tag{10.4.7}$$

is:

$$y = e^{-a(x)}\left\{\int g(x)\, e^{a(x)}\, dx + C_1\right\} \tag{10.4.8}$$

where C_1 is the constant of integration, and

$$a(x) = \int f(x)\, dx \tag{10.4.9}$$

Example 10.3 shows how we would integrate by hand and also symbolically a first order differential equation, by applying the integrating factor method.

Example 10.3: The general solution of first order linear ODEs

Consider the first order differential equation with constant coefficients

$$y'(x) + b\,y(x) = 1$$

where b is a constant. Solve for $y(x)$ using the integrating factor method, and verify the result using SymPy.

Solution

This equation is particularly relevant to first order systems such as circuits and certain oscillating mass systems.

We compare directly the given equation with the general form:

$$y'(x) + f(x)\,y(x) = g(x),$$

In this case, $f(x) = b$, $g(x) = 1$, therefore:

$$a(x) = \int f(x)\,dx = b\,x$$

and

$$\int g(x)\,e^{a(x)}\,dx = \int 1\,e^{b\,x}\,dx = \frac{e^{b\,x}}{b}$$

Using (10.4.8), the solution is

$$y = e^{-a(x)}\left(\int g(x)\,e^{a(x)}\,dx + C_1\right) = e^{-b\,x}\left(\frac{e^{b\,x}}{b} + C_1\right) = \frac{1}{b} + C_1 e^{-b\,x}.$$

The constant C_1 can be found if we are given the initial condition for this problem. The code uses dsolve() and produces the same result as the analytical solution.

```
from sympy import symbols, Function, Derivative as D, dsolve

print('-'*28,'CODE OUTPUT','-'*29,'\n')

# define function y and various symbols
y = Function('y')
b, x = symbols('b, x',real=True)

# use dsolve to find the solution
soln = dsolve(D(y(x), x) +b* y(x)-1, y(x),simplify=True).rhs

print('The symbolic solution is y(x) =',soln)


-------------------------- CODE OUTPUT ----------------------------

The symbolic solution is y(x) = C1*exp(-b*x) + 1/b
```

10.5   ODES FOR OSCILLATING SYSTEMS: GENERAL CONSIDERATIONS

Oscillating systems are of fundamental importance in physics and engineering systems, but interest in their study extends to almost every branch of science. With the exception of the simple pendulum, in this chapter we will focus our attention on small amplitude oscillations which refer to small displacements from the oscillator's equilibrium position. As we will see, when no external drive forces act on the oscillator, small amplitude oscillations are described by second-order homogeneous linear ordinary differential equations of the general form:

$$A_2 \frac{d^2 x}{dt^2} + A_1 \frac{dx}{dt} + A_0 x = 0 \tag{10.5.1}$$

where A_0, A_1 and A_2 are constants. If $x_1(t)$, $x_2(t)$ are two solutions of this equation, then the general solution $x_c(t)$ of (10.5.1) is a linear combination of these two solutions in the form:

$$x_c(t) = C_1 x_1(t) + C_2 x_2(t) \tag{10.5.2}$$

where C_1, C_2 are two arbitrary constants. The constants C_1, C_2 can be determined by the initial conditions of the system, which are usually given in the form of the initial position $x(0)$ and the initial speed $v(0) = \dot{x}(0)$ (dots denote differentiation with respect to time).

When external forces $f(t)$ are applied to the oscillator, we use a homogeneous linear second-order differential equation to describe the motion:

$$A_2 \frac{d^2 x}{dt^2} + A_1 \frac{dx}{dt} + A_0 x = f(t) \tag{10.5.3}$$

where again A_0, A_1 and A_2 are constants. In the general theory of differential equations, the solution of this equation is given by the sum of the solution of the homogeneous equation $x_c(t)$, sometimes called the *complementary solution*, and a *particular solution* $x_p(t)$ of the full nonhomogeneous equation. The particular solution $x_p(t)$ is *any* solution of the equation:

$$A_2 \frac{d^2 x_p}{dt^2} + A_1 \frac{dx_p}{dt} + A_0 x_p = f(t) \tag{10.5.4}$$

The general solution $x(t)$ of a nonhomogeneous second order differential equation then takes the form:

$$x(t) = x_c(t) + x_p(t) = C_1 x_1(t) + C_2 x_2(t) + x_p(t) \tag{10.5.5}$$

10.6   THE SIMPLE HARMONIC OSCILLATOR

In this section we study one of the most important problems in classical mechanics, the simple harmonic oscillator (SHO). Simple harmonic motion occurs when an object displaced a small distance from a stable equilibrium position, experiences a restoring force toward the equilibrium state.

An important example of simple harmonic motion is a mass m attached to a massless, horizontal spring with no friction, as shown in Figure 10.2.

When the mass is at the equilibrium position $x = 0$ (Figure 10.2A), the spring is neither stretched nor compressed, and the net force acting on the mass is zero. When the mass is moved from the equilibrium position (as shown in Figure 10.2B and C), the spring exerts a force $F(x)$ which is linearly proportional to the extension or compression of the spring in a direction toward the equilibrium position. The force F is called a *restoring force* because it attempts to return the system back to its equilibrium state. For small displacements x, we can expand $F(x)$ as a Taylor series about the equilibrium position $x = 0$, and the restoring

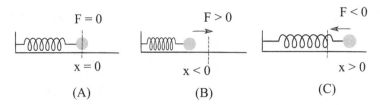

Figure 10.2 A mass m attached to a horizontal, massless, frictionless spring with spring constant k.

force takes on the form of *Hooke's Law* $F(x) = -kx$. We can then write Newton's second law, $F = ma = -kx(t)$, as:

$$m\frac{d^2x}{dt^2} + kx = 0 \tag{10.6.1}$$

Dividing by m we obtain:

$$\frac{d^2x}{dt^2} + \omega_0^2 x = 0 \tag{10.6.2}$$

$$\omega_0 = \sqrt{\frac{k}{m}} \tag{10.6.3}$$

where $\omega_0^2 \equiv k/m$ is called the *natural frequency* of the oscillator and is the frequency of small amplitude oscillations. Equation (10.6.2) is the differential equation of motion for the so-called *simple harmonic oscillator* (SHO). The SHO is an important equation in physics because it is the generic equation of motion for an object experiencing small amplitude oscillations about a stable equilibrium. Simple harmonic motion results when a particle experiences only a restoring force that is linearly proportional to the particle's displacement from equilibrium, *i.e.* when there are no drag forces or other external forces acting on the particle.

To find an analytical solution to (10.6.2), we need a function $x(t)$ whose second derivative gives itself back after differentiation, with a multiplicative constant and a minus sign. In other words, $x(t)$ must satisfy: $\ddot{x} = -\omega_0^2 x$, where the time derivatives are represented by dots above a function. While one obvious choice is either a sine or a cosine function, we will find that an exponential solution will be more useful (and still equivalent). We look for solutions that have the form $x = Ce^{\lambda t}$, where C is a constant. By substituting this into (10.6.2) we find:

$$\lambda^2 Ce^{\lambda t} + \omega_0^2 Ce^{\lambda t} = 0 \tag{10.6.4}$$

or by simplifying:

$$\lambda^2 + \omega_0^2 = 0 \tag{10.6.5}$$

or

$$\lambda_\pm = \sqrt{-\omega_0^2} = \pm i\,\omega_0 \tag{10.6.6}$$

and thus λ must be one of the complex numbers $i\,\omega_0$ or $-i\,\omega_0$. The general solution, then, will be a linear combination of the two solutions:

$$x(t) = C_1 e^{i\,\omega_0 t} + C_2 e^{-i\,\omega_0 t} \tag{10.6.7}$$

Note that in general C_1 and C_2 are arbitrary constants.

By using Euler's formula for relating complex exponentials to trig functions, we can change the exponentials in this equation into trigonometric functions. We find that the general solution can be written in two equivalent mathematical forms:

$$x(t) = A \cos(\omega_0 t) + B \sin(\omega_0 t) \tag{10.6.8}$$

$$x(t) = C \cos(\omega_0 t + \phi) \tag{10.6.9}$$

$$\text{with} \quad C = \sqrt{A^2 + B^2} \qquad \tan \phi = B/A \tag{10.6.10}$$

Example 10.4 shows the symbolic evaluation of the solution of the ODE for the SHO, with specified initial conditions.

Example 10.4: The general symbolic solution of the SHO

Solve (10.6.2) by hand and using SymPy, for the initial conditions $x(0) = x_0$ and $v(0) = x'(0) = v_0$.

Solution:
When solving by hand, we can use (10.6.8) or (10.6.9), where A, B and C are constants determined by the initial conditions. Inserting the initial condition $x(0) = x_0$ at $t = 0$ into (10.6.8) we find:

$$x(0) = A \cos 0 + B \sin 0 = A = x_0 \tag{10.6.11}$$

The speed of the mass $v(t)$ is found by taking the derivative of (10.6.8) with respect to time:

$$v(t) = \frac{dx}{dt} = -A\omega_0 \sin(\omega_0 t) + B\omega_0 \cos(\omega_0 t) \tag{10.6.12}$$

By using the given initial condition $v(0) = v_0$ at $t_0 = 0$:

$$v(0) = \omega_0 \left[-A \sin 0 + B \cos 0\right] = \omega_0 B = v_0 \tag{10.6.13}$$

and so $B = v_0/\omega_0$. Therefore the complete solution for these initial conditions is:

$$x(t) = x_0 \cos(\omega_0 t) + \frac{v_0}{\omega_0} \sin(\omega_0 t) \tag{10.6.14}$$

We can obtain the same result by using dsolve() in SymPy. The output of the symbolic evaluation is

```
C1*sin(sqrt(k)*t/sqrt(m)) + C2*cos(sqrt(k)*t/sqrt(m))
```

which after collecting the square root terms becomes the same as (10.6.14).
In the final answer, notice that SymPy does not combine the square-root terms, because of the symbolic nature of this library.

```
from sympy import symbols, Function, Derivative as D, dsolve

print('-'*28,'CODE OUTPUT','-'*29,'\n')

# define function x and various symbols
x = Function('x')
k, m, t = symbols('k, m, t', real=True, positive=True)
x0, v0 = symbols('x0, v0', real=True)

soln = dsolve(m*D(x(t), t, t) +k*x(t), x(t),\
simplify=True).rhs
print('The symbolic solution is:\n', 'x(t) =')
print(soln)

initCondits = {x(0): x0, D(x(t),t).subs(t, 0): v0}

soln = dsolve(D(x(t), t, t) +(k/m)*x(t), x(t),\
simplify=True,ics=initCondits).rhs

print('\nThe solution with the initial conditions x(0)=x0, v(0)=v0 \
 is:', '\nx(t) =')
print(soln)

------------------------- CODE OUTPUT --------------------------

The symbolic solution is:
 x(t) =
C1*sin(sqrt(k)*t/sqrt(m)) + C2*cos(sqrt(k)*t/sqrt(m))

The solution with the initial conditions x(0)=x0, v(0)=v0  is:
x(t) =
x0*cos(sqrt(k)*t/sqrt(m)) + sqrt(m)*v0*sin(sqrt(k)*t/sqrt(m))/sqrt(k)
```

10.7 NUMERICAL INTEGRATION OF THE ODE OF A SIMPLE PLANE PENDULUM

The plane pendulum is one of the simplest physical systems which can exhibit harmonic oscillations for small displacements from an equilibrium position. In this section, we calculate the period τ of the simple pendulum for oscillations with a small amplitude, and find expressions for the kinetic and potential energies as functions of time.

Figure 10.3 shows a pendulum consisting of a mass m which can swing freely in a vertical plane. A massless rod of length L connects the mass to the pendulum's pivot point. When the mass is displaced sideways from its equilibrium position, it is subject to a restoring force due to gravity. This force will accelerate the mass back toward the equilibrium position. The time for one complete cycle of oscillation is the period τ.

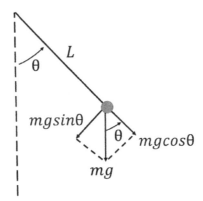

Figure 10.3 The simple plane pendulum.

We can find the equation of motion for the simple plane pendulum by applying Newton's second law to the tangential axis only, $F = -mg\sin\theta = ma$ and $a = -g\sin\theta$. We obtain the ODE for the simple plane pendulum:

$$\frac{d^2\theta}{dt^2} + \frac{g}{L}\sin\theta = 0 \tag{10.7.1}$$

For small displacements from equilibrium, we can approximate $\sin\theta \approx \theta$, and the equation becomes:

$$\frac{d^2\theta}{dt^2} + \frac{g}{L}\theta = 0 \tag{10.7.2}$$

Equation (10.7.2) is exactly similar to the equation for a mass-spring system $d^2x/dt^2 + \omega_0^2 x = 0$, with the angle θ replacing the position x, and with the natural frequency $\omega_0 = \sqrt{k/m}$ being replaced by $\omega_0 = \sqrt{g/L}$. Because the two equations (SHO and pendulum) are mathematically similar, we can use the solution of the SHO to solve (10.7.2). The solution to (10.7.2) is therefore given by:

$$\theta(t) = \theta_0 \cos(\omega_0 t + \phi) \tag{10.7.3}$$

where θ_0 is the initial angle of the pendulum, and ϕ is a constant phase angle value dependent on the initial conditions. The period of oscillation for small displacements is found using $\tau = 2\pi/\omega_0$ and is:

$$\tau = 2\pi\sqrt{\frac{L}{g}} \tag{10.7.4}$$

Notice that the period of oscillation is independent of the amplitude of oscillation. However, this condition holds only for small amplitude oscillations. Once the small angle approximation $\sin\theta \approx \theta$ no longer holds, then the period of oscillation will be related to the amplitude of oscillation. Such a relationship between the amplitude and the period is common in nonlinear oscillators.

The ODE for the simple pendulum cannot be solved in closed form without the use of elliptical functions, and must be solved numerically using SciPy.

Example 10.5 shows how to numerically evaluate and plot the solution of (10.7.1), using SciPy.

Example 10.5: The numerical solution of the ODE for simple pendulum

Integrate numerically and plot the solution of $(10.7.1)$ with $L = 0.5$ m for the initial conditions $\theta(0) = 2$ rad and $\dot{\theta}(0) = 0$ rad/s, using the Scientific Python library SciPy.

Solution:
To solve $(10.7.1)$, we will use the function `odeint()` from the `scipy.integrate` library which numerically solves first-order ODEs of the form:

$$\frac{dy}{dt} = f(y, t) \tag{10.7.5}$$

where y and t are the dependent and independent variables respectively, and $f(y, t)$ is a known function of the variables y, t.

To use `odeint()` on $(10.7.1)$, we need to rewrite $(10.7.1)$ as a system of two first-order ODEs, a common procedure needed for some ODE solvers. We begin by creating a new variable $\omega = \dot{\theta}$ and $(10.7.1)$ can be rewritten as the system:

$$\left.\begin{aligned}
\frac{d\theta}{dt} &= \omega \\
\frac{d\omega}{dt} &= -\frac{g}{L}\sin\theta
\end{aligned}\right\} \tag{10.7.6}$$

In the code we define a function `deriv(y,time)` whose arguments are a vector y and the time variable `time`. In our example, the first component `y[0]` of the vector y represents the angle variable θ, and the second component `y[1]` represents the angular velocity $\omega = d\theta/dt$. The function `deriv` uses $(10.7.6)$ to evaluate and return a vector with components $(\omega,\, d\omega/dt)$.

The function `odeint(deriv, yinit, t)` is called with three arguments. The first argument is the function `deriv` which contains the information on the first order ODE to be solved, the second argument is the initial conditions vector `yinit`, and the third argument is the time variable t.

The result of calling `odeint` is an array which is assigned to the variable y. The line `t = np.linspace(0, 10, 100)` defines the time interval over which the ODE will be integrated, in this case between $t = 0$ and $t = 10$ s. The variable `yinit = (2,0)` defines the initial conditions $\theta(0) = 2$ rad, and $\omega(0) = \theta'(0) = 0$ rad/s. The last 6 lines in the code are the graphics commands which plot the two functions $\theta(t)$ and $\omega(t) = \theta'(t)$, as shown in Figure 10.4.

```
from scipy.integrate import odeint

import numpy as np
import matplotlib.pyplot as plt


L = 0.5 # length of pendulum
g = 9.8 # gravitational acceleration


# function to define the ODE for pendulum
def deriv(y, time):
    return (y[1], - (g/L)* np.sin(y[0]))


# define times t
t = np.linspace(0, 10, 100)


# initial conditions theta(0)=2 rad,  theta'(0)=0
```

```
yinit = (2, 0)

# solve numerically using scipy odeint() function
y = odeint(deriv, yinit, t)

# plot angle and anglar velocity as functions of time
plt.plot(t, y[:, 0], 'o-',label=r'$\theta$(t)')
plt.plot(t, y[:, 1], '^-',label=r"$\theta$'(t)")

plt.legend(loc='best')
plt.xlabel('Time [s]')
plt.ylabel(r"$\theta$(t), $\theta$'(t)")
plt.title('Numerical solutions for the plane pendulum')
plt.show()
```

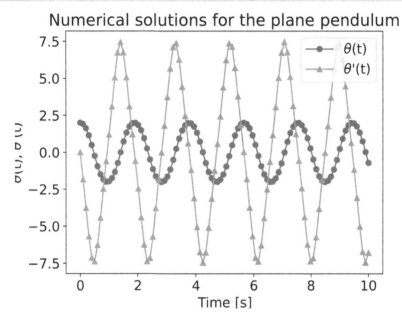

Figure 10.4 Graphical output from Example 10.5, showing a plot of the numerical solution of (10.7.1) from SciPy.

10.8 DAMPED HARMONIC OSCILLATOR

Next, we add a linear resistive force to the simple harmonic oscillator. The resistive force is sometimes called *damping* and is present, for example, when the oscillator is moving through a fluid such as air. The equation for the damped harmonic oscillator contains an additional friction term which is assumed to be linearly dependent on the speed in the form $F = -bv$, where b depends on the size and shape of the oscillator, as well as other physical parameters such as the type of fluid the oscillator moves through, and the coefficient of friction between the oscillator and a supporting surface.

The complete equation for the damped harmonic oscillator is obtained now from $F = ma = -kx - bv$. After dividing by the mass m we obtain:

$$\frac{d^2x}{dt^2} + \frac{b}{m}\frac{dx}{dt} + \frac{k}{m}x = 0 \tag{10.8.1}$$

We now define a constant γ called the *damping parameter*, in order to express the amount of damping present in the system:

$$\gamma = b/2m \tag{10.8.2}$$

The differential equation becomes:

$$\frac{d^2x}{dt^2} + 2\gamma\frac{dx}{dt} + \omega_0^2 x = 0 \tag{10.8.3}$$

where we used again $\omega_0 = \sqrt{k/m}$ for the natural frequency of the undamped oscillator.

Once more we look for solutions that have the form $Ce^{\lambda t}$, where C is a constant. By substituting this into (10.8.3) and dividing all terms by $Ce^{\lambda t}$, we find:

$$\lambda^2 + 2\gamma\lambda + \omega_0^2 = 0 \tag{10.8.4}$$

which is solved using the quadratic formula:

$$\lambda_\pm = -\gamma \pm \sqrt{\gamma^2 - \omega_0^2} \tag{10.8.5}$$

Because there are two solutions to (10.8.4), we have a pair of solutions $e^{\lambda_+ t}$ and $e^{\lambda_- t}$, corresponding to the two roots of the quadratic. The general solution is a linear combination of these two solutions:

$$x(t) = A_0 e^{\left(-\gamma + \sqrt{\gamma^2 - \omega_0^2}\right)t} + A_1 e^{\left(-\gamma - \sqrt{\gamma^2 - \omega_0^2}\right)t} \tag{10.8.6}$$

or by introducing the parameter $\omega \equiv \sqrt{\gamma^2 - \omega_0^2}$ we obtain the general solution $x(t)$ for the damped harmonic oscillator:

$$x(t) = e^{-\gamma t}\left(A_0 e^{\omega t} + A_1 e^{-\omega t}\right) \tag{10.8.7}$$

$$\omega = \sqrt{\gamma^2 - \omega_0^2} = \sqrt{(b/2m)^2 - (k/m)} \tag{10.8.8}$$

The exact mathematical form of this equation $x(t)$ depends on the numerical values of γ and ω_0, which determine whether the parameter $\omega = \sqrt{\gamma^2 - \omega_0^2}$ is real, imaginary, or zero.

There are three physically different behaviors of the damped harmonic oscillator as follows: overdamped oscillations ($\gamma > \omega_0$), underdamped oscillations ($\gamma < \omega_0$), and critically damped oscillations ($\gamma = \omega_0$).

The next three subsections examine each type of oscillation.

10.8.1   CASE I: OVERDAMPED OSCILLATIONS

In the case of *overdamped oscillations* we have $\gamma > \omega_0$, which means that the square root appearing in the expression $\omega = \sqrt{\gamma^2 - \omega_0^2}$ is real, and (10.8.7) will be written as the sum of two real exponential functions. The resulting motion is that the system returns to equilibrium by an exponential decay, without oscillating. Alternatively, one can write the solution as a linear combination of hyperbolic functions:

$$x(t) = \left(A_0 e^{\omega t} + A_1 e^{-\omega t}\right) e^{-\gamma t} \tag{10.8.9}$$

$$x(t) = [B_0 \sinh(\omega t) + B_1 \cosh(\omega t)]\, e^{-\gamma t} \tag{10.8.10}$$

$$\omega = \sqrt{\gamma^2 - \omega_0^2} \qquad \gamma > \omega_0 \tag{10.8.11}$$

The constants A_0, A_1, B_0, B_1 can be found from the initial conditions of the oscillator.

10.8.2   CASE II: UNDERDAMPED OSCILLATIONS

In the case of *underdamped oscillations*, we have $\gamma < \omega_0$, and the frequency $\omega = \sqrt{\gamma^2 - \omega_0^2}$ is imaginary. Mathematically, it is preferable to get rid of the imaginary quantities by using the real frequency parameter $\omega_d = \sqrt{\omega_0^2 - \gamma^2}$, expressing the general solution as a linear combination of trigonometric functions:

$$x(t) = \left[A_0 \sin(\omega_d t) + A_1 \cos(\omega_d t) \right] e^{-\gamma t} \tag{10.8.12}$$

or equivalently:

$$x(t) = A e^{-\gamma t} \cos(\omega_d t - \phi) \tag{10.8.13}$$

$$\omega_d = \sqrt{\omega_0^2 - \gamma^2} \qquad \gamma < \omega_0 \tag{10.8.14}$$

The solution of the underdamped case oscillates repeatedly through the equilibrium point as the amplitude of oscillation decays exponentially to zero. The amplitude's exponential decay is described by the $A e^{-\gamma t}$ term in (10.8.12) and (10.8.13). The underdamped behavior is unlike the overdamped case, where the oscillator may not pass through equilibrium at all, as its amplitude decays to zero. This is an important difference in the behaviors of the two oscillators. Also notice from (10.8.12) and (10.8.13) that the parameter $\omega_d = \sqrt{\omega_0^2 - \gamma^2} = \sqrt{k/m - \gamma^2}$ is characterizing the underdamped oscillator (with units of s^{-1}) and is smaller than the natural frequency $\omega_o = \sqrt{k/m}$.

10.8.3   CASE III: CRITICALLY DAMPED OSCILLATIONS

In the case of *critically damped oscillations*, we have $\gamma = \omega_0$, and $\omega = \sqrt{\gamma^2 - \omega_0^2} = 0$. A critically damped system returns to equilibrium as quickly as possible without oscillating. This is often desired for the damping of systems like doors and automobile suspensions. In this case, there is only one root of the quadratic equation, i.e. $\lambda = -\omega_0$. In addition to the solution $x(t) = e^{\lambda t}$, a second solution is given by the function $x(t) = t e^{\lambda t}$. The complete solution is a linear combination of these two solutions:

$$x(t) = (A + Bt) e^{-\omega_0 t} \tag{10.8.15}$$

where A and B are determined by the initial conditions of the system.

Example 10.6 illustrates the symbolic solution of the differential equation for the three cases of overdamped, underdamped and critically damped motion of an oscillator.

Example 10.6: The symbolic solution of the damped oscillator

Use SymPy to obtain the general solution of (10.8.3) for overdamped oscillations ($\gamma > \omega_0$), underdamped oscillations ($\gamma < \omega_0$), and critically damped oscillations ($\gamma = \omega_0$).

Solution:
We use the function `dsolve()` in SymPy to obtain the symbolic general solution of the differential equation for the three cases.
Note that in this and other examples in this chapter, the Python output is too long to be shown in one printed line, so we use the `textwrap` library to break up the output into smaller segments which appear in several lines.

```
from sympy import symbols, Function, Derivative as D, dsolve

import textwrap

print('-'*28,'CODE OUTPUT','-'*29,'\n')

# define function x and various symbols
x = Function('x')
x0, v0 = symbols('x0, v0', real=True)
k, m, gamma, t ,omega, C1, C2 = symbols('k, m, gamma, t,\
omega, C1, C2', real=True, positive=True)

# overdamped and underdamped oscaillation
soln = dsolve(D(x(t), t, t)+2*gamma*D(x(t), t) +omega**2*x(t), x(t),\
    simplify=True).rhs

print('\nFor overdamped and underdamped oscillations the \
solution :\nx(t) =')
print(textwrap.fill(str(soln),63))

# critically damped oscillation:  gamma=omega
soln = dsolve(D(x(t), t, t)+2*omega*D(x(t), t) +omega**2*x(t), x(t),\
    simplify=True).rhs

print('\nFor critically damped oscillations, the symbolic \
solution is:\nx(t) =\n',str(soln))

--------------------------- CODE OUTPUT ---------------------------


For overdamped and underdamped oscillations the solution :
x(t) =
C1*exp(t*(-gamma + sqrt(gamma - omega)*sqrt(gamma + omega))) +
C2*exp(-t*(gamma + sqrt(gamma - omega)*sqrt(gamma + omega)))

For critically damped oscillations, the symbolic solution is:
x(t) =
 (C1 + C2*t)*exp(-omega*t)
```

Example 10.7 illustrates plots of the symbolic solution of the differential equation for the three cases of overdamped, underdamped and critically damped motion of an oscillator.

Example 10.7: Graphing the three types of damped oscillations

Use SymPy to plot on the same graph the position $x(t)$ for an oscillator undergoing overdamped oscillations ($\gamma > \omega_0$), underdamped oscillations ($\gamma < \omega_0$), and critically damped oscillations ($\gamma = \omega_0$). Use the same numerical parameters $k = 0.1$ N/m, $m = 1$ kg for the three graphs and vary the friction coefficient b for the system with the values $b = 1, 0.2, 0.632$ Ns/m. The initial conditions are $y(0) = 1$ m, $v(0) = 0$ m/s.

Solution:

We define a function `plotx()` with variables k, m, b and a different label and marker for each graph in Figure 10.5.

For overdamped motion, we use the parameters $m = 1\,\text{kg}$, $k = 0.1\,\text{N/m}$, $y(0) = 1\,\text{m}$, $v(0) = 0\,\text{m/s}$ and $b = 1.0\,\text{Ns/m}$. For these parameters, $\omega^2 = \gamma^2 - \omega_0^2 = (b/2m)^2 - (k/m) = (1/2)^2 - 0.1 = 0.15 > 0$. The analytical solution is a linear combination of exponential terms.

For underdamped motion, we change only the value of $b = 0.2\,\text{Ns/m}$, and $\omega^2 = \gamma^2 - \omega_0^2 = (b/2m)^2 - (k/m) = (0.1/2)^2 - 0.1 = -0.05 < 0$. The analytical solution is a linear combination of exponentially decaying sine and cosine terms.

For critically damped motion we change only the value of $b = 0.632\,\text{Ns/m}$, and $\omega^2 = \gamma^2 - \omega_0^2 = (b/2m)^2 - (k/m) = 0$. The analytical solution is now a linear combination of the 2 possible solutions for critically damped oscillations, i.e. the $e^{-\omega_0 t}$ and $t\,e^{-\omega_0 t}$ terms.

```python
from sympy import symbols, Function, Derivative as D, dsolve, lambdify

import numpy  as np
import matplotlib.pyplot as plt

print('-'*28,'CODE OUTPUT','-'*29,'\n')

x = Function('x')
t = symbols('t',real=True,positive=True)

# function to plot the three solutions x(t) for given k, m, b
# and different markers and labels

def plotx(k, m, b, labl, mrker):
    # solve differential equation with initial conditions
    soln = dsolve(m*D(x(t), t, t)+b*D(x(t), t) +k*x(t),
      x(t),simplify=True,ics=initCondits).rhs

    # lambdify the solution to get a function x(t)
    xsoln = lambdify(t, soln,'numpy')

    # evaluate x(t) values from t=0 to t=50 s
    xvals = np.arange(0,50,.1)
    yvals = xsoln(xvals)

    # plot x(t) using different markers abnd labels
    plt.plot(xvals,yvals, mrker, label=labl);
    plt.plot(xvals,[0]*xvals)

# initial conditions x(0)=1  and v(0)=0
initCondits = {x(0): 1, D(x(t),t).subs(t, 0): 0}

# plot the solution x(t) for overdamped motion
k, m, b = .1, 1, 1
labl = 'Overdamped'
plotx(k,m,b,labl,'k-.')

# plot the solution x(t) for underdamped motion
b = .2
```

```
labl = 'Underdamped'
plotx(k,m,b,labl,'r--')

# plot the solution x(t) for critically damped motion
b = 0.632
labl = 'Critically damped'
plotx(k,m,b,labl,'b')

plt.title('The three types of damped oscillations')
plt.ylabel('x(t) [m]')
plt.xlabel('Time t [s]')
leg = plt.legend()
plt.show()
```

------------------------- CODE OUTPUT -----------------------------

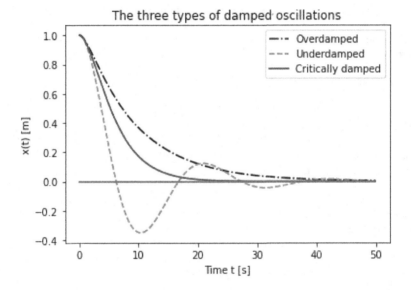

Figure 10.5 Graphical output from Example 10.7, showing the three cases of overdamped, under-damped and critically damped motion of an oscillator.

10.9   FORCED HARMONIC OSCILLATOR

Next, we consider the case where an external force $F_0 \cos(\omega t)$ is acting on a damped oscillator. Notice that the driving force has an amplitude of F_0 and a frequency of ω, sometimes referred to as the *drive frequency*. This case is referred to as the *forced* or *driven harmonic oscillator*. The net force is then:

$$m\frac{d^2x}{dt^2} = -kx - b\frac{dx}{dt} + F_0 \cos(\omega t) \tag{10.9.1}$$

The general equation for the driven system is then:

$$\frac{d^2x}{dt^2} + 2\gamma\dot{x} + \omega_0^2 x = D \cos(\omega t) \tag{10.9.2}$$

where $\gamma = b/2m$, $\omega_0^2 = k/m$, and $D = F_0/m$. You might be wondering why we limit ourselves to studying driving forces in the form of cosine functions. Clearly not all driving forces in nature are in the form of cosine or sine functions. As we have seen in our discussion of Fourier series, any periodic force (under certain conditions, which are often met in physical problems) can be approximated using a series of sines and cosines. Therefore, the case of cosine or sine driving forces is applicable widely and for many physical systems.

Note that (10.9.2) is a nonhomogeneous ordinary differential equation. As we discussed in the introduction of this chapter on the general theory of differential equations, the solution of nonhomogeneous equations is given by the sum of two parts, the homogeneous solution $x_c(t)$, plus the particular solution $x_p(t)$ of the full nonhomogeneous equation:

$$x(t) = x_c(t) + x_p(t) \tag{10.9.3}$$

We already saw in the previous section that the general solution $x_c(t)$ of the homogeneous equation is given by (10.8.7):

$$x_c(t) = e^{-\gamma t} \left(A_0 e^{\omega t} + A_1 e^{-\omega t} \right) \tag{10.9.4}$$

$$\omega = \sqrt{(b/2m)^2 - (k/m)} = \sqrt{\gamma^2 - \omega_0^2} \tag{10.9.5}$$

The solution $x_c(t)$ is commonly referred to as the *transient solution*, because the exponential decay term $e^{-\gamma t}$ causes $x_c(t)$ to decay to zero. After the transient solution has decayed to zero, the only solution left is $x_p(t)$ and is therefore known as the *steady state solution*. Many physical systems exhibit both a transient and steady state behavior. Physicists are often (but not always!) more interested in the steady state behavior of the system, because that is the system's long term behavior.

Next, we need a particular solution to the homogeneous ODE (10.9.2). As a trial solution we will choose an oscillation with the same frequency as the external force, but with a phase difference ϕ:

$$x(t) = A e^{i(\omega t - \phi)} \tag{10.9.6}$$

The choice of a complex exponential simplifies significantly the algebra. It must be kept in mind however, that at the end of the calculation we have to take the real part of the solution $x(t)$, since the final solution cannot contain complex numbers. After inserting (10.9.6) as a trial solution into the ODE (10.9.2), we obtain:

$$-A\omega^2 e^{i(\omega t - \phi)} + 2\gamma\omega i A e^{i(\omega t - \phi)} + \omega_0^2 A e^{i(\omega t - \phi)} = D e^{i\omega t} \tag{10.9.7}$$

By canceling out the $e^{i\omega t}$ from all terms and multiplying by $e^{i\phi} = \cos\phi + i\sin\phi$ on both sides, we obtain:

$$-A\omega^2 + 2\gamma\omega i A + \omega_0^2 A = D e^{i\phi} = D \left(\cos\phi + i\sin\phi \right) \tag{10.9.8}$$

By equating the real parts on the two sides of this equation, and also equating the imaginary parts on the two sides to each other, we obtain:

$$A \left(\omega_0^2 - \omega^2 \right) = D \cos\phi \tag{10.9.9}$$

$$2\gamma\omega A = D \sin\phi \tag{10.9.10}$$

Solving this system of equations for the amplitude A and for the phase angle ϕ, we obtain the particular solution $x_p(t)$ of the nonhomogeneous differential equation (10.9.2) for the driven harmonic oscillator. Finding ϕ from the above equations is simple, all one needs to do is divide the second equation by the first. Finding A involves squaring each equation and adding them, and recalling that $\sin^2\phi + \cos^2\phi = 1$. The complete solution $x(t) = x_c(t) + x_p(t)$ is then given by:

$$x(t) = x_c(t) + A\cos(\omega t - \phi) \tag{10.9.11}$$

$$x_c(t) = e^{-\gamma t}\left(A_0 e^{\omega t} + A_1 e^{-\omega t}\right) \qquad \omega = \sqrt{(b/2m)^2 - (k/m)} \tag{10.9.12}$$

$$A = \frac{F_0/m}{\sqrt{\left(\omega_0^2 - \omega^2\right)^2 + 4\gamma^2\omega^2}} \tag{10.9.13}$$

$$\tan\phi = \frac{2\gamma\omega}{\omega_0^2 - \omega^2} \tag{10.9.14}$$

Example 10.8 solves (10.9.2) symbolically in SymPy, using `dsolve()`, which was described previously in this chapter. The complete solution is a long algebraically complex expression, so we use again `textwrap` to format the output of the code.

Example 10.8: The symbolic solution of the ODE for a driven oscillation

Integrate symbolically (10.9.2) for the driven harmonic oscillator using SymPy, and plot the solution $x(t)$. Use the numerical values $F_0 = 1$, $\gamma = 2$, $\omega_0 = 1$, $\omega = 1.5$ with all parameters in SI units, and with the conditions $x(0) = 1$ and $x(1) = 2$.

Solution:
In the code we use `dsolve()` as before to obtain the symbolic solution, which is rather long when printed out on the output. After substituting the numerical values of the parameters and using `lamdify()`, the plot of the solution is shown in Figure 10.6.

```
from sympy import symbols, Function, solve, Derivative as D,\
    dsolve, lambdify, cos

import numpy as np
import matplotlib.pyplot as plt
import warnings
warnings.filterwarnings("ignore")
import textwrap

print('-'*28,'CODE OUTPUT','-'*29,'\n')

x = Function('x')
C1, C2 = symbols('C1,C2')
t, F, w, wo, gm  = symbols('t, F, w, wo, gm ',real=True,\
positive=True)

# initial conditions x(0)=1  and v(0)=0

soln = dsolve(D(x(t), t, t)+2*gm*D(x(t), t) +wo**2*x(t)-F*cos(w*t),
  x(t),simplify=True).rhs

print('The analytical solution is complicated!')
print('x(t) =')
```

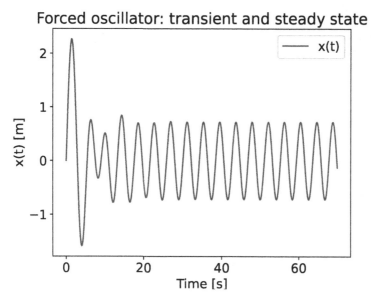

Figure 10.6 Graphical output from Example 10.8, showing a plot of the symbolic solution of the driven damped SHO, including the transient and steady state solutions.

```
print(textwrap.fill(str(soln),60))

# substitute numerical values for the parameters
u1 = soln.subs({F:1,w:1.5,wo:1,gm:.2,t:0})

u2 = soln.subs({F:1,w:1.5,wo:1,gm:.2,t:1})

C1C2 = solve([u1, u2-2],(C1,C2))

u = soln.subs({F:1,w:1.5,wo:1,gm:.2}).subs(C1C2)

xsoln = lambdify(t, u,'numpy')

# evaluate x(t) values from t=0 to t=70 s
xvals = np.arange(0,70,.02)
yvals = xsoln(xvals)

# plot x(t) using different markers abnd labels
plt.plot(xvals,yvals,'b',label='x(t)')
plt.xlabel('Time [s]')
plt.ylabel('x(t) [m]')
plt.title('Forced oscillator: transient and steady state')
leg = plt.legend()
plt.show()


------------------------- CODE OUTPUT -------------------------

The analytical solution is complicated!
x(t) =
C1*exp(t*(-gm + sqrt(gm - wo)*sqrt(gm + wo))) +
```

```
C2*exp(-t*(gm + sqrt(gm - wo)*sqrt(gm + wo))) +
2*F*gm*w*sin(t*w)/(4*gm**2*w**2 + w**4 - 2*w**2*wo**2 +
wo**4) - F*w**2*cos(t*w)/(4*gm**2*w**2 + w**4 - 2*w**2*wo**2
+ wo**4) + F*wo**2*cos(t*w)/(4*gm**2*w**2 + w**4 -
2*w**2*wo**2 + wo**4)
```

The following Example 10.9 numerically solves (10.9.2) using the `odeint` numerical integration method in SciPy, which was described previously in Example 10.5 of this chapter.

Example 10.9: The numerical solution of ODE for driven oscillation

Integrate numerically (10.9.2) for the driven harmonic oscillator using SciPy, and plot the solution $x(t)$. Use the numerical values $F_0 = 1$, $\gamma = 2$, $\omega_0 = 1$, $\omega = 1.5$ (with all quantities in SI units), and the initial conditions $x(0) = 1$ and $v(0) = \dot{x}(0) = 0$.

Solution:

In the code, the first component y[0] of the vector y represents the position variable x, and the second component y[1] represents the derivative dx/dt. The function `odeint` evaluates and returns a vector with components $(x, dx/dt)$. Note that the code in this example is much simpler than the code in the previous example, where we evaluated the symbolic solution of the ODE. The graph of $x(t)$ in Figure 10.7 shows clearly two different time regions with a different behavior: the transient solution $x_c(t)$ is dominant for $t < 40$ s, and that the sinusoidal steady state solution $x_c(t)$ dominates for $t > 40$ s.

```python
from scipy.integrate import odeint
import numpy as np
import matplotlib.pyplot as plt

# SI values of mass, spring constant, force amplitude
b ,m, k, Fo, om = 0.2,1,1,10, 1.5

def deriv(y, time):
    return (y[1], -(b/m)*y[1]- (k/m)*y[0]+Fo/m* np.cos(om*time))

t = np.linspace(0, 70, 200)
yinit = (1, 0)

soln = odeint(deriv, yinit, t)

plt.plot(t, soln[:, 0], '-',label=r'x(t)')
plt.ylabel('x(t)')
plt.xlabel('Time [s]')
plt.title('Externally Driven oscillator using SciPy')
leg = plt.legend()
plt.show()
```

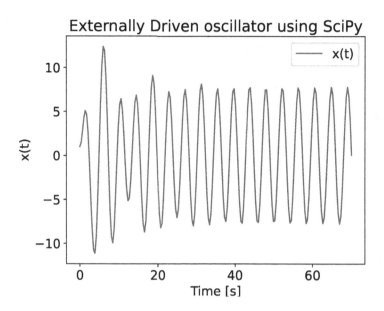

Figure 10.7 Graphical output from Example 10.9, showing a plot of the *numerical* solution of the driven damped SHO from SciPy.

Example 10.10 plots the amplitude A as a function of the external frequency ω, using (10.9.13). This example demonstrates the important physical phenomenon of resonance.

Example 10.10: The amplitude A as a function of the external frequency

Plot the amplitude $A(\omega)$ as a function of the external frequency ω, using (10.9.13) for different values of the damping constant b. Discuss the physical meaning of the graphs.

Solution:

One of the most striking results shown in this plot is the peak shape of the amplitude $A(\omega)$ as a function of frequency, which is known as the *resonance* phenomenon. As the damping constant b decreases in Figure 10.8, the resonance curve becomes higher and narrower. These graphs demonstrate that resonance represents the maximum amplitude response of the system to the external frequency.

In the code we define a function `ampl(omega,b)` of the frequency ω and of the damping constant b, and call it 3 times for a different value of b, in order to produce the three plots.

```
import numpy as np
import matplotlib.pyplot as plt

# SI values of mass, spring constant, force amplitude
m, k, Fo = 1, 1, 1

# natural frequency (undamped oscillation)
```

```
omega0 = np.sqrt(k/m)

# function to evaluate the amlitude for different (omega,b) values
def ampl(omega,b):
    gamma=b/(2*m)
    return (Fo/m)/((omega0*2-omega**2)**2+4*(gamma**2)*(omega**2))

omega = np.linspace(0.9,2,100)

# 3 plots for different values of b
plt.plot(omega,ampl(omega,.1),'b',label=r'b=0.1')
plt.plot(omega,ampl(omega,.2),'r--',label=r'b=0.2')
plt.plot(omega,ampl(omega,.3),'g-.',label=r'b=0.3')

plt.ylabel(r'Amplitude A($\omega$)')
plt.xlabel(r'External frequency $\omega$')
plt.title('Amplitude of Driven oscillator: Resonance')
leg = plt.legend()
plt.show()
```

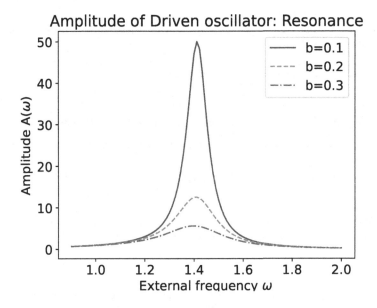

Figure 10.8 Graphical output from Example 10.10, showing a plot of the amplitude A of the driven SHO as a function of the external frequency ω, with the maximum amplitude occurring at the resonance frequency.

10.10   THE PRINCIPLE OF LINEAR SUPERPOSITION

As mentioned in the last section, focusing our attention on a single sinusoidal term as the external drive for the driven damped harmonic oscillator is not as restrictive as it may first appear. In our discussion of Fourier series, we will show that physically relevant periodic functions can be represented by an infinite series of sines and cosines. In the previous section, we examined the case where only one function appeared in the right-hand side of the damped harmonic oscillator. In this section, we will demonstrate how to solve *linear* ODEs with multiple terms on the right-hand side. What follows is true for *any* linear ODE, but we will focus on the damped harmonic oscillator as a concrete example.

Consider a damped harmonic oscillator experiencing n external drive forces $f_i(t)$:

$$\frac{d^2x}{dt^2} + 2\gamma\frac{dx}{dt} + \omega_0^2 x = \sum_{i=1}^{n} f_i(t) \tag{10.10.1}$$

The general solution to (10.10.1) is:

$$x(t) = x_c(t) + \sum_{i=1}^{n} x_i(t) \tag{10.10.2}$$

where $x_c(t)$ is the complementary solution of (10.10.1) (where the right-hand side equals zero) and $x_i(t)$ solve the equation:

$$\frac{d^2x_i}{dt^2} + 2\gamma\frac{dx_i}{dt} + \omega_0^2 x_i = f_i(t) \tag{10.10.3}$$

In other words, the general solution (10.10.2) is found by solving n ODEs and the homogeneous equation and then adding the solutions. Hence, (10.10.2) is an example of the *principle of linear superposition* where the solution of an ODE is a sum (or linear superposition) of the complementary solution and multiple particular solutions.

In practice, finding (10.10.2) can be a labor intensive task, especially for large n. After finding the homogeneous solution to (10.10.1), we would need to solve (10.10.1) n times, once for each $f_i(t)$ isolated on the right-hand side. But the fact that linear superposition holds, tells us that the steady state behavior of the driven system is the sum of the effects of each drive term. As we will see in Chapter 12, linear superposition does not hold for nonlinear systems, which is one reason they are difficult to solve in closed form.

We illustrate the principle of superposition for (10.10.1) in Example 10.11.

Example 10.11: The principle of linear superposition

Solve the following differential equation for a damped harmonic oscillator of mass m, spring constant k, damping parameter γ, and experiencing two sinusoidal drive forces:

$$\frac{d^2x}{dt^2} + 2\gamma\frac{dx}{dt} + \omega_0^2 x = \frac{F_1}{m}\cos(\omega_1 t) + \frac{F_2}{m}\cos(\omega_2 t) \tag{10.10.4}$$

by hand using the principle of linear superposition. Then, using the parameters $m = 1$, $\gamma = 0.1$, $\omega_0 = 1$, $F_1 = 0.5$, $F_2 = 2.0$, $\omega_1 = \omega_0/2$, $\omega_2 = \omega_0/3$ (all in SI units) solve the equation using SciPy and plot the solution for the initial conditions $x(0) = 1$ and $\dot{x}(0) = 0$.

Solution:

We can use much of the work done in previous sections to obtain a solution by hand:

$$x_c = De^{-\gamma t}\cos(\omega_0 t + \delta)$$
$$x_1 = A_1\cos(\omega_1 t - \phi_1)$$
$$x_2 = A_2\cos(\omega_2 t - \phi_2)$$

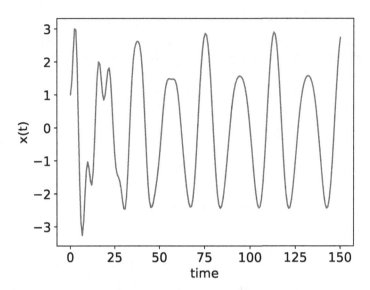

Figure 10.9 Graphical output from Example 10.11, showing a plot of the amplitude $x(t)$ of the damped driven harmonic oscillator.

where D and δ are found using initial conditions and:

$$A_1 = \frac{F_1/m}{\sqrt{(\omega_0^2 - \omega_1^2) + 4\gamma^2\omega_1^2}}$$

$$A_2 = \frac{F_2/m}{\sqrt{(\omega_0^2 - \omega_2^2) + 4\gamma^2\omega_2^2}}$$

$$\tan\phi_1 = \frac{2\gamma\omega_1}{\omega_0^2 - \omega_1^2}$$

$$\tan\phi_2 = \frac{2\gamma\omega_2}{\omega_0^2 - \omega_2^2}$$

Notice that the above values were obtained by plugging the drive force amplitude and frequency of each individual term on the right-hand side of (10.10.4), into the solutions we found for (10.9.2).

Below we solve the computational part of the example, using methods similar to those used in previous example problems in this chapter, with the result shown in Figure 10.9.

```
from scipy.integrate import odeint
import numpy as np
import matplotlib.pyplot as plt

m, gm, omega0, F1, F2 = 1.0, 0.1, 1.0, 0.5, 2.0
om1 = omega0/2
om2 = omega0/3

def deriv(y, t):
    return (y[1],-gm*y[1] - omega0**2*y[0] + \
        F1/m*np.cos(om1*t) + F2/m*np.cos(om2*t) )

time = np.linspace(0,150,200)
```

```
yint = (1,0)
soln = odeint(deriv,yint,time)

plt.plot(time, soln[:,0])
plt.xlabel('time')
plt.ylabel('x(t)')
plt.show()
```

10.11 ELECTRICAL CIRCUITS

The equation of motion for the forced harmonic oscillator can also be used to describe a driven RLC circuit. In this section, we develop the equations of motion for the RLC circuit and we explore what is sometimes referred to as the *electrical-mechanical analogy*. To begin, consider an RLC circuit as shown in Figure 10.10.

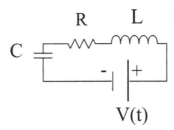

Figure 10.10 RLC series circuit with external voltage V, resistance R, inductance L, and capacitance C.

The governing differential equation can be found by substituting into Kirchhoff's voltage law (KVL), the constitutive equation for each of the three elements. From the KVL,

$$V(t) + V_R + V_L + V_C = 0 \tag{10.11.1}$$

where $V_R = -IR$ is the voltage drop across the resistor, $V_L = -L\frac{dI}{dt}$ is the voltage drop across the inductor, $V_C = -\frac{Q}{C}$ is the voltage drop across the capacitor, and $V(t)$ is the time varying voltage from the source. Note that we are using the normal conventions that I is the current in the circuit and Q is the charge on the capacitor. Furthermore, L is the inductance of the inductor, R is the resistance of the resistor, C is the capacitance of the capacitor and the current is $I = \dot{Q}$. For simplicity, we will assume that $V(t) = V_0 \cos(\omega t)$. Substituting these expressions in the KVL equation:

$$L\frac{dI}{dt} + IR + \frac{Q}{C} = V_0 \cos(\omega t) \tag{10.11.2}$$

Dividing by L leads to the second order differential equation for a series RLC circuit:

$$\frac{d^2Q(t)}{dt^2} + \frac{R}{L}\frac{dQ(t)}{dt} + \frac{1}{LC}Q(t) = \frac{V_0}{L}\cos(\omega t) \tag{10.11.3}$$

This equation has the exact same mathematical form as equation (10.9.2) for a driven damped harmonic oscillator:

$$\frac{d^2x}{dt^2} + \frac{b}{m}\frac{dx}{dt} + \frac{k}{m}x = \frac{F_0}{m}\cos(\omega t) \tag{10.11.4}$$

where (10.9.2) is written without the substitution $\gamma = b/2m$.

Because (10.11.3) and (10.9.2) have the same mathematical form, we can reuse the solution of (10.9.2) to obtain the solution of (10.11.3). The mathematical similarities imply that the physical behaviors of the two systems are similar. The forced damped harmonic oscillator has two solutions: a transient solution which decays as $t \to \infty$ and a steady state solution consisting of a sinusoidal oscillation with a frequency equal to the drive frequency. Because of the similarities between (10.11.3) and (10.9.2), we know that the RLC circuit will have two solutions, a transient which will also decay, and a steady state sinusoidal solution which has the same frequency as the voltage source in the circuit. In the case of the RLC circuit, it is the value of the charge on the capacitor that is oscillating.

Furthermore, we can then draw exact analogies between mechanical systems and electrical systems. These analogies are shown in Table 10.1 and are found by comparing the locations of the variables in (10.11.3) to those in (10.11.4). The simplest analogy is of that between the displacement x of the mass and the charge Q on the capacitor. A more interesting analogy exists between L and m. The inductance L in (10.11.3) appears in the same location as m in (10.8.3). This implies that L is taking on a role similar to that of mass. This analogy makes physical sense if you recall that the inductance L measures the inductor's ability to resist changes in current ($\dot{I}$), while the mass m is a measure of inertia, an object's ability to resist changes in velocity ($\dot{v}$).

It is also possible to use the analogies to identify qualitative characteristics of the RLC circuit's behavior. For example, we know that the coefficient of x in (10.8.3) is the angular frequency ω_0^2 of oscillation for an undamped oscillator. We can conclude from (10.11.3) that the oscillation frequency of a LC circuit (where $R = 0$) will be given by:

$$\omega_0 = \frac{1}{\sqrt{LC}} \tag{10.11.5}$$

because $1/(LC)$ is the coefficient of Q in (10.11.3).

Table 10.1

Analogies between electrical series RLC circuit and mechanical harmonic oscillator.

Mechanical system	Electrical system
Position x	Charge Q
Velocity $\frac{dx}{dt}$	Current $I = \frac{dQ}{dt}$
Mass m	Inductance L
Spring Constant k	Inverse Capacitance $1/C$
Damping γ	Resistance R
Natural Frequency $\frac{1}{2\pi}\sqrt{\frac{k}{M}}$,	Natural Frequency $\frac{1}{2\pi}\sqrt{\frac{1}{LC}}$
ODE $M\ddot{x} + \gamma\dot{x} + Kx = F$	ODE $L\ddot{Q} + R\dot{Q} + Q/C = V$

Analogies can be a powerful tool in physics, which can be used to take an understanding of one system and apply it to another mathematically similar system. As with any analogy (physics-related or not), be careful not to take the analogy too far. Doing so can lead to confusion at best, and to a complete misunderstanding of the system at worst.

10.12   PHASE SPACE

Phase space diagrams are a very useful concept in physics and other disciplines. A phase space diagram usually plots the position $x(t)$ of a particle on the x-axis and its corresponding speed $v(t)$ on the y-axis. Alternatively, one can also use the momentum $p = mv(t)$ on the y-axis, instead of the speed $v(t)$. The result is a graph that gives a qualitative description of the particle's motion. As we will see in Chapter 12, phase space diagrams are important in analyzing nonlinear systems. For now, we will explore the phase space diagrams associated with the harmonic oscillator.

Let us consider an undamped simple harmonic oscillator whose position and speed are described by:

$$x(t) = A \cos(\omega_0 t) \tag{10.12.1}$$

$$v(t) = dx/dt = -A\omega_0 \sin(\omega_0 t) \tag{10.12.2}$$

We can eliminate the time t in these equations by using the trig identity $\sin^2(\omega_0 t) + \cos^2(\omega_0 t) = 1$:

$$\frac{x^2}{A^2} + \frac{v^2}{A^2\omega_0^2} = 1 \tag{10.12.3}$$

This clearly represents an ellipse on the xv − plane. We conclude that the phase space representation of a simple harmonic oscillator is an ellipse, with semi-major axis equal to the amplitude of oscillation $a = A$, and the semi-major v-axis equal to $b = A\omega_0$.

Phase space diagrams can tell us something about the motion of the system. The curve in a phase diagram is sometimes referred to as a *trajectory in phase space*. An image of all possible trajectories in a phase space is sometimes called a *phase portrait* or *phase diagram*. Each point along a trajectory gives the system's position and velocity. The presence of closed-loop trajectories in a phase space diagram is evidence of oscillatory motion. A closed-loop trajectory tells us that after some time τ, the system returns to its initial position and velocity. The elliptical trajectories in (10.12.3) tells us that the SHO displays oscillatory motion.

If the oscillator is underdamped, we expect that motion will consist of oscillatory motion whose amplitude decays exponentially in time. In other words, the trajectory is an inward spiral.

Example 10.12 evaluates and plots the phase space diagrams for several types of oscillations of the damped harmonic oscillator.

Example 10.12: Phase space diagrams of damped oscillators

Plot the phase space of a damped harmonic oscillator, and discuss the differences between the phase spaces of overdamped and underdamped oscillations.

Solution:

The code is a modified version of Example 10.7. The main difference is the line
`dersoln=lambdify(t, diff(soln,t),'numpy')`
which evaluates symbolically the derivative $x'(t)$ and creates a new function using `lambdify`.
The phase space in Figure 10.11 for underdamped oscillations is a spiral from the initial point $(x(0), x'(0))$ toward the origin, where both the position and the speed of the particle become zero.
For overdamped and critically damped oscillations, the phase space trajectory does not cross the x-axis, in agreement with the plots in Example 10.7.

```
from sympy import symbols, Function, Derivative as D, dsolve, \
   lambdify, diff

import numpy  as np
import matplotlib.pyplot as plt
print('-'*28,'CODE OUTPUT','-'*29,'\n')

x = Function('x')
t = symbols('t',real=True,positive=True)

# function to plot the three phase spaces for given k, m, b

def plotx(k, m, b, labl, mrker):
    # solve differential equation with initial conditions
    soln = dsolve(m*D(x(t), t, t)+b*D(x(t), t) +k*x(t),
       x(t),simplify=True,ics=initCondits).rhs

    # lambdify the solution to get a function x(t), x'(t)
    xsoln = lambdify(t, soln,'numpy')
    dersoln = lambdify(t, diff(soln,t),'numpy')

    # evaluate x(t), x'(t) values from t=0 to t=50 s
    tvals = np.arange(0,50,.2)
    xvals = xsoln(tvals)
    yvals = dersoln(tvals)

    # plot x(t) using different markers abnd labels
    plt.plot(xvals,yvals, mrker, label=labl);
    plt.plot(xvals,[0]*xvals)

# initial conditions x(0)=1  and v(0)=0
initCondits = {x(0): 1, D(x(t),t).subs(t, 0): 0}

k, m, b = .1, 1, 1      # trajectory for overdamped motion
labl = 'Overdamped'
plotx(k,m,b,labl,'k-.')

b = .2    # trajectory for underdamped motion
labl='Underdamped'
plotx(k,m,b,labl,'r--')

b = 0.63
labl = 'Critically damped'  # trajectory for Critically damped
plotx(k,m,b,labl,'b:')

plt.title('Phase space trajectories for damped oscillations');
plt.xlabel('x(t) [m]')
plt.ylabel('v(t) [m/s]')
leg = plt.legend()
leg.get_frame().set_linewidth(0.0)
plt.tight_layout()
plt.show()
```

---------------------------- CODE OUTPUT ----------------------------

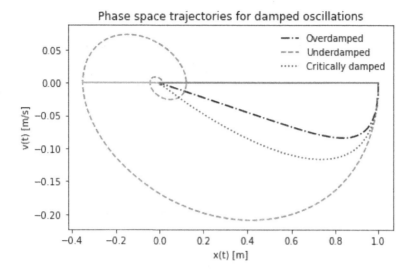

Figure 10.11 Graphical output from Example 10.12, showing the phase space diagrams for several types of oscillations of the harmonic oscillator.

The above algorithm can of course be used to generate the phase diagram for any system whose position $x(t)$ is known. While the phase space diagram does not provide new information about the behavior of the harmonic oscillator, we will find in Chapter 12 that phase space diagrams will become a critical tool for understanding the possible behaviors of a nonlinear system. In Chapter 12, we will develop a method of finding a system's phase space diagram without solving analytically the system's equations of motion. It is often the case in nonlinear systems that the equations of motion cannot be solved analytically. Therefore the phase space diagram may be the only tool available in getting information about the behavior of the system.

10.13   SYSTEMS OF DIFFERENTIAL EQUATIONS

Many physical systems are described by systems of *coupled differential equations*. Two or more ODEs form a coupled system if one or more of the equations in the system contain more than one dependent variable. For example

$$\left.\begin{array}{l} \dfrac{dx}{dt} = \sigma\,(y - x) \\[2mm] \dfrac{dy}{dt} = x\,(r - z) - y \\[2mm] \dfrac{dz}{dt} = xy - bz \end{array}\right\} \tag{10.13.1}$$

is a coupled system of ODEs called the *Lorenz equations* where the functions x, y, and z depend on time and σ, r, and b are constants. In this section we will give examples of how to solve linear coupled systems of ODEs symbolically in SymPy, and numerically using SciPy.

As a concrete example, consider two masses m_1 and m_2 attached to three springs with spring constants k_1, k_2, and k_3 and to the two fixed walls, as shown in Figure 10.12. We previously solved this system by using a matrix method in Chapter 7, in which we found the eigenvalues and eigenvectors of a 2×2 matrix. In this chapter we will solve directly the system of differential equations describing these coupled oscillations, by using both a symbolic calculation using SymPy, and also using a numerical integration of the equations using SciPy.

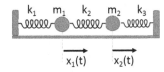

Figure 10.12   A system of two coupled harmonic oscillators, consisting of two masses m_1 and m_2, connected with three springs with constants k_1, k_2, and k_3.

Let us denote by $x_1(t)$ and $x_2(t)$ the horizontal displacements of the two masses from their respective equilibrium points. The force on the first mass due to the first spring is $-k_1 x_1$. The middle spring will be stretched by a distance $(x_1 - x_2)$, and the force on the first mass due to this middle spring will be $-k_2(x_1 - x_2)$. The total force on the first mass must then be $F_1 = -k_1 x_1 - k_2(x_1 - x_2)$. Similarly, the force on the second mass due to the middle spring is $-k_2(x_2 - x_1)$, and the force on the second mass due to this third spring will be $-k_3 x_2$.

The equations of motion from Newton's second law $F = ma$ for the two masses are:

$$\left. \begin{array}{c} m_1 \dfrac{d^2 x_1}{dt^2} = -k_1 x_1 - k_2(x_1 - x_2) \\[2mm] m_2 \dfrac{d^2 x_2}{dt^2} = -k_2(x_2 - x_1) - k_3 x_2 \end{array} \right\} \qquad (10.13.2)$$

In general, it is rather tedious to obtain the solutions $x_1(t)$ and $x_2(t)$ of the system of equations (10.13.2) analytically, and most of the time the solutions are obtained by numerically integrating the equations for the given initial conditions of the system. The initial conditions are usually given as the initial positions and initial speeds of the two masses.

The following Example 10.13 solves symbolically the special case of equal masses and equal spring constants in the two coupled oscillating masses using SymPy. However, the same code can be used for any values of the masses and spring constants.

Example 10.13: The symbolic solution of a system of ODEs

Solve the system of coupled differential equations (10.13.2) using SymPy, for the case of equal masses m and equal spring constants k. The initial conditions are $x_1(0) = 0$, $x_2(0) = 0$ m, $\dot{x}_1(0) = 0$ and $\dot{x}_2(0) = 1$. The frequency $\omega^2 = k/m = 1$ s^{-1}. Plot the numerical solutions $x_1(t)$ and $x_2(t)$, and discuss the physical behavior of the two oscillating masses.

Solution

Using equal spring constants $k_1 = k_2 = k_3 = k$ and equal masses $m_1 = m_2 = m$, and using $\omega^2 = k/m$, the system of equations becomes:

$$\left.\begin{aligned}\frac{d^2 x_1}{dt^2} &= -\omega^2 x_1 - \omega^2 (x_1 - x_2)\\\frac{d^2 x_2}{dt^2} &= -\omega^2 (x_2 - x_1) - \omega^2 x_2\end{aligned}\right\} \qquad (10.13.3)$$

This situation corresponds to the case where the first mass m_1 is initially at rest at its equilibrium position, and the second mass is pulled a distance from its equilibrium and released from rest. In the code the parameter om represents the frequency $\omega = \sqrt{k/m}$, and the analytical solution obtained by SymPy is:

$$x_1(t) = \frac{\sin(\omega t)}{2\omega} - \sqrt{3}\frac{\sin(\sqrt{3}\,\omega t)}{6\omega}$$

$$x_2(t) = \frac{\sin(\omega t)}{2\omega} + \sqrt{3}\frac{\sin(\sqrt{3}\,\omega t)}{6\omega}$$

These equations show that the motion of the system is a linear combination of two frequencies $\omega_1 = \omega$ and $\omega_2 = \sqrt{3}\omega$. These natural frequencies were obtained previously in Chapter 7 using matrix methods and are the *normal modes* of the system.

The plots of $x_1(t)$ and $x_2(t)$ in Figure 10.13 show that the motion of the two masses are coupled to each other. However, if one plots the sum $x_1(t) + x_2(t)$ and the difference $x_1(t) - x_2(t)$ as in the four graph panels shown here, it is possible to uncouple the two motions and obtain a pure sinusoidal oscillation of the system with the two frequencies ω_1 and ω_2.

In the Python code, we import and use the function dsolve_system to obtain the symbolic solution of the system of ODEs for the coupled oscillations. As usual, we use lamdify and .subs to evaluate and plot these solutions, as shown in Figure 10.13.

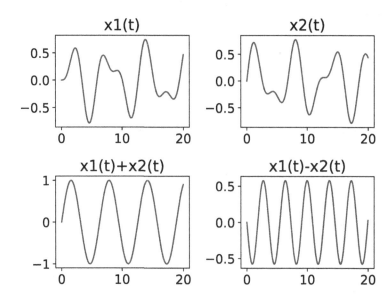

Figure 10.13 Graphical output from Example 10.13, for the SymPy solutions of the oscillating system of two equal masses and equal spring constants.

```
from sympy import Function,  Eq, Derivative as D, symbols, lambdify

from sympy.solvers.ode.systems import dsolve_system

import numpy as np
import matplotlib.pyplot as plt
print('-'*28,'CODE OUTPUT','-'*29,'\n')

k, m, t, om, x0, v0 = symbols('k, m, t, om, x0, v0',real=True,\
positive=True)

x1, x2 = symbols('x1, x2', cls=Function)

# function to plot x1,x2 etc
def plotx(y,ttle):
    plt.plot(xvals,y,'b');
    plt.title(ttle);

# Neton's law for the two masses
eq1 = Eq(D(x1(t),t,t), -om**2*x1(t)-om**2*(x1(t)-x2(t)))
eq2 = Eq(D(x2(t),t,t), -om**2*x2(t)-om**2*(x2(t)-x1(t)))

# initial conditions x1(0)=0, v1(0)=0, x2(0)=0  and v2(0)=1
initCondits = {x1(0): 0, D(x1(t),t).subs(t, 0): 0,\
               x2(0): 0, D(x2(t),t).subs(t, 0): 1 }

# solve the system of two differential equations symbolically
soln = dsolve_system((eq1, eq2), [x1(t),x2(t)], t, initCondits)

# extract the solutions x1(t) and x2(t), print x1, x2, x1+x2, x1-x2
X1 = soln[0][0].rhs
X2 = soln[0][1].rhs

print('x1(t) =',X1)
print('x2(t) =',X2)
print('\nx1(t)+x2(t) =',X1+X2)
print('x1(t)-x2(t) =',X1-X2)

# make x1(t) and x2(t) functions using lambdify, and insert omega=om=1
x1soln = lambdify(t, X1.subs(om,1),'numpy')
x2soln = lambdify(t, X2.subs(om,1),'numpy')

# evaluate x(t) values from t=0 to t=20 s
xvals = np.arange(0,20,.02)

# craete 4 plot panels for x1, x2, x1+x2 and x1-x2
plt.subplot(2,2,1)
plotx(x1soln(xvals),'x1(t)')
plt.subplot(2,2,2)
plotx(x2soln(xvals),'x2(t)')
plt.subplot(2,2,3)
plotx(x1soln(xvals)+x2soln(xvals),'x1(t)+x2(t)')
plt.subplot(2,2,4)
plotx(x1soln(xvals)-x2soln(xvals),'x1(t)-x2(t)')
```

```
plt.tight_layout()
plt.show()

-------------------------- CODE OUTPUT ----------------------------

x1(t) = sin(om*t)/(2*om) - sqrt(3)*sin(sqrt(3)*om*t)/(6*om)
x2(t) = sin(om*t)/(2*om) + sqrt(3)*sin(sqrt(3)*om*t)/(6*om)

x1(t)+x2(t) = sin(om*t)/om
x1(t)-x2(t) = -sqrt(3)*sin(sqrt(3)*om*t)/(3*om)
```

Example 10.14 solves *numerically* the more general case of two coupled oscillating masses using SciPy, with the given initial conditions for the system.

Example 10.14: The numerical solution of system of ODEs

Solve (10.13.2) numerically using SciPy, for the more general case with parameters $m_1 = 1$, $m_2 = 2$, $k_1 = 10$, $k_2 = 20$, $k_3 = 5$ (all in SI units). The initial conditions are $x_1(0) = 0$, $x_2(0) = 1$ m, $\dot{x}_1(0) = 0$ and $\dot{x}_2(0) = 0$. Plot the numerical solutions $x_1(t)$ and $x_2(t)$, and the speeds $\dot{x}_1(t)$ and $\dot{x}_2(t)$.

Solution:

This situation corresponds to the case where the first mass m_1 is initially at rest at its equilibrium position $(x_1(0) = 0$ and $\dot{x}_1(0) = 0)$, and the second mass is pulled a distance from its equilibrium and released from rest.
The plots of $x_1(t)$ and $x_2(t)$ are obviously complex, and it is not possible to give a simple physical description of the motion of the two masses. The key physical component which creates this complex behavior, is of course the middle spring in Figure 10.12, since this is the component that couples the motion of the two masses.

As in our previous examples, the function odeint(deriv, yinit, t) is called with three arguments. The first argument is the function deriv which contains the information on the second order ODEs to be solved, the second argument is the initial conditions vector yinit, and the third argument is the time variable t. In the code we define a function deriv(Y) whose argument is the vector Y, and the time variable time.

In our example, the first component Y[:,0] of the vector Y represents the variable $x_1(t)$, and the second component Y[:,1] of the vector Y represents the corresponding velocity $\dot{x}_1(t) = dx_1/dt$. Similarly, the third component Y[:,2] of the vector Y represents the variable $x_2(t)$, and the fourth component Y[:,3] represents the corresponding velocity dx_2/dt.
Note that the solver odeint also evaluates the derivatives of the positions, $v_1 = dx_1/dt$ and $v_2 = dx_2/dt$, and these are also plotted in the code output in Figure 10.14.

```
from scipy.integrate import odeint

import numpy as np
import matplotlib.pyplot as plt

# function defines the differential equations
```

```
# Y[0]=x1, Y[1]=x1', Y[2]=x2, Y[3]=x2'

def deriv(Y,t):
    return np.array([ Y[1], (- k1 * Y[0] + k2 * (Y[2] - Y[0] )) / m1,
                      Y[3], (- k3 * Y[2] + k2 * (Y[0] - Y[2] )) / m2])

# t= array of times to be evaluated
t = np.linspace(0,6,100)

# spring constants and masses
k1, k2, k3 = 10, 20, 5
m1, m2 = 1, 2

# yinit = initial conditions [x1(0), x1'(0), x2(0), x2'(0)]
yinit = np.array([0, 0, 1, 0])

# solve the odes using odeint
Y = odeint(deriv, yinit, t)

# plot x1(t), x2(t)
plt.subplot(2,1,1)
plt.plot(t,Y[:,0],label='x1(t)')
plt.plot(t,Y[:,2],label='x2(t)')
plt.legend(loc='lower right')

# plot x1'(t), x2'(t)
plt.subplot(2,1,2)
plt.plot(t,Y[:,1],label="x1'(t)")
plt.plot(t,Y[:,3],label="x2'(t)")

plt.legend(loc='lower right')
plt.show()
```

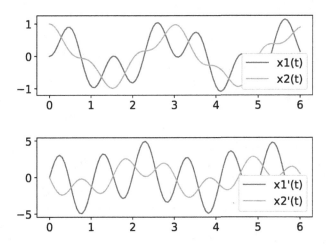

Figure 10.14 Graphical output from Example 10.14, showing the numerical solutions and their derivatives from SciPy, for the general case of two coupled oscillating masses.

10.14   THE LEGENDRE EQUATION

An important equation in physics and engineering that arises when working with problems in spherical coordinates is the Legendre equation:

$$\left(1 - x^2\right)\frac{d^2y}{dx^2} - 2x\frac{dy}{dx} + \ell\left(\ell + 1\right)y = 0 \tag{10.14.1}$$

where $y = y(x)$ and ℓ is a constant. Here, we will assume ℓ is an integer, as that is most common in physical applications. We will solve (10.14.1) using the method of series solution.

Series solutions are useful when solving an ODE and are often applicable to problems in science. We will focus on ODEs of the form

$$\frac{d^2y}{dx} + f(x)\frac{dy}{dx} + g(x)y = 0 \tag{10.14.2}$$

Under conditions that are generally met in physical applications, we can assume the solution can be written as a power series (a common assumption in physics), insert the series into the ODE, and find the series coefficients. We will demonstrate the method of series solution for (10.14.1). But first, you might be wondering why, with all of the powerful ODE solvers and numerical methods available, would we want a power series solution to an ODE? One reason is that power series solutions can give scaling laws for the function $y(x)$, something that might not be obvious from a closed form solution or graph given by a computer. For example, knowing that for small values of x, $y(x) \approx 1/x^2$ can be invaluable in interpreting the physics of a given problem.

Let us assume a solution to (10.14.1) in the form:

$$y(x) = \sum_n a_n x^n \tag{10.14.3}$$

If we insert (10.14.3) into (10.14.1), we have:

$$\left(1 - x^2\right)\left[\sum n(n-1)a_n x^{n-2}\right] - 2x\left[\sum na_n x^{n-1}\right] + \ell\left(\ell + 1\right)\left[\sum a_n x^n\right] = 0 \tag{10.14.4}$$

Let us expand the first term:

$$\left[\sum n(n-1)a_n x^{n-2}\right] - \left[\sum n(n-1)a_n x^n\right] - 2\left[\sum na_n x^n\right] + \ell\left(\ell + 1\right)\left[\sum a_n x^n\right] = 0 \tag{10.14.5}$$

Next, we collect the coefficients of x^n. For the first term, this means substituting $n \to n+2$:

$$\sum \left\{(n+2)\left(n+1\right)a_{n+2} + \left[-n(n-1) - 2n + \ell(\ell+1)\right]a_n\right\}x^n = 0 \tag{10.14.6}$$

The term in curly brackets must be zero. After simplification of the term in square brackets), we have:

$$a_{n+2} = -\frac{(\ell + n + 1)(\ell - n)}{(n + 2)(n + 1)}a_n, \tag{10.14.7}$$

Let us calculate the value of the first few a_n's starting with a_2:

$$a_2 = -\frac{\ell(\ell + 1)}{2}a_0 \tag{10.14.8}$$

$$a_3 = -\frac{(\ell + 2)(\ell - 1)}{6}a_1 \tag{10.14.9}$$

$$a_4 = -\frac{(\ell + 3)(\ell - 2)}{12}a_2 \tag{10.14.10}$$

$$a_5 = -\frac{(\ell + 3)(\ell - 4)}{20}a_3 \tag{10.14.11}$$

Notice that a_4 depends on a_2 and a_2 depends on a_0. Continuing the calculation of coefficients, the a_n's with even subscripts will all depend on a_0 and not on a_1. Likewise, the a_n's with odd subscripts will depend on a_1 and not a_0.

Therefore, we have a solution with two arbitrary constants a_0 and a_1. This is expected because (10.14.1) is a second order ODE. The solution (after simplification) is:

$$y = a_0 \left(1 - \frac{\ell(\ell+1)}{2!} x^2 + \frac{\ell(\ell+1)(\ell-2)(\ell+3)}{4!} x^4 + \cdots \right) \tag{10.14.12}$$

$$+ a_1 \left(x - \frac{(\ell-1)(\ell+2)}{3!} x^3 + \frac{(\ell-1)(\ell+2)(\ell-3)(\ell+4)}{5!} x^5 + \cdots \right)$$

The series (10.14.12) converges for $x^2 < 1$. Notice that if ℓ is odd, the a_1 series terminates at x^ℓ term but the a_0 series diverges. Likewise, when ℓ is even, the a_0 series terminates at the x^ℓ term and the a_1 series diverges. However, in physics, we often want (10.14.12) to converge at $x = \pm 1$. We accomplish this by choosing a value of ℓ and discarding the diverging series. If the value of a_0 or a_1 is chosen such that y is unity when $x = 1$, then the resulting polynomials are *Legendre polynomials*, $P_\ell(x)$. The first three Legendre polynomials are:

$$P_0(x) = 1 \tag{10.14.13}$$

$$P_1(x) = x \tag{10.14.14}$$

$$P_2(x) = \frac{1}{2} \left(3x^2 - 1 \right) \tag{10.14.15}$$

Legendre polynomials are most often generated using a computer. However, we can also write the Legendre polynomials using the *Rodrigues formula*:

$$P_\ell(x) = \frac{1}{2^\ell \ell!} \left(\frac{d}{dx} \right)^\ell \left(x^2 - 1 \right)^\ell \tag{10.14.16}$$

Legendre polynomials have useful relationships among them. The following recursion relationships can be useful in derivations and other closed form analyses.

$$\ell P_\ell(x) = (2\ell - 1)\, x P_{\ell-1}(x) - (\ell - 1)\, P_{\ell-2}(x) \tag{10.14.17}$$

$$\ell P_\ell(x) = x \frac{dP_\ell}{dx} - \ell \frac{dP_{\ell-1}(x)}{dx} \tag{10.14.18}$$

$$\ell P_{\ell-1}(x) = \frac{dP_\ell(x)}{dx} - x \frac{dP_{\ell-1}}{dx} \tag{10.14.19}$$

$$\left(1 - x^2 \right) P_\ell(x) = \ell P_{\ell-1}(x) - \ell x P_\ell(x) \tag{10.14.20}$$

$$(2\ell + 1) P_\ell(x) = \frac{dP_{\ell+1}}{dx} - \frac{dP_{\ell-1}}{dx} \tag{10.14.21}$$

$$\left(1 - x^2 \right) \frac{dP_{\ell-1}}{dx} = \ell x P_{\ell-1}(x) - \ell P_\ell(x) \tag{10.14.22}$$

There is an orthogonality relationship between Legendre polynomials of different orders.

$$\int_{-1}^{1} P_\ell(x) P_m(x)\, dx = \begin{cases} 0 & \ell \neq m \\ \frac{2}{2\ell+1} & \ell = m \end{cases} \tag{10.14.23}$$

The Legendre polynomials are complete on the interval $x \in [-1, 1]$. Therefore, on the interval $x \in [-1, 1]$, we can write a function $f(x)$ as

$$f(x) = \sum_{\ell=0}^{\infty} c_\ell P_\ell(x) \tag{10.14.24}$$

Using orthogonality methods we learned in Chapter 9, we can find c_ℓ. We begin by multiplying (10.14.24) by $P_m(x)$ and integrating

$$\int_{-1}^{1} P_m(x) f(x) dx = \int_{-1}^{1} \sum_{\ell=0}^{\infty} c_\ell P_\ell(x) P_m(x) dx \tag{10.14.25}$$

$$= c_m \frac{2}{2m+1} \tag{10.14.26}$$

Therefore (changing the index back to ℓ)

$$c_\ell = \frac{2\ell+1}{2} \int_{-1}^{1} P_\ell(x) f(x) dx \tag{10.14.27}$$

We will use the completeness of Legendre polynomials in the next chapter when solving partial differential equations. Example 10.15 demonstrates how to use the built in libraries for the Legendre polynomials in SciPy and SymPy, by displaying them and listing some of them.

Example 10.15: Legendre polynomials

(a) Using SymPy, display the first six Legendre polynomials
(b) Use SymPy to demonstrate the recursion relation between Legendre polynomials for $\ell = 6$:

$$\ell P_\ell(x) = (2\ell - 1) x P_{\ell-1}(x) - (\ell - 1) P_{\ell-2}(x) \tag{10.14.28}$$

(c) Using SciPy, plot the Legendre polynomials for $l = 1, 2, 3, 4$.

Solution:

(a) The SymPy command `legendre(l,x)` produces $P_\ell(x)$. We used a `for` loop to print the Legendre polynomials.
(b) We implement the right and left-hand side of (10.14.28) (variables `lhs` and `rhs` in the code), and test its validity using the logical double equal sign `==`, yielding `True`.
(c) The SciPy command `eval_legendre(l,x)` produces $P_\ell(x)$. We again used a `for` loop to evaluate and plot the Legendre polynomials between $x = -1$ and $x = +1$, with the result shown in Figure 10.15.

```
# Evaluate and plot the Legendre polynomials in SciPy
import numpy as np
from scipy.special import eval_legendre
from sympy import legendre, Symbol
import matplotlib.pyplot as plt


print('-'*28,'CODE OUTPUT','-'*29,'\n')
```

```
x = Symbol('x')

print('The first few Legendre polynomials are:\n')
for l in range(0,6):
    print('P_'+str(l)+' =', legendre(l,x))

l = 6

lhs = l*legendre(l,x)
print('\nThe left hand side = ',lhs.expand())

rhs = (2*l-1)* x * legendre(l-1,x)-(l-1)*legendre(l-2,x)
print('\nThe right hand side = ',rhs.expand())

print('\nThe identity is ',lhs.expand() == rhs.expand())

xc = np.linspace(-1, 1, 100)  # x-values between -1 and +1

s=['.',':','-','--','-.']
for j in range(1, 5):
    plt.plot(xc, eval_legendre(j, xc), str(s[j]),label=r'$P_{}(x)$'.format(j))

plt.title("Plots of Legendre polynomials")
plt.xlabel("x")
plt.ylabel(r'$P_l(x)$')
plt.ylim(-1.4,1);
plt.xlim(-1.2,1.2);
leg = plt.legend()
leg.get_frame().set_linewidth(0.0)
plt.tight_layout()
plt.show()
```

```
-------------------------- CODE OUTPUT --------------------------

The first few Legendre polynomials are:

P_0 = 1
P_1 = x
P_2 = 3*x**2/2 - 1/2
P_3 = 5*x**3/2 - 3*x/2
P_4 = 35*x**4/8 - 15*x**2/4 + 3/8
P_5 = 63*x**5/8 - 35*x**3/4 + 15*x/8

The left hand side =  693*x**6/8 - 945*x**4/8 + 315*x**2/8 - 15/8

The right hand side =  693*x**6/8 - 945*x**4/8 + 315*x**2/8 - 15/8

The identity is  True
```

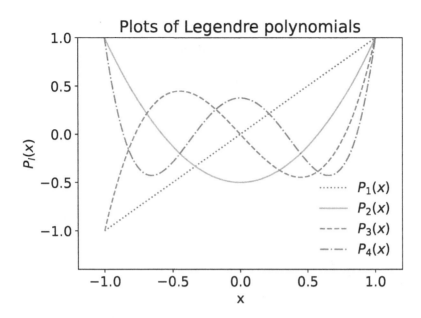

Figure 10.15 Graphical output from Example 10.15 showing Legendre polynomials $P_\ell(x)$ with $\ell = 1, \ldots 4$.

10.14.1 ASSOCIATED LEGENDRE FUNCTIONS

Another ODE that appears in problems with spherical symmetry is the *general Legendre equation*

$$\frac{d}{dx}\left[(1 - x^2)\frac{dy}{dx}\right] + \left[\ell(\ell + 1) - \frac{m^2}{1 - x^2}\right]y = 0 \qquad (10.14.29)$$

where $m^2 \le \ell^2$. The general Legendre equation is especially common in problems where the scalar function $y(r, \theta, \phi)$ is not independent of ϕ.

When $m > 0$, the solution $y = P_\ell^m(x)$ are called associated Legendre functions

$$P_\ell^m(x) = (-1)^m (1 - x^2)^{m/2}\left(\frac{d}{dx}\right)^m P_\ell(x) \qquad (10.14.30)$$

If $m < 0$, the solutions are still called associated Legendre functions, however we now use

$$P_\ell^{-m}(x) = (-1)^m \frac{(\ell - m)!}{(\ell + m)!} P_\ell^m(x) \qquad (10.14.31)$$

which is the standard used in many computer algebra systems, including SymPy.

A few of the associated Legendre functions are shown below. We use $\cos\theta$ as the argument because that is most often used in physics.

$$P_0^0 = 1 \qquad P_2^2 = 3\sin^2\theta$$
$$P_1^1 = -\sin\theta \quad P_2^1 = -3\sin\theta\cos\theta$$
$$P_1^0 = \cos\theta \qquad P_2^0 = \frac{1}{2}\left(3\cos^2\theta - 1\right)$$

As we will show in the next example, the associated Legendre functions are implemented in both the SymPy and SciPy libraries.

In quantum mechanics, the associated Legendre functions are often combined with an exponential to form a new function called *spherical harmonics*, $Y_\ell^m(\theta, \phi)$

$$Y_\ell^m(\theta, \phi) = \sqrt{\frac{(2\ell+1)}{4\pi}\frac{(\ell-m)!}{(\ell+m)!}}P_\ell^m(\cos\theta)\,e^{im\phi} \qquad (10.14.32)$$

The constant out front is chosen such that

$$\int_0^{2\pi}\int_0^\pi [Y_\ell^m]^*\,[Y_k^n]\sin\theta d\theta d\phi = \delta_{\ell k}\delta_{mn} \qquad (10.14.33)$$

where δ_{ij} is the *Kronecker delta* defined as

$$\delta_{ij} = \begin{cases} 1 & i = j \\ 0 & i \neq j \end{cases} \qquad (10.14.34)$$

Spherical harmonics are an important part of the wave function of the electron in a Hydrogen atom.

Example 10.16 demonstrates how to use the built in SymPy and SciPy functions for the associated Legendre functions $P_l^m(x)$ with $l = 0, 1, 2$.

===

Example 10.16: Associated Legendre functions

(a) Using SymPy, display the associated Legendre functions for $\ell = 0, 1,$ and 2.
(b) Using SciPy, plot the $P_l^m(x)$ from part (a), from $x = -1$ to $x = +1$.

Solution:

(a) The SymPy command `assoc_legendre(l,m,x)` produces the expressions for the $P_\ell^m(x)$ functions. We used a `for` loop to print the associated Legendre functions.
(b) The SciPy command `lpmv(l,m,x)` produces $P_\ell^m(x)$. We again used a `for` loop to evaluate and plot the functions between $x =$ -1 and $x =$ +1, with the plot shown in Figure 10.16.

```
import numpy as np
import matplotlib.pyplot as plt

print('-'*28,'CODE OUTPUT','-'*29,'\n')

from sympy import assoc_legendre, symbols
from scipy.special import lpmv

x = symbols('x')

print('\nl = 0')
```

```
l = 0
m =[ 0]
for i in range(len(m)):
    fn = assoc_legendre(l,m[i],x)
    print('P(0,',m[i],') = ',fn)

print('\nl = 1')
l = 1
m = [-1,0,1]

for i in range(len(m)):
    fn = assoc_legendre(l,m[i],x)
    print('P(1,',m[i],') = ',fn)

print('\nl = 2')
l = 2
m = [-2,-1,0,1,2]

for i in range(len(m)):
    fn = assoc_legendre(l,m[i],x)
    print('P(2,',m[i],') = ',fn)

xc = np.linspace(-1, 1, 100)  # x-values between -1 and +1

# define different markers for line plots
s = ['.',':','-','--','-.']

for j in range(1, 5):
    plt.plot(xc, lpmv(m[j],l, xc), str(s[j]),label='P(2,'+str(m[j])+')')

plt.title(r"Associated Legendre functions P$_l^m$(x)")
plt.xlabel("x")
plt.ylabel(r'$P_l^m(x)$')
plt.xlim(-1.1,1.4);

leg = plt.legend()
leg.get_frame().set_linewidth(0.0)

plt.show()

--------------------------- CODE OUTPUT ---------------------------


l = 0
P(0, 0 ) =  1

l = 1
P(1, -1 ) =  sqrt(1 - x**2)/2
P(1, 0 ) =  x
P(1, 1 ) =  -sqrt(1 - x**2)

l = 2
P(2, -2 ) =  1/8 - x**2/8
P(2, -1 ) =  x*sqrt(1 - x**2)/2
P(2, 0 ) =  3*x**2/2 - 1/2
```

```
P(2, 1 ) =  -3*x*sqrt(1 - x**2)
P(2, 2 ) =  3 - 3*x**2
```

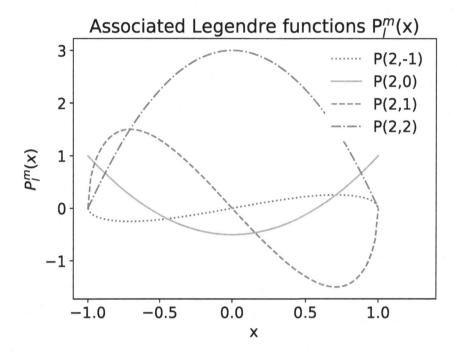

Figure 10.16 Graphical output from Example 10.16 showing the associated Legendre functions $P_l^m(x)$ with $l = 2$ and $m = -2, -1, 0, +1, +2$.

Example 10.17 demonstrates how to produce polar plots for the associated Legendre functions $P_l^m(x)$, when $x = \cos\theta$, and with $l = 2, 1, 0$. The graphs produced in this example are related to the atomic orbitals encountered in Quantum Mechanics.

Example 10.17: Polar plots of associated Legendre polynomials

Create polar plots of the associated Legendre functions with $m \geq 0$ found in Example 10.16.

Solution:

We create a grid of polar plots in Figure 10.17. Notice that we tell python each plot in the grid uses polar axes in the command subplots, so we use the plot command to make each individual plot.

```
from sympy import symbols, lambdify, cos
from sympy.functions.special.polynomials import assoc_legendre
import numpy as np
```

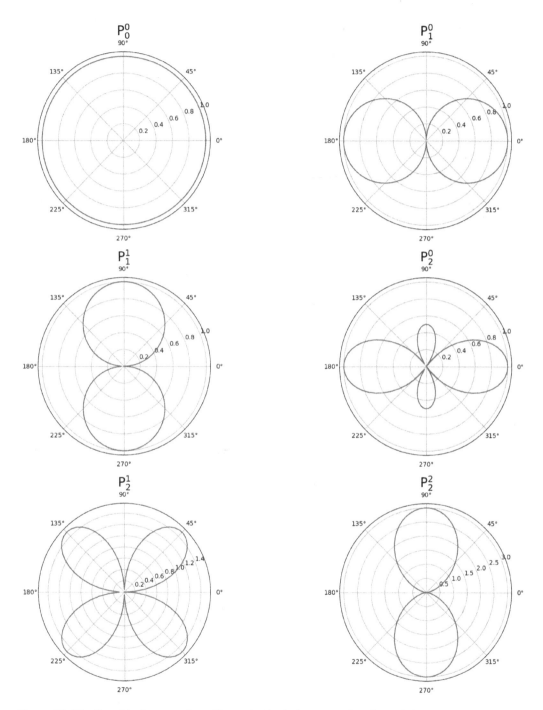

Figure 10.17 Graphical output from Example 10.17 showing polar plots of the associated Legendre functions with $\ell = 0, 1, 2$.

```
import matplotlib.pyplot as plt
from scipy.special import lpmv

theta = np.linspace(0,2*np.pi,100)
p00 = abs(lpmv(0,0,np.cos(theta)))
p10 = abs(lpmv(0,1,np.cos(theta)))
p11 = abs(lpmv(1,1,np.cos(theta)))
p20 = abs(lpmv(0,2,np.cos(theta)))
p21 = abs(lpmv(1,2,np.cos(theta)))
p22 = abs(lpmv(2,2,np.cos(theta)))

fig, axs = plt.subplots(3, 2, subplot_kw=dict(projection='polar'), figsize=(15,15))
axs[0,0].plot(theta,p00)
axs[0,0].set_title('P$_0^0$', fontsize=20)

axs[0,1].plot(theta,p10)
axs[0,1].set_title('P$_1^0$', fontsize=20)

axs[1,0].plot(theta,p11)
axs[1,0].set_title('P$_1^1$', fontsize=20)

axs[1,1].plot(theta,p20)
axs[1,1].set_title('P$_2^0$', fontsize=20)

axs[2,0].plot(theta,p21)
axs[2,0].set_title('P$_2^1$', fontsize=20)

axs[2,1].plot(theta,p22)
axs[2,1].set_title('P$_2^2$', fontsize=20)
fig.tight_layout()
plt.show()
```

10.15   THE BESSEL EQUATION

Another important ODE in physics which arises when working in cylindrical coordinates is Bessel's equation:

$$x^2 \frac{d^2y}{dx^2} + x \frac{dy}{dx} + \left(x^2 - p^2\right) y = 0 \qquad (10.15.1)$$

where p is a constant (but need not be an integer).

Notice that (10.15.1) is in a form amenable to series solutions. We will solve (10.15.1) using the Frobenius method which is a generalized method for series solutions. We assume

$$y(x) = \sum_{n=0}^{\infty} a_n x^{n+s} \qquad (10.15.2)$$

which will allow for the series expansion of $y(x)$ to have negative or fractional powers of x. We allow s to be positive, negative, or a fraction and we assume a_0 is not zero because $a_0 x^s$ is the first term in the series. We will demonstrate the Frobenius method to find the solution to (10.15.1). However, the method used here can be applied to many ODEs.

Before we insert (10.15.2) into (10.15.1), we rewrite (10.15.1) in a more convenient form

$$x\frac{d}{dx}\left(x\frac{dy}{dx}\right) + x^2y - p^2y = 0 \tag{10.15.3}$$

Next, we insert (10.15.2) into (10.15.3)

$$\sum_{n=0}^{\infty} a_n (n+s)^2 x^{n+s} + \sum_{n=0}^{\infty} a_n x^{n+s+2} - p^2 \sum_{n=0}^{\infty} a_n x^{n+s} = 0 \tag{10.15.4}$$

To find the value of s, we examine the coefficient of x^s in (10.15.4). Notice the middle term has no x^s term and therefore, the coefficient for the middle term is zero.

$$a_0 s^2 - p^2 a_0 = 0 \tag{10.15.5}$$

Or, $s = \pm p$. We find two values of s, one for each solution to (10.15.1).

Next, we examine the coefficient of x^{s+1}. As with the coefficient of x^s, the coefficient of x^{s+1} in the middle term of (10.15.4) is also zero. Therefore,

$$a_1 (1+s)^2 - p^2 a_1 = 0 \tag{10.15.6}$$

and $a_1 = 0$. Notice that if we examine the coefficient of x^{s+2}, then we find that the middle term of (10.15.4) has a nonzero coefficient, a_0. In general, we can write:

$$a_n (n+s)^2 + a_{n-2} - s^2 a_n = 0 \tag{10.15.7}$$

which can be solved for a_n

$$a_n = -\frac{a_{n-2}}{(n+s)^2 - p^2} \tag{10.15.8}$$

We will find the solution for $s = p$.

$$a_n = -\frac{a_{n-2}}{n(n+2p)} \tag{10.15.9}$$

Recall that $a_1 = 0$. Therefore all odd a_n's are zero. We can then replace n with $2n$

$$a_{2n} = -\frac{a_{2n-2}}{2^2 n(n+p)} \tag{10.15.10}$$

Leaving the 2^2 term as written will make more sense later. To simplify the solutions to (10.15.10), we will use the Γ function, $\Gamma(p+1) = p!$. The Γ function also has the property that $\Gamma(p+1) = p\Gamma(p)$. This is helpful for recursion relations. For example

$$\Gamma(p+3) = (p+2)\Gamma(p+2) = (p+2)\left[(p+1)\Gamma(p+1)\right] \tag{10.15.11}$$

Using (10.15.10), we find

$$a_2 = -\frac{a_0}{2^2(1+p)} = -\frac{a_0\Gamma(1+p)}{2^2\Gamma(2+p)} \tag{10.15.12}$$

$$a_4 = -\frac{a_2}{2^3(2+p)} = \frac{a_0}{2!2^4(1+p)(2+p)} = \frac{a_0\Gamma(1+p)}{2!2^4\Gamma(3+p)} \tag{10.15.13}$$

We can now write out the first three terms of the series solution with $s = p$

$$y(x) = a_0 x^p - \frac{a_0\Gamma(1+p)}{2^2\Gamma(2+p)}x^{p+2} + \frac{a_0\Gamma(1+p)}{2!2^4\Gamma(3+p)}x^{p+4} + \cdots \tag{10.15.14}$$

Using $\Gamma(1) = \Gamma(2) = 1$ and $x^p = 2^p (x/2)^p$, we can write the series solution as

$$y(x) = a_0 2^p \left(\frac{x}{2}\right)^p \Gamma(1+p) \left[\frac{1}{\Gamma(1)\Gamma(1+p)} - \frac{1}{\Gamma(1)\Gamma(2+p)} \left(\frac{x}{2}\right)^2 + \frac{1}{\Gamma(3)\Gamma(3+p)} \left(\frac{x}{2}\right)^4 + \cdots \right]$$

$$(10.15.15)$$

If we use

$$a_0 = \frac{1}{2^p \Gamma(1+p)} = \frac{1}{2^p p!} \tag{10.15.16}$$

then $y(x)$ is called the *Bessel function* of order p, $J_p(x)$:

$$J_p(x) = \sum_{n=0}^{\infty} \frac{(-1)^n}{\Gamma(n+1)\Gamma(n+1+p)} \left(\frac{x}{2}\right)^{2n+p} \tag{10.15.17}$$

If $p < 0$, then we can either perform the Frobenius method again, or simply insert $-p$ for p in the formulas above

$$J_{-p}(x) = \sum_{n=0}^{\infty} \frac{(-1)^n}{\Gamma(n+1)\Gamma(n+1-p)} \left(\frac{x}{2}\right)^{2n-p} \tag{10.15.18}$$

From (10.15.17), we see that if p is not an integer, then $J_p(x)$ is a series starting with the x^p term, while $J_{-p}(x)$ starts with x^{-p}. In this case, the solution to (10.15.1) is a linear combination of J_p and J_{-p}. But, if p is an integer, then it can be shown that $J_{-p} = (-1)^p J_p(x)$ and is therefore not an independent solution to (10.15.1). In this case, the second solution is not $J_{-p}(x)$ but instead has a logarithm which we typically ignore in physics because it is infinite at the origin.

Instead of using J_{-p}, it is common to use the Neumann function

$$N_p(x) = \frac{\cos(\pi p) J_p(x) - J_{-p}(x)}{\sin(\pi p)} \tag{10.15.19}$$

as the second solution to (10.15.1). It, is undefined when p is an integer, but it has a limit for any $x \neq 0$.

There are several useful relationships between Bessel functions of various orders. A search online will yield many of them. Below, we present only a few:

$$\frac{d}{dx}\left[x^p J_p(x)\right] = x^p J_{p-1}(x) \tag{10.15.20}$$

$$\frac{d}{dx}\left[x^{-p} J_p(x)\right] = -x^{-p} J_{p+1}(x) \tag{10.15.21}$$

$$J_{p-1}(x) + J_{p+1}(x) = \frac{2p}{x} J_p(x) \tag{10.15.22}$$

$$J_{p-1}(x) - J_{p+1}(x) = 2\frac{dJ_p}{dx} \tag{10.15.23}$$

Finally, there is a relationship of orthogonality between Bessel functions:

$$\int_0^1 x J_p(ax) J_p(bx)\,dx = \begin{cases} 0 & a \neq b \\ \frac{1}{2}J_{p+1}^2(a) = \frac{1}{2}J_{p-1}^2(a) = \frac{1}{2}\left[J_p'(a)\right]^2 & a = b \end{cases} \tag{10.15.24}$$

where a and b are zeros of $J_p(x)$. There are many zeros of a given Bessel function, which require a computer to find. The condition (10.15.24) will be used in the next chapter when we examine solutions to partial differential equations.

Example 10.18 shows how the Bessel equation can be solved using SymPy and how to plot Bessel functions of various orders p. Note that in SymPy the Bessel functions $J_p(x)$ are implemented using the function `besselj(p,x)`, while in SciPy they are implemented with the functions `jv(p,x)`.

Example 10.18: Bessel functions

(a) Solve the general Bessel ODE (10.15.1) using the SymPy function `dsolve()`.
(b) Starting with the series expansion (10.15.17), prove (10.15.20) by hand.
(c) Plot Bessel functions for $p = 0$, 1, 2 and 3 and discuss the result.

Solution:

(a) The Python code shows that the solution of the Bessel ODE using `dsolve()` is:

`Eq(y(x), C1*besselj(p, K*x) + C2*bessely(p, K*x))`

which represents a linear combination of the Bessel function `besselj(p,x)` and the Neumann function `bessely(p,x)`.

(b) We begin by proving (10.15.20). Starting with the series expansion (10.15.17):

$$x^p J_p(x) = \sum_{n=0}^{\infty} \frac{(-1)^n}{\Gamma(n+1)\Gamma(n+1+p)} \left(\frac{1}{2}\right)^{2n+p} x^{2n+2p}$$

we take the derivative:

$$\frac{d}{dx}\left[x^p J_p(x)\right] = \sum_{n=0}^{\infty} \frac{(-1)^n}{\Gamma(n+1)\Gamma(n+1+p)} \left(\frac{1}{2}\right)^{2n+p} \left[(2n+2p)x^{2n+2p-1}\right]$$

$$= \sum_{n=0}^{\infty} \frac{(-1)^n (n+p)}{\Gamma(n+1)\left[(n+p)(n+p-1)\cdots\right]} \left(\frac{1}{2}\right)^{2n+p-1} x^{2n+2p-1}$$

$$= x^p \sum_{n=0}^{\infty} \frac{(-1)^n (n+p)}{\Gamma(n+1)\left[(n+p)(n+p-1)\cdots\right]} \left(\frac{x}{2}\right)^{2n+p-1}$$

$$= x^p \sum_{n=0}^{\infty} \frac{(-1)^n}{\Gamma(n+1)\Gamma(n+p)} \left(\frac{x}{2}\right)^{2n+p-1}$$

$$= x^p J_{p-1}(x)$$

(c) Next, we use Python to plot the Bessel functions in Figure 10.18. We use the command `jv(n,x)` from the `scipy.special` library. The first argument of `jv` is the order of the Bessel function and the second argument is the variable. Notice that the Bessel functions have multiple zeros and appear similar to underdamped oscillations. As we will see in the next subsection, the fact that Bessel functions have more than one zero will lead to an infinite number of product solutions, when it comes to solving the Laplace equation in cylindrical coordinates.

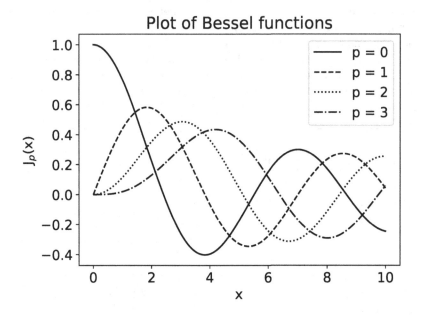

Figure 10.18   Graphical output from Example 10.18 showing Bessel functions $J_p(x)$ of orders $p = 0$, 1, 2 and 3.

```python
from scipy.special import jv

import matplotlib.pyplot as plt

import numpy as np
from sympy import dsolve, Symbol, symbols, Function, diff

y = Function('y')
x, K = symbols('x, K')
p = Symbol('p', positive=True)

# The general form of Bessel ODE
ode = x**2*y(x).diff(x,x)+x*y(x).diff(x)+(K**2*x**2-p**2)*y(x)
soln = dsolve(ode)

print('-'*28,'CODE OUTPUT','-'*29,'\n')
print('The general solution of the Bessel ODE is:\n')
print( soln)

# plot the Bessel functions for p=0,1,2,3
xs = np.linspace(0,10,1000)

plt.plot(xs, jv(0,xs), 'k', label = 'p = 0')
plt.plot(xs, jv(1,xs), 'k--', label = 'p = 1')
plt.plot(xs, jv(2,xs), 'k:', label = 'p = 2')
plt.plot(xs, jv(3,xs), 'k-.', label = 'p = 3')
plt.legend()
plt.xlabel('x')
plt.ylabel(r'J$_p$(x)')
```

```
plt.title('Plot of Bessel functions')
plt.tight_layout()
plt.show()

-------------------------- CODE OUTPUT --------------------------

The general solution of the Bessel ODE is:

Eq(y(x), C1*besselj(p, K*x) + C2*bessely(p, K*x))
```

Example 10.19 shows how to evaluate symbolically and how to demonstrate some of the properties of the Bessel functions using the implementation of these functions in SymPy.

Example 10.19: Properties of the Bessel functions

Use SymPy to demonstrate that following identities for Bessel functions:

$$J_0(x) + J_2(x) = \frac{2}{x} J_1(x) \tag{10.15.25}$$

$$J_{p-1}(x) - J_{p+1}(x) = 2\frac{dJ_p}{dx} \tag{10.15.26}$$

$$J_1(x) = -\frac{dJ_0}{dx} \tag{10.15.27}$$

$$J_0(x) - J_2(x) = 2\frac{dJ_1}{dx} \tag{10.15.28}$$

Solution:

In SymPy the Bessel functions $J_p(x)$ are implemented using the function `besselj(p,x)`, while in the previous example they were implemented with the SciPy functions `jv(p,x)`.
In the Python code, the derivatives are evaluated symbolically with `diff`, and the third equation is shown to hold true by using the logical double equal sign `==`, yielding `True`.

```
from sympy import  besselj, simplify, symbols, diff
x, p = symbols('x, p')

print('-'*28,'CODE OUTPUT','-'*29,'\n')

print('J(0,x)+J(2,x) =',((besselj(0,x)+besselj(2,x)).simplify()))

print("\nJ'(0,x) = ",diff(besselj(0,x)))

print("\nJ'(p,x) = ",diff(besselj(p,x),x))

ident1=(besselj(0,x)-besselj(2,x))==2*diff(besselj(1,x))
print("\nThe identity    J(0,x) - J(2,x) = 2 . J'(1,x) is:  ", ident1)

-------------------------- CODE OUTPUT --------------------------
```

```
J(0,x)+J(2,x) = 2*besselj(1, x)/x

J'(0,x) =  -besselj(1, x)

J'(p,x) =  besselj(p - 1, x)/2 - besselj(p + 1, x)/2

The identity    J(0,x) - J(2,x) = 2 . J'(1,x) is:    True
```

Example 10.20 shows how to find the zeros of Bessel functions.

Example 10.20: Zeros of the Bessel functions

The zeros of Bessel functions must be found using a computer. Unlike zeros of sine and cosine, the zeros of a Bessel function $J_p(x)$ are not evenly spaced. Use Python to show that as x gets large, the difference between two successive zeros approximates π. Use $p = 1$, 2 and 3.

Solution:

In SciPy, the zeros of $J_p(x)$ can be found using the `jn_zeros(p,n)` command, where p is the order of the Bessel function and n is the number of zeros to return. Note that we use the command `diff` from the NumPy library, which finds the successive differences between elements in a list. The result is shown in Figure 10.19.

```
from scipy.special import jn_zeros
import numpy as np

import matplotlib.pyplot as plt

zeros_1 = jn_zeros(1,40)
zeros_2 = jn_zeros(2,40)
zeros_3 = jn_zeros(3,40)

diff_zeros_1 = np.diff(zeros_1)
diff_zeros_2 = np.diff(zeros_2)
diff_zeros_3 = np.diff(zeros_3)

plt.plot(range(0,39),diff_zeros_1, 'ko--', label='p = 1')
plt.plot(range(0,39),diff_zeros_2, 'b^-.', label='p = 2')
plt.plot(range(0,39),diff_zeros_3, 'gd:', label='p = 3')
plt.legend()
plt.ylabel('Difference in zeros')
plt.xlabel('Order of zero')
plt.tight_layout()
plt.show()
```

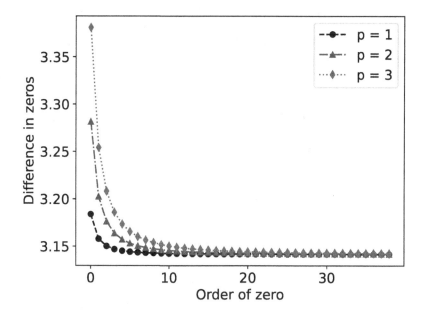

Figure 10.19 Graphical output of Example 10.20 showing the successive differences of the zeros of Bessel functions for $p = 1$, 2 and 3.

10.16 END OF CHAPTER PROBLEMS

1. **Numerical solution of ODE using odeint()** – Numerically solve the differential equation:

$$\frac{df}{dx} = 5\sin(x) - 4e^{-x}$$

for the initial condition $f(0) = 0$. Plot the solution for $x = 0$ to 3, with a step size of 0.1.

2. **Newton's law with time dependent force F(t) using integrate() in SymPy** – Find the velocity $v(t)$ and position $x(t)$ of a particle starting at position x_0 with speed v_0, experiencing the following forces (assume all constants a, b, A, ω are positive). DO this problem by hand and verify the result using the `integrate()` in SymPy,

    a. $F = a + bt$

    b. $F = A\cos(\omega t)$

3. **Newton's law with time dependent force F(t), using dsolve() in SymPy** – A 1.0 kg particle experiences a force $F = -ae^{-\beta t}$, where $a = 0.5\,\text{N}$, $\beta = 0.25\ s^{-1}$, and the minus sign denotes that the force opposes the particle's motion. The mass $m = 1$ kg. If the particle's initial position and velocity are $x_0 = 0$ m and $v_0 = 10$ m/s, find the position and velocity of the particle as a function of time. Do this problem by hand and also using `dsolve()` in SymPy.

4. **Newton's law with linear air resistance F(v) using dsolve()** – The resistive force on a mass m moving along the x-axis is given by $F = m\,dv/dt = -bv + F_0$ where b is a air resistance coefficient and F_0 is a constant force. Find the position $x(t)$ and the speed $v(t)$ of the mass m using three different methods: (a) by hand (b) Using the symbolic capabilities in Python and (c) by numerically integrating the ODE using `dsolve`. Plot the solutions for $m = 1.0\,\mathrm{kg}$, $b = 1\,\mathrm{Ns/m}$, $x_0 = 0\,\mathrm{m}$ and $v_0 = 2\,\mathrm{m/s}$.

5. **Newton's law with speed dependent force F(v)** – An object experiences a force, $F(v) = -c\,v^2$. If the particle starts at the origin with an initial velocity, $v(0) = v_0$, find the particle's velocity and position as a function of time. Do this problem by hand, and use SymPy to verify your results.

6. **Newton's law with linear and quadratic air resistance** -An object is thrown vertically upward with an initial velocity v_0.

   a. Find the amount of time it takes for the object to reach its highest point when there is no air resistance.

   b. How much time is required for the object to reach its maximum height if there is linear air resistance $F = -b\,v$?

   c. How about the case of a quadratic air resistance $F = -b\,v^2$?

7. **Newton's law with conservative force F(x)** – Find the velocity $v(x)$ and position $x(t)$ of a particle starting at rest at the origin experiencing the force $F = a + bx$ with $a, b > 0$.

8. **Newton's law with position and speed dependent force F(x,v)** – A particle experiences a force $F = cvx$, where $c > 0$. If at $t = 0$, the particle is passing through the origin with a velocity $v = v_0$, find the particle's position as a function of time.

9. **Series RC circuit** – Consider a series RC circuit consisting of a resistor R, a capacitor C and a switch S.

   a. The differential equation satisfied by the voltage V, the current I and the charge Q in this circuit is given by the equation

   $$R\frac{dQ}{dt} + \frac{Q}{C} = 0$$

   Assume that the capacitor has a charge $Q = Q_o$ at time $t=0$. Find the charge $Q(t)$ and explain the results physically.

   b. Repeat part (a) when the RC circuit includes a constant voltage source V_0, so that the charge $Q(t)$ follows the ODE:

   $$R\frac{dQ}{dt} + \frac{Q}{C} + V_0 = 0$$

10. **A general ODE** – Find the solution of the differential equation

$$4\frac{dx}{dt} + 3\frac{d^2x}{dt^2} + 25x = 0$$

What does this solution represent physically? Plot the solution with initial conditions $x(0) = 0$ m and $x'(0) = 1$ m/s.

11. **A series RLC circuit** – An RLC circuit has the values R=10 Ohms, L=0.5 Henry and C=0.01 Farad. Find the current $I(t) = dQ/dt$ in the system when $Q(0) = Q_0 = 1\,C$ and $I(0) = 0$. The differential equation in this case is

$$R\frac{dQ}{dt} + L\frac{d^2Q}{dt^2} + \frac{Q}{C} = 0$$

12. **System of coupled differential equations** – Solve the coupled differential equations, both numerically using SciPy and symbolically using SymPy:

$$\frac{dx}{dt} = 0.6x - 1.2xy,$$
$$\frac{dy}{dt} = xy - y,$$

using the initial conditions $x(0) = 40$ and $y(0) = 7$. Plot x as a function of time, and y as a function of time.

13. **Numerical solution of the pendulum** – A pendulum has length $L = 1$ m, mass $m = 0.1\,kg$, and the air resistance is given by $F = -m\,b\,(d\theta/dt)$ where $\theta(t)$ is the angle of the mass $m = 0.1$ kg, and $b = 0.2\,s^{-1}$ is a coefficient representing the air resistance. The pendulum is released from rest at a large angle of 20 degrees from equilibrium; in this case the small angle approximations are not valid.

  a. Evaluate and plot the angle $\theta(t)$ for the motion, by using SciPy.

  b. Calculate and plot the kinetic energy K as a function of time t.

14. **Externally driven oscillator** – The position $x(t)$ of a forced harmonic oscillator satisfies the equation

$$d^2x/dt^2 - 6dx/dt + 8x = 10\cos(2t)$$

At time $t = 0$ the particle is at the origin and at rest.

  a. Find the transient and steady state solutions for the position and velocity of the oscillator as functions of time t. Do this by hand, and also by using Python.

  b. Find and plot the kinetic energy $K = m\,v^2/2$ as a function of time t using Python, and plot this quantity over many *periods of oscillation*.

15. **System of coupled ODEs** – The position $x(t), y(t)$ of a particle on the xy-plane satisfies the equations
$$d^2x/dt^2 = -8y$$
$$d^2y/dt^2 = -2x$$

At time $t = 0$, the particle is at the origin and at rest. Do this problem by hand, and also using SymPy. The particle is initially at the point $(1, 1)$ and moves with a speed of 2 m/s along the positive x-axis.

  a. Find the position as a function of time t

  b. Find the period of the motion.

16. **Charge and power in RLC circuit** – An electrical circuit consists of a 10 kΩ resistor, a 80 mH inductance, a 10 μF capacitor, and an AC voltage of the form $V = 100\cos(10t)$ (where V is measured in volts) connected in series.

a. Find the current $I(t)$ in the circuit at any time t. Plot the current $I(t)$ and the charge $q(t)$, and identify the steady state part and the transient part of the graph.

b. Find the average power $P = VI$ dissipated in the resistor R, between $t = 0.2\ s$ and $t = 0.4\ s$. In general, the average power dissipated between two times t_1 and t_2 is given by

$$< P >= \frac{1}{t_2 - t_1} \int_{t_1}^{t_2} V\,I\mathrm{dt} = \frac{1}{t_2 - t_1} \int_{t_1}^{t_2} I^2\,R\mathrm{dt}$$

17. **System of two masses and two springs** – Consider two equal masses m attached to one wall and to each other with springs of spring constant k, as shown in Figure 10.20. Find and plot $x_1(1)$ and $x_2(t)$ with the initial conditions of positions $x_1(0) = 1$ and $x_2(0) = 0$, and initial speeds $v_1(0) = 1$ and $v_2(0) = 0$ (all quantities in SI units). Solve this problem by hand and also using SymPy.

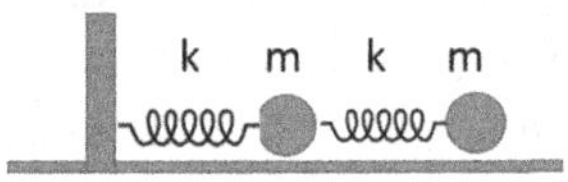

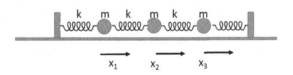

Figure 10.20 A system of two masses and two springs attached to a single wall

18. **System of three masses connected with springs** – Three equal masses m are connected to each other with identical springs of constant k, as shown in Figure 10.21. The masses are also attached to the walls. The spring constants k are the same. Find and plot $x_1(1)$, $x_2(t)$, $x_3(t)$ with the initial conditions of positions $x_1(0) = 1$, $x_2(0) = 0$, $x_3(0) = -1$ and initial speeds $v_1(0) = 1$, $v_2(0) = 0$, $v_3(0) = 0$ (all quantities in SI units).

Figure 10.21 A system of three masses and four springs attached to two walls

19. **Series expression for Legendre polynomials using Sum() and summation() in SymPy** – Use SymPy to show that the following series expression produces the Legendre polynomials $P_l(x)$:

$$P_l(x) = \frac{1}{2^n} \sum_{k=0}^{l} \binom{l}{k}^2 (x-1)^{l-k} (x+1)^k$$

You may need to use the symbolic commands Sum(), summation() and doit() in SymPy.

20. **The logistic equation** – The logistic equation

$$dx/dt = r\,x(1 - x/k)$$

is used to model population sizes in various branches of science and is a one-dimensional system. Note that r is a constant called the *linear growth rate* and k is a constant called the *carrying capacity*. Solve this differential equations using SymPy and by hand, and plot the result for two values of the parameter $r = 0.3$, -0.3 and $k = 10$, and the initial condition $x(0) = 1$. Discuss the physical meaning of the results.

21. **Coupled RLC circuits with external voltage** – Solve numerically the following the following system of differential equations for the currents $I_1(t)$ and $I_2(t)$ in two coupled RLC circuits. Plot the solutions $I_1(t)$ and $I_2(t)$ for the numerical values $R = 100\ \Omega$, $L = 20$ H, $C = 0.1$ F, $E = 10$ V. The initial conditions are $I_1(0) = 0$ A and $I_2(0) = 2$ A.

$$L\frac{dI_1}{dt} + R\left(I_1 - I_2\right) - E = 0 \tag{10.16.1}$$

$$R\frac{dI_1}{dt} - R\frac{dI_2}{dt} + \frac{I_2}{C} = 0 \tag{10.16.2}$$

22. **The Euler-Cauchy equation** – An important class of ODEs encountered in Physics is the Euler-Cauchy equation:

$$x^2\frac{d^2y}{dx^2} + a\,x\,\frac{dy}{dx} + b\,y = 0 \tag{10.16.3}$$

a. By substituting $y = x^m$, show that one obtains the equation $m^2 + (a-1)m + b = 0$.

b. Verify by hand and using SymPy that the linear combination $y = c_1 x^{m_1} + d x^{m_2}$ is a solution of the Euler-Cauchy equation, where m_1 and m_2 are the roots of the equation obtained in (a).

11 Partial Differential Equations

In this chapter we will study three partial differential equations (PDEs), which are an integral part of the education of all physicists and engineers. Specifically, we will learn how to use SymPy to obtain the general solutions of the heat equation, the Laplace equation, and the wave equation. We chose these three equations because of their importance in physics. However, the methods we demonstrate can be applied to any linear homogeneous PDE.

Once the general solutions are obtained using SymPy, the exact solutions for each physical situation are determined based on the boundary conditions and the initial conditions. We will study how Fourier analysis can help us obtain the analytical solutions to the PDEs as an infinite series and will learn how to plot these solutions by summing the corresponding Fourier series.

11.1 SEPARATION OF VARIABLES

Separation of variables is a powerful method for solving ordinary and partial differential equations, in which algebra allows one to rewrite an equation so that each of two variables occurs on a different side of the equation. In Chapter 10, we encountered separation of variables as means of solving first-order ODEs. However, when solving PDEs, separation of variables is used to separate the PDE into several ODEs, which can then be solved with the usual techniques of solving ODEs. For example, we will solve the ODEs resulting from the separation of variables, using the `dsolve()` function in SymPy.

We will look at two types of physical problems which are encountered often in physics and engineering. In the first type of problem, we are given specific values that the solution of the PDE must satisfy. For example, we may be given how the temperature varies on the surfaces of a cube and will be asked to find the solution of the PDE that gives the temperature inside the cube and also satisfies the conditions on the surfaces. This type of problem is called a *Dirichlet boundary condition problem*.

In the second type of physical situation, we are not given the values of the solution at the boundary, but we are given instead the values of the normal derivatives of the solution. These types of problems are called *Neumann boundary condition* problems.

In addition to the boundary conditions, the solution must also satisfy the initial conditions of the system, which can be, for example, the behavior of the system at time $t = 0$.

In most of the examples of this chapter we will use the `pde_separate()` function in SymPy, which allows us to separate symbolically the PDE into several ODEs. Subsequently, we will use `dsolve()` to obtain the general solution of the ODEs, and combine these solutions to obtain the general solution of the PDE.

Once the general solutions are obtained, we will apply the boundary conditions and the initial conditions of the system. Applying the boundary and initial conditions will usually determine the integration constants in the general solution, and thus will produce the specific solution to the physical situation. A valuable tool in obtaining the *complete* solution of the PDE is Fourier analysis, and the solution of the PDE will be expressed often as an infinite series.

If any two functions are solutions to a linear homogeneous PDE, any linear combination of them is also a solution. This is known as the *principle of superposition* and is of fundamental importance in the theory of ODEs and PDEs. This principle is used in practice to obtain the solutions to a complex problem, by constructing linear combinations of simple solutions to the problem.

11.2 THE HEAT EQUATION

The heat equation

$$\nabla^2 T = \frac{1}{a^2} \frac{\partial T}{\partial t} \tag{11.2.1}$$

is a rather famous partial differential equation, whose theory and solutions were developed by Joseph Fourier (1768 - 1830), in order to explain how heat diffuses through materials. The function $T = T(x, y, z, t)$ gives the temperature at a point in space and time. The parameter a^2 is called the *thermal diffusivity* and is determined by the thermal properties of the material whose temperature we wish to model with (11.2.1). The thermal diffusivity may or may not be constant. However, in this chapter, we will focus on the case where a^2 is constant. The heat equation appears in several varieties, for example in the description of diffusion and random walks of particles, like the Brownian motion. When studying diffusion, a^2 is called the *diffusion coefficient* and, if constant, the *diffusion equation* is mathematically the same as (11.2.1). In addition, the Schrodinger equation for a free particle in quantum mechanics, is mathematically very similar to the heat equation.

As a first example of applying SymPy to solve PDEs, we discuss the propagation of heat across a rod, as a function of time. Let us consider a uniform thin rod with length L, and with its axis along the x-direction. We assume that the rod is thermally insulated along its length, except at the two end points $x = 0$ and $x = L$.

We further assume that the initial condition for the temperature along the rod at time $t = 0$ is given by:

$$T(x, 0) = \frac{25\,x}{L} \qquad \text{for } 0 \le x \le L \tag{11.2.2}$$

Hence, the two ends of the rod are *initially* at 0°C and 25°C, respectively, as shown in Figure 11.1. In addition, the temperature at $t = 0$ varies linearly with the distance across the rod. We want to describe mathematically the heat flow during the cooling process across the rod.

The temperature $T(x, t)$ of the rod can be described as a function of the position x and of time t. Under certain physical assumptions, $T(x, t)$ will satisfy the heat equation in one dimension:

$$\frac{\partial^2 T}{\partial x^2} = \frac{1}{a^2} \frac{\partial T}{\partial t} \tag{11.2.3}$$

Physically the right-hand side of (11.2.3) is proportional to the change of temperature with time $\partial T / \partial t$, and the left-hand side of this equation describes the heat flow process until the rod reaches equilibrium at its final state.

11.2.1 THE PRODUCT SOLUTION TO THE ONE-DIMENSIONAL HEAT EQUATION

We separate the variables using a solution in the form:

$$T(x, t) = X(x)F(t) \tag{11.2.4}$$

Substituting into (11.2.3) and dividing by $T(x, t)$, we obtain:

$$\frac{1}{X(x)} \frac{\partial^2 X}{\partial x^2} = \frac{1}{a^2} \frac{\partial F}{\partial t} \frac{1}{F(t)} \tag{11.2.5}$$

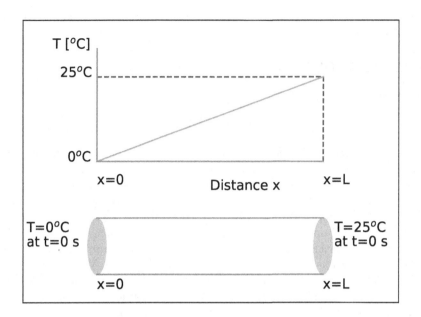

Figure 11.1 The two ends of a rod are initially at different temperatures 0°C and 25°C. The temperature across the rod increases linearly with distance L.

so that the two variables x and t have been separated. For this equation to be satisfied, each side of the equation can be equal to a negative constant $-k^2$, so that:

$$\frac{1}{X(x)} \frac{d^2 X}{dx^2} = -k^2 \tag{11.2.6}$$

$$\frac{1}{a^2} \frac{\partial F}{\partial t} \frac{1}{F(t)} = -k^2 \tag{11.2.7}$$

We have chosen in this example the constant to be negative $-k^2$, so that as time t increases, the temperature is decreasing (i.e. $dT/dt < 0$). Integrating (11.2.7), we find:

$$F(t) = C \exp\left(-a^2 k^2 t\right) \tag{11.2.8}$$

We also recognize that (11.2.6) is the equation for a simple harmonic motion, so that we can write the solution as:

$$X(x) = C_1 \sin(k\,x) + C_2 \cos(k\,x) \tag{11.2.9}$$

By combining (11.2.9) and (11.2.8), we obtain the *product solution* of the diffusion equation:

$$T(x,t) = \{C_1 \sin(k\,x) + C_2 \cos(k\,x)\} \exp\left(-a^2 k^2 t\right) \tag{11.2.10}$$

where the constant of integration C in (11.2.8) was absorbed in the two arbitrary constants C_1, C_2.

Example 11.3 demonstrates how to obtain the general solution of the diffusion equation using SymPy.

Example 11.1: The general solution of the heat flow equation in 1D

Write a Python code to obtain the product solution of (11.2.3).

Solution:

We use pde_separate() function with the option strategy='mul', which tells SymPy to use a solution of (11.2.3) in the form of a product. The function dsolve() in SymPy is used to obtain the solution $F(t)$ and $X(x)$.

In order to absorb the extra constant C appearing in (11.2.8), we substitute $C = 1$ in the code line solF.subs(C1,1). Note that SymPy used C1 to represent the integration constant C in (11.2.8). On the last line of the code we verify that the answer from SymPy is indeed a solution of the heat flow equation, by evaluating the partial derivatives of the solution $T(x,t) = X(x)F(t)$ and substituting on both sides of (11.2.3).

```
print('-'*28,'CODE OUTPUT','-'*29,'\n')

from sympy import  Eq, Function, pde_separate, diff,\
Derivative as D, dsolve, symbols

# Define symbols including integration constants C1, C2
a, t, k = symbols('a, t, k',positive=True, real=True)
x, y, C1, C2 =symbols('x, y, C1, C2')

# define function T(t), X(x) and F(x,t)
T, X, F = map(Function, 'TXF')

# Solve the heat flow equation using sparataion of variables
eq = Eq(D(T(x, t), x, 2), (1/a**2)*D(T(x, t), t))
soln = pde_separate(eq, T(x, t), [X(x), F(t)], strategy='mul')

print('The separated form of the Laplace equations is:\n')
print(soln)

solX = dsolve(soln[0]+k**2, X(x))

print('\nThe solution X(x) =', solX.rhs)

solF = dsolve(soln[1]+k**2, F(t))
# Absorb the extra integration constant C, by setting C=1
solF = solF.subs(C1,1)
print('The solution F(t) =', solF.rhs)

sol = solX.rhs*solF.rhs
print('\nThe general solution is:')
print('      T(x,t) =', sol)

# Verify that symbolic solution satisfies heat flow equation
print('\nIs this T(x,t) a solution of the diffusion equation? ',\
      diff(sol,x,2) == (1/a**2)*diff(sol,t))
```

------------------------ CODE OUTPUT ----------------------------

```
The separated form of the Laplace equations is:

[Derivative(X(x), (x, 2))/X(x), Derivative(F(t), t)/(a**2*F(t))]

The solution X(x) = C1*sin(k*x) + C2*cos(k*x)
The solution F(t) = exp(-a**2*k**2*t)

The general solution is:
    T(x,t) = (C1*sin(k*x) + C2*cos(k*x))*exp(-a**2*k**2*t)

Is this T(x,t) a solution of the diffusion equation?  True
```

In the next subsection we apply the boundary and initial conditions of the system to the product solution, in order to describe quantitatively how the temperature of the rod changes with distance x and the elapsed time t.

11.2.2   THE PARTICULAR SOLUTION OF THE HEAT EQUATION: THE COOLING PROCESS IN A UNIFORM ROD

Once the product form of the solution of the diffusion equation is found as in the previous subsection, then we can turn our attention to determining the constants C_1, C_2. This is usually done using the initial conditions at $t = 0$, and also by applying the boundary conditions given for the problem.

The boundary conditions in this example are the temperatures at the two ends of the rod for *any time* $t > 0$. For example, suppose that at time $t > 0$ we keep the temperature at the two ends $x = 0$ and $x = L$ at 0°C, by using an ice bath. We require that the solution $T(x, t)$ must satisfy the *boundary conditions*:

$$T(0, t) = T(L, t) = 0 \tag{11.2.11}$$

Substituting these conditions in (11.2.10) we obtain:

$$T(0, t) = C_2 = 0 \tag{11.2.12}$$
$$T(L, t) = C_1 \sin(k L) = 0 \tag{11.2.13}$$

The values of the parameter k can be found using (11.2.13):

$$k = \frac{n \pi}{L} \qquad n = \text{integer} \tag{11.2.14}$$

The product solution (11.2.10) becomes:

$$T(x, t) = C_1 \sin\left(\frac{n \pi}{L} x\right) \exp\left(-a^2 \frac{n^2 \pi^2}{L^2} t\right) \tag{11.2.15}$$

We now see that there are an infinite number of solutions to the problem, corresponding to different integers n, and that all of these solutions satisfy the boundary conditions (11.2.11). According to the superposition principle, the most general solution then must be a linear combination of this type:

$$T(x, t) = \sum_{n=1}^{\infty} B_n \sin\left(\frac{n \pi}{L} x\right) \exp\left(-a^2 \frac{n^2 \pi^2}{L^2} t\right) \tag{11.2.16}$$

where the constant coefficients B_n are to be determined next, by using the initial conditions of the problem; as stated above, we assume that the initial condition for the temperature along the rod at time $t = 0$ is given by:

$$T(x, 0) = \frac{25\,x}{L} \qquad \text{for } 0 \le x \le L \tag{11.2.17}$$

This equation tells us that the two ends of the rod are *initially* at 0°C and 25°C, respectively. Physically we expect that the temperature along the rod will start decreasing until the temperature of the whole rod reaches 0°C.

Substituting this initial condition in (11.2.16) we find:

$$\frac{25\,x}{L} = \sum_{n=1}^{\infty} B_n \sin\left(\frac{n\,\pi}{L}\,x\right) \tag{11.2.18}$$

which tells us that the constant coefficients B_n are the Fourier *sine* coefficients of the function $25\,x/L$. The values of the B_n can be found using Fourier's trick, which was previously discussed in Chapter 3 of this book:

$$B_n = \frac{2}{L} \int_0^L F(x, 0) \sin\left(\frac{n\,\pi}{L}\,x\right) dx = \frac{2}{L} \int_0^L \frac{25\,x}{L} \sin\left(\frac{n\,\pi}{L}\,x\right) dx \tag{11.2.19}$$

$$B_n = -50\,\frac{(-1)^{n-1}}{n\,\pi} \tag{11.2.20}$$

and by substituting in (11.2.16), we find the particular solution, which describes how the temperature varies with time t and the distance x across the rod during the experiment:

$$T(x, t) = \frac{50}{\pi} \sum_{n=1}^{\infty} \frac{(1)^{n-1}}{n} \sin\left(\frac{n\,\pi}{L}\,x\right) \exp\left(-a^2\,\frac{n^2\,\pi^2}{L^2}\,t\right) \tag{11.2.21}$$

Example 11.2 demonstrates how to sum and plot the series shown in (11.2.21), for different times t.

===

Example 11.2: Plotting the particular solution of the heat equation

(a) Use SymPy to verify (11.2.20) for the sine terms in the Fourier series.
(b) Plot (11.2.21), for different times t and from $x = 0$ to $x = L$. Discuss the physical meaning of the plots.

Solution:

The code evaluates the Fourier coefficients using the `integrate()` function in SymPy, and the resulting symbolic expression B_n is turned into a function of the integer value n using the `lambdify()` function. A function `T(t)` is defined and is used to evaluate the temperature $T(x, t)$ at all positions x between $x = 0$ and $x = L$. The parameter `numTerms` specifies how many terms of the series should be summed, and the sums are plotted for a series of times `times=[0,0.5,1,2,5]` using a simple loop.

The parameter `suma` stores the sum of the terms in the series for all x-values and is initially set to a zero array using `suma = [0]*len(x)`.

The graphical output is shown in Figure 11.2. At $t = 0$ we see that the temperature $T(x, 0)$ is a straight line representing the initial condition of the problem $T(x, 0) = 25\,x/L$, and the temperature at the two ends of the rod are 0°C and 25°C. As time t increases, we are keeping the two ends of the rod at 0°C, so the temperature curves decrease at all positions x along the rod. For large times $t = 5$ s, the temperatures across the whole rod start approaching 0°C, as expected.

Also notice the jagged line near $x = L$, this is the well known Gibbs phenomenon which occurs when we are evaluating a Fourier series.

```python
from sympy import symbols, integrate, sin, pi, lambdify
import numpy as np
import matplotlib.pyplot as plt


x = symbols('x')        # define symbols
n = symbols('n', integer = True, positive=True)


print('-'*28,'CODE OUTPUT','-'*29,'\n')


# function to evaluate temperature T(x,t) by summing the series T(x,t)
def T(t):
    suma = [0]*len(x)
    for i in range(1,numTerms):
        suma = suma+bn(i)*np.sin(i*np.pi*x/L)*np.exp(-(a*i*np.pi/L)**2*t)
    plt.plot(x,suma)


L = 5      # length of the rod
a = 1      # coefficient a in the diffusion equation (SI units)

# Find the sine Fourier coefficients of the functions 25*x/L
Bn = (2/L)* integrate(25*x/L*sin(n*pi*x/L),(x,0,L))
print("Fourier coefficients bn =",Bn)


# turn Bn into a function of n
bn = lambdify(n, Bn)


# Number of terms to be summed in the truncated Fourier series
numTerms = 200


# positions along the x-axis
x = np.linspace(0,L,200)


# call function to evaluate F(x,t) for different times t
times = [0,1,2,5]
for t in range(len(times)):
    T(times[t])

plt.text(3,21,'Time t=0 s')
plt.text(3.2,12,'t=1 s')
plt.text(3.2,8,'t=2 s')
plt.text(3.2,3,'t=5 s')
plt.title('Heat flow equation: solution for different times t')
plt.xlabel('Position x [m]')
plt.ylabel(r'Temperature [$^{o}$C]')
plt.tight_layout()
```

```
plt.show()

---------------------------- CODE OUTPUT ----------------------------

Fourier coefficients bn = -50.0*(-1)**n/(pi*n)
```

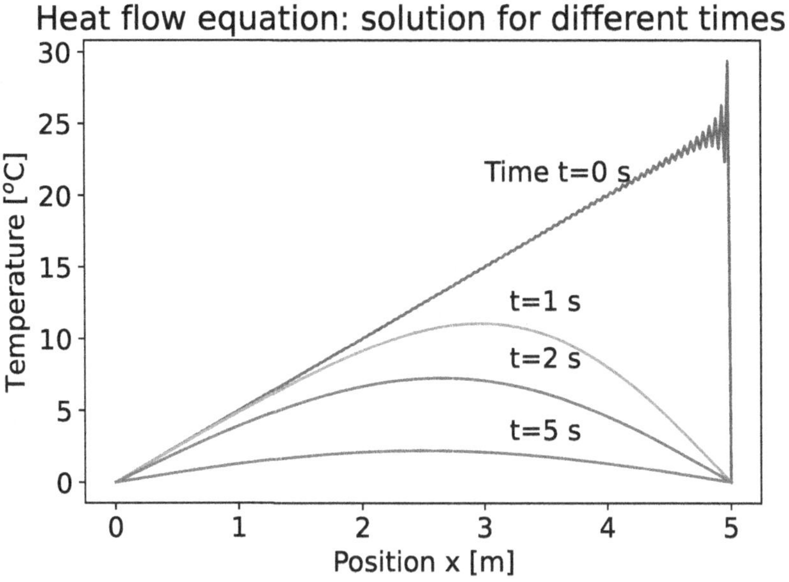

Figure 11.2 Graphical output from Example 11.2 showing a plot of the series shown in (11.2.21), for different times t.

11.3 THE LAPLACE EQUATION

The Laplace equation is a homogeneous, second-order, linear PDE with constant coefficients:

$$\nabla^2 u = 0 \tag{11.3.1}$$

where the scalar function $u(x, y, z)$ depends on the position variables (x, y, z), and ∇^2 is the Laplace operator or *Laplacian*. The Laplace equation finds applications in electrostatics, gravitation, and fluid dynamics. In general, the Laplace equation describes situations of equilibrium. For example, the electric potential $V(x, y, z)$ in spaces where there is no charge, satisfies the Laplace equation $\nabla^2 V = 0$.

If the right-hand side of (11.3.1) is equal to a specific function $f(x, y, z)$, this becomes the Poisson equation. For example, in the presence of a charge density ρ, we obtain Poisson's equation for electrostatics:

$$\nabla^2 V = -\frac{\rho}{\varepsilon_0} \tag{11.3.2}$$

Similarly, the gravitational field potential $\Phi(x, y, z)$ in empty space satisfies the Laplace equation:

$$\nabla^2 \Phi = 0 \tag{11.3.3}$$

Mathematical Methods using Python: Applications in Physics and Engineering

In this section we will obtain the general solution of the Laplace equation in two dimensions using Cartesian coordinates:

$$\nabla^2 V = \frac{\partial^2 V}{\partial x^2} + \frac{\partial^2 V}{\partial y^2} = 0 \tag{11.3.4}$$

Again we suppose that the solution $V(x,y)$ can be written in the form

$$V(x,y) = X(x)Y(y) \tag{11.3.5}$$

Substituting and dividing by $V(x,y)$ we obtain:

$$\frac{1}{X(x)} \frac{\partial^2 X}{\partial x^2} + \frac{1}{Y(y)} \frac{\partial^2 Y}{\partial y^2} = 0 \tag{11.3.6}$$

In order for both terms to sum to zero, they must each equal the same constant but with opposite signs. The choice of assigning the negative constant to either the X or Y term is often physically motivated. For example, the boundary conditions may lead one to expect the electric potential $V(x,y)$ to decay in one direction but not the other. In this case, we will make the following assignment (which will be motivated by the problem in the next section):

$$\frac{1}{X(x)} \frac{d^2 X}{dx^2} = k^2 \tag{11.3.7}$$

$$\frac{1}{Y(y)} \frac{d^2 Y}{dy^2} = -k^2 \tag{11.3.8}$$

These equations have the solutions:

$$X(x) = C_1 \exp(k\,x) + C_2 \exp(-k\,x) \tag{11.3.9}$$

$$Y(y) = C_3 \sin(k\,y) + C_4 \cos(k\,y) \tag{11.3.10}$$

The product solution is:

$$V(x,y) = \{C_1 \exp(k\,x) + C_2 \exp(-k\,x)\} \cdot \{C_3 \sin(k\,y) + C_4 \cos(k\,y)\} \tag{11.3.11}$$

where C_1, C_2, C_3, C_4 are arbitrary constants.

Example 11.3 demonstrates how to obtain the general solution of the Laplace equation.

Example 11.3: The product solution of Laplace equation in 2D

Write a Python code to obtain the product solution (11.3.11) of the Laplace equation in two dimensions.

Solution:

We use again the `pde_separate()` function, and the structure of this code is very similar to the code for the heat flow equation in Example 11.3. The main mathematical difference here is that the general solution for the Laplace contains both exponential and trigonometric functions. We use .subs() to replace the second set of integration constants C_1,C_2 with the symbols C_3, C_4 for clarity. In the last line of the code we verify that the product $X(x)Y(y)$ is indeed a solution of the Laplace equation, by evaluating the derivatives and substituting in both sides of (11.3.4).

```
print('-'*28,'CODE OUTPUT','-'*29,'\n')

from sympy import  Eq, Function, pde_separate, diff,\
Derivative as D, dsolve, symbols

# define symbols and functions
k = symbols('k', positive=True, real=True)
u, X, Y = map(Function, 'uXY')
x, y, C1, C2, C3, C4 = symbols('x, y, C1, C2, C3, C4')

# The Laplace PDE
eq = Eq(D(u(x, y), x, 2), -D(u(x, y), y, 2))

# use pde_separate() to separate the x and y variables
soln = pde_separate(eq, u(x, y), [X(x), Y(y)], strategy='mul')
print('The separated form of the Laplace equation is:')
print(soln)

# Solve the ODE for X(x) using dsolve()
solX = dsolve(soln[0]-k**2, X(x))
print('\nThe solution X(x) =', solX.rhs)

# Solve the ODE for Y(y) using dsolve()
solY = dsolve(soln[1]-k**2, Y(y))

# Replace constants C1, C2 in Y(y) with C3, C4 to avoid confusion
solY = solY.subs({C1:C3,C2:C4})

# print the function Y(y)
print('\nThe solution Y(y) =', solY.rhs)

# print combined solution X(x)*Y(y)
sol = solX.rhs*solY.rhs
print('\nThe general solution is')
print('u(x,y) =', sol)

print('\nIs this u(x,y) function a solution of the Laplace equation? ',\
diff(sol,x,2) == -diff(sol,y,2))

-------------------------- CODE OUTPUT ---------------------------

The separated form of the Laplace equation is:
[Derivative(X(x), (x, 2))/X(x), -Derivative(Y(y), (y, 2))/Y(y)]

The solution X(x) = C1*exp(-k*x) + C2*exp(k*x)

The solution Y(y) = C3*sin(k*y) + C4*cos(k*y)

The general solution is
u(x,y) = (C1*exp(-k*x) + C2*exp(k*x))*(C3*sin(k*y) + C4*cos(k*y))

Is this u(x,y) function a solution of the Laplace equation?  True
```

Once we obtained the general solution of the Laplace equation as above, we apply once more the boundary conditions in the next section, and seek the specific solution to an electrostatics problem.

11.4   APPLICATION OF THE LAPLACE EQUATION IN ELECTROSTATICS

We now look at a classic application of the Laplace equation to electrostatics, found in most undergraduate textbooks.

Consider a thin plate on the xy-plane, with the two sides at $y = 0$ and $y = a$ consisting of grounded wires as shown in Figure 11.3. The third side at $x = 0$ is kept at a constant electric potential $V = V_0$. We wish to find the potential $V(x, y)$ at all points on the xy-plane.

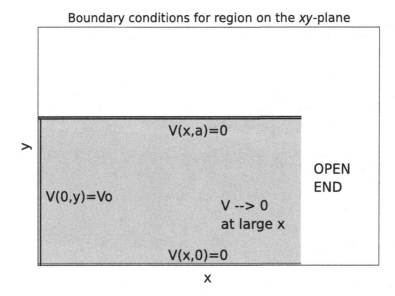

Figure 11.3 The boundary conditions for a thin plate on the xy-plane, with the two sides at $y = 0$ and $y = a$ consisting of grounded wires, and the third side at $x = 0$ is kept at a constant electric potential.

The boundary conditions in this problem are given as:

$$V(x, 0) = 0 \tag{11.4.1}$$

$$V(x, a) = 0 \tag{11.4.2}$$

$$V(0, y) = V_0 \tag{11.4.3}$$

$$V \to 0 \text{ as } x \to \infty \tag{11.4.4}$$

The mathematical condition in (11.4.4) makes physical sense, since we expect the potential far away along the x-axis to be zero. Note that conditions involving limits are not always explicitly stated in a problem, but are often inferred from the situation. Be on the lookout for them! Applying (11.4.4) to (11.3.11) yields $C_1 = 0$, because the potential must decay at large x-values. Hence,

$$V(x, y) = C_2 \exp\left(-k\, x\right) \left\{ C_3 \sin\left(k\, y\right) + C_4 \cos\left(k\, y\right) \right\} \tag{11.4.5}$$

Substituting the boundary conditions (11.4.1), (11.4.2) in (11.4.5), we obtain $V(x, 0) = C_4 \cos(k\,y) = 0$ and $V(x, a) = C_3 \sin(k\,L) = 0$. These conditions result in $C_4 = 0$ and in the values of the parameter k:

$$k = \frac{n\,\pi}{L} \qquad n = \text{integer} \tag{11.4.6}$$

so that the solution of the Laplace equation becomes:

$$V(x, y) = C \exp\left(-\frac{n\,\pi}{L}x\right) \sin\left(-\frac{n\,\pi}{L}y\right) \tag{11.4.7}$$

There are an infinite numbers of solutions, corresponding to different integers n. The general solution then must be a linear combination them:

$$V(x, y) = \sum_{n=1}^{\infty} B_n \exp\left(-\frac{n\,\pi}{L}x\right) \sin\left(\frac{n\,\pi}{L}y\right) \tag{11.4.8}$$

where the constant coefficients B_n are to be determined next, by using the boundary condition (11.4.3) of the problem. Substituting this boundary condition (11.4.3) in (11.4.8) we find:

$$V(0, y) = V_0 = \sum_{n=1}^{\infty} B_n \sin\left(\frac{n\,\pi}{L}y\right) \tag{11.4.9}$$

The constant coefficients B_n are the Fourier sine coefficients of the constant function V_0, and they can be found using Fourier's trick:

$$B_n = \frac{2}{L}\int_0^L V(0, y) \sin\left(\frac{n\,\pi}{L}y\right) dy = \frac{2}{L}\int_0^L V_0 \sin\left(\frac{n\,\pi}{L}y\right) dy \tag{11.4.10}$$

$$B_n = \frac{2\,V_0}{n\,\pi}(1 - (-1)^n) \qquad \text{where } n = \text{integer } = 1, 2, 3.. \tag{11.4.11}$$

or equivalently $B_n = 4\,V_0/(n\,\pi)$ with $n = \text{odd} = 1, 3, 5...$ By substituting in (11.4.8), we find the particular solution:

$$V(x, y) = \frac{2V_0}{\pi} \sum_{n=1}^{\infty} \frac{1 - (-1)^n}{n} \exp\left(-\frac{n\,\pi}{L}x\right) \sin\left(\frac{n\,\pi}{L}y\right) \tag{11.4.12}$$

Example 11.4 demonstrates how to create a wireframe plot of the electric potential $V(x, y)$ from (11.4.12) and how to plot the equipotential curves, $V(x, y) = c$.

Example 11.4:   Plotting the solution of the Laplace equation

Write a Python code to plot (11.4.12) and its contours for $V = 0.2$, 0.4, 0.6, and 0.8.

Solution:
The code here uses ax1.plot_wireframe(X, Y, Z) in MatPlotLib to produce the three dimensional plot shown in Figure 11.4. The command ax2.contour() is used to create the contour plot. The specific values of the contours are specified using the keyword argument levels.
    A for loop is used to sum the 200 terms of the Fourier terms for $V(x, y)$.

```python
from sympy import symbols, integrate, sin, pi, lambdify, simplify
import numpy as np
import matplotlib.pyplot as plt

print('-'*28,'CODE OUTPUT','-'*29,'\n')

x = symbols('x ')        # define symbols
n = symbols('n', integer = True, positive=True)

Vo = 1    # potential Vo=V(0,y)

L = 5     # length along the x-axis (SI units)

# Find the sine Fourier coefficients of the functions 25*x/L
Bn = (2/L)* integrate(Vo*sin(n*pi*x/L),(x,0,L))
print("Fourier coefficients bn=",simplify(Bn))

# turn Bn into a function of n
bn = lambdify(n, Bn)

# Number of terms to be summed in the truncated Fourier series
numTerms = 200

#Define the function V(x,y)
def V(x, y):
    suma = [0]*len(y)
    for i in range(1,numTerms):
        suma = suma+bn(i)*np.exp(-i*np.pi*x/L)*np.sin(i*np.pi*y/L)
    return suma

# positions along the x-axis and the y-axis
x = np.linspace(0,L,50)
y = np.linspace(0,L,50)
X, Y = np.meshgrid(x, y)
Z = V(X,Y)

# define the objects fig and axes for 3D plotting
# plot 3D surface as wireframe

fig = plt.figure()
ax1 = fig.add_subplot(121,projection='3d')
ax2 = fig.add_subplot(122)

ax1.plot_wireframe(X, Y, Z, rstride=3, cstride=3)
ax1.set_xlabel('x')
ax1.set_ylabel('y')
ax1.set_zlabel('V(x,y)')
ax1.set_title('Series Solution')

#plot the contours
cp = ax2.contour(X,Y,Z, levels=[ 0.2,0.4,0.6,0.8], colors='k')
ax2.clabel(cp, inline=True, fontsize=8, colors = 'k')
ax2.set_title('Contour plot')
ax2.set_ylabel('y')
ax2.set_xlabel('x')
```

```
#ax2.set_xbound(-0.5,6)

plt.tight_layout(pad = 6.0)
plt.show()
```

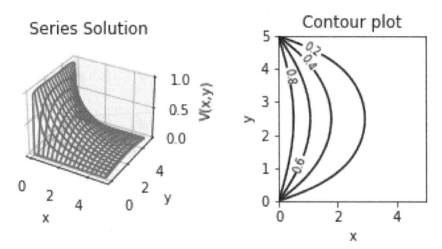

Figure 11.4 Graphical output from Example 11.4 showing a three-dimensional plot of the electric potential $V(x,y)$ and its contours.

11.5 THE WAVE EQUATION

The wave equation is a second-order linear partial differential equation which can describe different types of waves propagating in different types of media. For example, the wave equation is used in applied acoustics, in electromagnetism, continuum mechanics, quantum mechanics, plasma physics, general relativity, geophysics and fluid dynamics. The solutions of the wave equation are in general traveling waves, however linear combinations of traveling waves can yield standing waves.

In three dimensions, the wave equation is:

$$\nabla^2 u = \frac{1}{v^2} \frac{\partial^2 u}{\partial t^2} \tag{11.5.1}$$

where $u(x,y,z,t)$ is a displacement from equilibrium (such as height of water or the pressure of a gas above or below an equilibrium state) and v is a positive real constant.

In physics, one dimensional waves (such as a wave on a string) are common. The wave equation in one dimension is:

$$\frac{\partial^2 u}{\partial x^2} = \frac{1}{v^2} \frac{\partial^2 u}{\partial t^2} \tag{11.5.2}$$

where $u(x,t)$ is the displacement at time t and at the position x. Here v is the constant speed of the wave. We again assume the solution can be written in the form

$$u(x,t) = X(x)T(t) \tag{11.5.3}$$

Substituting and dividing by u we obtain:

$$\frac{1}{X(x)}\frac{\partial^2 X}{\partial x^2} = \frac{1}{Tv^2}\frac{\partial^2 T}{\partial t^2} \tag{11.5.4}$$

so that the two variables x and y have been separated. We set both sides equal to a constant $-k^2$:

$$\frac{1}{X(x)}\frac{d^2 X}{dx^2} = -k^2 \tag{11.5.5}$$

$$\frac{1}{T(t)}\frac{d^2 T}{dt^2} = -k^2 v^2 \tag{11.5.6}$$

These equations have the solutions:

$$X(x) = C_1 \sin(k\,x) + C_2 \cos(k\,x) \tag{11.5.7}$$

$$T(t) = C_3 \sin(k\,v\,t) + C_4 \cos(k\,v\,t) \tag{11.5.8}$$

After setting $\omega = k\,v$, the product solution is:

$$u(x,t) = \{C_1 \sin(k\,x) + C_2 \cos(k\,x)\}\{C_3 \sin(\omega\,t) + C_4 \cos(\omega\,t)\} \tag{11.5.9}$$

where C_1, C_2, C_3, C_4 are arbitrary constants and $\omega = kv$ represents an angular frequency.

Example 11.5 demonstrates how to find the general solution to the wave equation by using SymPy.

Example 11.5: Solving the wave equation

Use SymPy to obtain the product solution (11.5.9).

Solution:

We use again the `pde_separate()` function as in the previous examples of this chapter, with the only substantial difference that the solution $u(x,t) = X(x)T(t)$ is now a function of the position and time variables (x,t). The method `.subs(C1:C3,C2:C4)` is again used to clarify the symbols used in the final answer for the integration constants.

```
from sympy import  Eq, Function, Derivative as D,dsolve,\
   symbols,  pde_separate, diff

print('-'*28,'CODE OUTPUT','-'*29,'\n')

# define functions and symbols
u, X, T = symbols('y, X, T', cls=Function)

v, k, omega =symbols('v, k, omega',positive=True)

t, x, C1, C2, C3, C4 =symbols('t, x, C1, C2, C3, C4')

# this is the 1D wave equation
eq = Eq(v**2*D(u(x,t), x,2)-D(u(x,t),t,2), 0)
```

```
# use pde_separate() to separate the x and y variables
soln = pde_separate(eq, u(x, t), [X(x), T(t)], strategy='mul')
print('The separated form of the Laplace equation is:')
print(soln)

# Solve the ODE for X(x) using dsolve()
solX = dsolve(soln[0]+k**2, X(x))
print('\nThe solution X(x) =', solX.rhs)

# Solve the ODE for Y(y) using dsolve()
solY = dsolve(soln[1]+k**2, T(t))

# Replace constants C1, C2 in Y(y) with C3, C4 to avoid confusion
solY = solY.subs({C1:C3,C2:C4})

# print the function Y(y)
print('\nThe solution Y(y) =', solY.rhs)

# print combined solution X(x)*Y(y)
sol = solX.rhs*solY.rhs
print('\nThe general solution is')
print('u(x,y)=', sol)

print('\nIs this u(x,t) function a solution of the wave equation? ',\
      diff(sol,x,2) == diff(sol,t,2)/v**2)

--------------------------- CODE OUTPUT ---------------------------

The separated form of the Laplace equation is:
[Derivative(X(x), (x, 2))/X(x), Derivative(T(t), (t, 2))/(v**2*T(t))]

The solution X(x) = C1*sin(k*x) + C2*cos(k*x)

The solution Y(y) = C3*sin(k*t*v) + C4*cos(k*t*v)

The general solution is
u(x,y)= (C1*sin(k*x) + C2*cos(k*x))*(C3*sin(k*t*v) + C4*cos(k*t*v))

Is this u(x,t) function a solution of the wave equation?  True
```

11.6 APPLICATION OF THE WAVE EQUATION FOR A PLUCKED STRING: TRAVELING WAVES

In this section we study the solution of the wave equation for a string which is plucked near one of its fixed ends, as shown in Figure 11.5. The string is released from rest. We want to find out how this triangular shape will change over time and along the length of the string x.

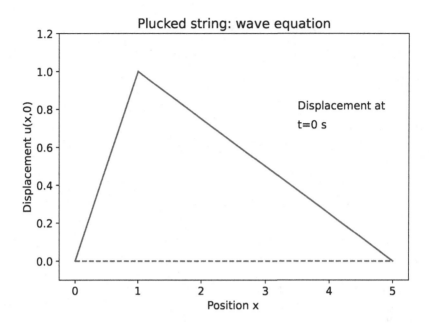

Figure 11.5 At time $t = 0$ the string is plucked into this triangular shape and is released from rest. The triangular shape is defined between $x = 0$ and $x = L$.

We start with the general solution derived in the last section:

$$u(x,t) = \{C_1 \sin(k\,x) + C_2 \cos(-k\,x)\} \{C_3 \sin(k\,v\,t) + C_4 \cos(k\,v\,t)\} \qquad (11.6.1)$$

where C_1, C_2, C_3, C_4 are arbitrary constants. In order to determine the integration constants we use the initial conditions at $t = 0$, and also apply the boundary conditions given for the problem.

For example, suppose that the initial displacement at the two ends at $x = 0, L$ is $y(0,t) = y(L,t)=0$. This leads to $C_2 = 0$ and $y(L,t) = C_1 \sin(k\,L) = 0$ (similar to the solutions we found for the Laplace equation). Therefore the values of the parameter k are:

$$k = \frac{n\,\pi}{L} \qquad n = \text{integer} \qquad (11.6.2)$$

so the solution becomes:

$$u(x,t) = \sin\left(\frac{n\,\pi}{L}\,x\right) \left\{C_3 \sin\left(\frac{n\,\pi}{L}\,v\,t\right) + C_4 \cos\left(\frac{n\,\pi}{L}\,v\,t\right)\right\} \qquad (11.6.3)$$

Once more there are an infinite numbers of solutions to the problem corresponding to different integers n, and the most general solution must be a linear combination of such functions:

$$u(x,t) = \sum_{n=1}^{n=\infty} A_n \sin\left(\frac{n\,\pi}{L}\,x\right) \left\{C_3 \sin\left(\frac{n\,\pi}{L}\,v\,t\right) + C_4 \cos\left(\frac{n\,\pi}{L}\,v\,t\right)\right\} \qquad (11.6.4)$$

where the constant coefficients A_n, C_3, C_4 are to be determined next, by using the initial conditions of the problem. We are given that the speed of the string when it is initially released is

$$\left.\frac{\partial u}{\partial t}\right|_{t=0} = 0 \qquad (11.6.5)$$

Writing out the partial derivative and setting $t = 0$:

$$\sum_{n=1}^{n=\infty} \left(\frac{n\pi}{L} v\right) A_n \sin\left(\frac{n\pi}{L} x\right) \left\{ C_3 \cos\left(\frac{n\pi}{L} vt\right) - C_4 \sin\left(\frac{n\pi}{L} vt\right) \right\}\Bigg|_{t=0} = 0 \qquad (11.6.6)$$

$$\sum_{n=1}^{n=\infty} \left(\frac{n\pi}{L} v\right) A_n \sin\left(\frac{n\pi}{L} x\right) C_3 = 0 \qquad (11.6.7)$$

Therefore $C_3 = 0$, so that (11.6.4) becomes:

$$u(x,t) = \sum_{n=1}^{\infty} B_n \sin\left(\frac{n\pi}{L} x\right) \cos\left(\frac{n\pi}{L} vt\right) \qquad (11.6.8)$$

In the final step, we require that this solution satisfy the initial condition $u(x,0) = f(x)$, where $f(x)$ is the initial shape of the string:

$$f(x) = \sum_{n=1}^{\infty} B_n \sin\left(\frac{n\pi}{L} x\right)$$

and the B_n are once more the coefficients of the Fourier sine series of $f(x)$.

Example 11.6 demonstrates how to find the solution to the wave equation for the given initial triangular shape of the plucked string.

Example 11.6: Plotting the solution of the wave equation at different times

Write a code to evaluate and plot the complete solution $y(x,t)$ for the plucked string with the initial triangular shape shown in Figure 11.5, for different time instants t. The equation for the initial triangular shape is:

$$f(x) = x \quad \text{for} \quad 0 \le x \le 1$$

$$f(x) = \frac{5-x}{4} \quad \text{for} \quad 1 \le x \le 5$$

Solution:

We evaluate the Fourier sine coefficients B_n by integrating the Fourier integrals using the `integrate()` function in SymPy. The integral consists of two parts, since the equation for the shape of the string is different in the intervals $0 \le x < 1$ and $0 \le x \le 5$.

We add 30 terms in the Fourier series, as determined by the parameter `numTerms=30`, and the function `F(t)` adds the Fourier terms for each of the times `t`. The result is shown in Figure 11.6.

Qualitatively the motion of the wave on the string can be described as the reflection of the wave at the two fixed ends of the string, while the initial triangular shape of the plucked string changes into a more complex shape consisting of two triangles.

```
from sympy import  pi, integrate, simplify, symbols, sin
from sympy.utilities.lambdify import lambdify
import numpy as np
import matplotlib.pyplot as plt

n = symbols('n', integer=True, positive=True)
```

```python
x = symbols('x')

print('-'*28,'CODE OUTPUT','-'*29,'\n')

# function to evaluate displacement by summing the series y(x,t)
def F(t):
    suma = [0]*len(x)
    for i in range(1,numTerms):
        suma = suma+bn(i)*np.sin(i*np.pi*x/L)*\
            np.cos((i*np.pi/L)*v*t)
    return suma


L = 5    # length of string
v = 1    # speed of the wave

# Find the sine Fourier coefficients of the triangular function
Bn = (2/L)*integrate(x* sin(n*pi*x/L),(x,0,1))+\
    (2/L)*integrate((1.25-.25*x)* sin(n*pi*x/L),(x,1,5))

print("Fourier coefficients bn=",simplify(Bn))

# turn Bn into a function of n
bn = lambdify(n, Bn)

# Number of terms to be summed in the truncated Fourier series
numTerms = 30

# positions along the x-axis
x = np.linspace(0,5,100)

# call function F(t) to evaluate series u(x,t) for times t=0-8s
times = np.arange(0,9,1)

for i in range(len(times)):
    plt.subplot(3,3,times[i]+1)
    plt.plot(x,F(times[i]))
    plt.ylim(-1.1,1.2)
    plt.plot(x,[0]*len(x))
    plt.text(3.4,.7,'t='+str(times[i])+' s')
    if i>5:
        plt.xlabel('x')
    if i==3:
        plt.ylabel('u(x,t)')

plt.tight_layout()
plt.show()
```

```
------------------------- CODE OUTPUT -------------------------

Fourier coefficients bn= 12.5*sin(pi*n/5)/(pi**2*n**2)
```

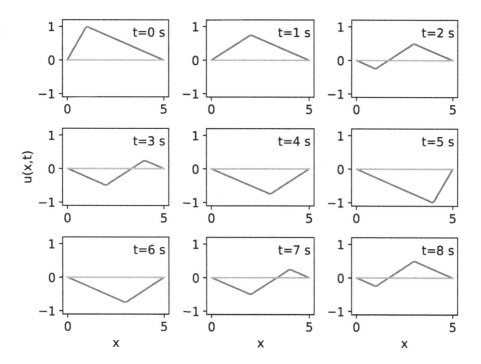

Figure 11.6 Graphical output from Example 11.6, showing the solution to the wave equation for the initial triangular shape of the plucked string, at different times between $t = 0$ s and $t = 8$ s.

In the next two sections, we revisit the Laplace equation, but in non-Cartesian coordinates.

11.7 THE LAPLACE EQUATION IN CYLINDRICAL COORDINATES

Problems with cylindrical symmetry are common in physics. For example, we may want to know the steady state temperature of a cylinder or the electric potential inside and outside a cylinder with a known surface charge density. The Laplace equation can be used to solve these problems, but for convenience we use cylindrical coordinates instead of Cartesian.

The Laplace equation in cylindrical coordinates is:

$$\nabla^2 u = \frac{1}{\rho}\frac{\partial}{\partial \rho}\left(\rho \frac{\partial u}{\partial \rho}\right) + \frac{1}{\rho^2}\frac{\partial^2 u}{\partial \phi^2} + \frac{\partial^2 u}{\partial z^2} = 0 \tag{11.7.1}$$

where $u(\rho, \phi, z)$ is a scalar function (such as electric potential). We will follow the method of separation of variables and assume that the product solution can be written in the form:

$$u(\rho, \phi, z) = R(\rho)\,\Phi(\phi)\,Z(z) \tag{11.7.2}$$

Substituting and dividing by $R(\rho)\,\Phi(\phi)\,Z(z)$ we obtain:

$$\frac{1}{\rho R}\frac{\partial}{\partial \rho}\left(\rho \frac{\partial R}{\partial \rho}\right) + \frac{1}{\rho^2 \Phi}\frac{\partial^2 \Phi}{\partial \phi^2} + \frac{1}{Z}\frac{\partial^2 Z}{\partial z^2} = 0 \tag{11.7.3}$$

Notice in (11.7.3) the separation of the variables is not as clean as in the case for the PDEs in Cartesian coordinates. A ρ^2 terms appears with the Φ term. However, we can perform the separation in two steps. In the first step, we isolate the Z term and set it equal to a constant.

$$\frac{1}{Z}\frac{d^2Z}{dz^2} = K^2 \tag{11.7.4}$$

$$\frac{1}{\rho R}\frac{\partial}{\partial\rho}\left(\rho\frac{\partial R}{\partial\rho}\right) + \frac{1}{\rho^2\Phi}\frac{\partial^2\Phi}{\partial\phi^2} = -K^2 \tag{11.7.5}$$

The choice to assign K^2 as a positive constant for (11.7.4) is often physically motivated. For example, we may want the electric potential along a semi-infinite cylinder to decrease if one end is kept at a fixed potential. We can solve (11.7.4) for Z:

$$Z(z) = A_1 e^{Kz} + A_2 e^{-Kz} \tag{11.7.6}$$

We can separate (11.7.5) by multiplying through by ρ^2:

$$\frac{\rho}{R}\frac{\partial}{\partial\rho}\left(\rho\frac{\partial R}{\partial\rho}\right) + \frac{1}{\Phi}\frac{\partial^2\Phi}{\partial\phi^2} = -K^2\rho^2 \tag{11.7.7}$$

Now we can separate (11.7.7) by introducing a new constant n^2 so that:

$$\frac{1}{\Phi}\frac{d^2\Phi}{d\phi^2} = -n^2 \tag{11.7.8}$$

$$\frac{\rho}{R}\frac{d}{d\rho}\left(\rho\frac{dR}{d\rho}\right) - n^2 + K^2\rho^2 = 0 \tag{11.7.9}$$

We chose to assign the constant $-n^2$ to the Φ-equation (11.7.8), because it is common for there to be periodic behavior in the angle ϕ. We can now solve (11.7.8):

$$\Phi(\phi) = B_1\sin(n\phi) + B_2\cos(n\phi) \tag{11.7.10}$$

Next, we rewrite (11.7.9):

$$\rho\frac{d}{d\rho}\left(\rho\frac{dR}{d\rho}\right) + \left(K^2\rho^2 - n^2\right)R = 0 \tag{11.7.11}$$

This is the Bessel equation, which we know from Chapter 10 has solutions $R = J_n(K\rho)$ called Bessel functions. We also know there is a second solution $R = N_n(K\rho)$ called the Neumann function. However, $N_n(K\rho)$ becomes infinite at $\rho = 0$. Therefore, like in most physical applications, we discard the Neumann solution.

Likewise, we often require exponential decay along the cylinder's axis. Therefore, we can write the product solution as:

$$u_n(\rho, \phi, z) = J_n(K\rho)\left(B_1\sin(n\phi) + B_2\cos(n\phi)\right)e^{-Kz} \tag{11.7.12}$$

The solution to the Laplace equation is a linear combination of u_n's.

We are not quite finished with the product solution (11.7.11). However, it is important to note that (11.7.11) is a typical starting place for problems which require a solution to the Laplace equation in cylindrical coordinates. In other words, when presented with a problem with cylindrical symmetry, you can start with (11.7.12) and apply boundary conditions. Unlike in the case of Cartesian coordinates, the ODEs have already been solved! In the next section, we will examine a specific example to see how boundary conditions are applied to (11.7.12) to solve a problem from electromagnetism.

11.7.1 APPLICATION OF THE LAPLACE EQUATION: ELECTRIC POTENTIAL INSIDE A CYLINDER

In this section, we will work through an example problem which will demonstrate how to apply (11.7.12) to a specific situation.

Consider an semi-infinite cylindrical shell of radius a whose axis runs along the z-axis. The bottom end of the cylinder is in the xy-plane and held at a potential of 10V. All other surfaces of the cylinder are grounded (held at 0V). Find the potential inside the cylinder.

We begin by stating the boundary conditions:

$$u(\rho, \phi, 0) = 10 \tag{11.7.13}$$

$$u(a, \phi, z) = 0 \tag{11.7.14}$$

$$\lim_{z \to \infty} u = 0 \tag{11.7.15}$$

For convenience, let us rewrite (11.7.12)

$$u_n(\rho, \phi, z) = J_n(K\rho)\left(B_1 \sin(n\phi) + B_2 \cos(n\phi)\right)e^{-Kz} \tag{11.7.16}$$

By including only the negative exponential in z, we have already accounted for (11.7.15). The boundary conditions are independent of ϕ. Therefore the potential must also be independent of ϕ. Hence, $n = 0$ and u includes only one Bessel function, J_0. Finally, (11.7.14) tells us

$$Ka = k_m \tag{11.7.17}$$

where k_m is the mth zero of J_0. As we saw in the last chapter, the zeros of any Bessel function can be found using Python. Because there are an infinite number of zeros for J_0 we have,

$$u = \sum_{m=1}^{\infty} b_m J_0\left(\frac{k_m \rho}{a}\right) e^{-k_m z/a} \tag{11.7.18}$$

The final boundary condition at $z = 0$, (11.7.13) will be used to get b_m

$$10 = \sum_{m=1}^{\infty} b_m J_0\left(\frac{k_m \rho}{a}\right) \tag{11.7.19}$$

We can get the values of b_m by using the orthogonality condition for Bessel functions

$$\int_0^1 x J_p(ax) J_p(bx) dx = \begin{cases} 0 & a \neq b \\ \frac{1}{2} J_{p+1}^2(a) = \frac{1}{2} J_{p-1}^2(a) = \frac{1}{2}\left[J_p'(a)\right]^2 & a = b \end{cases} \tag{11.7.20}$$

in a method analogous to Fourier's trick for Fourier series. We multiply both sides of (11.7.19) by $\rho J_0(k_\ell \rho/a)$ and integrate from $\rho = 0$ to $\rho = $a:

$$10 \int_0^a \rho J_0(k_\ell \rho/a) d\rho = b_\ell \int_0^a \rho \left[J_0(k_\ell \rho/a)\right]^2 d\rho \tag{11.7.21}$$

where $\ell = 0, 1, 2, \ldots$. We can use (11.7.21) to get each of the b_ℓ terms. We can evaluate the right-hand side of (11.7.21) using (11.7.20)

$$\int_0^a \rho \left[J_0(k_\ell \rho/a)\right]^2 d\rho = \frac{a^2}{2} J_1^2(k_\ell) \tag{11.7.22}$$

Next, we need to evaluate the left-hand side of (11.7.20), so we insert $x = k_\ell \rho/a$ into the Bessel function identity

$$\frac{d}{dx}\left[x^p J_p(x)\right] = x^p J_{p-1}(x) \tag{11.7.23}$$

to obtain

$$\frac{a}{k_\ell}\frac{d}{d\rho}\left[\frac{k_\ell \rho}{a}J_1(k_\ell \rho/a)\right] = \frac{k_\ell \rho}{a}J_0(k_\ell \rho/a) \tag{11.7.24}$$

Integrating both sides, we find

$$\int_0^a \rho J_0(k_\ell \rho/a)d\rho = \frac{a}{k_\ell}\rho\, J_1(k_\ell \rho/a)\big|_0^a = \frac{a^2}{k_\ell}J_1\left(k_m\right) \tag{11.7.25}$$

Inserting (11.7.22) and (11.7.25) into (11.7.21) we can get b_ℓ

$$b_\ell = \frac{20}{k_\ell J_1\left(k_\ell\right)} \tag{11.7.26}$$

Therefore, the solution is

$$u(\rho, \phi, z) = \sum_{k=1}^{\infty}\frac{20}{k_m J_1\left(k_m\right)}J_0\left(\frac{k_m\rho}{a}\right)e^{-k_m z/a} \tag{11.7.27}$$

Example 11.7 demonstrates how to plot (11.7.27).

Example 11.7: The electric potential of a charged cylinder

Using $a = 1$, plot (11.7.27). Include the first 10 terms.

Solution:

The methods used to plot (11.7.27) in Figure 11.7 are similar to what has been done in previous examples.

```
from scipy.special import jv, jn_zeros
import matplotlib.pyplot as plt
import numpy as np

num_terms = 10
a = 1

print('-'*28,'CODE OUTPUT','-'*29,'\n')

km = jn_zeros(0,num_terms)
print('The first few zeros of the Bessel function are: \n')
print(km)

def u(x,y):
    suma = [0]*len(x)
    for i in range(0,num_terms):
        suma = suma + 20/(km[i] * jv(1,km[i])) * jv(0, km[i]*x/a) * \
```

```
        np.exp(-km[i]*y/a)
    return suma

rho = np.linspace(0, a, 100)
z = np.linspace(0,6,100)
R,Z = np.meshgrid(rho,z)
U = u(R,Z)

fig = plt.figure()
ax = fig.add_subplot(projection='3d')

ax.plot_wireframe(R, Z, U, rstride = 3, cstride = 2)
ax.set_xlabel(r'$\rho$')
ax.set_ylabel('z')
ax.set_zlabel(r'Potential u($\rho$,z)')
ax.view_init(20, 50)

plt.show()

-------------------------- CODE OUTPUT --------------------------

The first few zeros of the Bessel function are:

[ 2.40482556  5.52007811  8.65372791 11.79153444 14.93091771 18.07106397
 21.21163663 24.35247153 27.49347913 30.63460647]
```

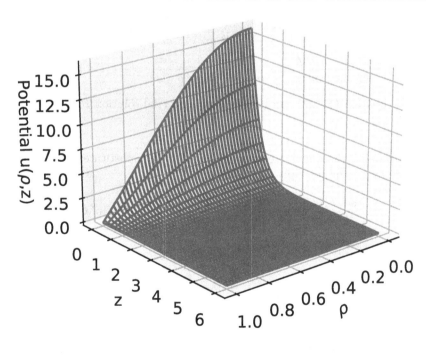

Figure 11.7 Graphical output from Example 11.7, demonstrating plots of the solution (11.7.27) of the Laplace equation in cylindrical coordinates.

11.8 THE LAPLACE EQUATION IN SPHERICAL COORDINATES

Problems with spherical symmetry are common in physics. For example, the electric potential due to a charged sphere is most easily modeled in spherical coordinates. In this section, we will examine the Laplace equation in spherical coordinates. As with cylindrical coordinates, we will be able to obtain a universally applicable product solution.

The Laplace equation in spherical coordinates for a scalar function $u(r, \theta, \phi)$ is

$$\frac{1}{r^2} \frac{\partial}{\partial r} \left(r^2 \frac{\partial u}{\partial r} \right) + \frac{1}{r^2 \sin \theta} \frac{\partial}{\partial \theta} \left(\sin \theta \frac{\partial u}{\partial \theta} \right) + \frac{1}{r^2 \sin^2 \theta} \frac{\partial^2 u}{\partial \phi^2} = 0 \tag{11.8.1}$$

Depending on the field of physics, we sometimes assume azimuthal symmetry. In the case of azimuthal symmetry, u does not depend on ϕ. Azimuthal symmetry is a common assumption in electromagnetism. However, in quantum mechanics, the scalar function (which is a particle's wave function) may not have azimuthal symmetry. In this section, we will study the more general case. However, we will examine the case of azimuthal symmetry in an example problem.

We begin by seeking solutions of the form $u = R(r)Y(\theta, \phi)$. After substitution and division by $u = RY$ and multiplication by r^2, we obtain

$$\frac{1}{R} \frac{\partial}{\partial r} \left(r^2 \frac{\partial R}{\partial r} \right) + \frac{1}{Y \sin \theta} \frac{\partial}{\partial \theta} \left(\sin \theta \frac{\partial Y}{\partial \theta} \right) + \frac{1}{Y \sin^2 \theta} \frac{\partial^2 Y}{\partial \phi^2} = 0 \tag{11.8.2}$$

For reasons that will become apparent later, we will set the constant for separation to be $\ell (\ell + 1)$:

$$\frac{1}{R} \frac{d}{dr} \left(r^2 \frac{dR}{dr} \right) = \ell (\ell + 1) \tag{11.8.3}$$

$$\frac{1}{Y \sin \theta} \frac{\partial}{\partial \theta} \left(\sin \theta \frac{\partial Y}{\partial \theta} \right) + \frac{1}{Y \sin^2 \theta} \frac{\partial^2 Y}{\partial \phi^2} = - \ell (\ell + 1) \tag{11.8.4}$$

The radial equation (11.8.3) can be solved using a trial solution $R(r) = a \, r^n$:

$$R = a \, r^\ell + b \, r^{-(\ell+1)} \tag{11.8.5}$$

Next, we examine the angular equation (11.8.4). We substitute $Y = \Theta(\theta)\Phi(\phi)$ into (11.8.4), divide by Y, and multiply by $\sin^2 \theta$:

$$\frac{\sin \theta}{\Theta} \frac{\partial}{\partial \theta} \left(\sin \theta \frac{\partial Y}{\partial \theta} \right) + \frac{1}{\Phi} \frac{\partial^2 \Phi}{\partial \phi^2} = -\ell (\ell + 1) \sin^2 \theta \tag{11.8.6}$$

We introduce the separation constant m^2:

$$\frac{1}{\Phi} \frac{\partial^2 \Phi}{\partial \phi^2} = -m^2 \tag{11.8.7}$$

$$\frac{\sin \theta}{\Theta} \frac{\partial}{\partial \theta} \left(\sin \theta \frac{\partial \Theta}{\partial \theta} \right) - m^2 = -\ell (\ell + 1) \sin^2 \theta \tag{11.8.8}$$

We chose to include the minus sign for m in (11.8.7) because typically, we want $\Phi(\phi)$ to be periodic, i.e. $\Phi(\phi + 2\pi) = \Phi(\phi)$, therefore we can solve (11.8.7):

$$\Phi(\phi) = e^{i m \phi} \tag{11.8.9}$$

where m must be an integer (to satisfy the periodicity requirement for Φ). Therefore, there are many solutions to (11.8.7), each indexed by m, and we will sum over them. Note that there are two solutions to (11.8.7), $\exp(\pm im\phi)$ but we will include the negative exponentials by summing over negative values of m.

Next, we will rewrite (11.8.8)

$$\sin\theta \frac{\partial}{\partial\theta}\left(\sin\theta \frac{\partial\Theta}{\partial\theta}\right) + \left(\ell(\ell+1)\sin^2\theta - m^2\right)\Theta = 0 \qquad (11.8.10)$$

Fortunately, we will not need to solve (11.8.10) because it is a form of a well-known differential equation called the *general Legendre equation*. We saw in Chapter 10 that the solutions to (11.8.10) are known as *associated Legendre functions* P_ℓ^m:

$$\Theta(\theta) = c\,P_\ell^m(\cos\theta) \qquad (11.8.11)$$

Note that there is a second solution to (11.8.10), but it blows up at $\theta = 0$ and/or $\theta = \pi$ and is therefore not physical.

Recall that for (11.8.11) to be a solution to (11.8.10), ℓ must be an integer. In addition, from Chapter 10, we know $P_\ell^m = 0$ for $|m| > l$. Therefore:

$$m = 0, \pm 1, \pm 2, \ldots \pm \ell \qquad (11.8.12)$$

Now that we have found $R(r)$, $\Theta(\theta)$, and $\Phi(\phi)$, we can write the full solution to (11.8.1), noting that now there are two separation constants to sum over

$$u(r, \theta, \phi) = \sum_{\ell=0}^{\infty} \sum_{m=-\ell}^{\ell} \left(A_{\ell,m}\, r^\ell + \frac{B_{\ell,m}}{r^{\ell+1}}\right) P_\ell^m(\cos\theta)\, e^{im\phi} \qquad (11.8.13)$$

Although (11.8.13) is not as clean as some of our other solutions, it can be used as the starting point for any problem involving the Laplace equation in spherical coordinates. One only needs to apply the problem's boundary conditions to find $A_{\ell,m}$ and $B_{\ell,m}$.

Next, we will work through a detailed example involving (11.8.13) to show how to apply boundary conditions to (11.8.13).

11.8.1   APPLICATION OF THE LAPLACE EQUATION: THE ELECTRIC POTENTIAL DUE TO A CHARGED SPHERE

In this subsection, we will examine an application of the Laplace equation in spherical coordinates where we will find the electric potential inside a charged sphere.

Consider a hollow sphere of radius a. The electric potential on the sphere's surface is

$$V(\theta) = V_0 \cos^2\theta \qquad (11.8.14)$$

The electric potential $V(r, \theta, \phi)$ inside the sphere is found using Laplace's equation in spherical coordinates. We know the general solution to be

$$V(r, \theta, \phi) = \sum_{\ell=0}^{\infty} \sum_{m=-\ell}^{\ell} \left(A_{\ell,m} r^\ell + \frac{B_{\ell,m}}{r^{\ell+1}}\right) P_\ell^m(\cos\theta)\, e^{im\phi} \qquad (11.8.15)$$

However, in electromagnetism, we expect V to be independent of ϕ because the boundary condition (11.8.14) is independent of ϕ. This mean that $m = 0$ and $P_\ell^0 = P_\ell$. Hence, we have (with $A_{l,0} = A_l$ and $B_{l,0} = B_l$):

$$V(r, \theta, \phi) = \sum_{\ell=0}^{\infty} \left(A_\ell r^\ell + \frac{B_\ell}{r^{\ell+1}}\right) P_\ell(\cos\theta) \qquad (11.8.16)$$

There is another boundary condition not explicitly stated in the problem. Because the area of interest is inside the sphere, the potential should not blow up at $r = 0$. Therefore, $B_\ell = 0$ for all values of ℓ. This leaves the potential to be

$$V(r, \theta, \phi) = \sum_{\ell=0}^{\infty} A_\ell r^\ell P_\ell (\cos \theta) \tag{11.8.17}$$

As mentioned earlier, when working in spherical coordinates, we only need to apply the boundary conditions to (11.8.13). Next, we need to find A_ℓ using (11.8.14) which at $r = a$ is

$$V_0 \cos^2 \theta = \sum_{\ell=0}^{\infty} A_\ell a^\ell P_\ell (\cos \theta) \tag{11.8.18}$$

Notice that we are attempting to express the boundary condition in terms of a linear combination of Legendre polynomials. That can be done because the Legendre polynomials are complete in the interval -1 ($\cos \pi$) to 1 ($\cos 0$).

To isolate the coefficients A_ℓ, we will need to modify the orthogonality condition for the Legendre polynomials first presented in Chapter 11

$$\int_0^\pi P_\ell (\cos \theta) P_m (\cos \theta) \sin \theta d\theta = \begin{cases} 0 & \ell \neq m \\ \frac{2}{2\ell+1} & \ell = m \end{cases} \tag{11.8.19}$$

Therefore, we will multiply (11.8.18) by $P_m (\cos \theta) \sin \theta$ and integrate

$$V_0 \int_0^\pi P_m (\cos \theta) \cos^2 \theta \sin \theta d\theta = \sum_{\ell=0}^{\infty} A_\ell a^\ell \int_0^\pi P_\ell (\cos \theta) P_m (\cos \theta) \sin \theta d\theta \tag{11.8.20}$$

$$V_0 \int_0^\pi P_m (\cos \theta) \cos^2 \theta \sin \theta d\theta = A_m a^m \frac{2}{2m + 1} \tag{11.8.21}$$

We can now solve for A_ℓ (note that we changed the dummy index from m to ℓ):

$$A_\ell = \frac{2\ell + 1}{2a^\ell} V_0 \int_0^\pi P_\ell (\cos \theta) \cos^2 \theta \sin \theta d\theta \tag{11.8.22}$$

We can insert (11.8.22) for A_ℓ into (11.8.17) to get the final answer:

$$V(r, \theta) = \sum_{\ell=0}^{\infty} \left[\frac{2\ell + 1}{2a^\ell} V_0 \int_0^\pi P_\ell (\cos \theta) \cos^2 \theta \sin \theta d\theta \right] r^\ell P_\ell (\cos \theta) \tag{11.8.23}$$

We need to find the A_ℓ's individually, which we can do as shown in Example 11.8.

Example 11.8: The potential due to a charged sphere

Using SymPy, calculate the coefficients A_ℓ from (11.8.22).

Solution:

Notice that in the code only two of the A_ℓ's are nonzero. That is because we can rewrite

$$\cos^2 \theta = \frac{2}{3} P_2 (\cos \theta) + \frac{1}{3} P_0 (\cos \theta)$$

Hence, only two terms survive the orthogonality condition (11.8.19). Therefore

$$V(r,\theta) = V_0 \left(\frac{1}{3} + \frac{2}{3a^2} r^2 P_2 \left(\cos\theta \right) \right) \tag{11.8.24}$$

or if we expand P_2:

$$V(r,\theta) = V_0 \left(\frac{1}{3} + \frac{r^2}{a^2} \left[\cos^2\theta - \frac{1}{3} \right] \right)$$

```python
from sympy import integrate, symbols, cos, sin, pi
from sympy.functions.special.polynomials import legendre

import numpy as np
import matplotlib.pyplot as plt

r, t,l = symbols('r,t,l')

a = 1
V0 = 1

def A(l):
    return (2*l + 1)/(2*a**l)*V0 *\
        integrate(legendre(l,cos(t))*(cos(t)**2)*sin(t), \
            (t,0,pi))

print('-'*28,'CODE OUTPUT','-'*29,'\n')

print('The Al coefficients for the potential V are:\n')
for l in range(0,5):
    print('A_'+str(l)+' =', round(A(l),2))


--------------------------- CODE OUTPUT ----------------------------

The Al coefficients for the potential V are:

A_0 = 0.33
A_1 = 0
A_2 = 0.67
A_3 = 0
A_4 = 0
```

11.9 GENERAL OUTLINE FOR SOLVING LINEAR PDES

We conclude this chapter with an outline of the methods used to solve the PDEs examined above.

1. Identify the independent variables relevant to the problem. For example, in the heat equation, the dependent function u depended on one spatial variable and time.

2. Write down a trial product function, which is the product of single variables functions. There is one term in the product function for each independent variable. For example, $u = X(x)T(t)$

3. Insert the trial function into the PDE and then divide by u. The result with be a sum of terms, each of which involves one independent variable.

4. Set each of the terms from Step 3 equal to a constant such that the constants are either equal or add to zero (depending on the PDE).

5. Solve the resulting ODEs from Step 4.

6. Apply the boundary conditions in order to get the integration constants from the solutions found in Step 5. Note that you may need to use the physics of the problem (such as electrostatic potentials tend toward zero at large distances from source charges) to identify all the boundary conditions. Expect multiple values for at least one of the constants.

7. Write the product solution as an infinite sum of terms, indexed by one of the constants found in Step 6.

8. Use the problem's initial condition and Fourier's trick to find the remaining unknown constants.

11.10   END OF CHAPTER PROBLEMS

1. **Solution of the heat flow equation** – Show using SymPy and by hand, that

$$T(x,t) = \cos(3\,x)\,\exp\left(-9\,a^2\,t\right) \qquad (11.10.1)$$

is a solution of the heat flow equation in 1D:

$$\frac{\partial^2 T}{\partial x^2} = \frac{1}{a^2}\frac{\partial T}{\partial t} \qquad (11.10.2)$$

2. **Solution of the Laplace equation in 2D** – Show using SymPy and by hand, that

$$V(x,y) = \exp(k\,x)\sin(k\,y) \qquad (11.10.3)$$

is a solution of the Laplace equation in 2D:

$$\frac{\partial^2 T}{\partial x^2} + \frac{\partial^2 T}{\partial y^2} = 0 \qquad (11.10.4)$$

3. **Solution of the wave equation** – Show using SymPy and by hand, that

$$y(x,t) = \sin(k\,x)\cos(\omega\,t) \qquad (11.10.5)$$

is a solution of the wave equation in 1D, with $\omega = k\,v$:

$$\frac{\partial^2 y}{\partial x^2} = \frac{1}{v^2}\frac{\partial^2 y}{\partial t^2} \qquad (11.10.6)$$

4. **Solution of the wave equation in spherical coordinates** – Show using SymPy and by hand, that

$$V(r) = \frac{f(r - v\,t)}{r} \qquad (11.10.7)$$

is a solution of the wave equation in spherical coordinates:

$$\nabla^2 V = \frac{1}{v^2}\frac{\partial^2 V}{\partial t^2} \qquad (11.10.8)$$

5. **General solution of the Laplace equation in 3D in Cartesian coordinates** – The Laplace equation in Cartesian coordinates in three dimensions is:

$$\nabla^2 V = 0$$

Show using SymPy and by hand, that the general solution of this equation in Cartesian coordinates in three dimensions is

$$V(x, y, z) = X(x)\, Y(y)\, Z(z) \tag{11.10.9}$$

$$X(x) = \{C_1 \sin(a\,x) + C_2 \cos(a\,x)\}$$

$$Y(y) = \{C_3 \sin(b\,y) + C_4 \cos(b\,y)\}$$

$$Z(z) = \{C_5 e^{f\,z} + C_6 e^{-f\,z}\}$$

where $f = \sqrt{a^2 + b^2}$ and a, b are positive constants.

6. **General solution of the Laplace equation in cylindrical coordinates with no z-dependence** – Show using SymPy and by hand, that the general solution of the Laplace equation $\nabla^2 V = 0$ in cylindrical coordinates when there is no z-dependence of V is:

$$V(\rho, \phi) = \left(C_3\, \rho^k + C_4\, \rho^{-k} + C_5 \ln \rho\right) \{C_1 \sin(k\,\phi) + C_2 \cos(k\,\phi)\} \tag{11.10.10}$$

7. **A solution of the Laplace equation in cylindrical coordinates with no z-dependence** – Show using SymPy and by hand, that

$$V(\rho, \phi) = (\rho + 1/\rho + \ln \rho)\, \sin(\phi) \tag{11.10.11}$$

is a possible solution of the Laplace equation $\nabla^2 V = 0$ in cylindrical coordinates, when there is no z-dependence of V.

8. **Solution of the heat flow equation with initial and boundary conditions** – Consider again the example of the heat flow across a rod in Section 11.2. Find and plot the solution of the heat flow equation in 1D, which satisfies the satisfy the *boundary conditions:*

$$T(0, t) = T(L, t) = 0 \tag{11.10.12}$$

and the initial condition at time $t = 0$:

$$T(x, 0) = \sin(\pi\,x/L) \qquad \text{for } 0 \le x \le L \tag{11.10.13}$$

Solve this problem using SymPy and also do by hand.

9. **Solution of the heat flow equation with a triangular initial condition** – Consider again the example of the heat flow across a rod in Section 11.2, with length $L = 2$ m, and a coefficient $a = 1$ (in SI units). Find and plot the solution of the heat flow equation in 1D, which satisfies the satisfy the *boundary conditions:*

$$T(0, t) = T(L, t) = 0 \tag{11.10.14}$$

and the initial condition at time $t = 0$ are such that the temperature profile is triangular:

$$T(x, 0) = 50\,x/L \qquad \text{for } 0 \le x \le L/2 \tag{11.10.15}$$

and

$$T(x, 0) = 50\,(L - x)\,/L \qquad \text{for } L/2 \le x \le L \tag{11.10.16}$$

Solve this problem using SymPy and also by hand.

10. **The Schrodinger equation** – The time-dependent Schrodinger equation for a free particle is the PDE:

$$i\hbar \frac{\partial \Psi(x,t)}{\partial t} = -\frac{\hbar^2}{2m} \frac{\partial^2 \Psi(x,t)}{\partial x^2} \tag{11.10.17}$$

where m is the mass of the particle, $\Psi(x,t)$ is the wave function which depends on time t and position x, and $\hbar = h/(2\pi)$ is the reduced Planck constant. Use SymPy to separate the variables and obtain the general solution, by using the solution in the form $\Psi(x,t) = \psi(x) T(t)$.

11. **Solution of the Schrodinger equation in 1D** – Find the relationship between the parameters E and k, so that the expression

$$\Psi(x,t) = \sin(k\,x) \exp(-i\,E\,t/\hbar) \tag{11.10.18}$$

is a solution of the time-independent Schrodinger equation for a free particle in 1D:

$$i\hbar \frac{\partial \Psi(x,t)}{\partial t} = -\frac{\hbar^2}{2m} \frac{\partial^2 \Psi(x,t)}{\partial x^2} \tag{11.10.19}$$

where m is the mass of the particle, $\Psi(x,t)$ is the wave function which depends on time t and position x, and $\hbar = h/(2\pi)$ is the reduced Planck constant. Solve this problem using SymPy and by hand.

12. **Solution of the Schrodinger equation in 3D** – Find the relationship between the parameters E, k_x, k_y, k_z so that the expression

$$\Psi(x,t) = \sin(k_x\,x) \sin(k_y\,y) \sin(k_z\,z) \exp(-i\,E\,t/\hbar) \tag{11.10.20}$$

is a solution of the time-independent Schrodinger equation for a free particle in 3D:

$$i\hbar \frac{\partial \Psi(x,y,z,t)}{\partial t} = -\frac{\hbar^2}{2m} \left(\nabla^2 \Psi(x,y,z,t) \right) \tag{11.10.21}$$

where ∇^2 is the Laplacian operator:

$$\nabla^2 \Psi(x,y,z,t) = \frac{\partial^2 \Psi(x,y,z,t)}{\partial x^2} + \frac{\partial^2 \Psi(x,y,z,t)}{\partial y^2} + \frac{\partial^2 \Psi(x,y,z,t)}{\partial z^2}$$

and $m=$ mass of the particle, $\Psi(x,y,z,t)$ is the wave function which depends on time t and position (x,y,z), and $\hbar = h/(2\pi)$ is the reduced Planck constant. Solve this problem using SymPy and by hand.

13. **General solution of the wave equation in 2D in Cartesian coordinates** – Show using SymPy and by hand, that the general solution of the wave equation in Cartesian coordinates in two dimensions is

$$V(x,y,t) = X(x)\,Y(y)\,T(t) \tag{11.10.22}$$

$$X(x) = C_1 \sin(k_x\,x) + C_2 \cos(k_x\,x)$$

$$Y(y) = C_3 \sin(k_y\,y) + C_4 \cos(k_y\,y)$$

$$T(t) = C_5 \sin(\omega\,t) + C_6 \cos(\omega\,t)$$

where $\omega = v \sqrt{k_x^2 + k_y^2}$, k_x, k_y are positive constants and v is the speed of the wave.

14. **Solution of the wave equation for a symmetric triangular initial pulse** – Consider the example of the plucked string in Section 11.6, in which we obtained the general solution with a triangular shaped pulse. Obtain and plot the general solution when the string is plucked in its middle point, while the end points are fixed. The string has length $L = 2$ m and a wave speed $v = 1$ m/s.

15. **Standing wave solution of the wave equation, for a sinusoidal initial condition** – Consider the example of the plucked string in Section 11.6, in which we obtained the general solution with a triangular shaped pulse. Obtain and plot the general solution when the string is plucked in the shape of the sine function $\sin(2\pi x/L)$, while the end points are fixed. The string has a length $L = 5$ m and the wave speed $v = 1$ m/s. Discuss the physical meaning of the results.

16. **A triangular pulse: example of the wave equation** – Obtain and plot the solution of the wave equation for the following triangular pulse on a string with length $L = 8$ m and a wave speed $v = 1$ m/s:

$$
\begin{aligned}
f &= 0 & 0 \le x \le 3 \\
f &= x - 3 & 3 \le x \le 4 \\
f &= 5 - x & 4 \le x \le 5 \\
f &= 0 & 5 \le x \le 8
\end{aligned}
$$

17. **A square pulse: example of the wave equation** – Obtain and plot the solution of the wave equation for the following square pulse, for a string with length $L = 8$ m and a wave speed $v = 1$ m/s:

$$
\begin{aligned}
f &= 0 & 0 \le x \le 3 \\
f &= 1 & 3 \le x \le 4 \\
f &= 0 & 4 \le x \le 8
\end{aligned}
$$

18. **Potential on a charged sphere** – In Section 11.8.1 we found the electric potential outside a charged hollow sphere of radius a. The electric potential on the sphere's surface was

$$
V(\theta) = V_0 \cos^2 \theta \tag{11.10.23}
$$

Find and plot the potential V as a function of the variables (r, θ), inside the sphere. Use Python to help with the algebra and integrals as necessary.

19. **Potential on a charged sphere** – Find the electric potential inside and outside a charged sphere with radius $R = 1$ and which has a potential $V_0 = 100$ Volts on the upper hemisphere and a potential $V = 0$ Volts on the lower hemisphere.

20. **Laplace equation for a sphere with boundary conditions depending on the angle** – Solve the Laplace equation for the potential inside and outside a hollow conducting sphere of radius $R = 1$, when the potential on the surface of the sphere is given by:

$$
V(\theta) = \cos 2\theta \tag{11.10.24}
$$

12 Analysis of Nonlinear Systems

Much of an undergraduate physics education focuses on modeling systems using linear ordinary or partial differential equations. The linear models work well in many situations. However, those models are often based on approximations. Consider the simple plane pendulum. In general, the angular acceleration of the pendulum depends on the sine of its angular displacement θ. If $\theta < \pi/12$ radians, the small angle approximation can be used and the pendulum's angular acceleration is well-described as a linear function of its angular displacement. However, if the angular displacement of the pendulum is greater than $\pi/12$ radians, then the small angle approximation is no longer valid, and the equation for the angular acceleration is nonlinear.

Examples of nonlinear systems abound in physics, engineering and the other sciences (both natural and social). In this chapter, we will examine how to analyze nonlinear differential equations which often are not solvable in closed-form. We will examine problems from a variety of fields to demonstrate the flexibility of the methods presented in this chapter.

12.1 THE DIFFERENCE BETWEEN LINEAR AND NONLINEAR SYSTEMS

In this chapter, we will consider a nonlinear system to be one which is modeled by a nonlinear ordinary differential equation. Recall that an ordinary differential equation is considered to be nonlinear if it contains nonlinear terms involving the dependent variable.

For example

$$\ddot{x} + \omega^2 x^2 = 0 \tag{12.1.1}$$

is a nonlinear ordinary differential equation because of the x^2 term. Note that we used dots to denote a second derivative of x with respect to time t. In this chapter, we will most often use time as the independent variable.

Let us now explore the consequence of the nonlinearity in (12.1.1). Suppose that we find two solutions to (12.1.1), $x_1(t)$ and $x_2(t)$. Is it true that $c_1 x_1(t) + c_2 x_2(t)$ is a solution? In other words, does linear superposition hold for nonlinear systems?

Inserting the superposition into (12.1.1), we find

$$(c_1\ddot{x}_1 + c_2\ddot{x}_2) + \omega^2 (c_1 x_1 + c_2 x_2)^2 = c_1\left(\ddot{x}_1 + \omega^2 x_1^2\right) + c_2\left(\ddot{x}_2 + \omega^2 x_2^2\right) + 2c_1 c_2 x_1 x_2$$
$$= 2c_1 c_2 x_1 x_2 \tag{12.1.2}$$

In other words, inserting $x = c_1 x_1(t) + c_2 x_2(t)$ into (12.1.1), does not produce zero. Therefore, the linear superposition of two solutions is not a solution to the differential equation. In general, linear superposition does not hold for nonlinear ordinary differential equations. That, combined with the difficulty in finding even one closed form solution for many nonlinear systems, means we will need other methods of extracting information about a system's behavior that do not involve finding a closed form solution.

The above may sound discouraging, but let us take a moment to consider what types of information do we typically need about a system.

1. Given a particular initial condition, what will be the long-term (steady-state) behavior of the system?

2. What are all the possible long-term (steady-state) behaviors of the system for a given collection of initial conditions?

DOI: 10.1201/9781003294320-12

3. What equilibria exist within the system, and what is their stability?

4. How far in the future are the predictions made by the model reliable?

The field of dynamical systems provides us a set of tools for addressing these questions, and more, even if we cannot solve the system's ODE in closed form. In this chapter, we will examine how each one of the above questions can be addressed for a variety of nonlinear systems.

Let us consider a nonlinear system with two dynamical variables x and y. For example, in physics, the dynamical variables are often a particle's position and velocity. The ODEs which model the system of interest will take the form

$$\left.\begin{aligned} \dot{x} &= f(x, y) \\ \dot{y} &= g(x, y) \end{aligned}\right\} \tag{12.1.3}$$

where f and g are nonlinear functions of the dynamical variables and are called the *equations of motion* for the system.

In physics, equations of motion are normally found using Newton's second law, the Euler-Lagrange equation, Hamilton's equations of motion, or other methods. The result is that we often work with second order ODEs, like in the case of the damped harmonic oscillator. Second order ODEs can be written in the form of (12.1.3). For example, we can rewrite (12.1.1) using $y = \dot{x}$, $f = y$, and $g = -\omega^2 x^2$.

Systems of the form (12.1.3) are called two-dimensional dynamical systems because they have two dynamical variables. In this chapter, we will focus primarily on two-dimensional systems, however, the methods we present can be easily extended to three and higher dimensions.

Before moving on, we should note that there are systems described by nonlinear partial differential equations. Solitons and nonlinear waves are examples of such systems. We will not examine nonlinear PDE's in this text, but we provide references for further reading at the end of this book.

12.2 PHASE PORTRAITS

Consider a physical system whose state can be modeled by two first order differential equations of the form (12.1.3). After reading the chapter on ODEs, your instinct may be to solve and plot $x(t)$ and $y(t)$. However, we can obtain a lot of information about the system if we instead plot its *phase portrait*. In the case of (12.1.3), the phase portrait is a plot of y vs. x.

We will typically need to find $x(t)$ and $y(t)$ numerically for nonlinear systems. To solve (12.1.3), we need to specify an initial condition $\mathbf{x}_0 = (x_0, y_0)$, which corresponds to the initial state of the system. The initial condition $\mathbf{x}_0$ is also a point in the phase portrait. Suppose the algorithm to solve (12.1.3) uses a step size Δt. Then at a time Δt later, the solution is $\mathbf{x}_1 = \mathbf{x}(\Delta t) = (x_1, y_1)$, which corresponds to a new point in the phase portrait (and a new state of the system). After another time step, we obtain a new point $\mathbf{x}_2$ and so on. Eventually we will have a curve in the phase portrait connecting one point to the next. A curve in the phase portrait is called a *trajectory*. Each trajectory corresponds to a solution to (12.1.3) and, therefore, illustrates the evolution of the system. Trajectories can take the form of curves, closed loops, or points. But in each case, the trajectory gives qualitative information about the system's behavior, without having a closed form analytical solution.

Example 12.1 demonstrates how to plot and interpret the phase portrait for the damped harmonic oscillator.

Example 12.1: Damped Harmonic Oscillator - revisited

Consider a massless horizontal spring with one end connected to a wall while the other is connected to a mass m. The mass is displaced from its equilibrium position and allowed to slide along a horizontal surface. A damping force, with damping parameter β, acts on the mass. From classical mechanics, we know the displacement from equilibrium x as a function of time satisfies the ODE

$$\ddot{x} + 2\beta\dot{x} + \omega_0^2 x = 0 \tag{12.2.1}$$

where $\omega_0^2 = k/m$ and k is the spring's spring constant. Using $\omega_0^2 = 1$, plot the phase portrait $\dot{x}$ vs x for $\beta = 0$ and $\beta = 0.2$ and interpret the result. Use as your initial conditions, $x(0) = 1$ and $\dot{x}(0) = 0$.

Solution:

Although we created the phase portrait for the damped harmonic oscillator in Chapter 10, here we revisit the problem with a focus on interpreting the result. We can rewrite $(12.2.1)$ as a system of two first order ODEs using $v = \dot{x}$

$$\left.\begin{array}{l} \dot{x} = v \\ \dot{v} = -2\beta v - \omega_0^2 x \end{array}\right\} \tag{12.2.2}$$

The code below uses Python to create the phase portrait v vs. x, shown in Figure 12.1.

Notice that when $\beta = 0$, the trajectory is a closed loop. After a period of time, the trajectory returns to its initial position,the point $(1,0)$, in the phase portrait. In other words, the system has returned to its initial station. Therefore, a closed loop trajectory in a phase portrait represents periodic behavior.

In the case of $\beta = 0.2$, the trajectory starts at the point $(1,0)$ but spirals in toward the origin. With damping present, the oscillation decays to the equilibrium position. Notice that after its first pass through equilibrium (to the left) the mass reaches a distance of only 0.5 from equilibrium. After the second pass (as it is moving to the right) the mass reaches a distance of 0.25 from the equilibrium. After which, the mass moves to the left coming to a rest at the equilibrium position (the origin in the phase portrait).

As you can see, we did not solve $(12.2.1)$ analytically, however, by plotting its phase portrait, we could obtain qualitative information about the behavior of the system.

```
import numpy as np
import matplotlib.pyplot as plt
from scipy.integrate import odeint

def dho(y, t, beta, omega):
    x, v = y
    dydt = [v, -2*beta*v - omega*x]
    return dydt

y0 = [1,0]
t = np.linspace(0,10,100)

sol0 = odeint(dho, y0, t, args = (0,1))
sol0_2 = odeint(dho, y0, t, args = (0.2,1))

fig, ax = plt.subplots(nrows = 1, ncols = 2)
```

```
ax[0].plot(sol0[:,0], sol0[:,1])
ax[0].set_title('Beta = 0')
ax[0].set_xlabel('x')
ax[0].set_ylabel('y')

ax[1].plot(sol0_2[:,0], sol0_2[:,1])
ax[1].set_title('Beta = 0.2')
ax[1].set_xlabel('x')
ax[1].set_ylabel('y')

fig.suptitle('Phase portrait of the damped harmonic oscillator')
fig.tight_layout()

plt.show()
```

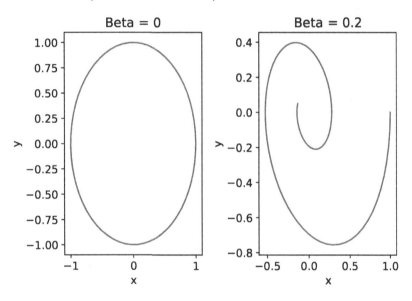

Figure 12.1 Graphical output from Example 12.1, demonstrating the phase portrait for the damped harmonic oscillator.

12.3 FIXED POINTS AND EQUILIBRIA

Notice that in Example 12.1 the trajectory ended at the origin in the case of $\beta = 0.2$ and orbited the origin in the case of $\beta = 0$. The origin is called a *fixed point* of the system. Fixed points correspond to equilibrium states. In this section, we will demonstrate how to calculate the fixed points of a system and how to determine their stability.

12.3.1 FINDING FIXED POINTS

We begin by considering a system of the form (12.1.3)

$$\left.\begin{aligned} \dot{x} &= f(x,y) \\ \dot{y} &= g(x,y) \end{aligned}\right\} \tag{12.3.1}$$

The fixed points (x^*, y^*) of (12.3.1) can be found by solving the system of equations

$$\left.\begin{aligned} f(x^*, y^*) &= 0 \\ g(x^*, y^*) &= 0 \end{aligned}\right\} \tag{12.3.2}$$

In other words, when $x = x^*$ and $y = y^*$, the dynamical variables x and y do not change because $\dot{x} = \dot{y} = 0$. A system whose initial state is a fixed point, will continue to stay in that state indefinitely.

Recall from your physics courses that a system is at equilibrium when it does not experience an acceleration. For example, if a particle is accelerating, its position and velocity (its dynamical variables) are changing. Hence, we can refer to fixed points as equilibrium states because at the fixed point, the dynamical variables are no longer changing.

For the damped harmonic oscillator, there is only one fixed point $(x^*, v^*) = (0,0)$, corresponding to the mass at rest at the spring's equilibrium position. Example 12.2 includes a model from population biology which has multiple fixed points.

Example 12.2: Predator-prey models

The Lotka-Volterra equations model the populations of two species who interact, one as predator and the other as prey. Suppose that $H(t)$ $(P(t))$ is the population of the prey (predator) species at time t. The Lotka-Volterra equations are:

$$\left.\begin{aligned} \dot{H} &= rH - aHP \\ \dot{P} &= -mP + bHP \end{aligned}\right\} \tag{12.3.3}$$

where r is the linear growth rate of the prey, m is the death rate of the predator, and a and b are parameters that describe the interaction of the two species. All parameters are real and positive. Find and interpret the fixed points of (12.3.3).

Solution:
We will use Python to solve the equations

$$rH^* - aH^*P^* = 0$$
$$-mP^* + bH^*P^* = 0$$

This system has two fixed points. The fixed point at the origin corresponds to the state where both species go extinct. The second fixed point has a fixed population of $H^* = m/b$ and $P^* = r/a$. The population of each species depends on the parameters of the other. For example, the equilibrium value of the prey depends on the death rate of the predator and how much the predator population benefits from predation. A high death rate for the predators means that more prey can avoid predation. Likewise, if b is large, then H^* is small. In other words, if b is high, each predation event makes more predators which can then consume more prey.
Note that equation-based population models are often used in cases where populations are large. For example, $H = 2$ could mean that there are 200 members of the population (or 2000, etc.) If the population gets to be too small, agent-based modeling, as opposed to differential equations, tends to be the preferred method of describing populations.

```
from sympy import solve
from sympy.abc import H,P, r, m, a, b

fps = solve((r*H-a*H*P, -m*P+b*H*P),(H,P))

print('-'*28,'CODE OUTPUT','-'*29,'\n')
print('The fixed points are: ', fps)


-------------------------- CODE OUTPUT ----------------------------

The fixed points are:  [(0, 0), (m/b, r/a)]
```

While (12.3.2) shows us how to calculate fixed points, it does not tell us anything about their stability. Recall from your classical mechanics course that equilibria have a stability associated with them. They can be stable, unstable, or neutrally stable. Fixed points in dynamical systems have similar stabilities, but in two-dimensional systems, their classifications are not as simple.

12.3.2   THE CLASSIFICATION OF FIXED POINTS

Let us return to the two-dimensional dynamical system (12.3.1)

$$\dot{x} = f(x, y)$$
$$\dot{y} = g(x, y)$$

To find the stability of a fixed point, we need to know how the system behaves near each equilibrium state (x^*, y^*). A Taylor series expansion of f and g would provide a polynomial approximation of the functions near a particular point. Hence, by Taylor expanding f and g, we can get a simple mathematical form for the equations of motion near the equilibrium states.

To perform the Taylor series expansion, we transform coordinates:

$$\left.\begin{aligned} x &= x^* + u \\ y &= y^* + v \end{aligned}\right\} \tag{12.3.4}$$

where u and v are small compared to x^* and y^*, respectively. Note that the origin of the (u, v) coordinate system is the fixed point (x^*, y^*). In other words, we will perturb (12.3.1) by starting the system just slightly off of the fixed point (equilibrium position) in the phase portrait.

Next, we substitute (12.3.4) into the equations of motion (12.3.1) and Taylor expand about the point (x^*, y^*). Note that $\dot{x} = \dot{u}$ because x^* is constant, similarly for $\dot{y}$.

$$\dot{u} = f(x^* + u, y^* + v) \tag{12.3.5}$$

$$= f(x^*, y^*) + u\frac{\partial f}{\partial x}\bigg|_{(x^*,y*)} + v\frac{\partial f}{\partial y}\bigg|_{(x^*,y*)} + \mathcal{O}\left(u^2, v^2\right) \tag{12.3.6}$$

$$= u\frac{\partial f}{\partial x}\bigg|_{(x^*,y*)} + v\frac{\partial f}{\partial y}\bigg|_{(x^*,y*)} \tag{12.3.7}$$

The term $f(x^*, y^*) = 0$ by the definition of fixed points and $\mathcal{O}\left(u^2, v^2\right)$ contains second-order and higher terms, which are ignored because u and v are small.

The same procedure can be repeated for $g(x, y)$:

$$\dot{v} = u \frac{\partial g}{\partial x}\bigg|_{(x^*, y*)} + v \frac{\partial g}{\partial y}\bigg|_{(x^*, y*)} \tag{12.3.8}$$

The Taylor expansion results in a linearized set of equations in u and v about the fixed point. The above process is sometimes called the *linearization* of the equations of motion.

By solving the linearized equations, we can identify the stability of the fixed point. Before solving the linearized equations, it is first helpful to cast them into matrix form:

$$\begin{pmatrix} \dot{u} \\ \dot{v} \end{pmatrix} = \begin{pmatrix} \frac{\partial f}{\partial x} & \frac{\partial f}{\partial y} \\ \frac{\partial g}{\partial x} & \frac{\partial g}{\partial y} \end{pmatrix} \begin{pmatrix} u \\ v \end{pmatrix} \tag{12.3.9}$$

where we have removed the notation for the evaluation of each partial derivative for simplicity in notation. The 2×2 matrix in (12.3.9) is called the *Jacobian matrix A*.

We can rewrite (12.3.9) in the form:

$$\dot{\mathbf{u}} = A\mathbf{u} \tag{12.3.10}$$

where $\mathbf{u} = (u, v)$.

Equation (12.3.10) represents a system of first-order differential equations which can be solved with the solution $\mathbf{u} = \mathbf{w}e^{\lambda t}$, where $\mathbf{w}$ is a constant vector. Inserting the trial solution into (12.3.10) yields

$$\lambda \mathbf{w} = A\mathbf{w} \tag{12.3.11}$$

which is the eigenvector equation for the matrix A. Hence, in our trial solution, $\mathbf{u} = \mathbf{w}e^{\lambda t}$, λ is the eigenvalue of A associated with the eigenvector, $\mathbf{w}$.

As we learned in Chapter 7, we can find the eigenvalues of A by solving

$$\det (A - \lambda \mathcal{I}) = \mathbf{0} \tag{12.3.12}$$

where $\mathcal{I}$ is the 2×2 identity matrix and $\mathbf{0}$ is the zero vector.

Because the matrix A is a 2×2 matrix, it will produce two eigenvalues and each eigenvalue can be inserted into $(A - \lambda \mathcal{I}) \mathbf{w} = \mathbf{0}$ to get its associated eigenvector, $\mathbf{w}$.

Using the principle of linear superposition, the solution to (12.3.10) is therefore:

$$\mathbf{u} = c_1 \mathbf{w}_1 e^{\lambda_1 t} + c_2 \mathbf{w}_2 e^{\lambda_2 t} \tag{12.3.13}$$

where λ_i is the eigenvalue of A associated with the eigenvector, $\mathbf{w}_i$ and c_i are constants determined by initial conditions.

Equation (12.3.13) tells us something about the growth or decay of the perturbation near the fixed point. If the eigenvalues are all real, then there are three possible cases.

1. $\lambda_1 < 0$ and $\lambda_2 < 0$: The perturbation decays exponentially along the directions $\mathbf{w}_1$ and $\mathbf{w}_2$ in the phase portrait. The trajectory settles onto the fixed point which is classified as a *stable node*. Stable nodes are an example of an attractor, an object in the phase portrait which nearby trajectories tend toward as $t \to \infty$.

2. $\lambda_1 > 0$ and $\lambda_2 > 0$: The perturbation grows exponentially along the directions $\mathbf{w}_1$ and $\mathbf{w}_2$ in the phase portrait. The trajectory repels away from the fixed point which is classified as an *unstable node*. Unstable nodes are unstable equilibrium states.

3. $\lambda_1 > 0$ and $\lambda_2 < 0$: The perturbation grows along the direction of $\mathbf{w}_1$ and decays along $\mathbf{w}_2$. The fixed point is still considered to be unstable, but it is now called a *saddle* point.

Saddles are important in dynamical systems. They sometimes partition phase portraits into different basins of attraction. An attractor's *basin of attraction* is the set of all points in the phase portrait, that when taken as initial conditions, have trajectories which lead to the fixed point.

The eigenvalues of A can also be complex. In this case $\lambda_2 = \lambda_1^*$. Suppose that $\lambda_1 = \alpha + i\beta$, then (12.3.13) becomes

$$\mathbf{u} = e^{\alpha t}\left(c_1\mathbf{w}_1 e^{i\beta t} + c_2\mathbf{w}_2 e^{-i\beta t}\right) \qquad (12.3.14)$$

The result is that trajectories near (x^*, y^*) oscillate around the fixed point. However, in the case of complex eigenvalues, there are three possibilities.

1. $\mathrm{Re}(\lambda_1) = 0$: In the case where $\alpha = 0$, the (x^*, y^*) is called a *center*. In Example 12.1, the fixed point at the origin is a center when $\beta = 0$. Centers are neutrally stable fixed points. Trajectories neither attract toward nor repel from a center.

2. $\mathrm{Re}(\lambda_1) < 0$: The perturbation decays exponentially while revolving around the fixed point. The fixed point is called a *stable spiral* (such as in the case of $\beta = 0.2$ in Example 12.1). Stable spirals are stable fixed points.

3. $\mathrm{Re}(\lambda_1) > 0$: The fixed point is called an *unstable spiral*, which corresponds to an exponentially growing oscillation. Unstable spirals are unstable fixed points.

If either eigenvalue is equal to zero, then we must rely on numerical solutions to understand the system's behavior. Degeneracies can also occur when $\lambda_1 = \lambda_2 = \lambda$. We will not examine degeneracies in this book, and we will focus only on non-zero eigenvalues for the Jacobian matrix.

The next two examples revisit the damped harmonic oscillator and the Lotka-Volterra equations, in order to calculate the stability of each fixed point.

Example 12.3: Stability of the damped harmonic oscillator

Classify the fixed point at the origin of the damped harmonic oscillator (12.2.2) with $\omega_0^2 = 1$, for $\beta = 0$ and $\beta = 0.2$

Solution:

We begin by writing (12.2.2) in the form of (12.3.1)

$$\dot{x} = f(x, y) = y$$
$$\dot{y} = g(x, y) = -\beta y - \omega_0^2 x$$

In the Python code we generate the Jacobian matrix A and its eigenvalues. Note the use of the variable ω in place of ω_0. The results show us that when $\beta = 0$, the fixed point is a center because the eigenvalues are purely imaginary. When $\beta = 0.2$, the eigenvalues are complex with a negative real part. In the case of $\beta = 0.2$, the fixed point is a stable spiral.

```
from sympy import diff, Matrix, symbols

x, v, beta, omega = symbols('x, v, beta, omega')
```

```
f = v
g = -beta * v - omega**2 * x

A = Matrix([[diff(f,x),diff(f,v)],[diff(g,x),diff(g,v)]])\
    .subs([(x,0),(v,0)])

print('-'*28,'CODE OUTPUT','-'*29,'\n')

print('The eigenvalues with beta = 0 are:')
print((A.subs([(beta,0),(omega,1)])).eigenvals())

print('\nThe eigenvalues with beta = 0.2 are:')
print((A.subs([(beta,0.2),(omega,1)])).eigenvals())
```

```
---------------------------- CODE OUTPUT ----------------------------

The eigenvalues with beta = 0 are:
{-I: 1, I: 1}

The eigenvalues with beta = 0.2 are:
{-0.1 - 0.99498743710662*I: 1, -0.1 + 0.99498743710662*I: 1}
```

Example 12.4: Stability and the Lotka-Volterra equations

Classify the fixed points of $(12.3.3)$.

Solution:

From $(12.3.3)$

$$f = rH - aHP$$
$$g = -mP + bHP$$

Although we could use SymPy's solve command to get the fixed points, we already know them. So we stored them each in their own array, fps_1 and fps_2.

Recall that the model parameters m, r, a, and b are positive. Hence, the origin with one positive and one negative eigenvalue is a saddle. The state where both species are extinct is unstable. That's good news for our animals! The other fixed point at $(H^*, P^*) = (m/b, r/a)$ has purely imaginary eigenvalues and is a center. Hence, the populations of both species will oscillate about that point.

```
from sympy import diff, Matrix, symbols, solve

r, a, m, b, H, P = symbols('r, a, m, b, H, P')

f = r*H-a*H*P
g = -m*P+b*H*P
```

```
fps_1 = [0,0]
fps_2 = [m/b, r/a]

A_1 = Matrix([[diff(f,H),diff(f,P)],[diff(g,H),diff(g,P)]]) \
    .subs([(H,fps_1[0]),(P,fps_1[1])])

A_2 = Matrix([[diff(f,H),diff(f,P)],[diff(g,H),diff(g,P)]]) \
    .subs([(H,fps_2[0]),(P,fps_2[1])])

print('-'*28,'CODE OUTPUT','-'*29,'\n')

print('The eigenvalues of the origin are:')
print(A_1.eigenvals())

print('\nThe eigenvalues of (m/b, r/a) are:')
print(A_2.eigenvals())

---------------------------- CODE OUTPUT ----------------------------

The eigenvalues of the origin are:
{r: 1, -m: 1}

The eigenvalues of (m/b, r/a) are:
{-sqrt(-m*r): 1, sqrt(-m*r): 1}
```

12.4   BIFURCATIONS OF FIXED POINTS

Notice that the location of the fixed point at $(H^*, P^*) = (m/b, r/a)$ in Example 12.4, depends on the value of the parameters, m, b, r, and a. For example, if $r \to 0$ and $m \to 0$, the fixed point will move toward the origin and eventually collide with it. This collision is an example of what is called a *bifurcation*.

Bifurcations are topological changes in the phase portrait. During a bifurcation, fixed points can collide and/or change stability. Sometimes, the fixed points disappear entirely. There are many different types of bifurcations with names like, saddle-node bifurcation (where a saddle and node collide and disappear), transcritical bifurcation (where two fixed points collide and survive but change stability), pitchfork bifurcation (where two new fixed points spring for a third), and many others. It is impossible to provide a survey of all possible bifurcations in one chapter. However, we provide opportunities to explore several of these bifurcations in the end of chapter problems.

Understanding the potential bifurcations in a dynamical system is of the utmost importance. As parameters change value, a once-stable and desirable equilibrium state may become unstable, as we demonstrate in Example 12.5.

Example 12.5: SIS model of disease in epidemiology

Mathematical models of disease spread are an important component to epidemiology, especially when it comes to public policy. The type of disease dictates the kind of model used. For example, the SIS model (susceptible-infected-susceptible) is appropriate for diseases that have repeat infections like the common cold.

Consider a fixed population of size $N = S + I$ where S is the number of people susceptible to a disease and I is the number of people infected by the disease. The SIS model is

$$\left.\begin{array}{l} \dot{S} = \mu I - \beta S I \\ \dot{I} = -\mu I + \beta S I \end{array}\right\} \tag{12.4.1}$$

where μ is the recovery rate from the disease and β is the disease's infection rate. Those who recover from the illness, can be infected again. Find the fixed points of (12.4.1) and identify any bifurcations that can occur as the models parameters μ and β change.

Solution:

Because $S = N - I$, we can focus on the $\dot{I}$ equation (the $\dot{S}$ equation does not provide extra information), thus reducing the two-dimensional model to a one-dimensional one. After substitution, the $\dot{I}$ equation in (12.4.1) becomes

$$\dot{I} = (\beta - \mu) I - \beta \frac{I^2}{N} \tag{12.4.2}$$

Solving $\dot{I} = 0$ for fixed points, I^* we find

$$I^* = 0, \left(1 - \frac{\mu}{\beta}\right) N$$

When $I^* = 0$, then $S^* = N$ and there is no disease in the population. However, when $I^* = \left(1 - \frac{\mu}{\beta}\right) N$, the disease is present in the population. For simplicity we will reserve the variable I^* for the fixed point at $I = \left(1 - \frac{\mu}{\beta}\right) N$.

Next, we need to identify the stability of each fixed point. In Problem 2, you will derive the stability condition for a one-dimensional system. For now, we will simply use its result. We define $\lambda = d\dot{I}/dI$ evaluated at the fixed point. If $\lambda < 0$, then the fixed point is stable. If $\lambda > 0$, then the fixed point is unstable.

Let us calculate $d\dot{I}/dI$ from (12.4.2)

$$\frac{d\dot{I}}{dI} = \dot{I}' = (\beta - \mu) - 2\beta \frac{I}{N} \tag{12.4.3}$$

Evaluating (12.4.3) at each of the fixed points gives the fixed points stability

$$\dot{I}'(0) = \beta - \mu$$
$$\dot{I}'(I^*) = -\beta + \mu$$

The stability and location of the fixed points along the one-dimensional phase portrait is summarized in Figure 12.2.

If $\beta < \mu$ then the disease-free state is stable. Note that when the disease-free state is stable, the other fixed point I^* is both unstable and less than zero (and therefore, also nonphysical). However, as β increases and passes through the value μ, I^* collides with 0. The bifurcation occurs at $\beta = \mu$, when the rate of infection is equal to the rate of recovery. The point becomes bistable (to see this, examine $\dot{I}$ for $\beta = \mu$). Once $\beta > \mu$, the disease-free state becomes unstable and I^* becomes stable. The disease is present in the population and the equilibrium value (i.e. the number of sick people) is $\left(1 - \frac{\mu}{\beta}\right) N$.

The stability analysis suggests an important model parameter is the *reproductive number* $R_0 = \beta/\mu$. When $R_0 < 1$, the disease-free state is stable and the disease dies out in the population. When $R_0 > 1$, the equilibrium $\left(1 - \frac{\mu}{\beta}\right) N$ is stable and, in time, the number of sick people will approach $\left(1 - \frac{\mu}{\beta}\right) N$.

For an initial number of infected people I_0, (12.4.2) can be solved in closed form to obtain

$$I(t) = \frac{I^*}{1 + I_1 \exp(-\alpha t)} \tag{12.4.4}$$

where $\alpha = \beta - \mu$, $I^* = \left(1 - \frac{\mu}{\beta}\right) N$, and $I_1 = (I^*/I_0 - 1)$.

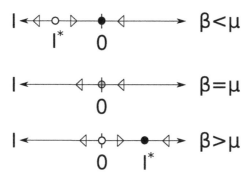

Figure 12.2 The one-dimensional phase portrait for (12.4.2). White (black) dots represent unstable (stable) fixed points. The gray dot at the origin of the middle graph represents a bistable point (attracts from one direction and repels from the other). The triangles represent the directions of trajectories on the line. Note that $I^* = \left(1 - \frac{\mu}{\beta}\right) N$.

Figure 12.3 is a plot of (12.4.4) which shows the qualitative difference for $R_0 < 1$ and $R_0 > 1$ with $I_0 = 40$ and $N = 100$.

```python
import numpy as np
import matplotlib.pyplot as plt

N = 100
I_0 = 40
t = np.linspace(0,40,100)

mu = 0.1
beta = 1.0
alpha = beta - mu
I_star =(1 - mu/beta)*N
I_1 = I_star/I_0 - 1

I_r_greater = I_star/(1 + I_1*np.exp(-alpha*t))

mu = 2.0
beta = 1.0
alpha = beta - mu
I_star =(1 - mu/beta)*N
I_1 = I_star/I_0 - 1

I_r_less = I_star/(1 + I_1*np.exp(-alpha*t))

plt.plot(t,I_r_greater, c = 'k', label = 'R_0 = 10')
plt.plot(t,I_r_less, c = 'k', linestyle='dashed', label = 'R_0 = 1/2')
plt.ylabel('I(t)')
plt.xlabel('time')
plt.title('Bifurcation in the SIR model')
plt.legend()
plt.show()
```

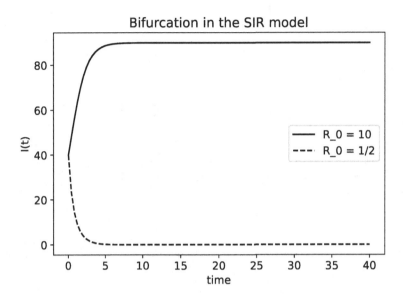

Figure 12.3 Graphical output from Example 12.5, showing the number of infected individuals $I(t)$ when the reproductive number $R_0 = 10$ (solid line) and when $R_0 = 1/2$ (dashed line). Notice that when $R_0 = 1/2$, the disease-free state is stable. However, when $R_0 = 10$, then the disease-free state is unstable and $I(t)$ quickly reaches a large value.

12.5 THE PHASE PORTRAIT, REVISITED

Although we presented phase portraits earlier this chapter, it is now time to see their full value when analyzing nonlinear systems. We begin by considering a two-dimensional dynamical system

$$\left.\begin{aligned}
\dot{x} &= f\,(x,y) \\
\dot{y} &= g\,(x,y)
\end{aligned}\right\} \tag{12.5.1}$$

where x and y are dynamical variables. The phase portrait is a plot of y vs. x. Note that in the case of three dynamical variables, a three-dimensional phase portrait can be plotted. For dimensions higher than three, we would need to visualize (typically two-dimensional) subspaces of the phase portrait.

Below, we outline the process of creating the phase portrait for a dynamical system of the form (12.5.1).

1. Find and classify the fixed points of (12.5.1).

2. For any saddle points, numerically solve (12.5.1) forward in time with an initial condition near the saddle point and in the direction of the eigenvector associated with the positive eigenvalue. The trajectory plotted is called the saddle's *unstable manifold*. There will be two unstable manifolds, generally on opposite sides of the saddle point.

3. For any saddle points, numerically solve (12.5.1) backward in time with an initial condition near the saddle point and in the direction of the eigenvector associated with the negative eigenvalue. The trajectory plotted is called the saddle's *stable manifold*. There will be two stable manifolds, generally on opposite sides of the saddle point. Stable manifolds often serve as important boundaries in the phase portrait.

4. Numerically solve and plot the trajectories of (12.5.1) using initial conditions that will lead to each of the attractors in the phase portrait such as stable nodes, spirals, limit cycles (see below), or strange attractors (see below).

Once completed, the phase portrait will provide a graphical representation of the transient and steady state behaviors of the system for many (and sometimes all) initial conditions. If there are bifurcations in the dynamical system, you should create several phase portraits, before and after each bifurcation.

The next two examples demonstrate the step-by-step procedure for plotting a phase portrait.

Example 12.6: A competition model

Consider two species, X and Y competing for the same limited resources. Let x (y) be the population size of species X (Y). A population model for the two species is sometimes called the *Lotka-Volterra model of competition*

$$\left.\begin{aligned}\dot{x} &= x\,(2 - x - 2y)\\ \dot{y} &= 3y\,(2 - 2x - y)\end{aligned}\right\} \qquad (12.5.2)$$

Plot the phase portrait of (12.5.2) and discuss the result. Note that as with predator-prey models, x and y are multiples of a large number (such as 100 or 1000).

Solution:
We begin by finding the fixed points of (12.5.2).

```
from sympy import symbols, solve

x, y = symbols('x,y')

f = x*(2 - x - 2*y)
g = 3*y*(2 - 2*x - y)

print('-'*28,'CODE OUTPUT','-'*29,'\n')
solve((f,g),(x,y))


-------------------- CODE OUTPUT -----------------------------

[(0, 0), (0, 2), (2/3, 2/3), (2, 0)]
```

Note that there are four equilibrium states: $x_0^* = (0,0)$ where both species are extinct, $x_1^* = (0,2)$ where species X is extinct, $x_2^* = (2,0)$ where species Y is extinct, and $x_3^* = (2/3, 2/3)$ where both species exist in equal numbers.

Next, we need to classify each fixed point. We define a function called JacEval which computes the Jacobian using the formula for the matrix in (12.3.9). The function returns the eigenvalues of the Jacobian matrix for each fixed point.

```
# This is a continuation of the previous code

from sympy import Matrix, diff

x0, y0 = [0,0]
x1, y1 = [0,2]
x2, y2 = [2,0]
x3, y3 = [2/3,2/3]

def JacEval(f,g, x_p, y_p):
    A = Matrix([[diff(f,x),diff(f,y)],[diff(g,x),diff(g,y)]])\
.subs([(x,x_p),(y,y_p)])
    return list(A.eigenvals().keys())

print('-'*28,'CODE OUTPUT','-'*29,'\n')
print('The eigenvalues of (0,0) are: ', JacEval(f,g,x0,y0))
print('The eigenvalues of (0,2) are: ', JacEval(f,g,x1,y1))
print('The eigenvalues of (2,0) are: ', JacEval(f,g,x2,y2))
print('The eigenvalues of (2/3,2/3) are: ', JacEval(f,g,x3,y3))

------------------------- CODE OUTPUT ---------------------------

The eigenvalues of (0,0) are:  [2, 6]
The eigenvalues of (0,2) are:  [-2, -6]
The eigenvalues of (2,0) are:  [-6, -2]
The eigenvalues of (2/3,2/3) are:  [1.07036751697599, -3.73703418364266]
```

We see that $\mathbf{x}_0^* = (0,0)$ is an unstable node, $\mathbf{x}_1^* = (0,2)$ is a stable node, $\mathbf{x}_2^* = (2,0)$ is a stable node, and $\mathbf{x}_3^* = (1,1)$ is a saddle. Fortunately for our animals, the state of extinction is unstable. However, there are only two attracting fixed points in the phase portrait. For each stable node, one of the species is extinct. The phase portrait will tell us which initial conditions lead to X being extinct and which ones lead to Y being extinct.

To understand the outcomes of various initial conditions, we need to understand the dynamics near the saddle point. The Jacobian matrix about the saddle is (see Problem 13)

$$J = \begin{pmatrix} -2/3 & -4/3 \\ -4 & -2 \end{pmatrix}$$

The eigenvectors of J will help us identify initial conditions for the saddle's stable and unstable manifold.

```
# This is a continuation of the previous code

import numpy as np
J = np.array([[-2/3,-4/3],[-4,-2]])
w, v = np.linalg.eig(J)

print('-'*28,'CODE OUTPUT','-'*29,'\n')
print('The eigenvalues of x_3 are: ', w)
print('The eigenvectors of x_3 are: ', np.transpose(v))

------------------------- CODE OUTPUT ---------------------------

The eigenvalues of x_3 are:  [ 1.07036752 -3.73703418]
The eigenvectors of x_3 are:  [[ 0.60889368 -0.79325185]
 [ 0.3983218   0.91724574]]
```

The command eig returns a list of eigenvalues and eigenvectors. The printed eigenvectors are the rows of the array v. We see that for the unstable manifold (the eigenvector associated with the positive eigenvalue), we will want to begin initial conditions along a line passing through the saddle with a negative slope. For the stable manifold (the eigenvector associated with the negative eigenvalue), the initial conditions should lie along a line passing through the saddle with a positive slope.

In the next block of code we solve $(12.5.2)$ for several initial conditions around the saddle and plot the phase portrait which appeared in Figure 12.4. The type of manifold is given in the variable name used with the command odeint. Notice that care must be used when integrating backwards in time to get the stable manifold. The solution can grow exponentially large if one integrates too far backward in time. We also include circles for the fixed points and a normalized vector plot to show the flow of the vector field.

Note that we normalized the vector field so that all of the arrows had the same length. The arrows in the vector field plot help guide the eye for all possible trajectories. We see that for initial conditions above the saddle's stable manifold, the attracting equilibrium is that only species Y survives. We say that the area of the phase portrait above the stable manifold is the *basin of attraction* for x_1^*. The basin of attraction is the set of initial conditions whose trajectories approach an attractor as $t \to \infty$. For initial conditions below the stable manifold, only species X survives. Hence, the phase portrait is split into two basins of attraction, one for each attracting fixed point.

```python
# This is a continuation of the previous code

from scipy.integrate import odeint
import matplotlib.pyplot as plt
import numpy as np
np.seterr(invalid='ignore');

def sys(vec, t):
    x, y = vec
    return [x*(2 - x - 2*y), 3*y*(2 - 2*x - y)]

#define times for integation
t_forward = np.linspace(0,100,1000)
t_backward_1 = np.linspace(0,-3,100)
t_backward_2 = np.linspace(0,-0.96,100)

#define initial conditions near saddle point
vec0 = [x3 + 0.01, y3 - 0.01 ]
vec1 = [x3 - 0.01, y3 + 0.01 ]
vec2 = [x3 - 0.01, y3 - 0.01]
vec3 = [x3 + 0.01, y3 + 0.01]

#compute the stable and unstable manifolds
unstable_1 = odeint(sys, vec0, t_forward)
unstable_2 = odeint(sys, vec1, t_forward)
stable_1 = odeint(sys, vec2, t_backward_1)
stable_2 = odeint(sys, vec3, t_backward_2)

#create a normalized vector field
x_v, y_v = np.meshgrid(np.linspace(0,2.25,20),np.linspace(0,2.25,20))
f_vec =x_v*(2 - x_v - 2*y_v)
g_vec = 3*y_v*(2 - 2*x_v - y_v)
r = (f_vec**2 + g_vec**2)**(0.5) #normalizes vectors
```

```
#create plot of phase portrait
fp0 = plt.Circle((x0,y0), 0.03, color='k')
fp1 = plt.Circle((x1,y1), 0.03, color='k')
fp2 = plt.Circle((x2,y2), 0.03, color='k')
fp3 = plt.Circle((x3,y3), 0.04, color='k')

fig, ax = plt.subplots()

ax.plot(unstable_1[:,0],unstable_1[:,1], 'k')
ax.plot(unstable_2[:,0],unstable_2[:,1], 'k')
ax.plot(stable_1[:,0],stable_1[:,1], 'k')
ax.plot(stable_2[:,0],stable_2[:,1], 'k')

ax.quiver(x_v,y_v,f_vec/r, g_vec/r)

ax.add_patch(fp0)
ax.add_patch(fp1)
ax.add_patch(fp2)
ax.add_patch(fp3)

plt.xlabel('x')
plt.ylabel('y')
plt.title('Two basins of attraction')
plt.show()
```

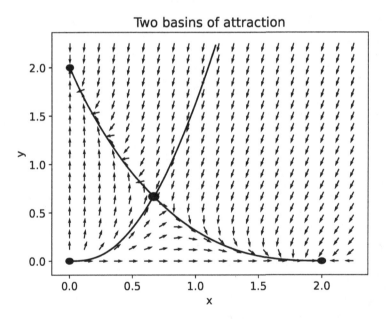

Figure 12.4 Graphical output from Example 12.6, showing how the phase portrait can split into two basins of attraction in the presence of several attracting fixed points.

In the next example, we will examine a system from physics which has no attractors, but has an interesting stable/unstable manifold structure called *homoclinic orbits*.

Example 12.7: The double-well potential

You are likely familiar with the elastic potential energy, $V = 1/2kx^2$ which has one equilibrium. Consider the double well potential,

$$V = -\frac{1}{2}x^2 + \frac{1}{4}x^4 \tag{12.5.3}$$

which, as we will see, has three equilibria. Using Newton's second law, we can find the equations of motion for unit mass in the double well potential

$$\ddot{x} = -\frac{dV}{dx} = x - x^3 \tag{12.5.4}$$

Plot and analyze the phase portrait for the double well potential.

Solution:

We begin by rewriting (12.5.3) as a two dimensional system,

$$\dot{x} = y$$
$$\dot{y} = x - x^3 \tag{12.5.5}$$

We can solve for the fixed points by inspection. Note that there are three equilibrium states: $\mathbf{x}_0^* = (0,0)$, $\mathbf{x}_1^* = (1,0)$ and, $\mathbf{x}_2^* = (-1,0)$. Next, we classify each fixed point.

```python
from sympy import Matrix, diff, symbols

x, y = symbols('x, y')

f, g = y, x-x**3

x0, y0 = [0,0]
x1, y1 = [1,0]
x2, y2 = [-1,0]

def JacEval(f,g, x_p, y_p):
    A = Matrix([[diff(f,x),diff(f,y)],[diff(g,x),diff(g,y)]])\
        .subs([(x,x_p),(y,y_p)])
    return list(A.eigenvals().keys())

print('-'*28,'CODE OUTPUT','-'*29,'\n')
print('The eigenvalues of (0,0) are: ', JacEval(f,g,x0,y0))
print('The eigenvalues of (1,0) are: ', JacEval(f,g,x1,y1))
print('The eigenvalues of (0,-1) are: ', JacEval(f,g,x2,y2))

------------------------ CODE OUTPUT ----------------------------

The eigenvalues of (0,0) are:  [-1, 1]
The eigenvalues of (1,0) are:  [-sqrt(2)*I, sqrt(2)*I]
The eigenvalues of (0,-1) are:  [-sqrt(2)*I, sqrt(2)*I]
```

We see that $\mathbf{x}_0^* = (0,0)$ is a saddle and the other two points are centers.

The Jacobian matrix about the saddle is

$$J = \begin{pmatrix} 0 & 1 \\ 1 & 0 \end{pmatrix}$$

The eigenvectors of J will help us identify initial conditions for the saddle's stable and unstable manifold.

```
# This is a continuation of the previous code

import numpy as np
J = np.array([[0,1],[1,0]])
w, v = np.linalg.eig(J)

print('-'*28,'CODE OUTPUT','-'*29,'\n')
print('The eigenvalues of x_0 are: ', w)
print('The eigenvectors of x_0 are: ', np.transpose(v))

------------------------- CODE OUTPUT ----------------------------

The eigenvalues of x_0 are:  [ 1. -1.]
The eigenvectors of x_0 are:   [[ 0.70710678   0.70710678]
 [-0.70710678   0.70710678]]
```

Next, we plot the phase portrait, which is shown in Figure 12.5. There is no need to plot the stable manifold because this saddle has a *homoclinic orbit*. Homoclinic orbits are trajectories which start and end at the same point. Notice that the saddle's manifolds serve as boundaries for each of the other equilibria. Initial conditions that start inside a homoclinic orbit, oscillate about the equilibrium inside the orbit. Trajectories starting outside the homoclinic orbits also oscillate, but they pass through all the equilibria. There are no basins of attraction in this case because there are no attractors. Conservative systems do not have attractors, only saddles and centers.

```
# This is a continuation of the previous code

from scipy.integrate import odeint
import matplotlib.pyplot as plt
np.seterr(invalid='ignore');

import warnings
warnings.filterwarnings("ignore")

def sys(vec, t):
    x, y = vec
    return [y, x - x**3]

t_forward = np.linspace(0,100,1000)

vec0 = [x0 + 0.01, y0 + 0.01 ]
vec1 = [x0 - 0.01, y0 - 0.01 ]

unstable_1 = odeint(sys, vec0, t_forward)
unstable_2 = odeint(sys, vec1, t_forward)


x_v, y_v = np.meshgrid(np.linspace(-1.5,1.5,20),np.linspace(-1,1,20))
```

```
f_vec = y_v
g_vec = x_v - x_v**3
r = (f_vec**2 + g_vec**2)**(0.5) #normalizes vectors

fp0 = plt.Circle((x0,y0), 0.03, color='k')
fp1 = plt.Circle((x1,y1), 0.03, color='k')
fp2 = plt.Circle((x2,y2), 0.03, color='k')

fig, ax = plt.subplots()
ax.plot(unstable_1[:,0],unstable_1[:,1], 'k')
ax.plot(unstable_2[:,0],unstable_2[:,1], 'k')

ax.quiver(x_v,y_v,f_vec/r, g_vec/r)

ax.add_patch(fp0)
ax.add_patch(fp1)
ax.add_patch(fp2)

plt.xlabel('x')
plt.ylabel('y')
plt.title('Phase portrait of the double well potential')
plt.show()
```

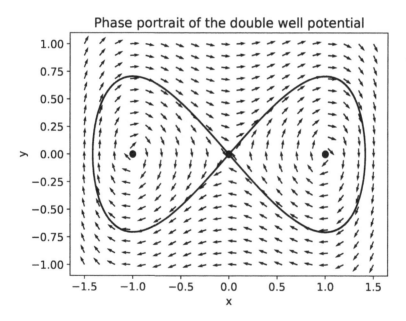

Figure 12.5 Graphical output from Example 12.7, showing the phase portrait of the double well potential.

12.6 NONLINEAR OSCILLATIONS AND LIMIT CYCLES

Oscillations are a common physical behavior. In a typical undergraduate physics education, the student focuses mostly on small amplitude oscillations. For example, a simple pendulum with mass m and length ℓ is described by the equation

$$\ddot{\theta} + \omega_0^2 \sin \theta = 0 \qquad (12.6.1)$$

where $\theta = \theta(t)$ is the angle the pendulum makes with the vertical (such that $\theta = 0$ corresponds to the lowest point of the pendulum) and $\omega_0^2 = g/\ell$. If one restricts the simple plane pendulum to small amplitude oscillations (approximately 15 degrees or less), then $\sin \theta \approx \theta$ and the motion can be approximated by the simple harmonic oscillator equation whose solutions are in the form of sines and cosines. For the simple harmonic oscillator, there is no relationship between the period of oscillation and the amplitude of oscillation. In the limit of small amplitude oscillations, the angular frequency of the pendulum oscillation is $\omega_0 = \sqrt{g/\ell}$ regardless of the pendulum's initial amplitude.

However, anyone who has ever been on a swing set knows that large amplitude oscillations have a period-amplitude relationship. As the amplitude of oscillation increases, so does the period. For larger amplitude oscillations, the small angle approximation is no longer accurate. The equation that best describes the pendulum's behavior (12.6.1) is nonlinear and we refer to the pendulum as a *nonlinear oscillator*. Nonlinear oscillators have period-amplitude relationships.

While linear oscillations can be expressed as a superposition of sines and cosines (see Fourier series and transforms), nonlinear oscillations do not obey the principle of superposition. While still periodic, the wave forms for nonlinear oscillators can often have very steep slopes. It would be impossible to provide a survey of all nonlinear oscillators in this section (or even this book!). Instead, we will examine the phase portrait and wave forms of a common nonlinear oscillator, the Duffing equation.

Recall Hooke's law which gives the restoring force for small displacements from a stable (in the physics-sense) equilibrium. Although in physics we refer to the equilibrium for Hooke's law as stable, it is actually a center.

You may have done a laboratory activity where you hang an increasingly large amount of mass from a rubber band to stretch it until Hooke's law no longer describes its elongation. In this case, we need a higher order correction for Hooke's law in order to describe the force needed to stretch the rubber band. The force can be given by

$$F = -kx - \epsilon x^3 \qquad (12.6.2)$$

The first correction is cubic in x so that the potential energy is not asymmetric about the equilibrium position. In other words, returning to thinking of a spring, the force exerted by the spring does not depend on which side of equilibrium the mass may sit.

Let us continue with the model system of a mass on a horizontal spring. If there is damping $-b\dot{x}$ and an external drive force $F_d = A \cos \omega t$, then the equation of motion for the mass is (via Newton's second law)

$$\ddot{x} + \delta \dot{x} + \beta x + \alpha x^3 = B \cos \omega t \qquad (12.6.3)$$

which is called the *Duffing equation*.

The Duffing equation exhibits many different nonlinear oscillations, even without changing its parameters. In phase portraits, nonlinear oscillations are represented by *limit cycles*, isolated closed trajectories. Example 12.8 examines several coexisting limit cycles in the Duffing equation's phase portrait.

Example 12.8: The Duffing equation

Plot the phase portrait and corresponding wave form $x(t)$ for the Duffing equation using parameters $\delta = 0.08$, $\beta = 0$, $\omega = 1.0$, $\alpha = 1.0$, and $B = 0.2$. For initial conditions use, $x(0) = -0.21$ and $\dot{x}(0) = 0.02$, $x(0) = -0.46$ and $\dot{x}(0) = 0.3$, and $x(0) = -0.43$ and $\dot{x}(0) = 0.12$.

Solution:

We begin by writing $(12.6.3)$ as a two dimensional system

$$\dot{x} = y$$
$$\dot{y} = -\delta y - \beta x - \alpha x^3 + B\cos\omega t$$

We can use Python to make the necessary plots. We are not interested in filling out the entire phase portrait, instead we will examine the trajectories which start with each initial condition. Notice however, that since the parameters of $(12.6.3)$ are unchanged, all the limit cycles exist in the same phase portrait.

In Figure 12.6, the left column is the phase portrait y vs. x and the right column is the wave form x vs. t. We omitted the axes labels to make the figure less cluttered. Notice that the first initial condition (corresponding to vec0) has an almost sinusoidal oscillation. The limit cycle is a closed loop. However, the other two initial conditions have non sinusoidal oscillations. Notice the loops in their limit cycles. These are not actual crossings of a trajectory (that would violate uniqueness of the solution). The phase portrait for a *nonautonomous system* (one that explicitly depends on time) requires a third axis corresponding to time.   Regardless, the loops identify

an important feature for the solution. The wave forms show that ic1, has a period-2 solution. Loosely speaking, the system goes through two oscillations before repeating itself. Likewise, ic2 leads to a period-3 solution.

```
from scipy.integrate import odeint
import matplotlib.pyplot as plt
import numpy as np

beta, omega, delta, B, alpha = 0, 1.0, 0.08, 0.2, 1.0

def sys(vec, t):
    x, y = vec
    return [y,-delta*y - beta*x - alpha*x**3 + B*np.cos(omega*t)]

t_forward = np.linspace(0,40,1000)

vec0 = [-0.21, 0.02]
vec1 = [-0.46, 0.3 ]
vec2 = [-0.43, 0.12]

ic0 = odeint(sys, vec0, t_forward)
ic1 = odeint(sys, vec1, t_forward)
ic2 = odeint(sys, vec2, t_forward)

fig, ax = plt.subplots(3,2)
ax[0,0].plot(ic0[:,0],ic0[:,1], 'k')
ax[0,0].set_title('Phase Portraits')
```

```
ax[0,0].text(0.17,0.15, 'IC0',bbox={'facecolor':'white'})

ax[0,1].plot(t_forward,ic0[:,0], 'k')
ax[0,1].text(36,0.15, 'IC0',bbox={'facecolor':'white'})
ax[0,1].set_title('Wave forms')

ax[1,0].plot(ic1[:,0],ic1[:,1], 'k')
ax[1,0].text(0.56,0.35, 'IC1',bbox={'facecolor':'white'})

ax[1,1].plot(t_forward,ic1[:,0], 'k')
ax[1,1].text(36,0.48, 'IC1',bbox={'facecolor':'white'})

ax[2,0].plot(ic2[:,0],ic2[:,1], 'k')
ax[2,0].text(0.37,0.22, 'IC3',bbox={'facecolor':'white'})

ax[2,1].plot(t_forward,ic2[:,0], 'k')
ax[2,1].text(36,0.33, 'IC3',bbox={'facecolor':'white'})
fig.tight_layout()
plt.show()
```

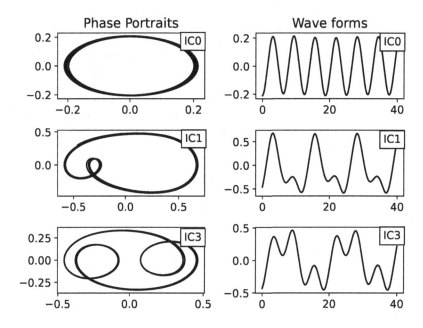

Figure 12.6 Graphical output from Example 12.8, showing phase portraits of the Duffing equation with several different types of nonlinear oscillations.

Limit cycles, like fixed points, have a stability. Furthermore, limit cycles can go through bifurcations, such as a change of stability. *Hopf bifurcations* are a particularly important bifurcation involving the appearance and disappearance of limit cycles that occurs as the eigenvalues pass through the imaginary axis as a parameter changes. In the end of chapter problems, you will explore the stability of limit cycles and examine some limit cycle bifurcations.

One of the most important bifurcations involving limit cycles are called *period-doubling bifurcations*. Period-doubling bifurcations occur when the period of the limit cycle doubles as a parameter is changed. A period-1 limit cycle becomes a period-2 limit cycle. As the parameter changes further, the period-2 becomes period-4 and so on. In Problem 18, you will examine a period doubling bifurcation in the Duffing equation. Eventually, period doubling bifurcations can lead to chaos. In the next section we examine chaos, an important nonlinear phenomenon.

12.7   CHAOS

Determinism is one of the hallmarks of Newtonian physics. If one knows the initial state for a given system (typically the position and velocity of each particle at $t = 0$) and the equations that govern the system's behavior (typically ODEs found from Newton's second law), then one can calculate the future or past state of said system. In principle, these predictions should be perfect and any errors that arise are due to a lack of knowledge of either the forces (or other influences) acting on the system, or the system's initial conditions.

For the linear systems you learn about in your undergraduate physics education, the above is generally true. Any errors in predictability can generally be explained by a lack of knowledge. However, as scientists began to look at more diverse nonlinear systems, such as weather models, it became clear that the break down of predictability wasn't due to a lack of knowledge of the forces acting on the system. It was found that while nonlinear deterministic systems can have a short term predictability, after a time t_h called the *horizon time*, the predictability breaks down. The observations and the model diverge exponentially.

The lack of long term predictability in nonlinear deterministic systems is called chaos. Note the word *deterministic* in the previous sentence. The loss of predictability is due to nonlinearities in the ODE describing the system, not randomness. Later, we will present a more formal definition for chaos which connects determinism to the lack of predictability through knowledge of the system's initial conditions.

12.7.1   THE LORENZ EQUATIONS

One of the most famous chaotic systems is the Lorenz equations

$$\left.\begin{aligned} \dot{x} &= \sigma\left(y - x\right) \\ \dot{y} &= rx - y - xz \\ \dot{z} &= xy - bz \end{aligned}\right\} \tag{12.7.1}$$

Edward Lorenz (1917 - 2008) developed (12.7.1) as a model of atmospheric convection. The dynamical variables x, y, and z are proportional to the rate of convection, the temperature diffusivity of ascending and descending convection rolls, and the vertical temperature variation, respectively. The model parameters σ and r are the Prandtl number and the Rayleigh number, respectively. The parameter b has no name.

Ellen Fetter (b.1940) and Margaret Hamilton (b.1936) worked with Lorenz creating numerical solutions to (12.7.1). What they found was bizarre behavior. For certain parameter values, the trajectories of (12.7.1) didn't settle on to a fixed point or limit cycle. However they remained bounded to a particular region of the phase portrait, eventually being attracted to a set of zero volume. This new type of attractor is called a *strange attractor* and the behavior of the system, once it has settled onto the strange attractor, came to be known as *chaos*.

Example 12.9: The phase portrait of the Lorenz equations

Plot the phase portrait of (12.7.1) using $\sigma = 10$, $r = 28$, and $b = 8/3$ with initial conditions $x(0) = 0.1$, $y(0) = 0$, and $z(0) = 0.1$.

Solution:

You will do the fixed point analysis of (12.7.1) in Problem 12. Here we plot the phase portrait of (12.7.1) and examine its strange attractor. In three dimensions, the phase portrait is a plot of the curve $(x(t), y(t), z(t))$.

Using the Python code below, we obtained the famous butterfly attractor, as shown in Figure 12.7. This structure is an attractor, in the sense that trajectories starting near the strange attractor approach it as $t \to \infty$. But, what makes it strange? First, strange attractors often are fractals and have a fractional dimension. In the case of the butterfly attractor, the fractal dimension is 2.05. Fractals are beyond the scope of this text, however, we include some sources for additional reading at the end of this book.

Another condition for the strange attractor deals with how the trajectories on the attractor behave. The trajectory appears to oscillate about two fixed points (at the center of each lobe). However, these are not limit cycles.

```python
from scipy.integrate import odeint
import matplotlib.pyplot as plt
import numpy as np

sigma, r, b = 10, 28, 8/3

def sys(vec, t):
    x, y, z = vec
    return [sigma*(y-x), r*x-y-x*z,x*y-b*z]

t = np.linspace(0,40,10000)

vec0 = [0.1,0,0.1]

soln = odeint(sys,vec0,t)

xs,ys,zs = soln.T

fig = plt.figure()
ax = plt.axes(projection='3d')
ax.plot(xs, ys, zs)
ax.set_xlabel("X")
ax.set_ylabel("Y")
ax.set_zlabel("Z")
plt.title('The strange attractor (Lorenz equations)')
plt.show()
```

The strange attractor (Lorenz equations)

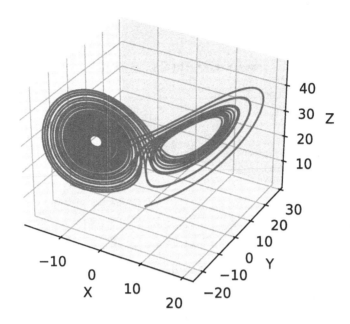

Figure 12.7 Graphical output from Example 12.9, demonstrating the phase portrait of the Lorenz equations with its strange attractor.

In the next example, we will examine the behavior of the dynamical variable x on the attractor, to better understand the nature of strange attractors.

Example 12.10: Trajectories on the butterfly attractor

Plot $x(t)$ of (12.7.1) using $\sigma = 10$, $r = 28$, and $b = 8/3$ with $y(0) = 0$, $z(0) = 0.1$ for $x(0) = 0.1$ and $x(0) = 0.101$ Discuss the results.

Solution:

Using code similar to Example 12.9, we plot $x(t)$ for each initial condition. The plot is shown in Figure 12.8. We plot one initial condition with a solid line and the other with a dashed line. You may want to download the code and change the colors in the plot command.

Although the two trajectories behave similarly early on, they noticeably diverge for $t > 22$. Note that the qualitative behavior of each trajectory is the same. Both trajectories are on the same attractor.

The divergence of two similar initial conditions is the second requirement for a strange attractor. The phenomenon is called *sensitivity to initial conditions* and we discuss it further below.

```
from scipy.integrate import odeint
import matplotlib.pyplot as plt
```

```
import numpy as np

sigma, r, b = 10, 28, 8/3

def sys(vec, t):
    x, y, z = vec
    return [sigma*(y-x), r*x-y-x*z,x*y-b*z]

t = np.linspace(0,40,10000)

vec0 = [0.1,0,0.1]
vec1 = [0.101,0,0.1]

soln0 = odeint(sys,vec0,t)
soln1 = odeint(sys,vec1,t)

x0 = soln0[:,0]
x1 = soln1[:,0]

plt.plot(t,x0, 'k')
plt.plot(t,x1,color='black', linestyle='dashed')
plt.ylabel('x')
plt.xlabel('time')
plt.title('Trajectories on the butterfly attractor')
plt.show()
```

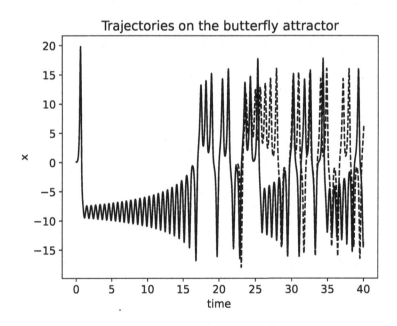

Figure 12.8 Graphical output from Example 12.11, demonstrating the sensitivity of initial conditions exhibited by the trajectories on the Lorenz equation's butterfly attractor.

12.7.2   SENSITIVITY TO INITIAL CONDITIONS

Let us begin by examining the simple harmonic oscillator with a unit natural frequency:

$$\ddot{x} + x = 0 \qquad (12.7.2)$$

In Figure 12.9, we plot the solution of (12.7.2) using initial conditions $x(0) = 1.0$ (solid line) and $x(0) = 1.1$ (dashed line), each with $\dot{x}(0) = 0$. Note that we used a larger difference in the initial conditions compared to Example 12.11 in order to see the effect on the graph.

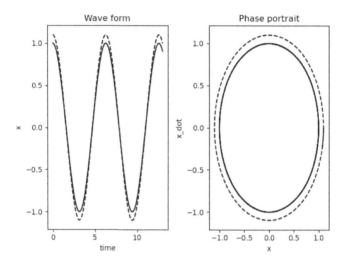

Figure 12.9 Two solutions to (12.7.2) with different initial conditions. Notice that the small difference in initial conditions remain small for $t > 0$.

Notice in Figure 12.9, the two solutions remain close both as a wave form and in the phase portrait. This is an example of a system which is not exhibiting sensitivity to initial conditions.

Sensitivity to initial conditions determines a system's predictability. Suppose the solid line in Figure 12.9 is the actual behavior of a system (the behavior that is observed) whereas the dashed line represents predictions from the mathematical model (12.7.2) used to simulate the system. Notice that the small initial error between the observation and the model's prediction (possibly due to rounding by instrumentation) continues to stay small as time progresses. Hence, the lack of sensitivity to initial conditions means that we have long term predictability, the model stays close to the observations.

Now let's consider $x(t)$ from Example 12.10, as shown in Figure 12.8. Again suppose the solid line is the observed state of the system and the dashed line is the state predicted by a model (in that case, the Lorenz equations). Notice that the model fails to track the true behavior of the system after $t = t_h = 22$. Beyond the horizon time t_h, the model can no longer make accurate predictions about the future state of the system. Sensitivity to initial conditions limits the predictability of the system. The Lorenz equations have no random element, the nonlinearities are responsible for the breakdown.

By returning to the phase portrait, we can provide an estimate for the horizon time and quantify how far in the future we can make reliable predictions about a chaotic system.

12.7.3 LYAPUNOV EXPONENTS AND THE HORIZON TIME

Formally, sensitivity to initial conditions means that initially close trajectories on the strange attractor diverge exponentially in time. Consider two trajectories on the chaotic attractor, $\mathbf{x}(t)$, and $\mathbf{x}(t) + \boldsymbol{\delta}(t)$, where $\boldsymbol{\delta}(t)$ is a vector that measures the distance between the two trajectories and $\boldsymbol{\delta}(0)$, the initial separation, is very small, $\delta(0) << 1$. Numerical studies show,

$$\|\boldsymbol{\delta}(t)\| \approx \|\boldsymbol{\delta}(0)\|\, e^{\lambda t} \tag{12.7.3}$$

where λ is called the *maximum Lyapunov exponent*. The maximum Lyapunov exponent measures the rate of exponential divergence between two initially nearby trajectories on the chaotic attractor. Chaotic systems have $\lambda > 0$. The maximum Lyapunov exponent is typically difficult to calculate from a system, however, there are methods of finding λ from measured data called *time series*.

The Lyapunov exponent and (12.7.3) can provide an estimate for the horizon time t_h. Let a serve as an error tolerance. In other words, when $\boldsymbol{\delta}(t_h) \approx a$, then the difference between the observations and the model system are too great for a reliable prediction. Setting $\boldsymbol{\delta}(t_h) = a$, (12.7.3) can be solved for t_h

$$t_h \approx \frac{1}{\lambda}\ln\left(\frac{a}{\delta(0)}\right) \tag{12.7.4}$$

where $\delta(0)$ is the magnitude of $\boldsymbol{\delta}(0)$.

Example 12.11: The horizon time

Suppose a chaotic system is known to have a maximal Lyapunov exponent of $\lambda = 1.50$. You wish to make future predictions about the state of the system and your error tolerance is one part in one hundred ($a = 0.01$). From your knowledge of chaotic systems, you know that the horizon time can be prolonged with more precise measurements of the system's initial state. Suppose you have two devices, one that measures the state of the system with a precision of $\delta_1(0) = 10^{-3}$ and another with $\delta_2(0) = 10^{-9}$, a million-fold improvement. Calculate the horizon time for each device.

Solution:
We use Python to calculate the horizon time.

```python
import numpy as np

def horizon_time(x):          #x = precision (initial error)
    a = 0.01                  #error tolerance
    lam = 1.5                 #maximal Lyapunov exponent
    t_h = np.log(a/x)/lam
    return t_h

t_h1 = horizon_time(10**(-3))
t_h2 = horizon_time(10**(-9))

print('-'*28,'CODE OUTPUT','-'*29,'\n')
print('When delta0 = 10^(-3), t_h =', t_h1)
print('When delta0 = 10^(-9), t_h =', t_h2)
```

```
print('Horizon time is increased by a factor of ', t_h2/t_h1)

--------------------------- CODE OUTPUT ----------------------------

When delta0 = 10^(-3), t_h = 1.5350567286626973
When delta0 = 10^(-9), t_h = 10.74539710063888
Horizon time is increased by a factor of  7.0
```

> Notice that the million-fold improvement in accuracy results in a less than one order-of-magnitude increase in horizon time. This is why predicting chaotic systems, such as the weather, is difficult.

12.8   HOW TO ANALYZE A NONLINEAR SYSTEM

In this chapter, we have seen that while we cannot necessarily solve nonlinear differential equations, we can obtain a lot of information about their dynamics by drawing their phase portrait. Before concluding the chapter, we present a guide which summarizes the methodology outlined in this chapter. While it is impossible to provide a comprehensive overview of nonlinear systems, what has been covered in this chapter should provide you with a start when you need to analyze a nonlinear system of your own.

1. Find the fixed points of the system.

2. Classify the fixed points by computing the Jacobian matrix, evaluating it at each fixed point, and calculating its eigenvalues and eigenvectors.

3. By examining the formulas for the fixed points and eigenvalues, identify any possible bifurcations in the system. Bifurcations can occur when fixed points collide or eigenvalues when eigenvalues change sign. Be on the look out for when eigenvalues become real or complex.

4. Using the eigenvectors of the Jacobian matrix, plot the stable and unstable manifolds of each saddle.

5. Plot a vector field plot as necessary, to fill in the phase portrait.

6. Plot the trajectories near any stable limit cycles.

7. In the case of a chaotic system, one trajectory in the phase space may be enough. To visualize the strange attractor, plot one trajectory that settles on to it.

12.9   END OF CHAPTER PROBLEMS

1. **Phase portrait of the simple pendulum** – Consider the simple pendulum equation

$$\ddot{x} + \sin x = 0$$

where x measures the angular displacement counter clockwise from the pendulum's lowest point. Find and classify the fixed points. Plot the phase portrait $\dot{x}$ vs. x.

2. **One-dimensional systems** – Consider the one-dimensional system

$$\dot{x} = f(x)$$

where x is the dynamical variable. What is the condition for x^* to be a fixed point of the system? By perturbing the fixed point, show how to classify a fixed point in a one-dimensional system. *This result is useful for the next five problems.*

3. **One-dimensional systems and potentials** – Consider a one-dimensional system, $\dot{x} = f(x)$. We can assign a potential function to the system, $V(x)$, such that

$$f(x) = -\frac{dV}{dx}$$

which makes the potential function similar to the potential energy. In this chapter, we discussed that the condition for fixed points (*i.e.* equilibrium points), x^*, $f(x^*) = 0$. What do fixed points correspond to when it comes to $V(x)$? How can one use $V(x)$ to identify the stability of a fixed point? Consider the system $\dot{x} = 1 - x^2$. Find the potential corresponding to this system. Using the potential, find the fixed points and identify their stability. Plot the phase portrait. *Consult Example 12.5 for the stability of fixed points in one-dimensional systems if you haven't solved Problem 2.*

4. **Bifurcations in one-dimensional systems** – One-dimensional models can be useful for studying the different types of bifurcations possible in nonlinear systems. In each of the cases below, find and classify the fixed points. Identify the parameter value when the bifurcation occurs. Note that all of these bifurcations can occur in higher dimensions. *Consult Example 12.5 for the stability of fixed points in one-dimensional systems if you haven't solved Problem 2.*

     a. Saddle-node bifurcation: $\dot{x} = \mu + x^2$.

     b. Transcritical bifurcation: $\dot{x} = \mu x - x^2$.

     c. Supercritical pitchfork bifurcation: $\dot{x} = \mu x - x^3$.

     d. Subcritical pitchfork bifurcation: $\dot{x} = \mu x + x^3$.

5. **The logistic equation** – The logistic equation, $\dot{x} = rx(1 - x/k)$, is used to model population sizes and is a one-dimensional system. Note that r is a positive constant called the *linear growth rate* and k is a positive constant called the *carrying capacity*. Find the fixed points for the logistic equation and give them a physical interpretation. Find the stability of each fixed point. Using Python, solve the logistic equation and plot the result for $r = 0.3$ and $k = 10$, let $x(0) = 1$. *Consult Example 12.5 for the stability of fixed points in one-dimensional systems if you haven't solved Problem 2.*

6. **Limit cycles in polar coordinates** – It can sometimes be convenient to describe limit cycles using polar coordinates. Consider the following system in polar coordinates:

$$\dot{r} = r(r - 1)(r - 2)$$
$$\dot{\theta} = 1$$

where r and θ are the usual polar coordinates with $r \geq 0$. Notice that one can find values of r that make $\dot{r} = 0$, but there are no values of θ or r for which $\dot{\theta} = 0$. That is OK! The limit cycles in this case are circles. What are the radii of the limit cycles? Plot the phase portrait vector field and several trajectories (you'll need to convert to Cartesian coordinates, use $\dot{x} = d/dt(r \cos \theta) = \dot{r} \cos \theta - r\dot{\theta} \sin \theta$ and similar for $\dot{y}$). Use the phase portrait to identify the stability of each limit cycle and the fixed point at the origin.

7. **The dynamics of love affairs** – In his book <u>Nonlinear Dynamics and Chaos</u>, Strogatz discusses the mathematics of love affairs, an interesting and fun way to

explore two-dimensional linear systems. Let $R(t)$ be Romeo's love (positive) or dislike (negative) for Juliet at time t, and $J(t)$ be Juliet's love or dislike for Romeo at time t. The love affair can be modeled using the equations:

$$\dot{R} = aR + bJ$$
$$\dot{J} = cJ + dR$$

All kinds of interesting relationships can be created by choosing different values of the parameters a, b, c, and d. Following the methods of Section 13.3 solve the love affair model in closed form for an eager Romeo ($a = 1$, $b = 2$) and a very cautious Juliet ($c = -1$, $d = -3$). Classify the fixed point at the origin and plot the phase portrait. Comment on the possible outcomes for various initial conditions of the relationship.

8. **Nonlinear love affairs** – In this problem, we will build on Problem 7 with a nonlinear romance! We will have both Romeo and Juliet be very sensitive to their own emotions:

$$\dot{R} = - 2R - 2J(1 - |J|)$$
$$\dot{J} = J + R(1 - |R|)$$

The nonlinear term $J(1-|J|)$ is sometimes called a *repair nonlinearity* and the number 1 in the last term is a measure of when Juliet's love becomes counter productive. Find the fixed points and classify them. Plot the phase portrait for this system and discuss the nature of the relationship between Romeo and Juliet.

9. **Lotka Volterra model of competition** – Consider the following Lotka-Volterra model of competition between two species:

$$\dot{x} = 4x - 3x^2 - xy$$
$$\dot{y} = 2y - 4y^2 - 3xy$$

For this model, identify the fixed points and their stability. Can the two species ever coexist? Plot the phase portrait.

10. **Predator-prey model** – Consider the predator-prey model

$$\dot{x} = 3x - 7xy$$
$$\dot{y} = - 2y + 5xy$$

where like the competition model, we will assume $x(t)$ and $y(t)$ measure the number of each species in the hundreds. Find the fixed points of this predator-prey model and identify their stability. Comment on the nature of the equilibrium state corresponding to each fixed point. Plot the phase portrait.

11. **SIS model with vital dynamics** – Consider two populations of individuals $s(t)$ and $i(t)$, representing the number of susceptible and infected individuals respectively. The SIS model is:

$$\dot{s} = \mu n - \mu s + \gamma i - \beta si$$
$$\dot{i} = - \mu i - \gamma i + \beta si$$

where the coefficients represent rates and $n = s + i$ is the total population. Comment on the physical meaning of each coefficient: μ, γ, and β (which are all positive). Find the fixed points for the system and identify their stability. Does this model have a parameter similar to R_0 from Example 12.5?

12. **Fixed points in the Lorenz equations** – Find the fixed points in the Lorenz equations. Describe the bifurcation that occurs near $r = 1$. There is another special bifurcation called a Hopf bifurcation when

$$r = r_H = \frac{\sigma(\sigma + b + 3)}{\sigma - b - 1}$$

Using $\sigma = 10$ and $b = 8/3$, calculate r_H. Describe what happens to the stability of each fixed point if r is increased through the value of r_H.

13. **Jacobian matrices** – Compute the Jacobian for the saddle point in Examples 12.6 and 12.7.

14. **The logistic map** – The *logistic map*, $x_{n+1} = rx_n(1 - x_n)$, is a discrete form of the logistic equation (see Problem 5). After choosing an initial value x_0 and parameter r, it is easy to compute successive values. For example, $x_1 = rx_0(1 - x_0)$ and $x_2 = rx_1(1 - x_1)$, and so on. Using $x_0 = 0.1$, plot x_n vs. n for $r = 2.9, 3.2, 3.5$, and 3.55.

15. **Logistic map and Poincaré plots** – Identify the periodicity of each solution from Problem 5 using a Poincaré plot. In its simplest form, the Poincaré for the logistic map is a plot of x_{n+1} vs. x_n. Plot the Poincaré plots for the logistic map using the values of r in Problem 5. Use the Poincaré plots to identify the periodicity of each realization of the logistic map. *Hint: Notice the transients in Problem 5. Use values of x_n after the transient time to get the best picture of the Poincaré plots.*

16. **The logistic map, revisited** – Plot the logistic map (see Problem 5) and its Poincaré plot using $r = 3.91$. What kind of behavior is the logistic map exhibiting when $r = 3.91$?

17. **The Lyapunov exponent and the logistic map** – As mentioned previously, the Lyapunov exponent is positive for chaotic systems. We can compute the Lyapunov exponent for the logistic map using the following

$$\lambda = \lim_{n \to \infty} \left[\frac{1}{n} \sum_{i=0}^{\infty} \ln|r - 2rx_i| \right] \tag{12.9.1}$$

where x_i is the ith value of the logistic map from Problem 5. The derivation of (12.9.1) is beyond the scope of this text. However, to compute λ, one would compute a large number of iterations of the logistic map (say 10,000, since we cannot computer an infinite number) and insert the values of x_i into (12.9.1). Using (12.9.1), calculate λ as a function of r. *Hint: Don't for get to remove the transients from your logistic map calculation. Calculate say 10,500 elements of the logistic map then remove the first five hundred, using the last 10,000 in (12.9.1).*

18. **Period doubling bifurcations in the Duffing equation** – Show that the Duffing equation, (12.6.3), undergoes a series of period doubling bifurcations by changing the amplitude of the cosine term on the right-hand side of the equation. Use as the parameters: $\beta = -1, \delta = 0.1$, $\omega = 1.4$, and $\alpha = 1$. You will need to remove the transients from the plot to see the limit cycles and strange attractor.

19. **Electronics and relaxation oscillators** – Electrical circuits provide excellent examples of nonlinear systems. One of the most famous circuit-inspired equations is the so-called *Van der Pol equation*

$$\ddot{x} - \epsilon \left(1 - x^2\right) \dot{x} + x = 0$$

which is a simple harmonic oscillator with a nonlinear damping term. Notice that the sign of the damping actually changes with the value of x. For example, if $\epsilon > 0$, then the damping is negative for $x < 1$ and is positive for $x > 1$. Thus small oscillations will grow due to negative damping, but once the oscillations are large enough, they will dampen out as the damping term becomes positive. The Van der Pol oscillator can also display so-called *relaxation oscillations*, a common form of oscillations in nonlinear systems where rapid changes of x are followed by slower variations. To see relaxation oscillations in action, solve the Van der Pol oscillator for $\epsilon = 8.0$. Using $x(0) = 0.1$ and $\dot{x}(0) = 0$, plot both $x(t)$ and the phase portrait. Describe the solution.

20. **Nonlinear Oscillators** – Weakly driven nonlinear oscillators can be excited to large amplitudes through a phenomenon called *autoresonance*. The oscillator is driven by a periodic forcing term whose amplitude is small, but its frequency changes (typically decreasing) with time. As the drive frequency decreases through the oscillator's resonant frequency, the oscillator becomes phase-locked with the driving force, resulting in the oscillator maintaining resonance even though the drive frequency is no longer the oscillator's natural frequency. Consider the following driven pendulum (note the pendulum equation on the right-hand side):

$$\ddot{x} + \omega_0^2 sin(x) = \epsilon \cos\left(\omega_0 t - {}^1\!/_2 \alpha t^2\right)$$

Notice that the drive frequency (the time derivative of the drive phase) is $f_d = \omega_0 - \alpha t$, so that at $t = 0$, the drive frequency is equal to the natural frequency of the pendulum. Starting with initial conditions, $x(-1000) = 0$ and $\dot{x}(-1000) = 0$, compute $x(t)$ for $\epsilon = 0.0459$ and $\epsilon = 0.0461$ for the pendulum with $\omega_0 = 2\pi$, $\alpha = 0.001$. For which value of ϵ is the pendulum displaying autoresonance?

21. **Chua's circuit** – Chua's circuit is a electrical circuit with a piece wise linear negative resistance N_R as shown in Figure 12.10. It is one of the simplest circuits to exhibit a variety of nonlinear behaviors. The circuit schematic is shown. Let x, y, and z be proportional to the voltages across the capacitors C_1 and C_2, and the current in the inductor L, respectively. The equations governing the circuit can be obtained from Kirchoff's rules and simplified to be:

$$\dot{x} = \alpha(y - x - g(x))$$
$$\dot{y} = x - y + z$$
$$\dot{z} = -\beta y$$

where $g(x) = m_1 x + \frac{1}{2}(m_0 - m_1)(|x + 1| - |x - 1|)$ with $\alpha = C_2/C_1$, $\beta = R^2 C_2/L$, and m_0 and m_1 being positive constants related to the electrical response of the resistor. Create a phase portrait (x-y-z space) for the equations of motion using $\alpha = 15.6$, $\beta = 25.58$, $m_0 = -8/7$, $m_1 = -5/7$. Choose as the initial conditions $x(0) = 0.1$, $y(0) = z(0) = 0$. The result will be the famous double-scroll attractor.

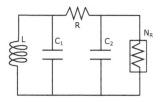

Figure 12.10 Problem 13.28: Chua's Circuit.

22. **Period doubling in Chua's circuit** – A period-doubling route to chaos can be observed in the equations for Chua's circuit (see Problem 21) by using the same parameters as used in Problem 21, but with $\beta = 50, 35, 33.8$ etc. For each value of β, create a graph of $x(t)$ and identify the periodicity of each solution. For what value of β is there a period-8 solution?

23. **Exponential divergence of nearby initial conditions** – Consider the Lorenz equations in Example 13.9. Numerically solve the equations using the initial conditions $(x(0), y(0), z(0)) = (0.1, 0, 0)$ and $(x(0), y(0), z(0)) = (0.1 + 10^{-9}, 0, 0)$. These are two initially very close initial conditions. Plot the natural log of the Euclidean distance between the two trajectories as as function of time. Fit the plot to a line to estimate the Lyapunov exponent (which has a known value near 0.9). *Hint: Don't integrate too long in time, about 30 time units should do. Keep only the part of the data that is increasing exponentially in time.*

24. **Arms race** – The Richardson arms race model models the expenditures $x(t)$ and $y(t)$ of two nations X and Y at war.

$$\dot{x} = ay - mx + r$$
$$\dot{y} = bx - ny + s$$

using $a = 2$, $m = 1$, $r = 5$, $b = 1$, $n = 1/2$, and $s = 3$, find and classify the fixed point. Plot the vector field phase portrait around the fixed point and interpret the result.

25. **Arms race with limited budget** – In the real world, a country's budget for warfare is limited. Consider the following modification to the Richardson arm race model,

$$\dot{x} = \left(1 - \frac{x}{x_c}\right)(ay - mx + r)$$
$$\dot{y} = \left(1 - \frac{y}{y_c}\right)(bx - ny + s)$$

where x_c and y_c are the carrying capacity (i.e. budget) for each country. The parameters a and b relate how each country responds to the other's spending on defense. Parameters m and n relate how each country responds to internal pressures to reduce spending on defense. Using the same parameters as in Problem 24 with $x_c = 7$ and $y_c = 4$, find the fixed points and plot the phase portrait. Interpret the result. If you did Problem 24, then compare your results to those of that problem.

26. **Rössler strange attractor** – Another chaotic system is called the Rössler equations

$$\dot{x} = -y - z$$
$$\dot{y} = x + ay$$
$$\dot{z} = b + (x - c)z$$

Using $a = b = 0.2$, $c = 5.7$, $x(0) = -1$, $y(0) = z(0) = 0.1$, plot the phase portrait and $x(t)$ for $t = 0 \ldots 200$.

27. **The Hénon Map** – The Hénon map

$$x_{n+1} = a - x_n^2 + b y_n$$
$$y_{n+1} = x_n$$

is a two-dimensional map computed in a fashion similar to the logistic map in Problem 5. Using $a = 1.29$, $b = 0.3$, $x_0 = y_0 = 0.5$, compute the phase portrait of the Hénon map, y_n vs. x_n. Compute to $n = 2000$. You'll get the famous Hénon attractor.

13 Analysis of Experimental Data

So far in this book we have discussed how to use mathematics for the purpose of modeling physical systems. Although not explicitly stated, we assumed perfect knowledge of values inserted into the equations. However, in the real world, perfect knowledge of a particular value is impossible. In this chapter, we will discuss how to propagate uncertainties in measured values through equations. In addition, we will present means of extracting information from measured data. The topics in this chapter are wide ranging and include the distributions of random numbers, scaling laws, linear and non-linear regression, and Fourier transforms. Although the topics vary, they all share a common theme. How can we transform data from an experiment and turn it into useful information about the system from which the data was measured?

13.1 ESTIMATING UNCERTAINTIES

How do we estimate errors in a series of measurements in the laboratory? In this section we discuss uncertainties in measurements.

The uncertainty in any one given measurement, can be estimated in several ways because there are several sources of potential error. For example, an important source of experimental error is the resolution of the instruments, which have a limited precision. Other factors are the uncertainty in reading the instrument, and the calibration of instruments which can lead to systematic errors. A detailed discussion of these types of errors and of how many significant figures one should use when reporting experimental results, is beyond the scope of this book.

When we have multiple measurements of the same quantity x, we can use the mean of the measurements to estimate the true value. The mean is found using

$$\bar{x} = \frac{1}{n} \sum_{i=1}^{n} x_i \tag{13.1.1}$$

where n is the number of measurements. The uncertainty in the measurement can be estimated using the standard deviation

$$\sigma = \sqrt{\frac{1}{n-1} \sum_{i=1}^{n} (x_i - \bar{x})^2} \tag{13.1.2}$$

The standard deviation σ measures the variation in the data and has the same dimensions as x. When using the standard deviation σ as an uncertainty, it is not uncommon to round it to one or two significant figures, depending on the precision of the experimental procedure. The best estimate of the mean value including its uncertainty can be written as:

$$x = \bar{x} \pm \sigma \tag{13.1.3}$$

We also evaluate the standard error or standard deviation of the mean $\sigma_{\bar{x}}$:

$$\sigma_{\bar{x}} = \frac{\sigma}{\sqrt{n}} = \sqrt{\frac{1}{n(n-1)} \sum_{i=1}^{n} (x_i - \bar{x})^2} \tag{13.1.4}$$

DOI: 10.1201/9781003294320-13

The coefficient of a variation $CV[\%]$ is defined as:

$$CV[\%] = 100\frac{\sigma}{\sqrt{n}} \qquad (13.1.5)$$

This dimensionless quantity is also known as the percent relative error.

Example 13.1 illustrates how to read data from a data file, and how to find the average and standard deviation of a data set.

Example 13.1: Estimating uncertainties in measured data

In a standard physics laboratory experiment to measure the acceleration g due to gravity, a series of measurements was obtained by releasing a ball from rest and timing how long it takes to drop from a given height.

The data obtained is stored as a *.csv* file at the web site

https://github.com/vpagonis/CRCbook/raw/main/data2.csv

The file contains two columns (t, y), the times t it takes the ball to drop a height y. The acceleration of gravity g can be obtained from the kinematic equation $g = 2y/t^2$.

(a) Write a code to read the *.csv* file from the website.

(b) What are the average measured values $\bar{y}$, uncertainties σ_y and $CV[\%]_y$ of the position y? Repeat for the times t.

(c) Which quantity was measured more precisely, t or y?

(d) Plot the data points y_i for the position and for the times t_i as a function of the data index i, and comment on the plot.

(e) Using the average values $\bar{y}$ and $\bar{t}$, find the average value $\bar{g}$.

(f) Another method of analyzing the data is by evaluating $g = 2y/t^2$ for each data point (x, y) and then averaging the resulting g values. Obtain the average $\bar{g}$ and its coefficient of variation using this alternative method, and compare with your answer in (e).

Solution:
(a) We import the **pandas** library and use **read_table** to read the data contained in the *.csv* file. Note that we tell Python that the data in the *.csv* file are separated by a comma, by using **delimiter=','**. The data is stored in a panda frame assigned to the variable **df**, and the function **pd.DataFrame.head(df)** is used to print the first few lines of the data in the file. It is always good to have a look at the data, to ensure that they are in a suitable form.

We can specify which column of the data to use with **df['t']**, where the argument in the square brackets is a string which is the header (first element) for the column of interest. The method **.iloc** can also be used to specify the column. For example, **df.iloc[:,0]** would specify the first column of the data file. Finally, the method **.to_numpy()** is used to convert the column of the data to a NumPy array.

(b) The code uses the NumPy functions **mean()** and **std()** to evaluate the mean and standard deviation of the data, and $CV[\%]$ is evaluated using (13.1.5).

(c) The result from (b) shows that the $CV[\%]$ for the time t is 0.92%, while the $CV[\%]$ for the position y is 3.0%. This shows that t was measured more precisely, and therefore may be contributing more to the overall uncertainty in the calculated value of g.

(d) The plot of the data points in Figure 13.1 shows that the measured values for t and y are located uniformly within narrow ranges.

(e-f) The two methods give similar average values $\bar{g} = 9.4227 \; m/s^2$ and $\bar{g} = (9.4256 \pm 0.3355)$ m/s^2, and the $CV[\%] = 3.56\%$.

```
import numpy as np
import matplotlib.pyplot as plt
import pandas as pd
print('-'*28,'CODE OUTPUT','-'*29,'\n')


# load the .csv file from GitHub
url='https://github.com/vpagonis/CRCbook/raw/main/data2.csv'

# read the contents of the file using pandas library
# The data in each row in the file are separated by commas
df = pd.read_table(url,delimiter=',')

# look at the contents of the first few lines of the file
print('The first few lines in the file are:\n')
print(pd.DataFrame.head(df))

# store columns #1 and #2 to NumPy arrays t_data, y_data
t_data = df['t'].to_numpy()
y_data = df['y'].to_numpy()

# find the number of data rows in the file
size = range(len(t_data))

#plot the data
plt.plot(size,t_data,'o',label='time')
plt.plot(size,y_data,'+',label='y')
plt.xlabel('Measurement number')
plt.ylabel('t [s],   y [m]')
plt.title('A set of data from a .csv file')
plt.legend()

# evaluate the average and standard deviation and CV[%]
tavg = np.mean(t_data)
dt = np.std(t_data)
print('\nMean t =',round(tavg,3),'+/-',round(dt,3),' s')
print('CV% of t values =',round(100*dt/tavg,2),' %')


yavg = np.mean(y_data)
dy = np.std(y_data)
print('\nMean y =',round(yavg,3),'+/-',round(dy,3),' m')
print('CV% of y values =',round(100*dy/yavg),' %')

print('\nMean g =',round(2*yavg/tavg**2,3),' m/s^2')

ratios = 2*y_data/(t_data**2)
print('\nMean g=',round(np.mean(ratios),3),'+/-',\
round(np.std(ratios),3),' m/s^2')
print('CV% of g =',round(100*np.std(ratios)/np.mean(ratios),2),' %')

plt.show()

-------------------------- CODE OUTPUT ----------------------------

The first few lines in the file are:
```

```
        t       y
0   0.644   1.938
1   0.647   1.944
2   0.655   1.927
3   0.647   2.027
4   0.654   1.913

Mean t = 0.65 +/- 0.006  s
CV% of t values = 0.92 %

Mean y = 1.991 +/- 0.057  m
CV% of y values = 3 %

Mean g = 9.422  m/s^2

Mean g= 9.425 +/- 0.334  m/s^2
CV% of g = 3.54 %
```

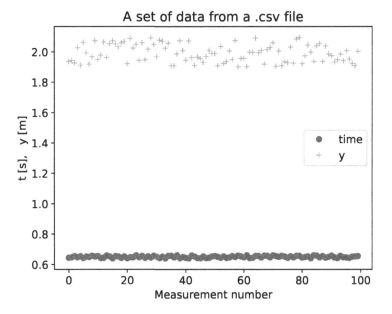

Figure 13.1 Graphical output from Example 13.1, showing a plot of time values t and position values y, from a *.csv* data file.

13.2 PROPAGATION OF UNCERTAINTIES

Now that we know how to represent the uncertainty of a measured quantity (either as the least count of a single measurement or the standard deviation of an ensemble of measurements) we consider how the uncertainty propagates through a calculation.

Consider a quantity $V(a, b)$ which is a function of the two quantities a and b with uncertainties σ_a and σ_b, respectively. If we consider the uncertainties to be small then we can calculate the change dV in V when we change a by an amount σ_a

$$dV \simeq \frac{\partial V}{\partial a}\sigma_a \tag{13.2.1}$$

Likewise, if we change b by an amount σ_b, then V would change by $(\partial V/\partial b)\,\sigma_b$. In the theory of error propagation, the rule for combining these uncertainties is:

$$\sigma_V = \sqrt{\left(\frac{\partial V}{\partial a}\sigma_a\right)^2 + \left(\frac{\partial V}{\partial b}\sigma_b\right)^2} \tag{13.2.2}$$

If there were more variables, then there would be more terms in (13.2.2).

For example, consider two measurements $a\pm\sigma_a$ and $b\pm\sigma_b$ which we wish to add together. Using $V = a + b$, $\partial V/\partial a = 1$ and $\partial V/\partial b = 1$, the above rule gives:

$$\sigma_{a+b} = \sqrt{\sigma_a^2 + \sigma_b^2} \tag{13.2.3}$$

Using (13.2.2) we can generate a list of rules for common mathematical relationships in physics. Table 13.1 shows the rules for finding σ_V for $V = V(a, b)$, when the measured uncorrelated quantities are $a \pm \sigma_a$ and $b \pm \sigma_b$.

Expressions for more complicated functions can be derived by combining simpler functions. For example, repeated multiplication, assuming no correlation, gives

$$f = ABC \qquad \left(\frac{\sigma_f}{f}\right)^2 \approx \left(\frac{\sigma_A}{A}\right)^2 + \left(\frac{\sigma_B}{B}\right)^2 + \left(\frac{\sigma_C}{C}\right)^2 \tag{13.2.4}$$

Table 13.1

Uncertainty propagation for common arithmetic operations in physics. Here a and b are measurements with corresponding uncertainties σ_a and σ_b.

Propagation of error	Function $V(a,b)$	Uncertainty σ_V of $V(a,b)$
addition/subtraction	$V = a \pm b$	$\sigma_V = \sqrt{\sigma_a^2 + \sigma_b^2}$
multiplication	$V = a\,b$	$\sigma_V = \lvert V\rvert \sqrt{\left(\frac{\sigma_a}{a}\right)^2 + \left(\frac{\sigma_b}{b}\right)^2}$
division	$V = a/b$	$\sigma_V = \lvert V\rvert \sqrt{\left(\frac{\sigma_a}{a}\right)^2 + \left(\frac{\sigma_b}{b}\right)^2}$
exponentiation	$V = a^y$	$\sigma_V = y\frac{\sigma_a}{a}$
power	$V = x^a$	$\sigma_V = a\,x^{a-1}\,\sigma_x$
multiplication by a constant c	$V = c\,a$	$\sigma_V = \lvert c\rvert\,\sigma_a$
logarithm	$V = \ln(a)$	$\sigma_V = 0.434\left(\frac{\sigma_a}{a}\right)$

Example 13.2: Uncertainty propagation in a free fall experiment

The acceleration due to gravity (g) is found in a laboratory free fall experiment by releasing a ball from rest, and timing how long it takes to fall from a given height. After several trials of the experiment, the initial height of the ball is found to be $y = (2.003 \pm 0.0005)$ m, and the time it takes the ball to fall is $t = (0.64 \pm 0.02)$s.

(a) What is the measured value of g and its uncertainty σ_g obtained used the rules of error propagation?

(b) Estimate the uncertainty Δg by calculating the range of values for g. How does this method of estimating Δg compare with the result from (a)?

Solution:

(a) We can use the kinematics equation for g:

$$g = \frac{2\Delta y}{t^2} \tag{13.2.5}$$

where t is the time it takes an object to fall a distance Δy.

To compute g, we need the uncertainty of the numerator $n = 2\Delta y$ and the uncertainty of the denominator $d = t^2$. We can use the rules in Table 13.1 to find each. Using the value of $t^2 = 0.41$:

$$n = 2(2.003) \pm 2(0.0005) \qquad = 4.006 \pm 0.001 \text{ m}$$
$$d = 0.41 \pm (0.41)(2)(0.02) \qquad = 0.41 \pm 0.02 \text{ s}^2$$

where we dropped units and rounded the uncertainties to one significant figure. Next, we calculate g

$$g = \frac{4.006}{0.41} \pm \frac{4.006}{0.41} \sqrt{\left(\frac{0.001}{4.006}\right)^2 + \left(\frac{0.02}{0.41}\right)^2} = (9.8 \pm 0.5) \text{ m/s}^2$$

(b) We can use Python to find the uncertainty Δg by calculating the minimum and maximum values of g, $gmin$ and $gmax$ respectively. The uncertainty is estimated using:

$$\Delta g = \frac{gmax - gmin}{2}$$

In the code we set up a simple function $g(y, t)$ that calculates the values of g using (13.2.5). The result is $g \pm \Delta g = 9.8 \pm 0.6$ m/s^2.

The result from parts (a) and (b) are similar.

```
y = 2.003
sigma_y = 0.0005

t = 0.64
sigma_t = 0.02

def g(y,t):
    return 2*y/t**2

g_mean = g(y,t)
g_min = g(y + sigma_y,t + sigma_t)
g_max = g(y - sigma_y,t - sigma_t)
sigma_g = (g_max - g_min)/2

print('-'*28,'CODE OUTPUT','-'*29,'\n')
print('g = ', round(g_mean,2),' m/s^2')
print('sigma_g = ', round(sigma_g,2),' m/s^2')


---------------------------- CODE OUTPUT ----------------------------

g =  9.78  m/s^2
sigma_g =  0.61  m/s^2
```

13.3   LINEAR REGRESSION AND THE METHOD OF LEAST SQUARES

Linear regression is a method of estimating the equation of a line that best fits a set of experimental data. Let $\{x_i, y_i\}$ represent the experimental data points. An algorithm is used to create a line of best fit for the data. You have likely used such algorithms either in Python or in programs like Microsoft Excel. For a given value of the independent variable x_i, there is a measured value y_i and a value $\hat{y}_i$ predicted by the best fit line. The difference $\hat{y}_i - y_i$ represents the error between the value predicted by the best fit line and the actual measured value.

The algorithm creating the line of best fit attempts to minimize the average of the squared error J:

$$J = \frac{1}{n} \sum_{i=1}^{n} (\hat{y}_i - y_i)^2 \tag{13.3.1}$$

where n is the number of data points.

The value of the best fit $\hat{y}_i$ depends on the slope of the best fit line m, and on the corresponding y-intercept b. Hence, we can write

$$J(m, b) = \frac{1}{n} \sum_{i=1}^{n} ([mx_i + b] - y_i)^2 \tag{13.3.2}$$

We want to find the values of m and b, denoted as m^* and b^*, that minimize J. Therefore we must calculate

$$\left. \frac{\partial J}{\partial m} \right|_{(m^*, b^*)} = 0 \quad \left. \frac{\partial J}{\partial b} \right|_{(m^*, b^*)} = 0 \tag{13.3.3}$$

After calculating the derivatives in (13.3.3) and solving for the best fit slope and y-intercept m^* and b^*, we find

$$m^* = \frac{\left(\sum x_i y_i \right) - n \bar{x} \bar{y}}{\sum x_i^2 - n \bar{x}^2} \tag{13.3.4}$$

$$b^* = \frac{\bar{y} \left(\sum x_i^2 \right) - \bar{x} \left(\sum x_i y_i \right)}{\sum x_i^2 - n \bar{x}^2} \tag{13.3.5}$$

where n is the number of data points, the sums range over all of the data points, and bars denote averages.

The R-squared value R^2, sometimes called the coefficient of determination, is found using

$$R^2 = 1 - \frac{\sum (y_i - \hat{y}_i)^2}{\sum (y_i - \bar{y})^2} \tag{13.3.6}$$

Loosely speaking, the dimensionless quantity R^2 gives the percentage of variability of the dependent variable y accounted for by the linear model. For example, if $R^2 = 1$ then all of the variation of the data is captured by a linear model. However, if $R^2 = 0.78$, then only 78% of the variation is accounted for.

Example 13.3 illustrates how to use Python to find the best fit line of a data set.

Example 13.3: Linear regression with Python

Suppose you measure the velocity of a toy car as a function of time (see Python code for the data). Find the line of best fit for the data and estimate the car's acceleration.

Solution:

The code below contains the velocity data as a function of time. The command `linregress` finds the line of best fit for the `velocity` data. Notice that the `linregress` command has five outputs, the slope, y-intercept, R-value, p-value, and standard error, respectively. The slope of the line in Figure 13.2 is an estimate of the car's acceleration. The data and the line of best fit are shown in Figure 13.2.

```python
import numpy as np
import matplotlib.pyplot as plt
from scipy.stats import linregress

time = np.array([0.0, 0.1, 0.2, 0.3, 0.4])
velocity = np.array([0.2, 0.38, 0.61, 0.77, 1.02])

reg = linregress(time, velocity)

slope = reg.slope                #slope
y_int = reg.intercept            #y-intercept
r_sq = reg.rvalue**2             #R-squared value
slope_err = reg.stderr           #standard error of slope
y_int_err = reg.intercept_stderr #standard error of y-int.

print('-'*28,'CODE OUTPUT','-'*29,'\n')
print('The slope is ', round(slope,3), '+/-', round(slope_err,3),' m/s^2')
print('The y-intercept is ', round(y_int,3), '+/-',\
round(y_int_err,3),' m/s')
print('The R^2 value is ', round(r_sq,4))

plt.plot(time, velocity,'o')
plt.plot(time, slope*time + y_int)
plt.xlabel('time [s]')
plt.ylabel('velocity [m/s]')
plt.title('Linear regression of data')
plt.show()

---------------------------- CODE OUTPUT ----------------------------

The slope is  2.03 +/- 0.074  m/s^2
The y-intercept is  0.19 +/- 0.018  m/s
The R^2 value is  0.9961
```

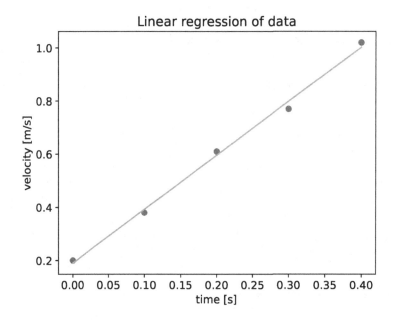

Figure 13.2 Graphical output from Example 13.3, illustrating the best fit line of a data set using Python.

13.4 DATA TRANSFORMATIONS

What happens if two quantities share a nonlinear relationship? For example, consider the drag force F which can depend quadratically on an object's velocity through a fluid (such as air). The magnitude of the quadratic drag force is:

$$F = b\,v^2 \tag{13.4.1}$$

where b is a constant which depends on the object's shape, the fluid's density, and other quantities. Suppose you measure the drag force on a given object as a function of velocity. How can you identify the nonlinear relationship between v and F?

There are two methods to answer this question. In the first method, we could perform feature engineering, similar to what is done in machine learning. The second method is to create a log-log or a log-linear plot of the data. We will examine both methods in the next two subsections.

13.4.1 FEATURE ENGINEERING

Let us begin with the *feature engineering* method, in which we plot the dependent variable (in this case F) vs. a function of some independent variable.

The difficulty with feature engineering is identifying the choice of v^2 as the independent variable over, say v^3 or $\sin v$ etc. Feature engineering works best when there is a theoretical justification for the choice of independent variable. In this case, (13.4.1) provides the theoretical justification for the choice of using v^2. The drag force equation (13.4.1) can be thought of as being linear in v^2, therefore if we plot F vs. v^2 then we should get a straight line. The slope of the best fit line would correspond to the value of b.

Example 13.4 shows how a set of data can be analyzed using a feature engineering method, and also how it can be analyzed assuming a linear relationship between the variables.

Example 13.4: Drag force and feature engineering

Suppose you measure the drag force on an object as a function of its velocity. The data is shown in the Python code below. Using feature engineering, find the drag force coefficient b. Repeat the analysis using linear regression of the data, and compare the two methods.

Solution:
The code below contains the data and the analysis. Notice that we use the square of the velocity data in both `linregress` and in the commands for the plots in Figure13.3a. Further note that we find that the y-intercept is close to zero, as expected from (13.4.1).
Now suppose, we repeat the analysis with performing a linear regression on the velocity data without squaring it, as in Figure 13.3b. The R-squared value is high, but not as high as the case where we performed feature engineering. In fact, the R-squared for the second case is high enough that a student might conclude the force is linear in the velocity.

```python
import numpy as np
import matplotlib.pyplot as plt
from scipy.stats import linregress

velocity = np.array([0.1, 0.2, 0.3, 0.4, 0.5])
force = np.array([0.0044, 0.014, 0.038, 0.060, 0.110])

# Use v^2 as the independent variable for feature engineering
# Use lingress to find the best fit parameters

reg = linregress(velocity**2, force)

# evaluate and print best fit parameters and R-squared
slope = reg.slope              #slope
y_int = reg.intercept          #y-intercept
r_sq = reg.rvalue**2           #R-squared value
slope_err = reg.stderr         #standard error of slope
y_int_err = reg.intercept_stderr   #standard error of y-int.

print('-'*28,'CODE OUTPUT','-'*29,'\n')

print('\nAnalysis with feature engineering:')
print('The drag force constant is ', round(slope,3),\
'+/-', round(slope_err,3))
print('The y-intercept is ', round(y_int,3), '+/-',\
round(y_int_err,3))
print('The R^2 value is ', round(r_sq,4))

# plot data and best fit to the data
plt.subplot(1,2,1)
plt.scatter(velocity**2, force)
plt.plot(velocity**2, slope*(velocity**2) + y_int)
plt.xlabel(r'v$^{2}$   [(m/s)$^{2}$]')
```

```
plt.ylabel('force [N]')
plt.title('(a)')

# Repeat assuming a linear relationship
reg = linregress(velocity, force)

slope = reg.slope                    #slope
y_int = reg.intercept                #y-intercept
r_sq = reg.rvalue**2                 #R-squared value
slope_err = reg.stderr               #standard error of slope
y_int_err = reg.intercept_stderr     #standard error of y-int.

print('\nAnalysis assuming a linear relationship:')
print('The drag force constant is ', round(slope,3),\
'+/-', round(slope_err,3))
print('The y-intercept is ', round(y_int,3),\
'+/-',round( y_int_err,3))
print('The R^2 value is ', round(r_sq,3))

plt.subplot(1,2,2)
plt.plot(velocity, force,'rv')
plt.plot(velocity, slope*(velocity) + y_int)
plt.xlabel('v [m/s]')
plt.ylabel('force [N]')
plt.title('(b)')
plt.tight_layout()
plt.show()

------------------------- CODE OUTPUT -------------------------


Analysis with feature engineering:
The drag force constant is  0.434 +/- 0.026
The y-intercept is  -0.002 +/- 0.004
The R^2 value is  0.9897

Analysis assuming a linear relationship:
The drag force constant is  0.257 +/- 0.041
The y-intercept is  -0.032 +/- 0.013
The R^2 value is  0.931
```

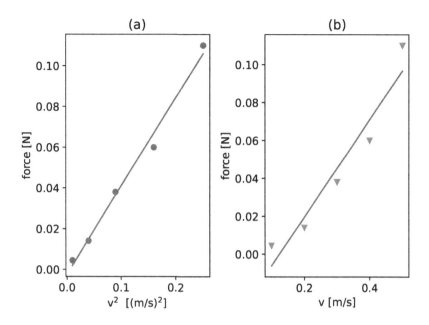

Figure 13.3 Graphical output from Example 13.4, illustrating (a) the feature engineering method (b) analysis of the same set of data, assuming a linear relationship between the two variables.

13.4.2   POWER LAW RELATIONSHIPS AND LOG-LOG PLOTS

Next we look at log-log plots. We begin by considering a power law relationship between a dependent variable y and an independent variable x.

$$y = ax^n \tag{13.4.2}$$

Let us take the natural logarithm of both sides

$$\ln y = \ln a + n \ln x \tag{13.4.3}$$

the result is that the natural logarithm of y is linear in the natural logarithm of x. The plot $\ln y$ vs. $\ln x$ is called a *log-log plot*. The slope of the best fit line is the power law n. The coefficient a is found by calculating the exponential of the best fit line's y-intercept.

In Example 13.5 we will analyze once more the data from Example 13.4, but this time we will use a log-log plot.

Example 13.5: Drag force and log-log plots

Find the power law relationship between the drag force and velocity from the data in Example 13.4 .

Solution:

The code below contains the data and the analysis. Notice that we use the logarithm of both `velocity` and `force` in `linregress` and in the commands for the plots in Figure 13.4.

The power law is correctly attained without making assumptions about the independent variable. The drag coefficient b is also the same as in the previous example.

```python
import numpy as np
import matplotlib.pyplot as plt
from scipy.stats import linregress

velocity = np.array([0.1, 0.2, 0.3, 0.4, 0.5])
force = np.array([0.0044, 0.014, 0.038, 0.060, 0.110])

reg = linregress(np.log(velocity),np.log(force))

slope = reg.slope                    #slope
y_int = reg.intercept                #y-intercept
r_sq = reg.rvalue**2                 #R-squared value
slope_err = reg.stderr               #standard error of slope
y_int_err = reg.intercept_stderr     #standard error of y-int.

print('-'*28,'CODE OUTPUT','-'*29,'\n')
print('The power law is n= ', round(slope,2), '+/-',\
round(slope_err,2))
print('\na = ', round(np.exp(y_int),2), '+/-', round(y_int_err,2))
print('\nThe R^2 value is ', round(r_sq,3))

plt.scatter(np.log(velocity),np.log(force))
plt.plot(np.log(velocity), slope*(np.log(velocity)) + y_int)
plt.xlabel('log(velocity)')
plt.ylabel('log(force)')
plt.title('Log-log plot')
plt.show()

---------------------------- CODE OUTPUT ----------------------------

The power law is n=  1.99 +/- 0.09

a =  0.4 +/- 0.14

The R^2 value is  0.993
```

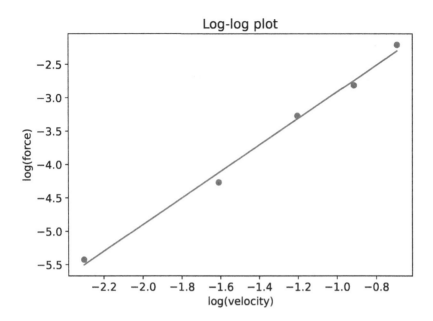

Figure 13.4 Graphical output from Example 13.5, illustrating the use of a log-log plot to fit the data.

13.4.3   EXPONENTIAL RELATIONSHIPS AND LOG-LINEAR (SEMI-LOG) PLOTS

Next, let us consider the case where two quantities are related by an exponential relationship:

$$y = ae^{\lambda x} \tag{13.4.4}$$

We take the natural logarithm of (13.4.4)

$$\ln y = \ln a + \lambda x \tag{13.4.5}$$

Notice now that the natural logarithm of y is linearly related to x. If we were to plot $\ln y$ vs. x, then the slope of the best fit line would be λ and the exponential of the y-intercept would be the quantity a. Such a plot is called a *log-linear plot*.

Example 13.6: Radioactive decay

The amount of radioactive material present in a sample is measured once per year over a period of four years, yielding the data included in the Python code below. Find the decay rate of the material.

Solution:
Let $N(t)$ be the number of radioactive particles present in a sample at time t. It can be shown

$$N(t) = N_0 e^{-\lambda t} \tag{13.4.6}$$

where $N_0 = 100$ is the initial number of particles in the sample and λ is the radioactive decay rate.

The code which includes the data and the analysis is below. In the code N is the number of radioactive particles left at time t.

Notice that in `linregress` we used the variables t and the natural logarithm of N. We used the same variables when plotting the data and the best fit line in Figure 13.5. The slope of the log-linear plot is

$$\lambda = -0.24 \pm 0.03 \, \text{s}^{-1}$$

which includes the negative sign in $(13.4.6)$. Furthermore, N_0 is found by taking the exponential of the y-intercept. We find that $N_0 = (96.5 \pm 0.07)$ particles. A log-linear plot of the data and line of best fit is shown in Figure 13.5.

```python
import numpy as np
import matplotlib.pyplot as plt
from scipy.stats import linregress

t = np.array([0, 1, 2, 3, 4])
N = np.array([100.0,  70.8,  64.6,  43.2, 38.7])

# find best fit line
reg = linregress(t,np.log(N))

slope = reg.slope                   #slope
y_int = reg.intercept               #y-intercept
r_sq = reg.rvalue**2                #R-squared value
slope_err = reg.stderr              #standard error of slope
y_int_err = reg.intercept_stderr    #standard error of y-int.

print('-'*28,'CODE OUTPUT','-'*29,'\n')
print(r'The decay constant is lambda =(', round(slope,2),\
'+/-', round(slope_err,2),') s^-1')
print('\nN_0 = ', round(np.exp(y_int),2), '+/-', round(y_int_err,2),\
' particles')
print('\nThe R^2 value is ', round(r_sq,3))

plt.scatter(t, np.log(N))
plt.plot(t, slope*(t) + y_int)
plt.xlabel('time t [s]')
plt.ylabel('ln(N[t])')
plt.title('Log-Linear plot (semilog)')
plt.show()

---------------------------- CODE OUTPUT ----------------------------

The decay constant is lambda =( -0.24 +/- 0.03 ) s^-1

N_0 =  96.5 +/- 0.07  particles

The R^2 value is  0.964
```

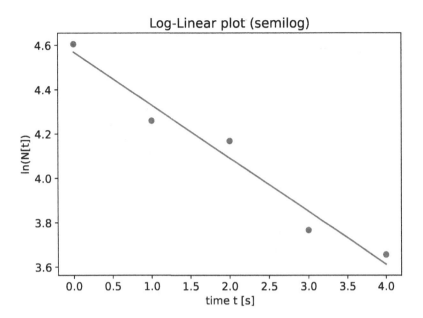

Figure 13.5 Graphical output from Example 13.10, illustrating the use of a log-linear plot to fit the data.

13.5 MULTIVARIATE LINEAR REGRESSION

It is not unusual to have more than one independent variable. We then need to perform multivariate linear regression. In such a case, we can think of the independent values creating a s-dimensional space (where s is the number of independent variables) and we attempt to fit a hyperplane instead of a line

$$\hat{y} = \sum_{i=1}^{s} m_i x_i + m_0 \tag{13.5.1}$$

Note that m_0 now serves as the intercept. Suppose we made n measurements.

The function J is now defined as

$$J(\mathbf{m}, b) = \frac{1}{n} \sum_{i=1}^{n} \left(\left[\sum_{i=1}^{s} m_i x_i + m_0 \right] - y_i \right)^2 \tag{13.5.2}$$

The m_i's values can be found by solving the system of equations

$$\frac{\partial J}{\partial m_i} = 0 \qquad i = 0, \ldots s \tag{13.5.3}$$

While there is a closed form solution to (13.5.3), in practice, we rely on software like Python to find the values of m_i. The formula for the R-squared value does not change.

Example 13.7: Predicting student understanding

At the end of each semester, a physics professor gives her students an assessment to measure their understanding of the course material (similar to the Force Concept Inventory). She hires you to create a model which will predict a student's performance based on their midterm and final exam grades. In order to do the job, you ask her for the grades of five students and their assessment score (provided in Python code below). Using multivariate linear regression, find a formula which predicts assessment performance. Calculate the predicted grade of a student who earns a score of 75% on their midterm exam and a 90% on their final exam.

Solution:
The code below contains the test data. The first array in the variable `test_data` is the list of midterm grades for each of the five students. The second array in the variable `test_data` is the list of final grades for each student. The variable `assessment_score` contains each student's assessment score.
We create the function `fit_func` which is the plane to which we will fit the data. The fit function has the form

$$\hat{y} = a\,x_0 + b\,x_1 + c \tag{13.5.4}$$

where x_0 is the midterm grade and x_1 is the final grade.
The command `curve_fit` fits the data to `fit_func`, and the best fit parameters a, b and c (similar to m_1, m_2, m_3 from (13.5.1)) are printed and the result is plotted in Figure 13.6. A plane of best fit is included in Figure 13.6 to show the trend in the data, similar to the best fit lines we saw in earlier examples.
    Using the fit parameters a, b and c, we find that the student is predicted to have a score of 0.87 on the assessment.

```
import numpy as np
import matplotlib. pyplot as plt
from mpl_toolkits.mplot3d import Axes3D

from scipy.optimize import curve_fit

test_data = np.array([
    [0.25,0.90, 0.80, 0.50, 0.63],
    [0.5,0.95,0.75,0.92,0.51]
    ])

assessment_score = np.array([0.43,0.97,0.80,0.77,0.60])

def fit_func(x,a,b, c):
    return a*x[0] + b*x[1] + c

popt, pcov = curve_fit(fit_func, test_data, assessment_score)

a, b, c = popt

print('-'*28,'CODE OUTPUT','-'*29,'\n')
print('a, b, c =', popt)
print('')
print('A student with a 0.75 on the midterm and a 0.90 on the final exam')
print('is predicted to have the grade: ', round(a*0.75 + b*0.90 + c,2))
```

```
xx, yy = np.meshgrid(np.linspace(0,1,10),np.linspace(0,1,10))

fig = plt.figure()
ax = fig.add_subplot(111,projection='3d')
ax.plot_surface(xx,yy,a*xx+b*yy+c,alpha=0.2)
ax.scatter(test_data[0],test_data[1],assessment_score)
ax.set_xlabel('mid-term')
ax.set_ylabel('final')
ax.set_zlabel('assessment')
ax.set_title('Multivariate linear regression')
plt.show()

------------------------ CODE OUTPUT -------------------------

a, b, c = [0.43487365 0.56213798 0.03800565]

A student with a 0.75 on the midterm and a 0.90 on the final exam
is predicted to have the grade:  0.87
```

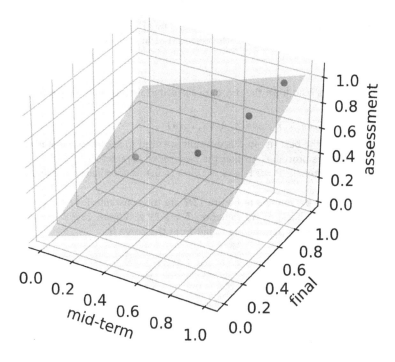

Figure 13.6 Graphical output from Example 13.7, illustrating a multivariate linear regression to a set of data.

13.6 NONLINEAR REGRESSION

In this section we show how to fit experimental data using nonlinear regression, when the data is expected to follow a previously known equation. The example is the least squares fit to the cosmic radiation data from the Cosmic background explorer project (COBE), using the Planck's radiation law equation:

$$I(\lambda) = A \frac{2c^2 h}{\lambda^5} \left(\frac{1}{e^{hc/\lambda kT} - 1} \right) \tag{13.6.1}$$

where $I(\lambda)$ is the intensity (power per unit area) of the emitted radiation at a wavelength λ, T is the black-body's temperature, $c = 2.008 \times 10^8$ m/s is the speed of light, $k = 1.381 \times 10^{-23}$ J/K is the Boltzmann constant, $h = 6.626 \times 10^{-34}$ J $\cdot$ s is Planck's constant and A is a constant in arbitrary units.

Example 13.8 illustrates how to use Python to find the best fit to the COBE data set.

Example 13.8: Nonlinear regression - Fitting the cosmic radiation data

The COBE experimental data analyzed here were adapted from the *hyperphysics* website *http://hyperphysics.phy-astr.gsu.edu/hbase/bkg3k.html#c1*

Find the best fit to this data using Planck's radiation law (13.6.1). Use as the fitting parameters the temperature T of the black-body and the constant A in (13.6.1).

Solution:
The experimental data is read from the .txt data file using the `pd.read_table` function, and the command

`optimize.curve_fit(Planck,x_data,y_data,inis)`

finds the parameters for the best fit. In this example we need to specify some starting parameters inis for the least squares procedure to find the best set of parameters.
The initial set of parameters is $T = 2$ K and $A = 400$ (a.u.), is specified in the line of code `inis=[2,400]`.
The values of the best fit parameters in Figure 13.7 are the constant $A = 1928$ (in a.u.), and the temperature of the cosmic background radiation $T = 2.71$ K. Figure 13.7 shows a graph of the data and the (13.6.1) with the parameters found from the curve fit.

```
import numpy as np
import matplotlib.pyplot as plt
from scipy import optimize
import matplotlib.pyplot as plt
import pandas as pd
print('-'*28,'CODE OUTPUT','-'*29,'\n')


# experimental data adapted from
# http://hyperphysics.phy-astr.gsu.edu/hbase/bkg3k.html#c1

# read the contents of the file using pandas library
# The data in each row in the file are separated by commas
url='https://github.com/vpagonis/CRCbook/raw/main/3K.txt'
df =pd.read_table(url,delimiter=',')
```

```
# look at the contents of the first few lines of the file
print('The first few lines of the data are:\n')
print(pd.DataFrame.head(df))


# store columns #1 and #2 to NumPy arrays x_data, y_data
x_data = df.iloc[:, 0].to_numpy()
x_data = 0.001*x_data  # wavelength is in mm, convert to m
y_data = df.iloc[:, 1].to_numpy()


plt.plot(x_data,y_data,'s')

h = 6.626e-34    # Planck constant in J s
k = 1.380649e-23 # Boltzmann constant in J/K
c = 2.998e8      # speed of light m/s


# T = black-body Temperature in K
# x = wavelength is in meters
# y = intensity
# A = scaling factor for intensity


# function for evaluating the Planck black-body equation
def Planck(x,T,A):
    return (A*2*h*c**2.0/x**5.0)*1/(np.exp(h*c/(x*k*T))-1)


inis = [2.0,400]  # starting values (T, A) for the fit

# find optimal parameters
# params are the best fit values for the parameters (T,A)
# cov is the covariance of the best fit parameters
params, cov =optimize.curve_fit(Planck,x_data,y_data,inis)

plt.scatter(x_data, y_data, label='COBE data')
plt.plot(x_data, Planck(x_data, *params),
c='r',linewidth=3, label='Planck equation')

plt.xlabel('Wavelength $\lambda$ [m]')
plt.ylabel(r'Intensity I(${\lambda}$)')
plt.title('Cosmic background radiation')
leg = plt.legend()
leg.get_frame().set_linewidth(0.0)

print('\nThe value of A from \
best fit =',round(params[1],2),' a.u.')


print('\nTemperature of background radiation from \
best fit =',round(params[0],2),' K')
plt.show()


------------------------- CODE OUTPUT -------------------------
                            •
The first few lines of the data are:

   Wavelength  Intensity
0    0.516338   0.216933
1    0.548758   0.282212
```

```
2    0.573032    0.343468
3    0.593078    0.408856
4    0.619386    0.461920

The value of A from best fit = 1928.5  a.u.

Temperature of background radiation from best fit = 2.71  K
```

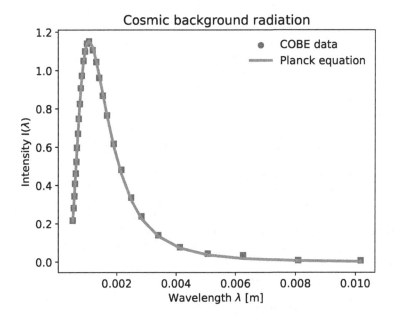

Figure 13.7 Graphical output from Example 13.8, illustrating the best nonlinear fit to the COBE data, using the Planck radiation law.

13.7 FOURIER TRANSFORMS

Let us return to the concept of a Fourier series, as presented in Chapter 3. Recall that any periodic function $f(t)$ with a period τ can be written as a sum of sines and cosines terms, resulting in a Fourier series:

$$f(t) = \frac{a_0}{2} + \sum_{n=1}^{\infty} a_n \cos(n\,\omega\,t) + b_n \sin(n\,\omega\,t) \tag{13.7.1}$$

where $\omega = 2\pi/T$. We can certainty think of using (13.7.1) as a means of approximating $f(t)$ by a finite sum of sines and cosines. However, we can also use (13.7.1) to identify what frequencies ω make up the function $f(t)$.

When a guitar string is plucked, the string does not vibrate with a single frequency. Instead, it vibrates with a fundamental frequency and its harmonics, frequencies which are integer multiples of the fundamental frequency.

Let us return to (13.7.1). The periodic function $f(t)$ has a period τ and a fundamental frequency $\omega = 2\pi/\tau$. The sum in (13.7.1) consists of sines and cosines which have frequencies consisting of the harmonics $n\omega$. For example, for a given $f(t)$ if a_2 and b_2 are both zero, then the function $f(t)$ does not have the frequency 2ω. We can make this idea concrete by revisiting Example 3.5.

In Example 3.5, we found that the Fourier series for the periodic function $f(x) = x$ (for $x = -\pi \le x \le \pi$) is

$$f(x) = 2 \sum_{n=1}^{\infty} \frac{(-1)^n}{n} \sin(n\,x) \tag{13.7.2}$$

We can interpret (13.7.2) as telling us that the function $f(x)$, which has a period $\tau = 2\pi$ contains all the harmonics, since there are no values of n such that $b_n = 0$. In this case the fundamental frequency is $\omega = 1$ and the amplitude of each harmonic $n\omega$ scales with $1/n$.

Why would we want to know what frequencies make up a given function? Consider a thin glass tube of Hydrogen gas. If we pass electric current through the gas, the gas will glow with a purple color. Purple light is an electromagnetic wave with a range of specific frequencies. If you viewed that light through a diffraction grating, you would find that in fact, the light consists of many more frequencies than that associated with purple. If we think of the light emitted by the Hydrogen gas as a periodic function, we could identify which frequencies make up the light, and therefore which colors appear in the spectrum. We know from the theory of atomic physics, that each of those frequencies correspond to an electron transition. Therefore, knowing the frequencies that make up a measured signal (a function) can tell us something about the physical phenomena responsible for producing that signal. Fourier series gives us a tool for identifying the frequencies in a periodic signal.

However, how can we get similar information from a nonperiodic function (signal) or from time series data? We can perform what is called a *Fourier transform*.

Let us return, temporarily, to periodic functions. The Fourier series (13.7.1) can be rewritten as a sum of complex exponentials. Consider the periodic function $f(x)$ with a period τ, then the complex Fourier series of f can be written as

$$f(t) = \sum_{n=-\infty}^{\infty} c_n e^{i n \omega t} \tag{13.7.3}$$

where the amplitudes c_n are

$$c_n = \frac{2}{\tau} \int_{-\tau}^{\tau} f(t) e^{-i n \omega t} dt \tag{13.7.4}$$

Notice that now the sum in (13.7.3) contains negative integral values of n, because of the identities that relate the complex exponential with trigonometric functions (see Chapter 6).

We can think of (13.7.4) as providing a formula for $c_n = c_n(\omega)$. If we compute $|c_n|^2$, we would have the relative intensity of the term in (13.7.3) with frequency $n\omega$. Colloquially, we can think of c_n as telling us how much the frequency $n\omega$ contributes to the periodic signal.

We can extend (13.7.3) to the case where we have a nonperiodic function with many frequencies, which are not harmonics of a fundamental. In this case, the sum becomes an integral (as we are no longer summing over discrete frequencies). The Fourier transform $F(\omega)$ of the function $f(t)$ is

$$F(\omega) = \frac{1}{2\pi} \int_{-\infty}^{\infty} f(t) e^{-i \omega t} dt \tag{13.7.5}$$

$$f(t) = \int_{-\infty}^{\infty} F(\omega) e^{i \omega t} d\omega \tag{13.7.6}$$

Notice that (13.7.5) corresponds to (13.7.4). The constant $1/2\pi$ is not universal in (13.7.5), the reader should consult the manual for any software when computing Fourier transforms to know which constant the software's author is using. Likewise, the *inverse Fourier transform* (13.7.6) corresponds to (13.7.3).

The Fourier transform is a complex function. Hence, we typically plot the *power spectrum*

$$\Phi(\omega) = \frac{1}{2\pi} F^*(\omega)\, F(\omega) \tag{13.7.7}$$

which is a real function and is sometimes multiplied by a constant (such as $1/2\pi$). The power spectrum provides the intensity of the frequency ω in the signal.

Example 13.9: Fourier transforms

Compute the Fourier transform of the function

$$f(t) = \begin{cases} t^2 - 1 & -1 \le t \le 1 \\ 0 & \text{otherwise} \end{cases} \tag{13.7.8}$$

and plot the power spectrum (13.7.7).

Solution:
We will use (13.7.5) with limits from -1 to 1 because $f(t)$ is zero otherwise.

$$F(\omega) = \frac{1}{2\pi} \int_{-1}^{1} \left(t^2 - 1\right) e^{-i\omega t} dt \tag{13.7.9}$$

We begin by breaking the integral up into two

$$\frac{1}{2\pi} \int_{-1}^{1} \left(t^2 - 1\right) e^{-i\omega t} dt = \frac{1}{2\pi} \left(\int_{-1}^{1} t^2 e^{-i\omega t} dt - \int_{-1}^{1} e^{-i\omega t} dt \right) \tag{13.7.10}$$

Using Euler's relationships between sine and exponentials we can show

$$\int_{-1}^{1} e^{-i\omega t} dt = \frac{2}{\omega} \sin \omega \tag{13.7.11}$$

The other integral can be done using parts:

$$\int_{-1}^{1} t^2 e^{-i\omega t} dt = t^2 \frac{1}{-i\omega} e^{-i\omega t} \Big|_{-1}^{1} - 2\left(\frac{1}{-i\omega}\right) \int_{-1}^{1} t e^{-i\omega t} \tag{13.7.12}$$

$$= \frac{2}{\omega} \sin \omega + \left(\frac{2}{i\omega}\right) \left\{ t \frac{1}{-i\omega} e^{-i\omega t} \Big|_{-1}^{1} - \left(\frac{1}{-i\omega}\right) \int_{-1}^{1} e^{-i\omega t} dt \right\} \tag{13.7.13}$$

$$= \frac{2}{\omega} \sin \omega + \frac{4}{\omega^2} \cos \omega - \frac{4}{\omega^3} \sin \omega \tag{13.7.14}$$

Next, we can simplify the Fourier transform by inserting (13.7.11) and (13.7.14) into (13.7.9)

$$F(\omega) = \frac{2}{\pi \omega^3} \left(\omega \cos \omega - \sin \omega\right) \tag{13.7.15}$$

However, this work can also be done in SymPy. The command _fourier_transform(f, t, omega, a, b computes the Fourier transform of the function f:

$$F(\omega) = a \int_{-\infty}^{\infty} f(t)e^{bi\omega t}dt \qquad (13.7.16)$$

There is an alternative command in SymPy, fourier_transform however, _fourier_transform has arguments a and b which can be used to match the constants in (13.7.5).

Notice that the first result from SymPy is not in the form of trigonometric functions. However, by using the .rewrite(cos).simplify() method, the SymPy answer becomes the same as in (13.7.15). Note also that we use the lambdify() function so that we can plot the result.

In Figure 13.8, we see that most of the power of the function is in frequencies between 0 and 4 rad/s.

```
from sympy.integrals.transforms import _fourier_transform
from sympy import  cos, pi, symbols, lambdify, simplify, Piecewise
import numpy as np
import matplotlib.pyplot as plt

print('-'*28,'CODE OUTPUT','-'*29,'\n')

t, omega = symbols('t, omega',real=True)
f = Piecewise((0,t<-1),(0, t> 1), (t**2 - 1,True))

F = _fourier_transform(f, t, omega, a = 1/(2*pi), b = -1,\
name ='ft', simplify =True)

u = F.args[0][0]
print('The Fourier transform is:\n',u)

# rewrite by replacing exponentials with trig functions

v = u.rewrite(cos).simplify()
print('\nThe Fourier transform in trig form is: \n', v)

ft = lambdify(omega, v)

omega = np.linspace(-8,8,100)

plt.plot(omega,ft(omega)**2)
plt.xlabel(r'Frequency $\omega$')
plt.ylabel(r'Power spectrum')
plt.title('Power spectrum of function f(t)')

plt.tight_layout()
plt.show()


------------------------ CODE OUTPUT -----------------------------

The Fourier transform is:
 (omega + (omega + I)*exp(2*I*omega) - I)*exp(-I*omega)/(pi*omega**3)

The Fourier transform in trig form is:
 (2*omega*cos(omega) - 2*sin(omega))/(pi*omega**3)
```

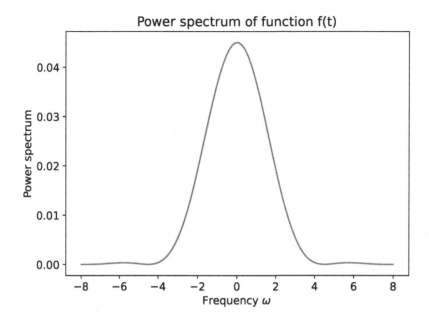

Figure 13.8 Graphical output from Example 13.10, showing the power spectrum $\Phi(\omega)$ of the function $f(t)$ in (13.7.8).

13.8   DISCRETE FOURIER TRANSFORMS

Now we move onto the case where we don't have a continuous function, but instead, have time series data. We can use the discrete Fourier transform (DFT) to obtain the frequency representation of the data.

Consider the time series

$$x = \{x_0, x_2, \ldots, x_{N-1}\} \tag{13.8.1}$$

where the index denotes the time at which the quantity x was measured and N is the number of elements in the time series. More concretely, the value x_n is the measurement of the quantity x taken at time $n\Delta t$, where Δt is called the sampling time.

Although we are no longer working with continuous functions, the insights of the previous sections still apply. The quantity x is measured as a function of time, although now it is a discrete function of time. By analogy to (13.7.4), we can write the DFT X as

$$X_k = \sum_{n=0}^{N-1} x_n e^{-2\pi i n k/N} \tag{13.8.2}$$

where $k = 1 \ldots N$. Note that because of the periodicity of the complex exponential, $X_{N-k} = (X_k)^*$. Further note that X_0 is always real.

To understand the connection between (13.8.2) and (13.7.4), let us rewrite (13.8.2)

$$X_k = \sum_{n=0}^{N-1} x_n \exp\left(-i\left[\frac{2\pi k}{N\,\Delta t}\right]\{n\Delta t\}\right) \tag{13.8.3}$$

The time of the measurement appears in the curly brackets in (13.8.3), therefore the frequency is

$$\omega_k = \frac{2\pi k}{N \Delta t} \tag{13.8.4}$$

From (13.8.2), we can plot the power spectrum

$$\Phi = \frac{2}{N} \left\{ |X_0|^2, |X_1|^2, \dots |X_{N-1}|^2 \right\} \tag{13.8.5}$$

which will give us information on the relative intensity of each frequency ω_k in the time series x. Note that we typically only plot Φ out to $k = N/2$ because $X_{N-k} = (X_k)^*$ and therefore the frequencies for $k > N/2$ are identical to the earlier frequencies.

The inverse DFT is also useful and parallels (13.7.3)

$$x_n = \frac{1}{N} \sum_{k=0}^{N-1} X_k e^{2\pi i n k/N} \tag{13.8.6}$$

The information above assumes that x is a regularly sampled time series. If the time between measurements is constant (i.e. there are no gaps in the data), the data is said to be regularly sampled. Fourier transforms can only be applied to regularly sampled time series.

If you are working with irregularly sampled time series (as is often the case in astronomy and many other fields), the Lomb-Scargle periodogram can be used to obtain a power spectrum from the data. The Lomb-Scargle periodogram is beyond the scope of this text, but there is an implementation of it in Python's SciPy library.

In Example 13.10, we demonstrate how to perform the DFT using Python, and how to interpret its results. Although, for convenience, we will use a mathematical equation to produce our data, the analysis is identical to the case where an experimental time series would be used (with the exception of needing to import the file that contains the data you wish to analyze).

Example 13.10: The discrete Fourier transform

Compute the DFT for a time series generated from the function

$$x(t) = 3\sin(2\pi t) + \sin\left[2\pi(3)t\right] + 2\sin\left[2\pi(7)t\right] \tag{13.8.7}$$

where we explicitly stated the frequency of each term.

Solution:
Python's SciPy library include an implementation of the fast Fourier transform (FFT) which can be used to efficiently compute the terms X_k.

We begin by plotting the data, which is always good practice. The plot is shown in Figure 13.9a.

Next, we compute the DFT using the command fft. The command fftfreq provides a list of frequencies for each of the values of k in (13.8.2). Note that we include only the first half of the lists x_ft and freqs to avoid the aforementioned symmetry. We show the power spectrum of the data in Figure 13.9b. Finally, we included only the plot out to $f = \omega/2\pi = 10$ to more clearly show the power spectrum.

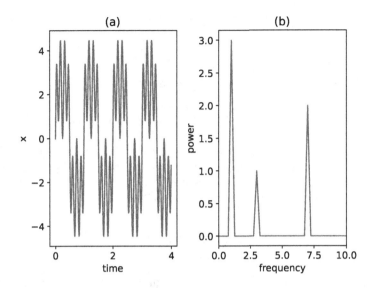

Figure 13.9 Graphical output from Example 13.10, showing (a) the function $x(t)$ from (13.8.7), and (b) its power spectrum evaluated using the DFT.

Notice that the location and amplitude of each spike in the graph in Figure 13.9 corresponds to the amplitudes and frequencies in the original formula of $f(t)$. Specifically, the ratio of the three amplitudes is 3:1:2 and the corresponding frequencies are 1,3 and 7, in agreement with (13.8.7).

```python
import numpy as np
import matplotlib.pyplot as plt

from scipy.fft import fft, fftfreq

f1, f2, f3 = 1.0, 3.0, 7.0
dt = 0.01

t = np.arange(0,4,dt)
x = 3.0*np.sin(2*np.pi*f1*t) + np.sin(2*np.pi*f2*t) +\
2*np.sin(2*np.pi*f3*t)

N = len(t)
plt.subplot(1,2,1)
plt.plot(t,x)
plt.xlabel('time')
plt.ylabel('x')
plt.title('(a)')

x_ft = fft(x)
freqs = fftfreq(N,0.01)[:N//2]
power_spec = 2.0/N * np.abs(x_ft[:N//2])

plt.subplot(1,2,2)
plt.plot(freqs, power_spec)
plt.xlim((0,10));
```

```
plt.xlabel('frequency')
plt.ylabel('power')
plt.title('(b)')
plt.tight_layout()

plt.show()
```

13.9 DISCRETE AND CONTINUOUS RANDOM VARIABLES

Physicists must deal with randomness not only as measurement noise, but also as a fundamental part of some physical processes. Stochasticity is inherent in quantum processes, where physicists can only predict probability distributions, not the values of single measurements. Sometimes, phenomena at a population level are best modeled as random numbers. For example, human height (as a population) can be modeled as a normal distribution. In this section, we discuss how to model and analyze random variables.

13.9.1 DISCRETE RANDOM VARIABLES

A discrete random variable is one that can take on only a countable number of distinct values. For example, consider an ensemble of 10 identically prepared spin 1 particles. Recall from quantum mechanics that a spin 1 particle has quantum numbers $s = 1$ and $m = -1, 0, 1$. Suppose that we measure m for each of the 10 particles and obtain

$$M = \{-1, -1, 0, 1, 0, 1, 1, 1, 0, 1\} \tag{13.9.1}$$

From this data, we can calculate the probability $p(m_i)$ of measuring each possible value $i = 0$, 1, or -1 by counting the number of occurrences of each value (N_i) and dividing by the total number of measurements $(N = 10)$

$$p(m_i) = \frac{N_i}{N} \tag{13.9.2}$$

Hence, in M we have $p(0) = 3/10$, $p(1) = 5/10$, and $p(-1) = 2/10$. Note that the probabilities add to unity.

Using the set M, we can calculate the average (or expectation value) of m by calculating

$$\langle m \rangle = \frac{1}{N} \sum_{i=0}^{9} M_i \tag{13.9.3}$$

where M_i is the ith element of the set M.

However, we can replace the sum in (13.9.3) with a sum over the possible allowed values of m in the set M, rewriting (13.9.3) as

$$\langle m \rangle = \sum_{m_i=-1}^{1} m_i \frac{N_i}{N} \tag{13.9.4}$$

which can be rewritten as

$$\langle m \rangle = \sum_{m_i=-1}^{1} m_i \, p(m_i) \tag{13.9.5}$$

We call $\langle m \rangle$ the expectation value of m. Likewise, the variance can be found using

$$\sigma_m^2 = \sum_{m_i=-1}^{1} (m_i - \langle m \rangle)^2 \, p(m_i) \tag{13.9.6}$$

Recall that the standard deviation is the square root of the variance

$$\sigma_m = \sqrt{\sigma_m^2} \tag{13.9.7}$$

For any data set $\{x_i\}$ the above functions can be generalized to

$$\langle x \rangle = \sum_{x_i} x_i \, p(x_i) \tag{13.9.8}$$

$$\sigma_x^2 = \sum_{x_i} (x_i - \langle x \rangle)^2 \, p(x_i) \tag{13.9.9}$$

where the sums are over the values of x_i that appear in the data set and $p(x_i)$ is the probability of the value x_i appearing in the data set.

Let us consider again the data set M. Suppose we wanted to know the probability of measuring a spin less than 1. We could find that by computing $p(-1) + p(0)$. In other words, we add the probabilities of each quantity that is less than 1. We can generalize this idea

$$p(x < x_i) = \sum_{x_j < x_i} p(x_j) \tag{13.9.10}$$

where the sum is over values in the data set that are less than x_i.

Note that if the data set is long enough, the experimentally obtained estimates of the mean and variance, (13.9.8), (13.9.9), and (13.9.10) will approximate the true values. Mathematically speaking, long enough means an infinite number of samples, however physicists never have that luxury. We have to learn to make due with finite samples and hope for significant representation.

Example 13.11: Discrete probabilities in Quantum Mechanics

Using Python, compute the expectation value and the variance of the spin data in (13.9.1):

$$M = \{-1, -1, 0, 1, 0, 1, 1, 1, 0, 1\} \tag{13.9.11}$$

Solution:
We use the command unique from the NumPy library to create a list of the unique elements in the array M and the number of times each element occurs in M. Because we are working with NumPy arrays, we can multiply arrays element by element and sum the result. As we have seen multiple times in this text, using NumPy arrays not only simplifies calculations, it also results in more easily understood code.

The expectation value of 0.3 reflects the tendency to measure a 1 for the particle's spin. In your quantum mechanics class you will learn how to compute the probabilities of each measurement using the particle's spinor, a column vector associated with the particle which describes its spin state.

```
import numpy as np

M = np.array([-1,-1,0,1,0,1,1,1,0,1])

unique, counts = np.unique(M, return_counts=True)
probs = counts/len(M)
avg = sum(unique*probs)
var = sum((unique-avg)**2*probs)

print('-'*28,'CODE OUTPUT','-'*29,'\n')
print('The elements in the set are ', unique)
print('\nThe probability of each element is ', probs)
print('\nThe expectation value is ', round(avg,2))
print('\nThe variance is ', round(var,2))

--------------------------- CODE OUTPUT ----------------------------

The elements in the set are  [-1  0  1]

The probability of each element is  [0.2 0.3 0.5]

The expectation value is  0.3

The variance is  0.61
```

13.9.2 CONTINUOUS RANDOM VARIABLES

Particle spin is a discrete random variable, since the spin can take on only a certain set of countable values. However, many variables in physical systems are not discrete. For example, human height can take on any value (even though we may round to the nearest centimeter, human heights can be 175.01 cm, 175.012 cm, and so on). In this section, we will generalize (13.9.8), (13.9.9), and (13.9.10) using what is called a *probability density function*.

The probability that a human male living in the United States has a height that is exactly 70 inches is almost zero. However, we can quantity the probability of an American male having a height between 69.1 and 70.1 inches. This is done by measuring the heights of a large number of people and fitting the data to a probability density function.

If we were to measure the height of every American and plot them, the data would fit curves that look like Figure 13.10. The horizontal axis in Figure 13.10 is height in inches while the vertical axis is proportional to the number of times a given height is measured. We will be more specific below, for now, the larger the value on the y-axis, the more common the measurement.

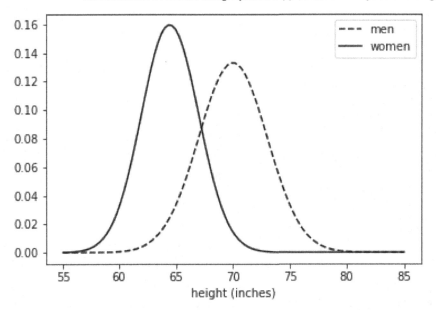

Figure 13.10 The probability density functions of the heights of Americans.

The mathematical form for the curves in Figure 13.10 is

$$p(x) = \frac{1}{\sigma\sqrt{2\pi}} \exp\left[-\frac{1}{2}\left(\frac{x - \langle x \rangle}{\sigma}\right)^2\right] \tag{13.9.12}$$

where x is the height of an individual, $\langle x \rangle$ is the mean height of the population (70 in for men and 64.5 in for women) and σ is the standard deviation of the heights (3.0 in for men and 2.5 in for women). The function (13.9.12) is a common probability density function and is known as a normal (or Gaussian) distribution. It serves as the probability of a measurement in the case of discrete random variables. However, because we now have continuous values, we need to replace the sums in (13.9.8), (13.9.9), and (13.9.10) with integrals.

For example, if we wish to know the probability of a woman being shorter than 60 inches, we would use the values of $< x >$ and σ for women in (13.9.12) and

$$p(x < 60) = \int_{-\infty}^{60} p(x)\, dx \tag{13.9.13}$$

Notice the limits of the integral serve as the bounds in the sum.

Likewise, if we wish to compute the mean or the variance, we could calculate

$$\langle x \rangle = \int_{-\infty}^{\infty} x\, p(x)\, dx \tag{13.9.14}$$

$$\sigma^2 = \int_{-\infty}^{\infty} (x - \langle x \rangle)^2\, p(x)\, dx \tag{13.9.15}$$

Example 13.12: Human height

Using Python, calculate the probability that an American man has a height less than 67.3 inches (the average global height of men) and the probability that an American woman has a height less than 62.8 inches (the average global height of women).

Solution:
The Python code for the calculation is shown below. Note that we could have used a variety of integration methods that we have covered in the text so far.

The result from the code shows that American men are less likely to be shorter than the global average, than American women.

```
from sympy import pi, sqrt, exp, oo, symbols, integrate

x = symbols('x')

mu_men = 70
s_men = 3
mu_women = 64.5
s_women = 2.5

f_men = 1/(s_men*sqrt(2*pi))*exp(-0.5*((x-mu_men)/s_men)**2)
f_women =  1/(s_women*sqrt(2*pi))*exp(-0.5*((x-mu_women)/s_women)**2)

p_men = integrate(f_men, (x,-oo,67.3)).evalf()
p_women = integrate(f_women, (x,-oo,62.8)).evalf()

print('-'*28,'CODE OUTPUT','-'*29,'\n')
print('The probability for men is ', round(p_men,3))
print('The probability for women is ',round( p_women,3))

--------------------------- CODE OUTPUT ---------------------------

The probability for men is  0.184
The probability for women is  0.248
```

In Example 13.13 we present an example of averaging over the distribution of a continuous variable.

Example 13.13: Probability distributions in Classical Mechanics

A car starting at rest accelerates with a constant acceleration a and travels a distance L. During the trip, a photographer takes many pictures of the car at random times. The photographer then measures the distance the car has traveled (from its starting position) on each image. What is the average distance measured? Plot the probability density function p and use SymPy to calculate the average and standard deviation.

Solution:

The car travels from a point x to a point $x + dx$ in a time dt. Therefore, the probability of a photograph being taken during that interval is dt/T, where T is the time it takes for the car to travel the full distance L.

Note that the probability density function here is $p(t) = 1/T$. However, this probability density function is time-dependent. We would like a position-dependent probability density function $p(x)$, given the nature of the problem. The two probability density functions are related by:

$$p(t)\, dt = p(x)\, dx \qquad (13.9.16)$$

Using kinematics, we know that the distance and time are related by $y = a\,t^2/2$, and in our example $L = a\,T^2/2$, so that:

$$T = \sqrt{\frac{2L}{a}}$$

Using $v = dx/dt$ or $dt = dx/v$, we find from (13.9.16) the probability density function $p(x)$:

$$p(x) = \frac{dt}{dx\,T} = \frac{1}{v\,T} = \frac{1}{\sqrt{2a\,x}}\sqrt{\frac{a}{2L}} = \frac{1}{2\sqrt{L\,x}}$$

The average distance is

$$\langle x \rangle = \int_0^\ell x\frac{dx}{2\sqrt{L\,x}} = \frac{L}{3}$$

The reason for the low expectation value of $<x>$ is that the car is moving at slower velocities early in its motion. Therefore, we are more likely to have photographs of the car early in its motion than later.

Using (13.9.15), the variance and the standard deviation are:

$$\sigma^2 = \int_{-\infty}^{\infty} (x - \langle x \rangle)^2\, p(x)\, dx = \int_{-\infty}^{\infty} \left(x - \frac{L}{3}\right)^2 \frac{dx}{2\sqrt{L\,x}} = \frac{4}{45}L^2$$

$$\sigma = \frac{2}{\sqrt{45}}L$$

```python
import numpy as np
from sympy import symbols, integrate, sqrt, simplify

print('-'*28,'CODE OUTPUT','-'*29,'\n')

x, L = symbols('x,L', positive=True)

p = 1/(2*sqrt(L*x))

meanx = integrate(x*p,(x,0,L))

print('\nThe mean is = ', simplify(meanx))

variance = integrate((x - meanx)**2*p,(x,0,L))
std = sqrt(variance)

print('\nThe standard deviation is = ', simplify(std))
```

```
--------------------------- CODE OUTPUT ---------------------------


The mean is =  L/3

The standard deviation is =  2*sqrt(5)*L/15
```

13.10   USEFUL PROBABILITY FUNCTIONS

In this section, we will examine some probability functions (both continuous and discrete) that are useful in physics and other sciences. In the interest of brevity, we present only the basic information about each distribution. The interested reader should consult textbooks and websites on statistics for more information.

13.10.1   THE BINOMIAL DISTRIBUTION

Consider a case where a measurement has one of two possible outcomes. For example, we could measure the spin of 10 identically prepared spin-1/2 particles. Each particle has as its potential spin either spin up ($m = 1/2$) or spin down ($m = -1/2$). If each particle has an even chance for each spin measurement, what is the probability that we get 7 particles with a spin up and 3 with a spin down (in any particular order)?

The probability for each spin measurement to be either spin up or spin down is $1/2$. Hence, the probability for 7 spin ups and 3 spin downs must be proportional to $(1/2)^{10}$ (a probability of $1/2$ for each measurement). However, out of the set of 10 particles, there are many ways we can select 7 of them to have spin up and 3 to have spin down. There is a mathematical formula for counting how many possible ways we can choose 7 of the particles to have spin up. It is called the *binomial coefficient*.

The binomial coefficient $C(n, k)$ is defined as

$$C(n, k) = \left(\begin{array}{c} n \\ k \end{array} \right) = \frac{n!}{k! \, (n - k)!} \tag{13.10.1}$$

is the number of ways a subset of k (unordered) items can be chosen from a fixed set of n elements. The value $C(n, k)$ is sometimes read as n choose k.

We can generalize to any problem with two possible outcomes. Suppose p is the probability of success and q is the probability of failure ($q = 1 - p$). For example, in the case of the spin-1/2 particle, p would be the probability of measuring $m = 1/2$ and q would be the probability of measuring $m = -1/2$. Then the probability of having exactly x successes in n trials is

$$\rho(x) = C(n, x) \, p^x \, q^{n-x} \tag{13.10.2}$$

Example 13.14: Spin probabilities in Quantum Mechanics

Consider an ensemble of 20 identically prepared spin-1/2 particle such that each particle has a 2/3 probability of producing a spin up ($m = 1/2$) measurement. What is the probability of finding 14 particles to have spin up?

Solution:
We use (13.10.2) to calculate this probability. In this case, $n = 20$, $p = 2/3$, and $q = 1/3$. We want to calculate

$$\rho(14) = C(20, 14) \left(\frac{2}{3} \right)^{14} \left(\frac{1}{3} \right)^{6}$$

We calculate the binomial coefficient with the `binom` command from SciPy's special functions library `special`.

```
from scipy.special import binom

n = 20
x = 14
p = 2/3
q = 1/3

C = binom(n,x)

print('-'*28,'CODE OUTPUT','-'*29,'\n')
print('The probability is ', round(C*p**14*q**6,2))

------------------------- CODE OUTPUT ---------------------------

The probability is  0.18
```

Example 13.15: The importance of repetition in experimentation

Plot the binomial distribution for $p = 1/2$ and $n = 10$ and for the two cases $n = 100$, and discuss the result.

Solution:
The Python code for the plot is shown below.
In Figure 13.11 we see that by increasing n, the binomial distribution tightens around the probability $1/2$. Hence, for large number of measurements the standard deviation of the distribution gets smaller, as expected from statistical considerations.

```
from scipy.stats import binom
import matplotlib.pyplot as plt
import numpy as np

p = 0.5

def dist(r,n,p):
    return [binom.pmf(item,n, p) for item in r]

dist1 = np.array(dist(np.arange(10),10,p))
dist2 = np.array(dist(np.arange(101),101,p))

fig, ax = plt.subplots(nrows = 1, ncols = 2)
fig.tight_layout(pad = 2.0)
fig.suptitle('Binomial Distribution, p(x) vs. x/n')
fig.subplots_adjust(top=0.88)


ax[0].bar(np.arange(10)/10,10*dist1, width = 0.07);
ax[0].set_title('n = 10, p = 1/2')
ax[0].set_xbound(0,1)
ax[0].set_xlabel('x')
```

```
ax[0].set_ylabel('Binomial distribution')

ax[1].set_xlabel('x')
ax[1].bar(np.arange(101)/101,101*dist2, width= 0.01);
ax[1].set_title('n = 100, p = 1/2')
ax[0].set_xbound(0,1)

plt.show()
```

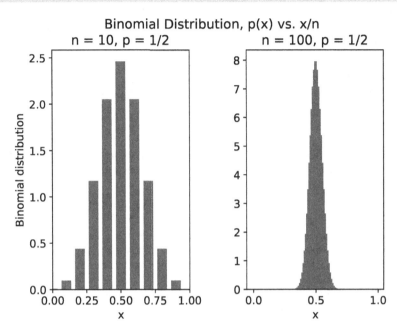

Figure 13.11 Graphical output from Example 13.15, showing the binomial distribution for $p = 1/2$ and two values of $n = 10$ and $n = 100$. $\rho(x)$.

13.10.2 NORMAL DISTRIBUTION

Another common random number distribution used in physics is the normal distribution already presented in (13.9.12)

$$p(x) = \frac{1}{\sigma\sqrt{2\pi}} \exp\left[\frac{1}{2}\left(\frac{x - \langle x \rangle}{\sigma}\right)^2\right] \qquad (13.10.3)$$

The normal distribution is commonly used, because of the Central Limit Theorem, which states (loosely) that if one randomly selects a large number of values from a population with a fixed mean and standard deviation, then the distribution of the randomly chosen samples approximates a normal distribution. Although beyond the scope of this text, the normal distribution is the distribution with the maximum entropy when the mean and standard deviation of a random population is known.

The normal distribution has the special property in that the probability of drawing a random value from a normally distributed data set depends on σ, the standard deviation of the distribution. For a normally distributed random numbers, 68% of all values fall within one standard deviation of the mean, 95% fall within two standard deviations, and 99.7% fall within three standard deviations. In particle physics, a measurement often needs to be 5σ from the mean to be considered significant, giving it a 1 in 3.5 million chance for the discovery of a particle to be false. Robert Wadlow (1918 - 1940), the tallest human in history was 8 ft 11 in and therefore 12.3 standard deviations from the current mean for American male heights. For all intents and purposes, the probability of encountering another human with Mr. Wadlow's height is zero.

Example 13.12 already examined how to use (13.9.12) to calculate probabilities. We can also use NumPy to numerically generate normally distributed numbers. The next example shows how large a random sample must be to faithfully reproduce the mean and standard deviation of the normal distribution.

Example 13.16: The mean and standard deviation of normally distributed random numbers

Plot the mean and standard deviation for sets of normally distributed random numbers with a mean of 2.0 and a standard deviation of 1.0 as a function of the population size (the number of elements in the randomly generated list).

Solution:
We will use NumPy's `random.normal` command which generated a list of normally distributed random numbers with a specified mean (first argument), standard deviation (second argument), and size (third argument). We then compute the mean and standard deviation of each population as a function of the population's size.
Notice from Figure 13.12 that even with small population sizes, the command `random.normal` produces a good representation of the actual mean and standard deviation.

```
import numpy as np
import matplotlib.pyplot as plt

mean = 2.0
stdev = 1.0

mean_list, std_list = [],[]

for s in range(5,505,5):
    data = np.random.normal(2.0, 1.0, size = s)
    mean_list.append(np.mean(data))
    std_list.append(np.std(data))

plt.plot(range(5,505,5), mean_list, 'r', label = 'mean')
plt.plot(range(5,505,5), std_list,  'b.', label = 'standard dev')
plt.ylabel('Mean, Standard deviation')
plt.legend()
plt.xlabel('data size')
plt.title('Mean and stadnard deviation of Gaussian distribution')
plt.show()
```

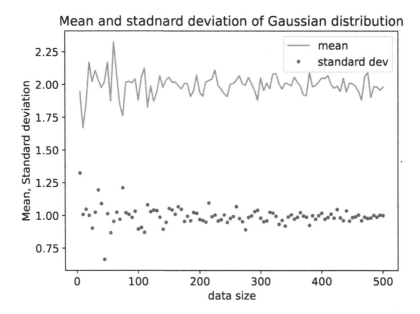

Figure 13.12 Graphical output from Example 13.16, showing the mean and standard deviation of sets of numbers chosen from a Gaussian distribution.

As a final note in this subsection, you will some times encounter the term *standard normal distribution*, which is a normal distribution with zero mean and unit standard deviation.

13.10.3   POISSON DISTRIBUTION

The Poisson distribution is a discrete probability distribution that gives the probability of a number of events occurring in a fixed interval of time. The events occur independently of the time since the last event and at a known constant mean rate. As an example, consider the process of radioactive decay. During a time interval Δt (much less than the element's half life), the probability that a particle will be emitted is $\mu \Delta t$ where μ is the mean rate of decay.

It can be shown that the probability of n events occurring per unit time P_n is

$$P_n = \frac{\mu^n}{n!} e^{-\mu} \tag{13.10.4}$$

To better understand how (13.10.4) can be used to solve problems, consider Example 13.17.

Example 13.17: Radioactive decay and the Poisson distribution

Suppose you observe the number of particles emitted by a radioactive substance to be 20 over a 10 minute period. During how many one-minute intervals should we expect to observe 3 particles?

Solution:
The average number of counts per minute is $20/10 = 2$. Hence, $\mu = 2$ for the Poisson distribution. We define (13.10.4) as a function and compute it for various interval durations.

From Figure 13.13, we see that we should observe 3 particles in one minute during 18% of the one minute intervals.

```python
import numpy as np
from scipy.special import factorial
import matplotlib.pyplot as plt

mu = 2

def p_n(n):
    return (mu**n)/factorial(n) * np.exp(-mu)

intervals = np.arange(0,11,1)
counts = []

for n in intervals:
    counts.append(p_n(n))

plt.bar(intervals,counts);
plt.ylabel('P_n')
plt.xlabel('interval')
plt.title('Histogram of data from a Poisson distribution')
plt.show()
```

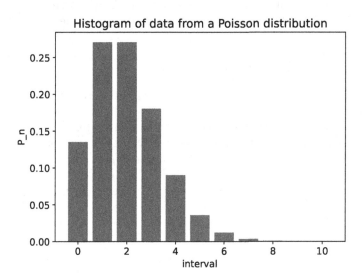

Figure 13.13 Graphical output from Example 13.17, showing a histogram of the number of one-minute intervals in which one would expect to observe 3 particles. The data is evaluated from a Poisson distribution.

13.11   END OF CHAPTER PROBLEMS

The data files in Table 13.2 are needed to solve some of the problems below. They can be found in the GitHub website: *https://github.com/vpagonis/CRCbook*.

Table 13.2
Table of data files for Chapter 13.

File name	Problem #	Description
dataset1.csv	3, 4	Data for Linear regression and noise analysis
dataset2.csv	5	Determining the power law followed by the data
dataset3.csv	6, 7	Determining which law the data follows
dataset4.csv	9	Linear superposition of sine waves
dataset5.csv	10	Linear superposition of sine and cosine waves
hub.txt	22	Hubble's 1929 data for distance and speed of galaxies
radioact.csv	23	Radioactivity data with two exponential components
qz.csv	24	Optically stimulated luminescence signal from quartz
einst.txt	26	Specific heat data for diamond $C_v(T)$
MB.txt	27	Experimental data from Maxwell-Boltzmann experiment

1. **Uncertainty in reading an instrument** –

    a. In the United States, many automobile speedometers have division of 5 mph. If while you are driving, the speedometer needle points to 55 mph, what is your car's speed including the uncertainty of its measurement. Should you share this with a police officer who pulled you over for speeding?

    b. An ammeter is connected in series to a resistor. The resistor an ammeter are connected to a 5.00 V power supply. The voltage from a the power supply is controlled by a dial and its output is displayed by a screen. The least count of the voltage is 0.01 V. You observe the ammeter for ten seconds. During that time, the largest current the ammeter measures is 2.35 A. The largest current measured is 2.70 A. Using Ohm's law $(V = IR)$, find the resistance of the resistor. Use Python to compute the uncertainty in the measurement.

2. **Propagation of uncertainties in thermodynamics** – A container holds 2.0×10^2 ml of liquid water. The volume of water was measured with a graduated cylinder that had as its smallest division, 5 ml. The water was initially at a temperature of $20°$ C. Its temperature was measured with a thermometer with a resolution of $1°$C. How much heat is needed to warm the water such that the thermometer reads $30°$C? Ignore the heat absorbed by the water's container and assume the specific heat of water is known exactly to be $c = 4186$ J kg^{-1} K^{-1}. Recall that the formula for the energy needed to increase the temperature of an object is $Q = mc\Delta T$.

3. **Linear regression** – Using linear regression, find the slope, y-intercept (and their respective standard errors), and the R-squared value of the linear fit for the data found in `dataset1.csv`. Plot the data and the best fit line.

4. **Noise analysis** – Using the regression parameters you found in Problem 3, find the random values added to the linear model used to create the data in `dataset1.csv`. Plot a histogram of the random numbers and make an educated guess about the distribution. Support your guess with evidence.

5. **Scaling law** – The dependent variable in `dataset2.csv` was generated by using a power law relationship for the independent variable. Find the power law. Justify your result through a plot and by calculating the appropriate R-squared value.

6. **Identifying relationships in data** – Identify whether the dependent variable in the data set `dataset3.csv` is linear in the dependent variable, or instead obeys a power law or exponential relationship.

7. **Noise analysis and filtering** – Using the regression parameters you found in Problem 6, estimate and plot the noise in the data set `dataset3.csv`. Do this by subtracting the best fit model from the data.

8. **Fourier Transforms** – Calculate and plot the Fourier transform and power spectrum of the following functions. To make the plots, choose $a = 1$ and $f = 2$. Plot the power spectra for the range $\omega \in [-4, 4]$.

     a. $f(t) = \begin{cases} 1 & -1 \le t < 1 \\ 0 & \text{otherwise} \end{cases}$

     b. $f(t) = \begin{cases} t+1 & -1 \le t < 0 \\ 1-t & 0 \le t \le 1 \\ 0 & \text{otherwise} \end{cases}$

     c. $f(t) = e^{-a|t|}$

9. **The discrete Fourier transform** – The data set in `dataset4.csv` contains a linear superposition of sine waves of different frequencies and amplitudes. The first column (with a the header t) contains the time and the second column (with the header y) contains the amplitude at time t. Using the DFT, identify the frequencies in the data and the amplitude of each frequency.

10. **The discrete Fourier transform 2** – Repeat Problem 9 using the data set `dataset5.csv` which consists of a linear superposition of sine and cosine waves.

11. **Maxwell-Boltzmann distribution** – The Maxwell-Boltzmann distribution

$$f(v) = \left(\frac{m}{2\pi \, k \, T}\right)^{3/2} 4\pi v^2 \exp\left(\frac{-m \, v^2}{2 \, k \, T}\right)$$

describes distribution of speeds of the particles making up an ideal gas of temperature T. Each particle in the gas has a mass m and k is the Boltzmann constant. If the ideal gas is Nitrogen (N2) at 300 K, what is the probability that gas particles will have a velocity between 350 m/s and 450 m/s?

12. **Normal distribution** – Show that the factor $1/\sigma\sqrt{2\pi}$ in (13.9.12) is required in order for the integral of the normal distribution from $x = -\infty$ to $x = \infty$ to be equal to one. Why would we require that the total area under the normal distribution be equal to one?

13. **Human height** – Suppose you randomly select 1000 men and 1000 women from the population of the United States. How many of the men can you expect to have a height between 70 and 75 inches? How many women can you expect to find in the same height range?

14. **Spam calls** – Suppose that, on average, you receive 5 spam calls per day on your phone. What is the probability that you receive no spam calls on a given day? How about ten calls?

15. **Airplane defects** – If there are, on average, three defects per 100 new airplanes built, what is the probability that a new airplane has 1 defect? What is he probability that all 100 airplanes have a defect?

16. **Coin tosses** – Consider an unfair coin which has a probability of 55% of landing with heads up when tossed. What is the probability of getting 5 heads in eight coin tosses? How does that compare to a fair coin (50% chance of landing heads up when tossed)?

17. **Wave functions as probability distributions** – The wave function in quantum mechanics is a continuous probability distribution function. Consider the first excited state of a particle of mass m trapped in a harmonic well ($V = 1/2 m\omega^2 x^2$)

$$\psi(x) = A \left(\frac{m\omega}{\pi\hbar}\right)^{1/4} \sqrt{\frac{2m\omega}{\hbar}} x \exp\left(\frac{-m\omega}{2\hbar}x^2\right)$$

The probability of a particle to be between $x = a$ and $x = b$ is found by

$$P\left(a \leq x \leq b\right) = \int_a^b |\psi(x)|^2 \, dx$$

Calculate:

a. The value of A such that $P(-\infty \leq x \leq \infty) = 1$. In other words, normalize the wave function.

b. The probability the particle will be between $x = 0$ and ∞.

c. The average position of the particle. In other words, the expectation of the position operator.

d. The standard deviation of the particle's position.

18. **Physics GRE** – In 2018, the Physics GRE score had a mean of 712 and a standard deviation of 160. Assuming the scores on the exam are normally distributed, what score was needed in 2018 to be in the 10 ten percentile?

19. **Human pregnancy** – The average length of a human pregnancy is 266 days with a standard deviation of 16 days. What percentage of all pregnancies last between 10 and 11 months?

20. **Life insurance** – An life insurance agent sells policies to five 45-year-old men, all of whom are healthy. According The US Social Security Administration, the probability of a person living in these conditions for 30 years or more is 71%. Calculate the probability that after 30 years:

    a. All 5 men are still alive.

    b. At least three of the men are still alive.

    c. Exactly two of the men are still alive.

21. **Euler's number** – Calculate $(2 - e)$ to 100 decimal digits. What are the probabilities of getting each of the digits 0 - 9? Plot the histogram. What is the most probable digit?

22. **Hubble's law and the age of the universe** – Edwin Hubble in 1929 analyzed a set of experimental data shown in Figure 13.14, relating the apparent distance of galaxies and their apparent speeds. The data file `hub.txt` contains some of the original data analyzed by Hubble. Plot the experimental data and find the best linear fit. From the slope of the best fit, estimate the apparent age of the universe.

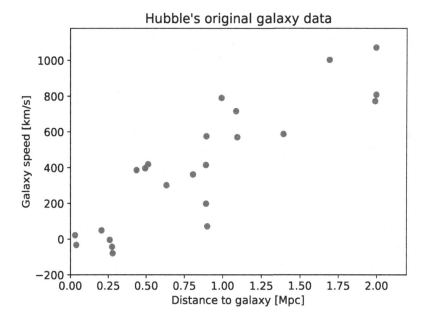

Figure 13.14   The original Edwin Hubble data relating the apparent distance of galaxies and their apparent speeds, contained in the data file `hub.txt`.

23. **Radioisotope generators and multiple radioactivity components** – Radioisotope generators are devices which often contain multiple components of radioactive material. In many cases the radioactivity from such generators can be described as the sum of two exponential functions:

$$y = A \exp\left(-t/\tau_1\right) + B \exp\left(-t/\tau_2\right)$$

where A, B are constants and τ_1, τ_2 are the characteristic decay times of the two components. Plot the experimental data in the file `radioact.csv` as shown in Figure 13.15 and find the best fit parameters A, B, τ_1, τ_2.

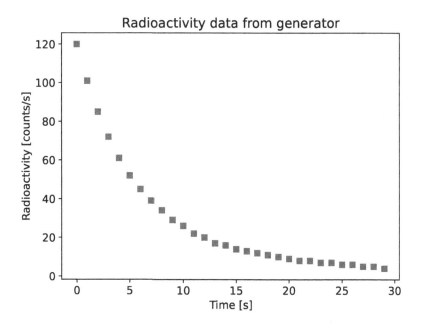

Figure 13.15   Radioisotope generator data containing multiple components of radioactive material, contained in the data file `radioact.csv`.

24. **Analysis of optically stimulated luminescence signals from quartz** – When quartz samples are exposed to brief pulses of blue light, they emit ultraviolet light. The experimental data shown below was obtained in such an optically stimulated luminescence experiment; when the light pulse is on, the luminescence signal increases, while it decreases after the light pulse is turned off. Plot the experimental data in the file `qz.csv` as shown in Figure 13.16, and find the best fit using the equation:

$$y = A\left[1 - \exp\left(-t/\tau_1\right)\right] + B\exp\left(-t/\tau_2\right)$$

where A, B are constants and τ_1, τ_2 are the characteristic decay times of the luminescence components during the pulse on/off time periods. Find the best fit parameters A, B, τ_1, τ_2. The increasing part of the signal corresponds to points #0-25 in the data file, and the decreasing part of the signal corresponds to points #25-98 in the data file.

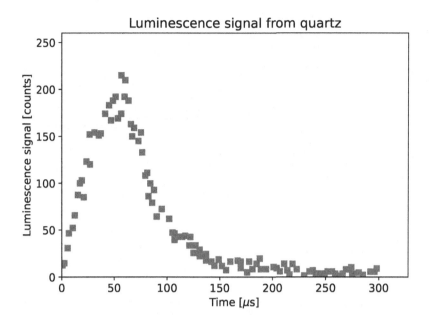

Figure 13.16 Optically stimulated luminescence signals from quartz, with the experimental data, contained in the data file `qz.csv`.

25. **Fitting a Poisson distribution to radioactivity data** – The following distribution of counts was obtained during a radioactivity experiment. The data points are
x = [0,1,2,3,4,5,6,7,8]
y = [0.4, 0.89, 0.77, 0.5, 0.3, 0.14, 0.05, 0.1, 0.1].
Find the least squares fit to this data by assuming a Poisson distribution.

26. **Einstein's theory of specific heat** –
Einstein treated the atoms in a crystal as N simple harmonic oscillators, all oscillating with the same frequency ν_E which depends on the strength of the chemical bonds within the solid. Even though Einstein's theory is an approximation, it explains several experimental characteristics of the specific heat C_v, as a function of the temperature T of the solid. The total energy of the oscillators in the Einstein model is given by:

$$U(T) = 3N \left(\frac{h\,\nu_E}{2} + \frac{h\,\nu_E}{e^{\frac{h\,\nu_E}{kT}} - 1} \right) \qquad (13.11.1)$$

where k is the Boltzmann constant and h is Planck's constant.

a. Evaluate the specific heat by hand and verify your result using Python:

$$C_v = - \left(\frac{\partial U}{dT} \right)_v \qquad (13.11.2)$$

b. Plot C_v as a function of the temperature T of the solid, by using the experimental value of the oscillator frequency ν_E for diamond, for which $\nu_E = 2.708 \times 10^{13}$ s^{-1}.

c. Plot the experimental data for diamond contained in the file `einst.txt`, as shown in Figure 13.17, and find the best fit to the experimental data $y(x)$ using the Einstein equation in the form:

$$y(x) = a \frac{e^{1/x}}{x^2 \left(e^{1/x} - 1\right)^2}$$

where a is a constant and $x = kT/(h \nu_E)$.

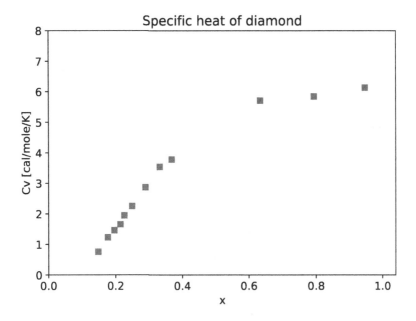

Figure 13.17 Experimental data of the specific heat C_v as a function of $x = kT/(h\nu_E)$ for diamond, with the experimental data contained in the data file `einst.txt`.

27. **Experimental data for Maxwell-Boltzmann distribution** – The data shown in the graph below were obtained during an experimental study of the distribution of speeds from a Maxwell Boltzmann distribution. The data was obtained using a velocity selector and is contained in the file `mb.txt`. For a detailed description of the experiment, the reader can consult the book *Concepts in Thermal Physics* by S. Blundell and K. Blundell (Oxford, 2006). The MB distribution is given by

$$f(v) = \sqrt{\frac{2}{\pi} \left(\frac{m}{kT}\right)^3} \, v^2 \exp\left(\frac{-mv^2}{2kT}\right)$$

a. Show that it is not possible to fit this data with the Maxwell-Boltzmann distribution

$$f(v) = a \, v^2 \exp\left(-b \, v^2\right)$$

with a, b are constants, and that the best fit curve is too broad.

b. Show that this data shown in Figure 13.18 can be fitted successfully using a *modified* Maxwell-Boltzmann distribution of the type:

$$f(v) = a' \, v^4 \exp\left(-b' \, v^2\right)$$

where a', b' are constants.

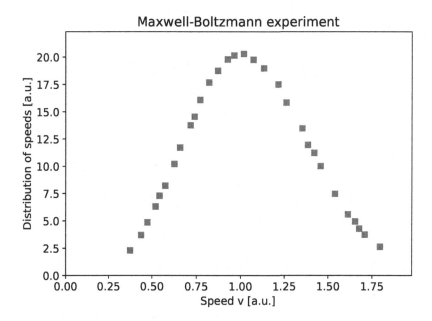

Figure 13.18 Experimental data for the distribution of speeds from a Maxwell Boltzmann distribution obtained using a velocity selector, with the data contained in the data file `mb.txt`.

Further Reading and Additional Resources

Here we provide a list of texts and other resources that the reader will find useful. We have annotated this list so that the reader can know what to expect from each resource.

TEXTBOOKS

1. M. Boas, Mathematical Methods in the Physical Sciences 3rd Edition, Wiley (2005).

   This comprehensive text is a classic for mathematical methods courses in the undergraduate physics curriculum.

2. C. W. Kulp and V. Pagonis, Classical Mechanics: A Computational Approach with Examples using Python and Mathematica, CRC Press (2020).

   This is a unique textbook which covers the traditional topics for an intermediate mechanics course, with a focus on using computer programming to solve problems.

3. W. H. Press, S. A. Teukolsky, W. T. Vetterling, and B. P. Flannery, Numerical Recipes: The Art of Scientific Computing 3rd Edition, Cambridge University Press (2007).

   This book presents algorithms used to solve numerical problems in physics, engineering, and more. Specific versions can be found for various languages.

4. J. M. Kinder and P. Nelson, A Student's Guide to Python33 for Physical Modeling 2nd Edition, Princeton University Press (2021).

   A no-nonsense guide for students to quickly get to speed in Python.

5. S. Strogatz, Nonlinear Dynamics and Chaos: With Applications to Physics, Biology, Chemistry, and Engineering 2nd Edition, CRC Press (2015).

   An introduction to the field of nonlinear dynamics. Readers interested in the material found in Chapter 12 of this book should consult this book for a more thorough introduction to the field.

6. R. H. Enns, It's a Nonlinear World, Springer (2010).

   We recommend this text to readers who are looking for a follow up to Strogatz's Nonlinear Dynamics and Chaos. It covers PDEs and has examples from a wide variety of fields.

7. H. Kantz and T. Schreiber, Nonlinear Time Series Analysis, Cambridge University Press (2003).

   For readers interested in extending the concepts from Chapter 12 to data measured from nonlinear systems. It covers topics including phase space reconstruction and Lyapunov exponent estimation from time series data.

WEBSITES

1. Anaconda Python distribution, https://www.anaconda.com

   Free and easy to install Python distribution.

2. Google Collaborator, https://colab.research.google.com

   Run Jupyter notebooks online without installing anything on your computer.

3. Microsoft Visual Studio Code (VS Code), https://code.visualstudio.com

   Free and easy install coding environment available on all platforms, MS Windows, MacOS, and Linux.

4. W. F. Trench, Elementary Differential Equations available online at:

   https://digitalcommons.usf.edu/oa˙textbooks/9/

   A freely available text on ODEs which expand upon the topics covered in Chapters 10 and 11 of this book.

5. Python library pages - Although a quick online search for Python commands is often the fastest way to get the help you need, the pages for the following libraries are useful

   a. NumPy - https://numpy.org

   b. SymPy - https://www.sympy.org

   c. SciPy - https://scipy.org

   d. Matplotlib - https://matplotlib.org

ADDITIONAL TEXTBOOKS

For students interested in learning more about the physics covered in any of the examples problems from this text, we recommend the following:

1. C.W. Kulp and V. Pagonis, Classical Mechanics: A Computational Approach with Examples using Python and Mathematica, CRC Press (2020)

2. D. J. Griffiths, Introduction to Electromagnetism 4th Edition, Cambridge University Press (2017).

3. S. Blundell and K. Blundell Concepts in Thermal Physics (Oxford, 2006)

4. D. J. Griffiths and D. F. Schroeter, Introduction to Quantum Mechanics 3rd Edition, Cambridge University Press (2018).

5. S. T. Thornton, A. Rex, and C. E. Hood, Modern Physics for Scientists and Engineers 5th Edition, Cengage Learning (2020).

Index

Printed in the United States
by Baker & Taylor Publisher Services